Traditional and Folk Herbal Medicine

Recent Researches

Traditional and Folk Herbal Medicine
Recent Researches

– Volume 1 –

Editor
Dr. V.K. Gupta
Deputy Director & Head,
Central Animal Facility,
Indian Institute of Integrative Medicine (CSIR),
Canal Road, Jammu – 180 001,
India

2012
DAYA PUBLISHING HOUSE®
New Delhi - 110 002

ISBN 9789351242154

Published by : **Daya Publishing House®**
A Division of
Astral International Pvt. Ltd.
– ISO 9001:2008 Certified Company –
4760-61/23, Ansari Road, Darya Ganj,
New Delhi - 110 002
Phone: 23245578, 23244987
Fax: (011) 23260116
e-mail : dayabooks@vsnl.com
website : www.dayabooks.com

Laser Typesetting : **Classic Computer Services**
Delhi - 110 035

Printed at : **Chawla Offset Printers**
Delhi - 110 052

PRINTED IN INDIA

— Dedication —

This book is dedicated
to my beloved parents

Editorial Board

Prof. Ian Fraser Pryme

Department of Biomedicine,
University of Bergen,
Jonas Lies vei 91, N-5009 Bergen, NORWAY
Tel: +47-55-586438; Fax: +47-55-586360
E-mail: ian.pryme@biomed.uib.no

Dr. George Qian Li

Herbal Medicines Research and Education Centre,
Faculty of Pharmacy,
The University of Sydney,
NSW 2006, AUSTRALIA
Tel: 612-9351; Fax: 9351-8638
E-mail: george.li@sydney.edu.au

Dr. Julia Serkedjieva

Institute of Microbiology,
Bulgarian Academy of Sciences,
26, Academician Georgy Bonchev St.,
1113 Sofia, BULGARIA
Tel: +359 2 979 31 85; Fax: +359 2 870 01 09
E-mail: jserkedjieva@microbio.bas.bg

电话：0872-2214251
传真：0872-2257401
地址：云南省大理白族自治州大理市下关万花路
大理学院药学院
邮编：671000

Tel: +86-872-2214251
Fax: +86-872-2257401
Address: College of Pharmacy, Dali University
Wanhua Rd., Xiaguan, Dali 671000,
Yunnan Province, P. R. China

Foreword

The history of human development is the history of the struggle with diseases. Although modern medicine has developed into the stage of sophisticated techniques like organ reconstruction, artificial cells, and cloned animals, people still have to facing the perhaps intolerable damages, both existing and potential, caused by so-called new stuffs such as superbugs (e.g. New Delhi Metallo-1), nuclear radiation, chemical and biological pollution, computer and mobile phone radiations, transgenic foods, etc.

Should we follow our ancients because we envy prehistoric humans who content with a natural way of life, that is, getting food and medicine from their natural surroundings to obtain the balance on the survival and to achieve harmony with the nature? Or should we keep in accordance with the modernization to "conquest" the nature for meeting our endless requests and pursuing luxury life and vanity, which caused costly invention for modern expensive antibiotics of generation after generation, in turn put ourselves into the risk of super bacteria, SARS or other variant (such as the MDR-HIV species) and varying viruses of strong epidemic? It is worthy for us modern citizens of the world with serious consideration.

From a historical perspective, there is no shortage of medicines nowadays categorized as complementary medicines, alternative medicines, or, in a fashioned word, ethnomedicines. In fact, the alternative therapies traced back from bloodletting, enema, witchcraft, and shaman ceremony in Paleolithic till the medieval church accommodating patients overnight as a treatment, are worthy of pondering over and deserve thorough investigation.

Ayurvedic and traditional Chinese medicine (TCM) both have splendid histories over thousands of years with the theoretical system prior to that of Western medicine.

E-mail: dryuzhao@126.com or dryuzhao@hotmail.com

Furthermore, the natural drugs from animals and botanical sources are anthropomorphic and abundant of local cultural connotations in many countries. The study of these ethnopharmacological species thus calls for more carefulness and comprehensiveness, and should be combined with its multi-faceted cultural factors.

Traditional medicines hold irreplaceable advantages and rightful place in the aspects of economics, culture, national crisis and many others. However, the uneven applications were found in the world. 80% of the populations of the African continent are using traditional medicines, while complementary and alternative medicines represents only as street markets in contrast of western medicine as a super market in north America and most of European countries. However, we are pleased to notice that, in recent years, it is surging again on every continent to carry out the researches of traditional medicine, of the collection of the pharmaceutical data, and of the construction work of the local specimen library of traditional medicines. Meanwhile, it is apparent that meditation, Qigong, Tai Chi, Yoga and other exercises of capable to balance the physical and mental health have become increasingly popular; which reflects the cultural diversity is hopeful to get passed down in the world. We hope that these healthy traditional medicines and therapies earn more understanding and supports worldwide.

On the other hand, it is a huge task to deeply investigating, to further explore and re-evaluate, and finally to evidence and identity the traditional medicines with the strategy of taking full advantage of the costly developed modern medical technology such as molecular pharmacology, molecular biology, modern pharmaceutics, computer-aided drug design, genomics, proteomics, glycomics and other emerging disciplines. It is nevertheless a huge contribution to human health and life, which merits the joint efforts of our and next generations. This book series is useful to bridge the divide between western and traditional medicines.

It is the first volume of the book series entitled, "*Traditional and Folk Herbal Medicine: Recent Researches*" which encompasses a total of 19 papers from 14 countries. These papers describe various utilizations of local natural drugs against many harmful diseases. Each paper demonstrated the summarized experiences and unique insights of the individual authors. I believe it is a book with rich content, of practice, and with a high reference value, whilst its publication will unambiguously provide useful supplementary information to the R&D departments of pharmaceutical companies, functional food companies, clinicians, pharmaceutical workers, pharmacological and toxicological researchers, and the undergraduate as well as graduate students at medical schools.

Dr. V.K. Gupta kindly took the codification work of this book, his team exhibited great enthusiastic about soliciting manuscripts, the layout, and follow-up publishing for this book. On behalf of all of the authors, I express the heartfelt thanks to them. Moreover, our gratefulness should also go to Daya Publishing House, New Delhi for jointing in the launching of this book series.

To the point of about ending this preface, we sadly figured out that the nuclear crisis caused by the earthquake in eastern Japan has not yet resolved; and Libya gunfire still ringing in our ears; while the legend 2012 of doomsday prediction is more rampant. We should perhaps sincerely revisit Einstein warned us: the need for Society is to begin to consider as soon as possible for the wisdom to find a balance between the access and its application of knowledge. Maybe we really ought to reflect on whether the rapid development of technology has brought a greater sense of security and comfort to human beings. It might be worth to apply the wisdom of ancient philosophers such as Lao Tzu, the Buddha, Hippocrates, etc. for the healing of the traumas of human body and mind caused by various diseases.

Finally, I'd like to take this opportunity to wish our planet, this unique blue planet, enjoy a permanent peace; also I wish the success of the publication and distribution of our book series.

Sincerely yours,

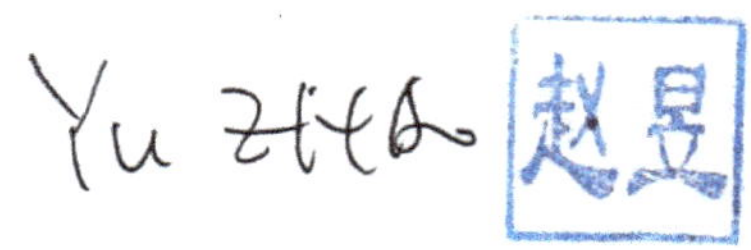

Yu

Yu ZHAO, PhD.

Cheung-kong Chair Professor of Chinese Ministry of Education
Assistant to the President of Dali University
Honorary Dean of the School of Pharmacy, Dali University
Wan Hua Road, Xia Guan, Dali 671000
Yunnan Province, CHINA
Tel: +86-872-2214251; Fax: +86-872-2257401
E-mail: dryuzhao@126.com

Preface

Traditional and folk medicines have been developed over millennia using the acquired experience and accumulated knowledge of man. Both beneficial and harmful effects of plant materials have been studied and theories about their use have been developed. Thus, man has succeeded in discovering cures using herbal, animal and mineral medicines. Many indigenous medicines began as a myth and then transferred to new generations as folk medicines and finally developed as modern drug treatments following a scientific analysis of their effects.

Plants have formed the basis of traditional medicine for thousands of years and continue to provide mankind with new remedies. Modern medicine has its roots in ancient medicine, and it is likely that many important new remedies will continue to be discovered and commercialized in future by following the leads provided by traditional and folklore knowledge and experience. Most of the secondary plant compounds employed in modern medicine were first discovered through ethnobotanical investigation.

The history of medicine is an account of mankind's efforts to deal with human illness and diseases ranging from primitive attempts of preliterate man to the present complex array of therapeutic specialties. The development of traditional medicines has been influenced by the different cultural and historic conditions in which they were first developed. Practices known as traditional medicines include Ayurveda, Siddha, Unani, Islamic, Chinese, Korean, African and other medical knowledge and practices all over the globe. Although modern medicine may be available in these countries, herbal medicines (phytomedicines) have often maintained popularity for historical and cultural reasons.

The use of traditional medicine has changed dramatically over the past thirty years. Due to its affordability, availability and accessibility, traditional medicine has played an important role in meeting the demands of primary health care in many developing countries, particularly in Africa and Asia. Since the 1990s, the use of traditional medicine has surged. It not only maintains its function in primary health care in developing countries (70-80% of the population in India and Ethiopia still depend on traditional medicine and practitioners for primary health care), but its use has expanded widely in many developed countries where it is referred to as Complementary or Alternative Medicine (CAM). Concurrently, many people in developed countries have begun to turn to alternative or complementary therapies, including medicinal herbs. For instance, 70% of the population in Canada and 80% in Germany have used traditional medicine as CAM treatment.

It is owing to a world wide and sustained effort of scientist's that an enormous information is being generated on traditional and folk medicinal plant researches. Based on this rational, the present volume of the book series entitled, "*Traditional and Folk Herbal Medicine: Recent Researches*" presents edited information on 19 research and review communications received from India and abroad. The chapters included in this volume are: The Potential Effectiveness of Medicinal Plants Used for the Treatment of Sickle Cell Disease in the Democratic Republic of Congo Folk Medicine: A Review; Ayurvedic Pharmaceutics and Insights on Personalized Medicine; Bioactivities of *Duabanga grandiflora* for Human Skin Cells; *Chromolaena odorata*, A Traditional Medicinal Herb with Hemostatic and Wound Curative Properties: An Overview; Chemistry, Pharmacology and HPTLC Standardisation of *Asparagus recemosus* (Shatavari), an Important Ayurvevic Herb; Ethnobotanical, Phytochemical, Pharmacological and Conservational Studies on *Glycyrrhiza glabra*: An Overview in Pakistan; Pharmacotherapeutic Potential of Medicinal Plants against Cardiometabolic Diseases; Recent Investigations on Therapeutic Potential of *Clitoria ternatea* Linn.; The Research Advance of Tibetan Medicine Dida; Euphorbia Latex: A Magic Potion; Traditional Medicine: Its Role in the Search for New Antiparasitic Candidates; A Simplified Procedure for Monitoring the Stability of Herbal Remedies for the Purpose of Manufacturing and Regulation; Effect of a Cysteine Proteinase from *Bauhinia forficata* Leaves on Thrombin-Induced Human Platelet Aggregation; Ajwain (*Tracyspermum ammi*) Seeds in Health Promotion and Disease Prevention; Quality Control, Standardization and Chemo Profiling of Herbal Drugs and ISM Preparations; *Ficus gnaphalocarpa, Ficus cordata* and *Ficus chlamydocarpa*: Potential Sources of Bioactive Compounds; Effects of *Holarrhena floribunda* Leaf Extract on Gastrointestinal Functions in Male Rats; Insulin Releasing Effects of Cumin (*Cuminum cyminum* L.) Seeds in Diabetic Rats; Phytochemical Analysis and *in vitro* Antioxidant Activity of Essential Oils and Methanolic Extracts of *Ocimum* spp. (Indian Basils) Growing in Uttarakhand Regions of Central Himalaya.

The studies included are likely to lead further researches in this direction and it is hoped that this publication would attract world wide audience of researchers and the academicians of allied disciplines engaged in the search of new drug from herbal resources.

V.K. Gupta

Contents

**Traditional and Folk Herbal Medicine:
Recent Researches Vol. 1 (2012)**
Editor: **V.K. Gupta**
Published by: **DAYA PUBLISHING HOUSE, NEW DELHI**

Pages **1–11**

1

The Potential Effectiveness of Medicinal Plants Used for the Treatment of Sickle Cell Disease in the Democratic Republic of Congo Folk Medicine: A Review

P.T. Mpiana[1*], K.N. Ngbolua[2], V. Mudogo[1], D.S.T. Tshibangu[1], E.K. Atibu[1], B.M. Mbala[1], B. Kahumba[3], M.T. Bokota[4] and L.K. Makelele[4]

ABSTRACT

In our continued efforts to search for novel and bioactive antisickling compounds from indigenous plants, we recently investigated crude extracts from 53 plants species growing spontaneously in the Democratic Republic of Congo which are prescribed by traditional healers or sold in markets for their potential against sickle cell disease.

The dried and powdered plant materials were repeatedly extracted by cold percolation three times with solvents of increasing polarity (n-hexane, dichloromethane, ethanol, methanol, water) for 72 hrs at room temperature. The specific combined extracts were filtered and concentrated to dryness under reduced pressure using a rotary evaporator. Dry crude extracts were screened for their biological activity using in vitro sickling suppressive test of Emmel.

1 Département de Chimie, Faculté des Sciences B.P. 190, Université de Kinshasa, Kinshasa XI, R. D. Congo.

2 Département de Biologie, Faculté des Sciences B.P. 190, Université de Kinshasa, Kinshasa XI, R. D. Congo.

3 Faculté de Pharmacie, Université de Lubumbashi, Lubumbashi, R.D. Congo.

4 Faculté des Sciences, Université de Kisangani, Kisangani, R.D.Congo.

* *Corresponding author*: E-mail: ptmpiana@yahoo.fr, Tel: +243 81 811 60 19.

The results of this preliminary test have indicated that the polar extracts of 50 on 53 studied plants possess a good antisickling activity in vitro (with a sickling suppressive rate ≥ 70 per cent in hypoxic conditions). Fourteen of these plants: Alchornea cordifolia, Bombax pentadrum, Centella asiatica, Crotalaria retusa, Ficus capensis, Hymenocardia acida, Jatropha curcas; Justicia secunda, Morinda lucida, Ocimum basilicum, Sterculia quinueloba, Trema orientalis, Vigna unguiculata, Zizyphus mucronata; for which the rate of normalization of sickle erythrocytes is very high were submitted to chemical screening (fingerprinting) and the chemical composition of plants were determined after which some secondary metabolites were screened. The antisickling activity of Congolese plants could be attributed to anthocyanins. This activity is dose dependent. Anthocyanins are recognized as potent antioxidants and may represent chemo protective agents for sicklers but they are sensitive to light and temperature. Anthocyanins are well known for their antioxidative properties but before our work, their antisickling activity had not been reported.

A discussion on how free radicals are built up in sickle erythrocytes and their roles in the pathophysiology of sickle cell disease and few additional putative modes of action of anthocyanins are proposed.

Keywords: Sickle cell disease, Medicinal plants, Anthocyanins, Dem. Rep. of Congo.

Sickle Cell Disease and Ethnopharmacology

Sickle Cell Disease (SCD) is one of the most neglected diseases of the 21st century with an important death rate in children living in the third world especially in Africa. Like malaria, SCD is also a major public health problem and it is the first genetic disease by the number of affected people in tropical Africa regions (Labie and Elion, 2003). Clinical symptoms of SCD are associated to the polymerization of unstable haemoglobin S under anoxic conditions. These symptoms include erythrocytes shape modification as well as anaemia. The intracellular polymerization of deoxy haemoglobin S occurs as a result of the activation of both KCl-cotransport system and Gardos channel, leading into sickle erythrocyte dehydration while anaemia occurs as a result of radical oxygen species (ROS) induced erythrocyte membrane damage. At molecular level, SCD is characterized by the low activity of the glutathione reductase system (Hebbel *et al.*, 1982) and the hexose metabolism in sickle erythrocytes is very reduced (Osuagwu and Mbeyi, 2007), so methaemoglobin and other ferriprotoporphyrin (containing Fe^{3+}) breakdown products builds up in the SCD patients' blood system as result of oxidative stress. Little progress had been made in suitable treatment of the disease all over the world. Some agents were developed by rational drug design to inhibit the red blood cells (RBCs) sickling process. Most of these agents, unfortunately, did not show promising success in terms of clinical use (Mehanna, 2002; Eliot *et al.*, 2006). Conventional treatments are expensive and unaffordable by the rural populations. So action needs to be taken quickly, notably against this genetic disease for which there is not yet an effective remedy.

These last few years, phytotherapy is explored as a therapeutic approach that can relieve tropical diseases. The Democratic Republic of Congo (DRC) which contains a large area of Congo basin forest is reputed for the extraordinary richness of its flora (qualified as one of the mega bio diversities of the world) and boasts a wide variety of medicinal plants species.

The use of medicinal plants is a simple, inexpensive and appropriate strategy for disease control in developing countries such as DRC. These plants represent an enormous reservoir of new molecules with potential therapeutic activities which are waiting to be discovered. Natural products have played a key role throughout the history of drug discovery as a source of both novel compounds and scaffolds for the development of blockbuster drugs. In the case of SCD, a number of *in vitro* biological screening studies have been undertaken in our laboratory in order to evaluate the antisickling activity of various plant extracts. Crude extracts of some Congolese plants were find to be promising therapeutic candidates in the hope of finding remedies that can be prepared and administered locally at very low cost and would increase the life expectancy of sicklers. Anthocyanins extracts from some Congolese plants used in traditional medicine for managing SCD had displayed good antisickling activity by alleviating or preventing all of the *in vitro* clinical phenotype of the SCD (Mpiana *et al.*, 2007-2010). So, anthocyanin-derived preparations could represent an alternative to the use of synthetic drugs, cheap and readily available for sicklers.

Epidemiology and Situation in the Democratic Republic of Congo

SCD is one of the serious health problems around the world, especially in the sub-Saharan Africa (Figure 1.1). Around 20 per cent of some African populations are heterozygous AS. In Central Africa, 25 to 30 per cent of populations are affected (Iyamu *et al.*, 2002; Moody *et al.*, 2003; Kaplan and Delpech, 1993; Voet and Voet, 1998). Two per cents of the Democratic Republic of Congo populations are homozygous SS, about a million of individuals. Children are the most concerned. About 80 per cent of anaemic children no medically treated die before the age of five years (Mpiana *et al.*, 2007a). SS anaemia is spilled today on the 5 continents because of populations migrations (Neuwinger, 2000; Ibrahim *et al.*, 2007).

Each year about 300 000 infants are born with major haemoglobin disorders – including more than 200 000 cases of sickle-cell anaemia in Africa. Globally, there are more carriers of thalassaemia than of sickle-cell anaemia, but the high frequency of the sickle-cell gene in certain areas leads to a high rate of affected newborns.

The sickle-cell gene has become common in Africa because the sickle-cell trait confers some resistance to falciparum malaria during a critical period of early childhood, favoring survival of the host and subsequent transmission of the abnormal haemoglobin gene. Although a single abnormal gene may protect against malaria, inheritance of two abnormal genes leads to sickle-cell anaemia and confers no such protection, and malaria is a major cause of ill-health and death of children with sickle-cell anaemia. There is increasing evidence that malaria not only influences outcome but also changes the manifestations of sickle-cell anaemia in Africa (WHO fifty-ninth world health assembly A59/9 Provisional agenda item 11.4 24 April 2006).

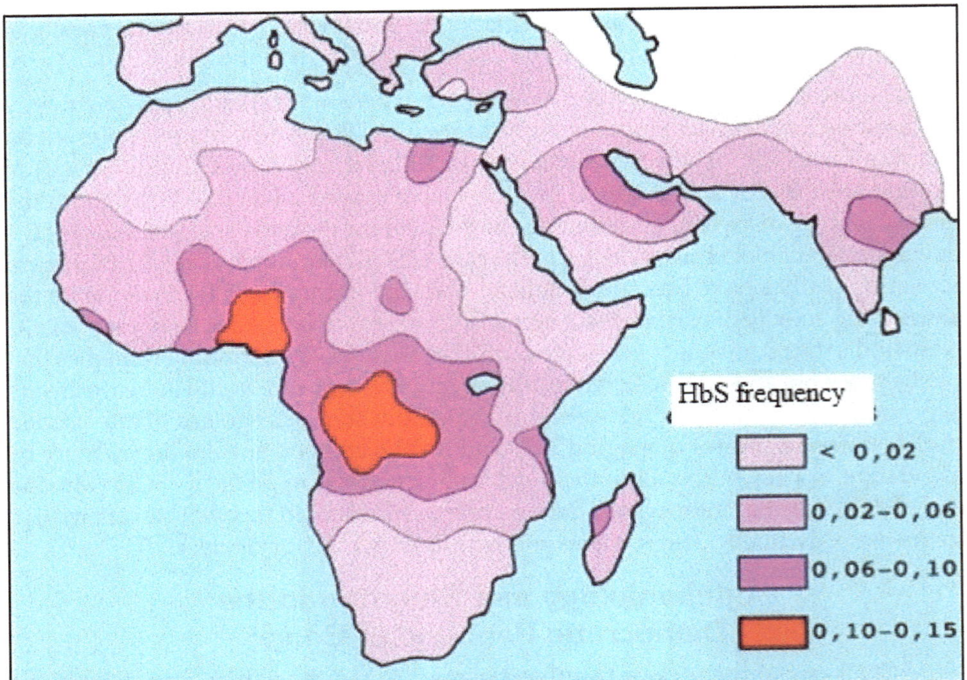

Figure 1.1: Geographic repartition of S haemoglobin.

Pharmacological Targets and Antisickling Assays Methodology

When searching for active antisickling agents, it is necessary to have relatively simple biological or pharmacological tests available in order to localise the chosen activity in the plant candidates. The great progress made during the last few years in biochemistry and cellular biology of SCD has indicated that three main targets are of particular importance for chemotherapy. These include erythrocyte membrane, unstable haemoglobin S and free radicals which are spontaneously produced within red blood cells (RBCs) (Mpiana *et al.*, in press; Buchanan *et al.*, 2004). It is possible therefore to act directly on each target involved in the pathophysiology of SCD. For example, plant extract which inhibit the sickling of RBCs or the polymerization of deoxy form of haemoglobin S in hypoxic conditions and free radicals scavengers will alleviate all the clinical phenotype of SCD such as cell dehydration, cell haemolysis and haemoglobin oxidative damage.

In our bio prospecting work, a RBCs sickling suppressive test of Emmel has been adapted and used successfully to test systematically Congolese plants for their antisickling properties. The testing procedure consists of exposing sickled erythrocytes to plant extracts and the number of the normalized RBCs is evaluated microscopically. A 6x zoom CANNON-type digital camera is used to convert the photonic micrograph image into a digital image, which is then digitalized with the help of MOTIC Images 2000 software (Mpiana *et al.*, 2007-2010).

A sample is considered to possess: very high activity (+++) if normalization >70 per cent; high activity (++) if 50 < normalization <70 per cent; weak activity (+) if 10<normalization <50 per cent; no activity (–) if normalization <10 per cent (Mpiana *et al.*, 2008). The modes of action were proposed by investigating the effects of anthocyanins extracts on inhibition of polymerization of deoxy HbS and erythrocyte membrane stability using Itano and osmotic fragility tests respectively (Mpiana *et al.*, in press). The solubility of the deoxygenated sickle cell haemoglobin (Itano test) is expressed as the increase of the optical density at 540 nm after its treatment with active extracts. Fragility of RBCs is evaluated by placing cells in graded series of hypotonic saline solutions (from 0.2 per cent to 0.8 per cent NaCl) buffered with phosphate at pH 7.4. The number of RBCs not lysed by saline concentration is determined microscopically. Extract is considered as active if the mean corpuscular fragility (which is the concentration of saline causing 50 per cent haemolysis of the RBCs) value of treated cells is greater than that of the untreated sickle erythrocytes and means that RBCs are rehydrated.

Results and Discussion

The ethnobotanical investigations carried out during the last few years by our research team have revealed that 53 medicinal plant species belonging to 30 different families are used by Congolese traditional healers for the treatment of SCD in the DRC (these plants are summarized in Table 1.1). A preliminary screening for the *in vitro* antisickling activity using Emmel test has indicated that the polar extracts of 50 on 53 studied plants possess a significant sickling suppressive rate in hypoxic conditions. The leaves of *Alchornea cordifolia, Bombax pentadrum, Centella asiatica, Crotalaria retusa, Ficus capensis, Hymenocardia acida, Jatropha curcas, Justicia secunda, Morinda lucida, Ocimum basilicum, Sterculia quiqueloba, Trema orientalis, Vigna unguiculata, Zizyphus mucronata* were submitted to chemical screening. The fingerprinting revealed the presence of some secondary metabolites such as alkaloids, saponins, tannins and anthocyanins. Bio guided extraction has indicated that anthocyanins extracts could be responsible for the antisickling activity of Congolese plants. Anthocyanins are sensitive to light and temperature (Mpiana *et al.*, 2008).

53 plant species belonging to 30 families were collected and screened for their antisickling properties. *Duvernoya splendens, Eleutherine plicata, Persea americana* were found to be no active while 50 plants were found to be active.

These results suggest a possible correlation between antisickling properties of the active plants and their use in the management of Sickle Cell Disease by traditional healers. This activity could be due to anthocyanins extracts that will act by various mechanisms: inhibition of deoxy Hb S gelation (antigelling agent), inhibition of RBCs sickling (antisickling agent), re-hydration of sickle erythrocytes, reduction of metHb level in sicklers blood (reducing agent) and protection of erythrocyte membrane against radical oxygen species (Mpiana *et al.*, 2007-2010). They are sensitive to light and temperature.

Apart from the inhibition of haemoglobin S polymerization, endothelial injury and the erythrocyte membrane, free radicals production have been defined as new target in SCD therapy (Buchanan *et al.*, 2004)· Because of the reduced glucose

Table 1.1

Sl.No.	Plant Family	Bionomial Latin Name
1.	**Acanthaceae**	1. *Justicia secunda* Valh
		2. *Duvernoya splendens*
		3. *Thomandersia hensii*
2.	**Annonaceae**	4. *Annona reticulanta* L.
		5. *Annona senegalensis* Pers
3.	**Apiaceae**	6. *Centella asiatica*
4.	**Bombacaceae**	7. *Adansonia digitata* L.
		8. *Bombax pentadrum*
5.	**Burseraceae**	9. *Dacryodes edulis* G. Don
6.	**Cannabaceae**	10. *Trema orientalis* L. Blum
7.	**Caricaceae**	11. *Carica papaya*
8.	**Chrysobalanaceae**	12. *Parinari mobola*
9.	**Clusiaceae**	13. *Harungana madagascariensis*
10.	**Connaraceae**	14. *Byrsocarpus orientalis* (Baill.) Baker
11.	**Euphorbiaceae**	15. *Alchornea cordifolia* Mill. Arg
		16. *Bridelia atroviridis*
		17. *Bridelia ferruginea* Benth
		18. *Bridelia ripicola*
		19. *Diplorrhynbchus condilocarpus*
		20. *Euphorbia heterophylla*
		21. *Jatropha curcas*
		22. *Macaranga monandra*
		23. *Ricinodendron heudelotii*
12.	**Fabaceae**	24. *Cajanus cajan*
		25. *Crotalaria retusa*
		26. *Dalberigia bochmintaub*
		27. *Vigna micrantha* Harms
		28. *Vigna unguiculata (L) Walg*
13.	**Flacourtiaceae**	29. *Caloncoba welwischii* Gilg
14.	**Hypericaceae**	30. *Harungana madagascariensis* Lam. ex Poir.
15.	**Hypoxydaceae**	31. *Hypoxis angustifolia* Lam
16.	**Iridaceae**	32. *Euletherine plicata*
17.	**Lamiaceae**	33. *Coleus kilimandscharis*
		34. *Ocimum basilicum* L.
18.	**Lauraceae**	35. *Persea americana* Mill
19.	**Malvaceae**	36. *Ceiba pentandra* (L) Gaertn
20.	**Mimosaceae**	37. *Acacia kirkii*

Contd...

Table 1.1–*Contd...*

Sl.No.	*Plant Family*	*Bionomial Latin Name*
21.	**Moraceae**	38. *Ficus capensis*
		39. *Ficus mucosa*
		40. *Treculia africana* Decne
22.	**Myrtaceae**	41. *Syzygium guineense* (Willd.)
23.	**Nyctaginaceae**	42. *Bourgainvillea* sp
24.	**Phyllanthaceae**	43. *Hymenocardia acida* Tul
25.	**Poaceae**	44. *Cymbopogon densiflorus* Stapf
26.	**Rhamnaceae**	45. *Ziziphus mucronata*
		46. *Maesopsis eminii*
27.	**Rubiaceae**	47. *Mitragyna stipulosa*
		48. *Morinda lucida Benth*
		49. *Rothmania witfchidii*
28.	**Rutaceae**	50. *Fagara macrophylla*
29.	**Sterculiaceae**	51. *Sterculia quinqueloba*
	Urticaceae	52. *Urera hypselodendron*
30.	**Zingiberaceae**	53. *Afromomum albo violaceum* Redley

metabolism and the low activity of both the glutathione reductase system and methemoglobin reductase, which are involved in the protection of hemoglobin and membrane from oxidative breakdown (Mills *et al.*, 1957), methemoglobin, a biomarker of oxidative stress, and radical oxygen species (ROS) are build up spontaneously in sickle erythrocyte (Hebbel *et al.*, 1982; Misra and Fridovich, 1972; Osuagwu and Mbeyi, 2007; Giulivi *et al.*, 1994; Balagopalakrishna *et al.*, 1996; Nagababu and Rifkind, 2000). These ROS act as biological nucleophile in the de-esterification of membrane lipids leading into the hyperhemolysis of sickle erythrocyte (Niehaus *et al.*, 1978; Carrell *et al.*, 1975). Cell-free plasma hemoglobin released during intravascular hemolysis leads to scavenging of nitric oxide (NO) (Reiter *et al.*, 2002). In addition, hemolysis releases red cell arginase which degrades arginine, the substrate for endothelial NO synthase, resulting in decreased NO production (Morris *et al.*, 2005). The NO pathway is one of the major pathways known to be perturbed in patients with SCD. NO is a potent vasodilator and plays an important role in vascular endothelial homeostasis. Its depletion leads to vasoconstriction, endothelial dysfunction, platelet activation, oxidative stress, and proliferative vasculopathy that can ultimately lead to PH in SCD and other hemolytic anemia (Rother *et al.*, 2005; Gladwin *et al.*, 2005). The oxidant properties of anthocyanins extracts (Kahkonen *et al.*, 2003) is an additional advantage protecting sickle RBCs against ROS induced precocious hemolysis and therefore will increase the bioavailability of NO in patients with SCD.

Conclusions

The anthocyanins extracts from Congolese plants exhibited antisickling activity and could, therefore, be used as drug leads in SCD chemotherapy.

The no cytotoxicity of both anthocyanins extracts and edible anthocyanins rich plants such as *Vigna unguiculata* and their efficacy in RBCs sickling inhibitory/ suppressive test validate the wide use of medicinal plants in traditional medicine for the palliative/symptomatic treatment of SCD in the DRC, where drepanocytosis is endemic in some individuals. For such edible plants, no toxicological testing is required before clinical trials. The use of the leaves by traditional healers for the preparation of receipts is particularly important for the conservation of these plants because leaves are affordable and their use does not damage the plants and limit the supply as the use of the roots would.

Acknowledgements

The authors are indebted to the International Foundation for Science (IFS) and the Organization for the Prohibition of Chemical Weapons (OPCW) for the Research Grant given to one of them for his PhD degree (NGBOLUA KOTO-TE-NYIWA, IFS Research Grant F/4921-1).

References

Balagopalakrishna, C., Manoharan, P.T., Abugo, O.O., and Rifkind, J.M. (1996). Production of superoxide from hemoglobin-bound oxygen under hypoxic conditions. *Biochemistry*, **35**: 6393-8.

Buchanan, G.R., De Baun, M.R., Quinn, C.T., and Steinberg, M.H. (2004). Sickle Cell Disease. *Hematology*, 35-47.

Carrell, R.W., Winterbourn, C.C., and Rachmilewitz, E.A. (1975). Activated oxygen and haemolysis. *Br J Haematol.*, **30:** 259–64.

Eliot, R., Davies, M., and Harmse, D.J. (2006). Dermatomyositis-like eruption with long-term hydrourea. *Dermatolog Treat.*, **17**(1): 56-60.

Giulivi, C., Hochstein, P., and Davies, K.J.A. (1994). Hydrogen peroxide production by red blood cells. *Free Radic.Biol.Med.*, **16**: 123–9.

Gladwin, M.T., and Kato, G.J. (2005). Cardiopulmonary complications of sickle cell disease: role of nitric oxide and hemolytic anemia. *Hematol (Am Soc Hematol Educ Program)*. 51-57.

Hebbel, R. P., Eaton, J. W., Balasingam, M., and Steinberg, M. H. (1982). Spontaneous Oxygen Radical Generation by Sickle Cell Erythrocytes. *J. Clin. Invest.*, **70**: 1253-1259.

Ibrahim, H., Sani, F.S., Danladi, B. H., and Ahmadu, A.A. (2007). Phytochemical and antisickling studies of the leaves of *Hymenocardia acida* Tul (Euphorbiaceae). *Pak. J. Biol. Sci.*, **10**(5): 788-791.

Iyamu, E. W., Turner, E.A. and Asakura, T. (2002). *In vitro* effect of NIPPRISAN (Mix-0690), and naturally occurring potent antisickling agent. *Br J. Haematol.*, **118**: 337-343.

Kahkonen, M.P., Heinamaki, J., and Heinonen, M. (2003). Antioxidant activity of anthocyanins and their aglycons. *J Agric Food Chem.*, **52**: 628–633.

Kaplan, J.C. and Delpech, M. (1993). *Biologie moléculaire et Médercine*, 2e éd., Flammarion Médecine-Sciences, Paris.

Labie, D. and Elion, J. (2003). Génétique et physiopathologie de la drépanocytose, in: *La drépanocytose* (R. Girot, P. Bégué, and F. Galacteros, eds.). *John Libbey Eurotext.*, 1-10.

Mehanna, A.S. (2002). Sickle Cell Anaemia and Antisickling Agents: Then and Now. *Curr Med Chem.*, **8(2)**: 79-88.

Mills, G.C.J. (1957). Hemoglobin catabolism. I. Glutathione peroxidase, an erythrocyte enzyme which protects hemoglobin from oxidative breakdown. *J Biol Chem.*, **229**(1): 189-97.

Misra, H. P and Fridovich, T. (1972). Super-oxide Ion Generation by Oxidation of Oxyhemoglobin to methaemoglobin. *J. Biol. Chem.*, **247**: 6960-6962.

Moody, J.O., Ojo, O.O., Omotade, O.O., Adeyemo, A.A., Olumese, P.E., and Ogundipe, O.O. (2003). Antisickling potential of Nigerean herbal formula (*Ajawaron HF*) and the major plant component (*Cissus populnea* L. CPK). *Phytother. Res.*, **17**: 1173-1176.

Morris, C.R., Kato, G.J., Poljakovic, M., *et al.* (2005). Deregulated arginine metabolism, hemolysis-associated pulmonary hypertension, and mortality in sickle cell disease. *JAMA*. **294**: 81-90.

Mpiana, P.T., Ngbolua, K.N., Bokota, M.T., Kasonga, T.K., Atibu, E.K., and Mudogo, V. In *vitro* Effetcts of Anthocyanins Extracts From *Justicia secunda* VAHL on the Solubility of Hemoglobin *S* and Membrane Stability of Sickle Erythrocytes, *Blood Transfusion* (In press; DOI 10.250/2009.0120-09).

Mpiana, P.T., Balangayi, E.K., Kanangila, A.B., Kalonda, E.M., Ngbolua, K.N., Tshibangu, D.S.T., Atibu, E.K., Lumbu, J.B.S. (2009). Activité antidrépanocytaire et thermodégradation des anthocyanes extraits de *Sterculia quinqueloba* et *Ficus capensis, Int. J. Biol.& Chem. Sci.*, **3(3):** 551-560.

Mpiana, P.T., Ngbolua, K.N., Bokota, M.T., Kasonga, T.K., Atibu, E.K., and Mudogo, V. (2009). Antisickling activity of *Justicia secunda Vahl, In:* Drug discovery from African rain forest, The 13th Symposium of the Natural Product Research Network for Eastern and Central Africa, August 10-14, University of Kinshasa, Democratic Republic of Congo, p.21 (SL-4).

Mpiana, P.T., Mudogo, V., Kabangu, Y.F., Tshibangu, D.S.T., Ngbolua, K.N., Atibu, E.K., Mangwala, K.P., Mbala, M.B., Makelele, L.K., and Bokota, M.T. (2009). Antisickling Activity and Thermostability of Anthocyanins Extract from a congolese plant, *Hymenocardia acida Tul.* (Hymenocardiaceae). *Int. J. Pharmacol.*, **5 (1):** 65 – 70.

Mpiana, P.T., Mudogo, V., Tshibangu, D.S.T., Ngbolua, K.N., Atibu, E.K., Kitwa, E.K., Kanangila, A.B., and Makelele, L.K. (2009). Activité antifalcémiante et thermodégradation d'une fraction d'anthocyanes extraits de *Zizyphus mucronata*. *Ann.Afr.Med.*, **2(2)** : 91-97.

Mpiana, P.T., Mudogo, V., Nyamangombe, L., Kakule, M. K., Ngbolua, K.N., Atibu, E.K., Mbala, M.B., Mbongo, A.K., and Ntumba, J.N. (2009). Antisickling activity and photodegradation effect of anthocyanins extracts from *Alchornea cordifolia* (SCHUMACH *and* Thonn.) and *Crotalaria retusa L. Ann.Afr.Med.*, **2(4):** 240-245.

Mpiana, P.T., Mudogo, V., Tshibangu, D.S.T., Shetonde, O.M., Ngbolua, K.N. and Mbala, M.B. (2007). Antisickling activity of some Congolese plants, *In*: Drug discovery from African flora, The 12th Symposium of the Natural Product Research Network for Eastern and Central Africa, July 22-26, University of Makerere, Kampala, Uganda, p.45 (PS-6).

Mpiana, P.T., Mudogo, V., Tshibangu, D.S.T., Ngbolua, K.N., Mangwala, P. K., Atibu, E.K., Kakule, M.K., Makelele, L.K., and Bokota, M.T. (2010). Antisickling activity and thermodegradation of an anthocyanin fraction from *Ocimum basilicum L.* (LAMIACEAE). *In*: Comprehensive Bioactive Natural Products, Vol. 3: Efficacy, Safety and Clinical Evaluation (Partt II), V.K. Gupta (Ed): Studium Press LLC, USA,: 278-287.

Mpiana, P.T., Mudogo, V., Ngbolua, K.N., Tshibangu, D.S.T., Shetonde, O.M., and Mbala B.M. (2007). *In vitro* antisickling activity of anthocyanins from *Ocimum basilicum* L. (Lamiaceae). *Int. J. Pharmacol.*, **3 (4):** 371-374.

Mpiana, P.T., Mudogo, V., Tshibangu, D.S.T., Kitwa, E.K., Kanangila, A.B., Lumbu, J.B.S., Ngbolua, K.N., Atibu, E.K., and Kakule, M.K. (2008). Antisickling activity of anthocyanins from *Bombax pentadrum, Ficus capensis, Ziziphus mucronata*: photodegradation effect. *J.Ethnopharmacol.*, **120**: 413-418.

Mpiana, P.T., Mudogo, V., Tshibangu, D.S.T., Ngbolua, K.N., Shetonde, O.M., Mangwala P.K., and Mavakala, B.K. (2007). *In vitro* antisickling activity of anthocyanins extracts of a Congolese plant: *Alchornea cordifolia. M.Arg.J.Med.Sci.*, **7(7):** 1182-1186.

Mpiana, P.T., Tshibangu, D.S.T., Shetonde, O.M., and Ngbolua, K.N. (2007). *In vitro* antidrepanocytary activity (anti-sickle cell anaemia) of some Congolese plants. *Phytomedicine*, **14:** 192-195.

Mpiana, P.T., Mudogo, V., Tshibangu, D.S.T., Ngbolua, K.N., Tshilanda, D.D., and Atibu, E.K. (2009). Antisickling activity of anthocyanins extract of *Jatropha curcas* L. *In:* Recent progress in Medicinal Plants: Chemistry and Medicinal Value, J.N. Govil, V.K. Singh (Eds): Studium Press LLC, USA, Vol.25, p. 104-108.

Mpiana, P.T., Mudogo, V., Ngbolua, K.N., Tshibangu, D.S.T., and Atibu, E.K. (2010). *In vitro* antisickling activity of anthocyanins extracts from *Morinda lucida* Benth (RUBIACEAE) *In* Medicinal Plants: Phytochemistry, Pharmacology and Therapeutics, V.K. Gupta, G.D. Singh, Surjeet Singh and A. Kaul (Eds): Daya Publishing House, New Delhi, Vol.1, p. 330-337.

Mpiana, PT., Mudogo, V., Ngbolua, K.N., Tshibangu, D.S.T., Atibu, E.K., Kitwa, E.K. and Kanangila, A.B. (2009). *In vitro* antisickling activity of anthocyanins extracts of *Vigna unguiculata* (L.) walp., *In* Recent progress *in* Medicinal Plants: Chemistry and Medicinal Value. J.N. Govil, V.K. Singh (Eds): Studium Press LLC, USA, Vol.25, p. 91-98.

Mudogo, V., Mpiana, P.T., Tshibangu, D.S.T., Ngbolua, K.N., Mangwala, K.P., Atibu, E.K., Mbala, M.B., Shetonde, O.M., Kakule, M.K., Makelele, L.K., and Bokota, M.T. (2009). Investigation into the structure, properties and reactivity of bioactive compounds. *In*: The Third World Academy of Science (TWAS) 11th General Conference, Durban, South Africa, 20-23 October, (p22).

Mudogo, V., Mpiana, P.T., Tshibangu, D.S.T., Shetonde, O.M., Ngbolua, K.N., and Mbala, M.B. (2009). Antisickling Activity of Some Medicinal Plants from Congolese Biodiversity. *African Issues Symposium: Food Security, Environmental Suitability and Human Health. Kansas State University (USA), March 30 - April 1, 2009.*

Nagababu, E., and Rifkind, J. M. (2000). Heme degradation during autoxidation of oxyhemoglobin. *Biochem.Biophys.Res.Comm.*, **273** : 839–45.

Neuwinger, H. D. (2000) *African Traditional Medicine*. Mepharm Scientific Publisher, Stuttgart.

Niehaus, W.G. Jr. (1978). A proposed role of superoxide anion as a biological nucleophile in the deesterification of phospholipids. *Bioorganic Chemistry*, **7**: 7-84.

Osuagwu, C. G., and Mbeyi C. U. (2007). Altered Plasma Hexose Sugar Metabolism in Sickle Cell Anaemia. *African Journal of Biochemistry*, **1(3):** 37-40.

Reiter, C.D., Wang, X., Tanus-Santos, J.E., *et al.* (2002). Cell-free hemoglobin limits nitric oxide bioavailability in sickle-cell disease. *Nat Med.*, **8:** 1383-1389.

Rother, R.P., Bell, L., Hillmen, P., and Gladwin, M.T. (2005). The clinical sequelae of intravascular hemolysis and extracellular plasma hemoglobin: a novel mechanism of human disease. *JAMA*, **293:** 1653-1662.

Voet, D., and Voet, J.G. (1998). *Biochimie*. De Boeck Université. 2 éd., Paris.

Traditional and Folk Herbal Medicine: Recent Researches Vol. 1 (2012) *Pages* **13–52**
Editor: **V.K. Gupta**
Published by: **DAYA PUBLISHING HOUSE, NEW DELHI**

2

Ayurvedic Pharmaceutics and Insights on Personalized Medicine

Venil. N. Sumantran[1*] and Girish Tillu[2]

ABSTRACT

The search for new drugs has led to renewed study of Ayurvedic drugs and their mechanisms of action. For the past two decades, studies have focused on pure compounds isolated from the medicinal plants listed in Ayurvedic texts. However, empirical studies perfected over thousands of years have established stringent protocols for preparation of Ayurvedic drugs as crude extracts with minimal adverse effects. Indeed, the concept of preparing standardized crude extracts of herbal drugs is gaining acceptance at the scientific and commercial level. As the advantages and disadvantages of purifying medicinal compounds from source herbs are becoming clear, we require a greater understanding of the biochemical and pharmacological bases underlying the preparation of traditional Ayurvedic drugs as crude extracts. Therefore, the first section of this chapter explains the principles and traditional methods for preparing the different types of basic Ayurvedic formulations. We also examine and compare the traditional and modern methods of quality control for Ayurvedic formulations. Using examples, we then discuss single and multi-herbal Ayurvedic formulations. Throughout this section, we emphasize latest amendments of Drug and Cosmetics Act and ICMR guidelines which recognize the importance and value of traditional methods of herbal processing and solvent extraction. The principles of Ayurvedic pharmaceutics explained in the first section, lay the foundation for exploring issues such as Ayurvedic diagnostics, and factors which determine what drug is used for a given disease in a given patient. Therefore, the second

1 Adjunct Professor, Depatment of Biotechnology, IIT-Madras, India.

2 Group Leader, High Performance Business Computing Group, Centre for Development of Advanced Computing, (C-DAC), Pune – 411 007, Maharashtra, India.

* *Corresponding author*: 7/1 Valli Ammai Aachi Road, Kotturpuram, Chennai – 600 085; E-mail: venil.sumantran@gmail.com.

section of this chapter describes Ayurvedic concepts and approaches for personalized medicine. Although personalized medicine is new to Modern Medicine, it is a familiar practice in Ayurveda. Thus, using an example, we explain how two individuals suffering from the same ailment may be treated with different Ayurvedic drugs. Personalized Ayurvedic drugs can also be used to treat a single disease as it progresses and changes within a single individual. In summary, this chapter addresses crucial aspects of Ayurvedic Pharmaceutics, diagnostics, and explains how they can be applied to the complex topic of personalized medicine.

Keywords: Ayurvedic pharmaceutics, Extraction methods, Ayurvedic therapeutics, Personalized Medicine.

Introduction

Ayurveda stands at a major crossroad in history. The renewed, worldwide interest and acceptance of alternative medicine, has sparked a major resurgence in the field. However, issues such as authenticity of source herbs, the lack of uniform standards, inadequate quality control, and high metal content of certain Ayurvedic drugs; are serious problems which must be addressed. In order to gain international acceptance, Ayurvedic drugs must also go through rigorous analysis by modern science in order to obtain certified proof of therapeutic efficacy. However, modern medicine also faces its own challenges of very high costs and adverse side-effects. Another major concern is the lack of sufficient numbers of new, approved drugs for many infectious, chronic, and lifestyle related diseases. These challenges in modern medicine provide Ayurveda a window of opportunity to resolve its problems and establish itself as a scientifically validated form of medicine.

Attempts have been made to investigate Ayurvedic drugs with the tools of modern science. Thus, the main approach for the past two decades has been the preparation of pure bioactive compounds from Ayurvedic herbs and substances whose therapeutic effects are well documented in traditional texts. Since traditional Ayurvedic methods of preparing crude drugs and formulations are difficult, time consuming, and require validation and standardization; scientists in research and industry have opted to use modern organic solvent extraction methods to isolate bioactive components from the original plants/substances. However, the extraction and isolation of pure chemical compounds from a medicinal plant is controversial, as Ayurveda and other systems of herbal medicine have always advocated holistic healing with single/multi-herbal formulations. Indeed, empirical studies perfected over thousands of years have established stringent protocols for preparation of therapeutic Ayurvedic formulations with minimal side effects. Furthermore, ICMR guidelines and latest Drug and Cosmetics rules state that Ayurvedic formulations prepared by traditional methods do not have to undergo tests for safety and efficacy if they are used for diseases specified in the original texts. ICMR guidelines also state that Ayurvedic drugs prepared by modern methods ('with deviations from classical methods') will be treated as new chemical entities (NCE), and must go through appropriate pre-clinical and clinical tests (ICMR, 2006). In this context, it is very important to closely examine the traditional methods of preparation of therapeutic Ayurvedic formulations. Therefore, the first section of this chapter will focus on the traditional methods used for preparing the most common types of Ayurvedic drug formulations. We will then discuss important types of Ayurvedic single and multi-herbal drug formulations. Quality control aspects related to Ayurvedic drug formulations will also be discussed. The principles of Ayurvedic pharmaceutics explained in the first section, lay the foundation for a discussion on Ayurvedic diagnostics and therapeutics. Then, it is also relevant to discuss the path from Ayurvedic pharmaceutics to actual clinical practice, wherein an Ayurvedic physician designs drugs according to the specific needs of individual patients. Therefore, the second section of this chapter describes Ayurvedic principles, perspectives, and approaches for personalized medicine. Although the concept of personalized/customized medicine is new to modern medicine, it is a familiar concept to those practicing Ayurveda. Personalized medicine requires designing drugs specifically for individual patients and patient sub-types. The second section uses real examples of specific diseases wherein Ayurveda uses personalized medicine.

SECTION I
Ayurvedic Drugs: Preparation, Properties, and Quality Control

The principles, rationale, and methods for preparing different types of traditional Ayurvedic drugs as crude extracts have been explained and described in great detail in a book entitled Extraction Techniques for Medicinal and Aromatic Plants (Swami *et al.*, 2008). In this section, we begin by defining the term 'Ayurvedic crude drug', and explain the value of such drugs. Then, we give an overview of the most commonly used Ayurvedic drugs prepared as crude extracts. This overview includes the processes used to prepare crude extracts, and their physical, chemical, and medicinal properties. We emphasize certain types of Ayurvedic formulations which are particularly relevant to the pharmaceutical industry. In addition, we examine issues of quality control which are of immense importance in the preparation of Ayurvedic formulations. In light of the commercial potential of Ayurvedic drugs, the regulatory bodies such as Department of AYUSH and ICMR, have drafted rules pertaining to the use of Ayurvedic source herbs and their methods of extraction. The Drugs and Cosmetics Act (6th amendment) 2010, is also relevant in this regard. These guidelines and rules will be discussed wherever applicable. Before we describe the preparation, properties, and actions of Ayurvedic formulations, we define the term "Ayurvedic crude drugs" and explain important principles of Ayurvedic pharmaceutics.

Ayurvedic Formulations as 'Crude Extracts'

Ayurvedic drugs have traditionally been prepared as extracts of single or multiple herbs/substances. Such traditional formulations are enriched in a class of chemicals (*e.g.* water soluble or oil soluble). They are termed 'crude extracts' since there is no deliberate attempt to isolate or purify a single chemical entity from the formulation. Therefore, our definition of a crude extract simply refers to a single/multi-herbal formulation prepared by the traditional Ayurvedic methods. Although the chemical composition of these 'crude extracts' is not completely known, it is clear that each of these traditional formulations contains a large number of phytochemicals with different medicinal properties. These 'crude extracts' are likely to have sophisticated and complex therapeutic properties which depend on their chemical composition and bio-availability. This is because the herbal constituents of 'crude extracts' have been prepared by strict, well-established traditional protocols which have withstood the test of time. Modern research also suggests that Ayurvedic drugs as 'crude extracts' have unique properties. In fact, three recent studies compared effects of 'crude extracts' of a given herb with that of pure compounds isolated from the same herb. Results indicated that there were major differences in the biological actions of 'crude extracts' versus that of the pure compounds isolated from a given source herb. The three studies are described in the next paragraph.

First, a recent review on Bael (*Aegle marmelos*), a medicinal plant in Ayurveda suggests that purified compounds from Bael have anti-cancer, anti-diabetic and cardioprotective activities, whereas the crude extract of Bael has antimicrobial and anti-ulcer activities (Maity *et al.*, 2009). The *second* study found that crude extract of *Anoectochilus formosanus* induced apoptosis (programmed cellular suicide) in a breast cancer cell line termed as MCF-7, whereas a purified compound from the same plant arrested cellular growth without inducing cell death (Yang *et al.*, 2004). *Third*, Deocaris *et al.*, tested effects of crude extract of leaves of Ashwagandha (*Withania somnifera*) versus purified withanoloides from this same plant. Using bioinformatic analysis of key genes which regulate viability of cancer cells, they found that purified withanoloides from Ashwagandha altered the expression of large number of genes involved in maintaining cell growth and survival. However, crude extract of leaves of Ashwagandha regulated the expression of a few crucial ("hub") genes which showed high connectivity in the 'gene network' which controls growth and proliferation of cancer cells (Deocaris *et al.*, 2008). Therefore, the authors infer that although there are many hundreds of genes controlling cell growth, proliferation, and cell death; the most effective anticancer drugs like (leaves of Ashwagandha), may work by targeting few genes at the 'heart' of the 'gene network' which controls the proliferation of cancer cells. The authors summarize their important study with the phrase *"No longer a crude view of the crude extract"*.

Renewed Interest in Ayurvedic Formulations as Crude Extracts

Deocaris *et al.*, also point out that the discovery of cellular and molecular networks with complex interactions and regulatory mechanisms, has caused mainstream (modern) medicine to shift its focus from a 'mono-molecular' or a single target approach to combinations and multiple target strategies (Deocaris *et al.*, 2008). We also note that the alarming rise of resistance to antibiotics and anti-cancer drugs in patients, has greatly contributed to this renewed interest in drugs with multiple therapeutic actions. In this context, Ayurvedic formulations as crude extracts become very relevant, as such chemically complex formulations are highly likely to have multiple therapeutic actions. Another reason for interest in Ayurvedic formulations/crude extracts, is because they are not associated with the adverse events which occur with intake of modern medicines. Indeed, the concept of preparing standardized Ayurvedic formulations as 'crude extracts' with strict quality control is gaining acceptance at the scientific and commercial level. For example, standardized crude extracts of Ginger and Guggulu, contain a fixed percentage of gingerols (Funk *et al.*, 2009) and boswellic acids respectively (Lalithakumari *et al.*, 2006). The renewed interest and acceptance of Ayurvedic crude extracts is also reflected in ICMR's rules pertaining to Ayurvedic crude drugs. Accordingly, the latest ICMR guidelines on ethical guidelines for biomedical research on human participants, clearly states that *'purification of the Ayurveda-Siddha-Unani medicines to the level of single chemical constituents is not required, as the purification may result in loss or degradation in synergistic action or properties of original drug'* (ICMR, 2006).

Due to the rising importance of Ayurvedic crude drugs, the remainder of this section will explain and discuss traditional methods of Ayurvedic drug preparation. This topic is also timely and relevant, in lieu of certain ICMR guidelines (ICMR, 2006). Thus, in addition to recognizing the importance of Ayurvedic drugs in their traditional form, and using them for diseases indicated in the original classics, ICMR guidelines also state that the traditional methods of Ayurvedic drug preparation cannot be set aside. Rather, the burden of proof of safety and efficacy rests on those who try to prepare Ayurvedic drugs by new methods, or those who try to use existing Ayurvedic drugs for diseases not specified in the original texts. To quote the ICMR rule: *"When an extract of a plant or a compound isolated from the plant and any compound formulation having plants, metals, minerals and animal products as ingredients has to be clinically evaluated for a therapeutic effect not originally described in the texts of traditional systems or, the method of preparation is different, it has to be treated as a new substance or new chemical entity (NCE) and the same type of acute, sub-acute and chronic toxicity data will have to be generated as required by the regulatory authority for synthetic products before it is cleared for clinical evaluation"* (ICMR, 2006).

Principles of Ayurvedic Pharmaceutics

According to Ayurveda, all matter is made up of *Panchamahabhootas*, (five primordial entities). Therefore, each Ayurvedic drug is also made up of these five primordial entities. These primordial entities are thought to confer Ayurvedic drugs with multiple targets and properties. Target actions of each drug depend on the drug's intrinsic properties and host factors. Intrinsic drug properties in turn, partially depend on how the drug is collected and processed. Factors such as soil type, agro-climatic conditions, cultivation techniques, harvesting, storage, packaging and processing, influence the combination of *Panchamahabhootas*; which in turn affects intrinsic properties and target actions of the drug. Maintaining the intrinsic properties of the drug requires strict quality control. Other relevant factors include correct identification and processing of the source herb before extraction with specific solvents. Finally, the host factors such as overall health status and 'body constitution' of the patient, are also taken into consideration.

Though Ayurvedic drugs can contain herbal, animal, metallic, and mineral components; this chapter focuses mainly on drugs of herbal origin. In the next two sections, we discuss the preparation and processing of the source herbs, medicinal compounds, and the different solvent extraction processes used for preparation of Ayurvedic drugs.

Processing of Ayurvedic Source Herbs and Medicinal Compounds

A key concept is that the methods of processing the source herb can change the properties of the final drug produced from it. Therefore, simple processes like cleaning, mincing, churning, or extracting fresh herbal juices from the source herb; are carefully done according to strict, prescribed methods. The various heating methods and specific utensils used for storage and processing of source herbs are also carefully selected, cleaned, and monitored, since they can exert unwanted effects on the final drugs.

A unique Ayurvedic practice is that of converting toxic source herbs into safe therapeutic drugs with specific medicinal properties. This practice is termed as *Shodhana*, and involves special protocols for processing toxic herbs and every metal and mineral based drug. *Shodhana* includes a range of special purification methods that aim to reduce toxic effects of the drug, and augment its therapeutic properties. For example, boiling of crude aconite with cow urine reduces its toxicity (Thorat and Dahanukar, 1991). The purified aconite is used in combination with Borax for further reduction of toxic effects (Sathe, 1991). Another example is the case of Bhallataka (*Semecarpus anacardium*), wherein a highly toxic source herb is boiled in cow's milk for three hours before being used in formulations (Gogte, 2000 A). Besides being used to mitigate toxicity of a poisonous herb or substance, the process of *Shodhana* can be used to convert a drug into a form that is more bio-available. The *Shodhana* process is essential for metal and mineral based drugs. Thus, metals are boiled with certain natural solvents (milk, cow urine, decoctions, fresh juices, oils, butter milk etc) for a fixed number of times in a specific sequence. For example, certain metal based drugs are heated and dipped in series of solvents (sesame oil, butter milk, cow urine, specially prepared vinegar, decoction of *Dolichos biflorus*) for 7 times each (Parshuramshastri, 2005 A). In summary, optimal processing of the source herb, mineral, metal, or substance; is an essential step before it is extracted with specific solvents in order to produce a final Ayurvedic crude drug. The *Shodhana* process ensures that the final Ayurvedic crude drug has the desired therapeutic actions without adverse effects.

Solvents and Vehicles Used for Herbal Extraction

Extraction of the source herb with specific solvents is used to produce Ayurvedic formulations of high quality. The extraction processes are based on natural and desired therapeutic properties of the final drug. Easy administration, tolerability, and suitability for long-term consumption, are also important attributes of a good Ayurvedic drug. Furthermore, the drug must also be effective at low doses. The types of solvents and vehicles used for extraction determine bioavailability, ease of digestion, speed of action, and the ability to reach the target organ.

Another unique Ayurvedic concept is that of "compatibility" between the properties of drug and the solvent used to extract it. This concept states that the best solvents are those which preserve/maintain the essential/major attributes of the herb. Thus, solvents with 'cooling' properties, will induce 'cooling' effects in the extracted drug. For example, Shatavari (*Asparagus racemosus*) is best extracted in milk/ghee because both drug and solvent posses similar properties (cooling, rasayana, reunion) (Bapalal, 1999 A). On the other hand, non-polar organic solvents like methanol with penetrating or corrosive properties are expected to induce similar (undesirable) properties in the final drug extract. Overall, the Ayurvedic theory believes that the ideal solvent for herbal extraction and preparation of the Ayurvedic drug, is one that maintains and even augments the medicinal properties of the original drug during the various steps of processing. This concept of "compatibility" between the properties of drug and the solvent used to extract it, has yet to be understand in terms of modern science and medicine. Such an understanding may help us better understand issues of Ayurvedic drug uptake, bio-availability, and distribution of the drug within the tissues of the body.

Since our aim is to better understand Ayurvedic crude drugs, we now list and discuss the most commonly used types of traditional Ayurvedic drug formulations (Tables 2.1 and 2.2).

Ayurvedic Formulations Produced by Simple Extraction Methods

Table 2.1 shows the powders, pastes, and 5 different types of aqueous extracts. According to the Ayusoft database, powders and decoctions (*Choornas* and *Kwathas*) are the most commonly used formulations (AyuSoft formulation data mining tool, 2006). For each type of Ayurvedic drug formulation, we explain its method of preparation, properties, and the dose which is administered. General Properties and target tissues of each type of formulation are listed. However, individual subtypes of each formulation, for example; a particular powder (*choorna*) or decoction (*kwatha*), would have additional properties which reflect its unique composition of source herbs, minerals, and other chemical compounds. Below, we discuss each type of formulation (from Table 2.1) in further detail. Examples of formulations and unique properties of formulations are also given.

Choornas/Powders of Crude Herbs

Powders of specific dried herbs are termed *Choornas*. Multiple dried herbs can be powdered and mixed in certain proportions to form more complex *Choornas*. In other cases, dried herbal powders from various herbs can be mixed with different minerals or metals to produce a *choorna*. For example, *Sitopaladi Choorna* contains a mixture of specific herbs and salts (*Joshi YG*, 2003 A). In some cases, *Choornas* are administered with an *Anupana* or vehicle. For example, Sariva (*Hemidesmus indicus*) *Choorna* with milk is prescribed for acid peptic disease (Amlapitta) (Bapalal, 1999 B).

Kalkas (Pastes)

Kalkas are pastes produced by grinding of fresh herbs to get a soft pulp. This dosage form, permits easy consumption, and is thought to have affinity to tissues such as muscle. Usually, *kalkas* are mixed honey, ghee, sugar, or jaggery to improve palatability and absorption. For example, for diseases associated with *Pitta dosha*; ghee is added to the *Kalka*. In contrast, honey is mixed with the *Kalka* for diseases associated with *Kapha dosha* (Paradkar, 2000).

Table 2.1 clearly shows that aqueous extraction is the most common type of method used to produce an Ayurvedic Formulation. Four types of aqueous extracts are discussed below (*Hima*, *Phanta Kwatha* and *Ghanasara*).

Table 2.1: Ayurvedic formulations produced by simple extraction methods

Type of Formulation	*Extraction Method (Shelf Life)*	*Properties and Targets*	*Number of Formulations in AyuSoft database*	*Dose (per day)*
Choorna (powders)	**Dried herb Powder** (Two months)	Dry Easy to digest	**1106**	10 gm
Kalka (paste)	**Fresh herb Paste** (Unstable. Use immediately)	Heavy to digest Properties depend on Source drugs	**752**	10 gm
Svarasa (juices)	**Juice of fresh herb** (Unstable. Use immediately)	**Alleviation of *Pitta*** Targets: *Rasa Dhatu, Rakta* (Blood)	227	20-40 ml
Hima (cold infusions)	**Aqueous.** (Unstable. Use immediately)	**Alleviation of *Pitta*** Easy to digest, Cooling	157	80 ml
Phanta (hot infusions)	**Aqueous** (Unstable. Use immediately)	Easy to digest **Improves *Agni*** (digestive and metabolic status)	116	80 ml
Kvatha (decoctions)	**Aqueous** (Unstable. Use immediately)	Easy to digest Good bioavailability **Targets: *Rasa, Rakta Dhatu* (Blood)**	**1465**	80 ml
Ghanasaras (pastes derived by boiling decoctions)	**Aqueous** Stable upto 1 year)	Generally heavy to digest. Properties depend on source drugs.	57	10 gm

Table 1.1 shows 2 types of solid formulations (*Choornas* and *Kalkas*) and 5 types of aqueous extracts (juices, hot infusions, cold infusions, decoctions, and boiled pastes). Large doses of each of these formulations must be consumed fresh.

Hima (Cold Infusions)

Hima is a type of aqueous extraction process which results in a Cold Infusion. *Hima* infusions are indicated for extracting delicate herbs and flowers which contain volatile oils. Hima infusions are also the preferred extraction method for source herbs with"cooling" properties.

Phanta (Hot Infusions)

Phanta is an Aqueous extract wherein hot water is directly poured onto the coarsely powdered source herb/substance. The resulting mixture should be consumed after it reaches a lukewarm temperature. Thus, the *Phanta* extraction process is shorter and milder than the process used to produce *Kwathas* (Decoctions). Therefore, when a drug has heat-sensitive active principles, its aqueous extract would be prepared as a *Phanta* rather than a *Kwatha*. For example, hot infusion of dry Ginger (*Shunthi Phanta*) is commonly used as *Anupana* (vehicle) for Castor oil (Bapalal, 1999 C).

Kwathas (Decoctions)

Kwathas are aqueous extracts which are specially prepared to yield decoctions or concentrated liquids. For this purpose, the raw material is cleaned, minced, boiled gently with fixed a ratio of water, and then concentrated such that the final volume is 1/8 of the original volume. The *Kwathas* can be consumed after filtration. Since the filtration process does not use filters of fixed pore size, it is very likely that *Kwathas* are enriched with water soluble compounds, but also contain some proportion of water insoluble compounds and suspended particles. The process of preparing *Kwathas* is shown as a Flow Chart 2.1.

Flow Chart 2.1: Protocol for Preparation of Kvatha (Decoction)

Coarse powder + Water **(1:16)**

↓

Boil gently to get 1/8th part of decoction. **(2 parts)**

↓

Filter, and use Filtrate

Ghanasaras (Semi-solid Pastes)

These aqueous extracts are produced by prolonged boiling (3-5 hours) of the *Kwatha* decoction, juices and *Hima* and *Phanta* preparations. The resulting semi-solid paste can be dried and stored as a powder. Unlike the *Kwatha, Hima* and *Phanta* preparations; *Ghanasaras* most likely contain a significant proportion of water insoluble compounds such as complex polysaccharides, and are heavier to digest. The process of preparing *Ghanasaras* is shown as a Flow Chart 2.2.

Flow Chart 2.1: Protocol for Preparation of Ghanasara

Coarse powder + Water **(1:16)**

↓

Overnight maceration

↓

Boil gently to get 1/8 th part of decoction. **(2 parts)**

↓

Filter, and use Filtrate

↓

Prepare Dry aqueous extract from Filtrate

Overall, Table 2.1 shows two types of solid formulations (*choornas kalkas*) and 5 types of aqueous extracts (juices, hot and cold infusions, decoctions, and boiled pastes). Table 2.1 also shows that *Choornas* and *Kwathas* are the most commonly used Ayurvedic formulations. Most of these formulations have a short shelf-life, and must be consumed fresh. In addition, the formulations shown in Table 2.1 are relatively easy to digest, although the required therapeutic doses are large (Parashuramshastri, 2005 B).

Ayurvedic Formulations Produced by Complex Extraction Methods

Table 2.2 shows Ayurvedic formulations which are more difficult to produce than those in Table 2.1. These include medicated formulations (medicated milk, ghee, and oil), pills, and 4 types of highly specialized formulations (*Avalehas, Asavas, Arishtas,* and *Bhasmas*). For each type of formulation, the Table lists the method of preparation, medicinal properties, shelf-life, and the therapeutic dose.

Ksheerpaka (Medicated Milk)

In *Ksheerpaka* or medicated milk, milk is the solvent. According to Ayurveda, this type of formulation has 'cooling' and anabolic properties. Apart from therapeutic action of the drug, *Ksheerpaka* extracts also nourish tissues and have anabolic effects. These extracts are also thought to reduce the hot (*Ushna*), penetrating (*Teekshna*) properties of the drug, while improving efficacy and specific target action. Other advantages of the *Ksheerpaka* extracts include better palatability, the ability to administer large doses of the drug, and its suitability for all patient types and age groups. Indeed, *Ksheerpaka* extracts also help patients comply with the prescribed drug dosing regimen. This, Pippali (*Piper longun*) is an effective drug for respiratory diseases, and can be administered in increasing doses only if it is prepared as a 'medicated milk' (Joshi, *2003 B*). The medicated milk avoids the hot and penetrating properties of Pippali and prevents adverse effects like hyperacidity, burning urination, and bleeding. The final dose of Pippali administered as a *Ksheerpaka* extract, could be as much as 20 gm per day. Such a high dose is difficult to achieve with other dosage forms. The process of preparing *ksheerpakas* is shown as a Flow Chart 2.3.

Flow Chart 2.3: Protocol for Preparation of Ksheerapaka (Medicated milk)

Coarse powder of Drug + Cow Milk + Water **(1:8:32)**

↓

Boil gently till water remains

↓

Filter the milk, and use Filtrate

Ghritha (Ghee Extract)

Ghrithas are formulations which use Ghee as solvent for extract. *Ghrithas* require the mixing of juices or decoctions with pastes of the drugs (Kalkas) and Ghee (or oils).

Table 2.2: Ayurvedic formulations produced by complex extraction methods

Type of Formulation	*Extraction Method (Shelf Life)*	*Properties and Targets*	*Number of Formulations in AyuSoft database*	*Dose (per day)*
Ksheerpaka (medicated milk)	**Milk** (Use within four hours)	**Alleviates Pitta** Anabolic properties	321	80 ml
Ghrita (medicated ghee)	**Ghee** (Use within Four months)	**Alleviates Pitta and Vata** Regenerative properties Targets: muscles, adipose, nervous system, and cartilage	**997**	5 to 50 ml
Thaila (medicated oils)	**Oil** (Use within Four months)	**Alleviates Kapha and Vata** Regenerative properties Targets: muscles, adipose, nervous system, bones and cartilage	**740**	5 to 50 ml
Guti, Vati (pills, tablets)	**Aqueous extracts** used as ingredients of pills and tablets (Use within One Year)	Properties and target action depend on ingredients and processes	390	0.25–1.0 gm
Avaleha (semisolid)	**Aqueous** (Use within One Year)	**Anabolic properties** Targets: Muscles, Adipose tissue, Shukra dhatu	418	40 gm
Asava and Arishta (Fermented Extracts)	**Self generated alcohol** (Medicinal properties increase over time)	**Alleviates *Kapha* and *Vata*** Increases Pitta Aids digestion/metabolism Targets: *Rasa*, *Rakta* dhatu (Blood)	120	40 ml
Bhasma (specially processed metals and minerals)	**Metallic or minerals** (Medicinal properties increase over time)	Fast acting, quick absorption, Medicinal properties depend on source drug and processing	32	30 to 250 mg

Table 2.2 shows 3 types of medicated formulations (milk, ghee, oils), pills, and 4 types of specialized formulations. (*Avaleha, Asava, Arishta,* and *Bhasmas*). These formulations are complex, relatively stable, and are effective at smaller doses than the formulations shown in Table 2.1.

This mixture is heated until most of the water is evaporated and the remaining paste becomes waxy so that it can be rolled with a finger. Therefore, the heated ghee fraction of the *Ghritha*, probably extracts drugs from the source herbs in the presence of water and other liquids. The process of preparing *Ghritas* is shown as a Flow Chart 2.4 (Parashuramshastri, 2005 G). The quantity and duration of heat affects properties and target actions of *Ghritas*. It is important to note that the Ghee itself can 'absorb' properties of the decoctions and pastes with which it is mixed. In this respect, the Ghee has adjuvant properties as it serves as the medium/carrier for the bioactives present in the decoctions and pastes. During the preparation of *Ghritas*, the Ghee may enhance the properties of the phytochemicals that it interacts with. This property of ghee is termed as *'yogavaahitwa'*. In modern terms, ghee is said to have adaptogenic action (Umretia *et al., 2009*).

Flow Chart 2.4: Protocol for Preparation of Ghritha or Thaila (Medicated Oil or Ghee)

Pulp of drug(s) + Oil/Ghee + Water/decoction or fresh juices of drug(s)
(1:4:16 parts)

↓

Boiled on mild fire
(Stir well continuously so that pulp is not charred or adheres to the vessel)

↓

Boil till the liquid part evaporates

↓

Carefully heat the mixture to avoid charring of pulp
(Retain some amount of moisture in the pulp)

↓

Filter the mixture to get the medicated Oil or Ghee

Thaila (Oil Extract)

Thailas (medicated oils) are prepared by the same procedure used for the preparation of Ghrithas. The major difference between *Thailas* and *Ghrithas* concern their properties and target actions. Generally, *Ghrithas* are cooling (*Sheeta*) whereas *Thailas* are hot (*Ushna*). Therefore, *Ghrithas* are used for their cooling effects in diseases associated with imbalace of *Pitta* and *Vata doshas* (Acharya, 2000 A). Since *Thailas* are hot, they are thought to permeate tissues rapidly, and are useful for alleviation of pain. Hence, *Thailas* are indicated for diseases associated with imbalance of *Kapha* and *Vata Doshas* (Acharya, 2000 A).

Guti/Vati (Pills)

The *Guti/Vati* dosage forms are basically tablets or pills. A convenient dosage form and long shelf-life are the two major advantages of Gutis/Vatis. These tablets or pills can be prepared from different *Choornas, Kalkas* or *Ghanasara* extracts. Though they are heavy to digest, they are administered with *Anupana* (vehicles), and can therefore reach their target site. Properties of individual *Gutis/Vatis* depend upon the unique ingredients and the extraction process originally used to generate the *Choorna, Kalka,* or *Ghanasara* extract.

Avaleha (Pastes with Regenerative Effects)

Avaleha or *Prasha* is a type of semisolid preparation obtained by heating decoctions, juices or pastes of drugs. *Avaleha* includes jaggery, sugar or sugar-candy, and ghee or oils; boiled with specific juices or decoction of drugs. The process of preparing *Avalehas* is shown as a Flow Chart 2.5. *'Chyavanprasha'* is well known example of *Avaleha*, it is popular for its anti-aging, regeneration, and immunomodulatory properties. Overall, Avalehas have anabolic and regenerative effects, and target muscle and adipose tissue (the *Mamsa, Meda* and *Shukra Dhatu* (Parashuramshastri, 2005 G) Therapeutic properties of a specific type *Avaleha* depend upon the unique mixture of ingredients used for its preparation. For example, *Pippalyavaleha* (main ingredient: *Piper longum*) is indicated in respiratory diseases. Another Avaleha termed *Agasti Haritaki Avaleha* (a rasayana formulation with main ingredient: *Terminalia chebula*), is indicated in obesity (Panditrao *et al.*, 1990 A). Although *Avalehas* (Table 2.2) and *Kalkas* (Table 2.1) have a similar consistency, they vary in certain aspects. Unlike *Avalehas, Kalkas* have a short shelf life, and must be consumed fresh. More importantly, the therapeutic effects of *Kalkas* are not as sustained as those of *Avalehas*, which can nourish tissues.

Flow Chart 2.5: Protocol for Preparation of Avaleha

Prepare Decoction or Juice of drugs

↓

Sugar or Jaggery is dissolved in the Decoction/Juice and filtered.

↓

This mixture is Gently boiled till it becomes a Semisolid.

↓

Ghee or oils are added as specified in the formula.

↓

Add fine powders of specific herbal drugs (*Piper longum, Piper nigrum,* and *Zingiber officinale*) and mix well to form a homogenous mixture.

↓

Honey maybe added to the Cooled mixture

Asavas and *Arishtas* (Fermented Extracts)

Many of the current studies on Ayurvedic drugs use hydro-alcoholic extracts of the source herbs for research and evaluation. In lieu of the popularity of hydro-alcoholic extracts, we now discuss the *Asavas* and *Arishtas*. *Asavas* and *Arishtas* are Ayurvedic drugs generated by a fermentation process outlined by Ayurvedic texts. Due to fermentation, these drugs contain alcohol, in addition to the medicinal phytochemicals. In other words, *Asavas* and *Arishtas* represent a natural form of hydro-alcoholic drug extract in Ayurveda. *Asavas* and *Arishtas* are used for a very wide range of ailments. Indeed, there are 79 such formulations, and some have been commercialized (Mishra, *et al.*, 2010). The methods of preparation of *Asavas* and *Arishtas*, and their unique properties are presented in the next few paragraphs.

Preparation and properties of Asavas and Arishtas

Asavas (fermentated infusions) and *Arishtas* (fermentated decoctions) are prepared by a slow, natural, fermentation process using certain herbs, jaggery, and a source of yeast. This fermentation process generates alcohol (5 -15 per cent), in a self-limiting manner. The process of preparing *Asavas* and *Arishtas* is termed *Sandhana kalpana*, and is shown as a Flow Chart 2.6. *Asavas* and *Arishtas* have certain unique and noteworthy properties. First, these formulations are very stable since the bioactive products of fermentation are continuously exposed to a low concentration of alcohol. Indeed, the medicinal properties *Asavas* and *Arishtas* are thought to increase with time. Second, the constant presence of alcohol in these formulations may detoxify certain phytochemicals, and increase the potency of others. Third, these fermented extracts exhibit rapid therapeutic effects at low doses. Thus, the aqueous-alcoholic nature of the extract may promote increased uptake of the drug by target organs (Mishra *et al.*, 2010). According to Ayurveda, 'no substance is inert'; and every substance or solvent, (for example, methanol) that is in contact with the drug, affects its properties. Hence, most Ayurvedic experts do not consider modern 'hydro-alcoholic extracts' as equivalent to the classical *Asavas* and *Arishtas*. Since hydro-alcoholic extracts of medicinal plants are commonly used in modern research on Ayurveda, and by the herbal drug industry, it is important to discuss the main differences between modern hydro-alcoholic extracts and *Asavas* and *Arishtas*.

Differences Between Asavas and Arishtas and Modern Hydro-alcoholic Extracts

- ☆ Various types of alcohols (methanol, ethanol, and butanol etc) could be used to prepare a modern hydro-alcoholic extract, whereas ethanol is the primary type of alcohol generated during the fermentation process used for preparation of *Asavas* and *Arishtas*.
- ☆ Modern hydro-alcoholic extracts of medicinal plants are generated by chemical extraction of the plant with a specific hydro-alcoholic solvent. In contrast, *Asavas* and *Arishtas* arise from the natural biological process of yeast-based fermentation. Therefore, the traditional process of preparing *Asavas* and *Arishtas* results in 'self-generation' of alcohol.
- ☆ The end product of a modern hydro-alcoholic extract does not contain any alcohol, while *Asavas* and *Arishtas* constantly maintain a fixed, low content of alcohol (5-15 per cent) which is thought to enhance bio-availability of these fermented extracts.
- ☆ Due to these major differences between *Asavas* and *Arishtas*, and modern hydro-alcoholic extracts, the phytochemical components and properties of an *Asava* or *Arishta* containing 85 per cent water and 15 per cent ethanol would be very different from a modern hydro-alcoholic extract containing the exact same ratio of water : alcohol.

Flow Chart 2.6: Protocol for Preparation of Asava/Arishta

Coarse powder + Water **(1:16)**

↓

Boil gently to get 1/8 th part of decoction. **(2 parts)**

↓

Filtrate

↓

Decoction #

↓

Add Sugar or Jaggery in mentioned proportion (20 – 40 per cent) according to formula

↓

Add additive **drugs*** according to formula

↓

Add fermenting agent

↓

Put the mixture in wooden or earthen container, and seal with a lid.

↓

Remove lid after specified period (3 to 6 months), and examine contents to ascertain completion of the fermentation process

↓

Filter the **Arishta** and fill in bottle. Seal the bottle.

\# **Asava**: Use fresh juices **Arishta**: Use decoctions
(*Drugs with Bio-enhancing property)

These major differences between modern hydro-alcoholic extracts and *Asavas / Arishtas* listed above are significant, and there are important rules on this issue. Thus, according to Drugs and Cosmetics (6th amendment) Rules 2010, *hydro-alcoholic extracts prepared by different processes than those mentioned in Ayurvedic classics (books mentioned in schedule 1 of Drug and Cosmetics act, 1940) would require proof of safety and efficacy for issue of license. Whereas extracts (aqueous or hydro-alcoholic) prepared by traditional method, mentioned in classics will not require proof for efficacy or safety, but if these extracts are to be used for 'new indication', they will require proof of 'effectiveness'.*

In summary, the modern hydro-alcoholic extracts are not considered to be equivalent to the classical *Asavas* and *Arishtas*. Therefore, pharmaceutical companies would have to conduct tests for chemical content, safety, and therapeutic efficacy; for all hydro-alcoholic drug extracts of Ayurvedic source herbs generated by modern methods (which have deviated from classical methods).

Bhasmas

Bhasmas are very complex metallic or mineral based formulations. During preparation of bhasmas, the toxic raw metals or minerals are converted into safe and absorbable forms by a special *Shodhana* process. This process of Shodhana includes special heating methods which aim to purify the metal/mineral. After *Shodhana,* the metals undergo a process termed as *Marana,* wherein the metals are triturated with herbal juices and heated to specified high temperatures in specialized furnaces to produce the final *bhasma*. It is possible that the *Marana* process converts the original metal into an organo-metallic form with reduced toxicity (*Rajeshwardattashastri*, 2005 A). This fascinating possibility needs to be investigated. Although modern science has established that free metals are toxic because they poison vital enzymes, little is known about organo-metals and their potential for curative action. The properties of *bhasmas* depend upon the source metal/mineral, the herbal drugs used for the *Marana* process, and the type and duration of heat cycles. Specially purified mercury and sulfur are used in the processing of many *bhasmas* (*Rajeshwardattashastri,* 2005 A). *Bhasmas* are fast acting, tasteless, and are thought to be effective in low doses (*Sharma D,* 1996). Hence *bhasmas* are used in combination with herbs or animal products. For example, *Saptamrut Loha* includes *Loha bhasma* (iron bhasma), *Triphala, Glycerrhiza glabra* and honey and Ghee (*Rajeshwardattashastri,* 2005 B).

In summary, Table 2.2 shows three types of medicated formulations (medicated milk, ghee, and oil), pills, and four types of highly specialized formulations (*Avalehas, Asavas, Arishtas,* and *Bhasmas*) According to the Ayusoft database, formulations made of medicated ghee and and oil (*Ghrithas* and *Thailas*) are the amongst the most commonly used formulations (AyuSoft formulation data mining tool, 2006). Most of these formulations are highly complex and difficult to prepare. However, these formulations have a longer shelf-life and are effective at far lower doses than the relatively simple formulations listed in Table 2.1. Table 2.2 also shows general properties, extraction methods, and targets for each type of formulation. Individual sub-types of each formulation (*e.g.* a particular *avaleha* or *ghritha*) would have additional properties which reflect its unique composition of source herbs, minerals, and other chemical compounds.

Modern Research on Extraction Methods for Ayurvedic Source Herbs

Thus far, we have discussed the different principles and methods governing the production of a wide range of Ayurvedic formulations (Tables 2.1 and 2.2). Indeed, the unique medicinal properties of formulations ranging from *choornas, kwaathas, ghrithas, guti/vati, avalehas, asava/arishtas,* and *bhasmas*; are in large part, due to the carefully controlled and diverse methods of processing and extracting the source herbs and compounds. The complex chemistry, biochemistry, and physiological effects of Ayurvedic formulations produced by these traditional methods has not been studied. However, a quick survey of the literature is heartening, because it shows that scientists are doing research on different types of extraction methods for drugs derived from Ayurvedic source herbs. Indeed, we found three types of interesting publications which we discuss in the next three paragraphs.

First, many studies are testing the effects of aqueous extracts of Ayurvedic drugs for different medicinal properties in cell and/or animal based models. This is encouraging, since it indicates a renewed interest in investigating aqueous extracts, which are one of the most commonly used types of extracts of Ayurvedic herbs (Table 2.1). In fact, studies on aqueous extracts of Ayurvedic herbs were relatively rare in the previous decade. An interesting and recent study was done to compare the bioactivity of *Ashwagandha* roots prepared by 2 different extraction methods. Thus, *Jain et al*, extracted Ashwagandha roots by hot continous percolation or maceration. Water, water-alcohol in different ratios, and alcohol alone; were used as solvents. Results showed that the Ashwagandha extracts produced by maceration with water contained a higher content of herbal matter and showed greater potency of anti-stress activity than Ashwagandha extracts produced by hot continuous percolation with water. The data also showed that the alcoholic extracts lacked anti-stress activity, whereas the hydro-alcoholic extracts produced by maceration had significant anti-stress activity. The authors conclude that the traditional method (maceration with water (1 part root : 8 parts water) could be the best method to extract Ashwagandha roots (Jain *et al.*, 2010).

Second, there are studies which have compared several different extraction methods for a single medicinal plant/substance. For *e.g. Bacopa monnieri* (Brahmi) has been used in Ayurvedic medicine as a tonic for promoting mental health and memory. The tetracyclic triterpenoid saponins in Brahmi are reported to be the molecules responsible for cognition enhancement. This study tested and compared 9 different methods of extraction of the Brahmi plant. Water, methanol, or hexane at different temperatures and pressure, were used as solvents; and the Soxhlet apparatus was used in some cases. The final HPLC data showed that the *Brahmi* extract containing the highest content of total saponins, was obtained by percolation with ethanol after soaking the plant material with water (Sakchai and Kornkanok, 2007).

The *third* type of interesting study in the literature, compares standard modern drugs for a particular disease with traditional ayurvedic medicines indicated for that same disease. For *e.g.* Lieu *et al.* compared standard drug (levodopa), used for treatment of Parkinson's disease with a water extract of *Mucuna pruriens,* (a legume extensively used in Ayurveda for treatment of Parkinson's disease). Both drugs showed long-term reduction of neurological symptoms, however, animals treated with the *Mucuna pruriens* extract did not show the adverse effects associated with administering levodopa (Lieu *et al.*, 2010).

Overall, the trends in modern research are addressing very pertinent issues on preparation and therapeutic actions of Ayurvedic drugs.

Single Herbal Formulations with Multiple Therapeutic Effects

Thus far, we have described basic types of Ayurvedic formulations (Tables 2.1 and 2.2), and emphasized the importance of the traditional extraction methods. In addition, the types of solvents and extraction methods chosen to prepare a drug from a given source herb are also extremely important. This point is best appreciated when we consider that in Ayurveda, a single plant can be extracted with different solvents to produce drugs with very different medicinal properties. We illustrate this simple but powerful concept with the case of Guduchi (*Tinospora cordifolia*), a well established *Rasayana* which is reported to have a wide range of properties, such as, anti-inflammatory, anti-diabetic, antipyretic, and anti-ulcer effects. Before we discuss the example of *Rasayana* of Guduchi (*Tinospora cordifolia*), we will define the term *Rasayana.* The word '*Rasayana*' means producing tissues of good quality. *Rasayana* therapy aims at producing tissues of optimum quality and quantity. *Rasayanas* could be general or tissue specific. For example, Amla (*Phyllathus emblic*) is a general *rasayana*, whereas the *Rasayana* of Sariva (*Hemidesmus indicus*) targets Blood.

According to Ayurvedic theory and practice, the medicinal effects of *Guduchi Rasayana* depend on its dosage form. Thus, Figure 2.1 shows that Guduchi (*Tinospora cordifolia*) in 6 different dosage forms has 6 different and specific medicinal effects. For example, the fresh juice of Guduchi with honey is indicated in Diabetes (*Prameha*) due to its *Rasayana* activity (*Parashuramshastri*, 2005 C). Guduchi decoction with Pippali powder is effective in inflammation due to its '*Pachana*' action. Cold infusion of Guduchi is effective in fever (*Parashuramshastri*, 2005 D). In this example, all three liquid forms of Guduchi (juice, decoction and cold infusion) target the blood (*Rasa Dhatu*). Medicated ghee extract of Guduchi is indicated in acid peptic disease as it alleviates the vitiated *Pitta Dosha* and initiates healing of ulcer due to its *Rasayana* property (Savrikar *et al.*, 2010) (Rajeshwardattashastri, 2005 C). Guduchi given in the form of a rectal enema (in medicated milk), is used for bone diseases such as Osteoporosis. The medicated milk extract of Guduchi is thought to alleviate the vitiated *Vata Dosha*, and strengthens bones due to its *Rasayana* properties (Acharya, 2000 B). *Arishta* of Guduchi is indicated in diseases of respiratory system caused by imbalances in *Kapha Dosha.* This is because the *Arishta* form of Guduchi has *Agnideepana* activity and can alleviate the vitiated *Kapha Dosha* (Gogte, 2000 B). To summarize, Figure 2.1 shows that the aqueous extracts of Guduchi (juice, decoction and cold infusion) are effective for inflammation and fever, whereas the other 3 different types of Guduchi formulations show varied *Rasayana* properties which can be used for treatment of diabetes, ulcers, and Osteoporosis respectively. Therefore, the *Rasayana* property of Guduchi can be targeted to different body tissues, by varying the solvents and types of extraction procedures used for the source herb: *Tinospora cordifolia.*

Multi-Component Formulations: Synergy and Target action

Ayurvedic pharmacology has also developed methods wherein combinations of specific drugs or drugs and vehicles, are used to obtain beneficial synergistic effects. For *e.g.* the combination of *Shunthi*/Dry Ginger (*Zingiber officianle*) + *Tinospora*

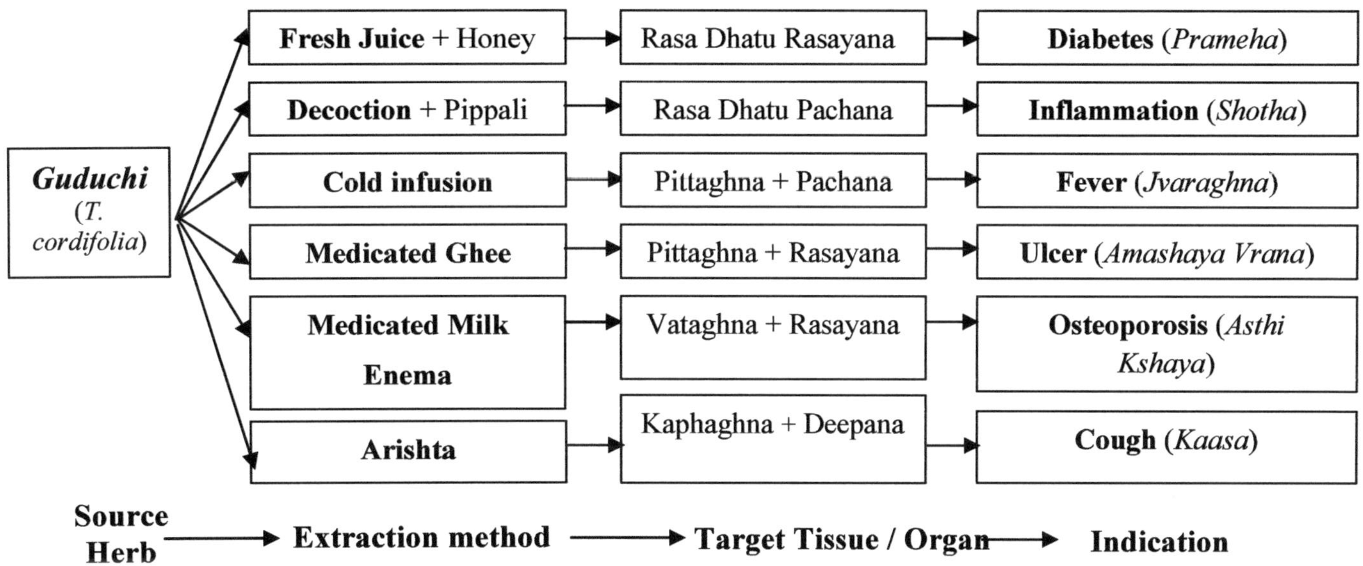

Figure 2.1: Specific formulations and therapeutic activity of *Guduchi* (*Tinospora cordifolia*).

Guduchi (*Tinospora cordifolia*) in different dosage forms has specific medicinal effects. Aqueous extracts of *Guduchi* (juice, decoction and cold infusion) target the blood, and are effective for inflammation and fever. The three different *Rasayana* properties of *Guduchi* depend on the type of *Guduchi* formulation. *Guduchi* with honey is indicated for diabetes. The medicated ghee form of *Guduchi* is effective for ulcers, whereas the medicated milk enema of *Guduchi* is effective for Osteoporosis. The *Arishta* form of *Guduchi* is effective for coughs.

cordifolia is synergistic and effective for *Amavata* (Rheumatoid Arthritis) (*Sharma*, 1982). The synergistic interactions between herbal components are determined by careful and specific choices of the starting herbs which are used. Other important factors include extraction procedures and maturation time for the final drug to be produced. Indeed, a multi-factorial consideration of all these aspects is necessary to ensure that the combinations of Ayurvedic drug have the desired medicinal effects. A good example of synergism comes from interactions between the Ayurvedic drug and its vehicle. The vehicle (*Anupana*) is a liquid/other substance, which is administered along with the drug. At the simplest level, the *Anupana* improves bio-availability, absorption, and distribution of the drug. Examples of routinely used *Anupanas*, are hot water, honey, ghee, milk, buttermilk, and oils. The *Anupana* which is used, partially depends on the *Dosha* which is vitiated in the patient. Thus, for diseases associated with *Vata Dosha*, ghee or oils are used as *Anupana*. For diseases associated with *Pitta dosha*, milk can be used as *Anupana*, whereas ginger juice is a suitable *Anupana* for diseases associated with *Kapha Dosha*. At a more complex level, the *Anupanas* can themselves be 'drugs' with specific target actions. For example, the components of the *Anupana* can synergize with the actions of the primary drug being used. We illustrate this concept with the example of *Mahayogaraja Guggulu* in the next paragraph.

Mahayogaraja Guggulu is a commonly used herbo-mineral drug. The main ingredients are *Guggulu* oleoresin (gum resin) of *Commifora mukul*, *Triphala*, jiggery, ghee, and *bhasmas* of mica, iron, tin, lead and silver. This formulation is also a *Rasayana* with broad spectrum activity, and it alleviates symptoms related to ailments caused by imbalance in one/all of three *Doshas*. *Mahayogaraja Guggulu* is effective in rheumatic, neuromuscular, respiratory and other diseases caused by abnormal *Agni* (digestive and metabolic functions) (Parashuramshastri, 2005 E). When the main drug *Mahayogaraja Guggulu* is administered with a particular *Anupana*, the actions and target site of *Mahayogaraja Guggulu* change dramatically. Thus, Figure 2.2 shows that *Rasnadi Kvath* + *Mahayogaraja Guggulu* is useful for treatment of diseases associated with *Vata dosha* (Parashuramshastri, 2005 E). The combination of *Mahayogaraja Guggulu* + *Kakolyadi Kvatha* is useful for treatment of diseases associated with *Pitta dosha* (*Parashuramshastri*, 2005 E). Finally, *Mahayogaraja Guggulu* + *Aragvadhadi Kvatha* is useful for treatment of diseases associated with *Kapha dosha* (Parashuramshastri, 2005 E). Figure 2.2 also shows that *Mahayogaraja Guggulu* can be combined with 3 different *Anupanas* (*Darvi* decoction, cow urine, and honey), in order to treat diabetes, anemia, and obesity, respectively. (Parashuramshastri, 2005 E). Similarly, a combination of *Mahayogaraja Guggulu* with decoctions of *Guduchi* or *Triphala* is indicated for gout or eye diseases respectively (Parashuramshastri, 2005 E). Therefore, the case of *Mahayogaraja Guggulu* serves as good example of synergistic interactions between the components of a multi-component Ayurvedic drug and its vehicle/*Anupana*.

Choice of Ayurvedic Drug and Type of Formulation

Having surveyed the various types of types of Ayurvedic drugs, drug combinations, and the wide variety of extractions method used to produce these

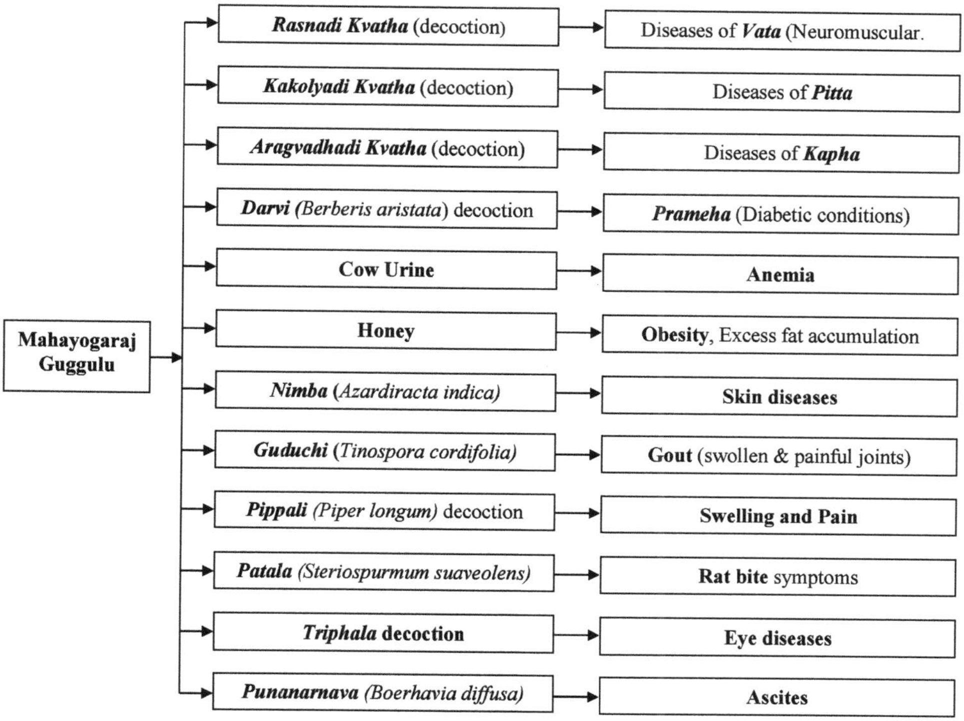

Figure 2.2: Different formulations of *Mahayogaraj Guggulu*

This diagram shows how *Mahayogaraj Guggulu* could be adminstered with 12 different vehicles (*Anupana*) for 12 specific clinical targets.

drugs, we are faced with an obvious question: *How do Ayurvedic physicians select the most appropriate drug and extraction process for a given patient with a specific illness?* The factors pertaining to the patient's overall health, body constitution, and the drug's intrinsic properties, are analyzed in totality. This type of analysis determines which source herbs and solvent extraction processes (aqueous, fermentation, or dietary solvents) are required. Finally, the sub-type of extraction process which will be used (aqueous decoction, aqueous infusion, fermented extract, ghee/milk based extracts etc.) is also chosen, because it is important in determining the bio-availability of the drug, and the time taken for it to exert its therapeutic effects. Thus, *Bhasmas, Asavas, Arishtas,* and *Kwathas;* have relatively rapid action, whereas drugs administered as milk, ghee, or oil based extracts; have a slower but more sustained effect. The choice of extraction process of the herb also depends on what the patient can tolerate and benefit from. A deeper discussion on the 'choice of drug' will be explained in the context of Ayurvedic diagnostics in section 2 of this chapter.

Quality Control of Ayurvedic Formulations:

Traditional and Modern methods of Quality Control

The most commonly used Ayurvedic formulations have been discussed in Tables 2.1 and 2.2. Although, the chemical composition of these traditional Ayurvedic crude drug formulations is usually not analyzed, there are parameters one can to use to verify whether a particular formulation is being produced correctly. Thus, juices and pulps should possess the colour, odour, and taste of the source herb. Decoctions and hot infusions should also maintain the colour, odour, and taste of the individual ingredients. Many extraction processes require a heating step. Since heating by flame can be inexact, traditional Ayurvedic texts describe the changes in colour, odour, viscosity, turbidity, frothiness, etc. that must take place in an extract heated with a particular solvent for a certain period of time in a certain type of utensil. Interestingly, even the changing sounds produced by the extract being heated are noted ! Following these detailed precautions would ensure that the material is heated to the right temperature and at the desired rate. The processes of *Shodhana* (purification) and *Marana* (conversion of metals and minerals into Bhasma form) described earlier, are extremely important, and must be performed according to strict protocols explained in the classical texts. Overall, a close observation of the physical properties of formulations during processing, serve as valuable indicators of the quality and therapeutic efficacy of the final crude drug.

Today, the quantitative physical and chemical parameters such as specific gravity, pH, total content of solids, water and alcohol soluble content, and ash values; are preferred. The HPLC profile of the final product now represents a crucial piece of data, and is widely used to establish the purity and quantity of important analytical and chemical markers within the Ayurvedic drug. The ICMR guidelines also emphasize the importance of standardization and quality control based on analysis of important chemical markers and biomarkers in the source herb. Chemical fingerprints of all compounds present within the Ayurvedic drug are also being analyzed.

Need for Novel Approaches to Improve Quality Control of Ayurvedic Drugs

Although traditional and modern quality control processes are rigorous, and are useful for obtaining reproducible batches of standardized Ayurvedic drugs, there are aspects of quality control which require additional research. For example, analytical markers for a given drug may not always be equivalent to the actual chemicals responsible for the drug's therapeutic activity. This important point has been clearly documented in the case of St. Johns Wort (Laakmann *et al.*, 1998). Therefore, there is an urgent need for novel approaches which link the HPLC (chemical) profiles of an Ayurvedic drug/formulation with its bioactivity. One biochemical approach has been suggested to link the HPLC profiles of the herbal components of *Triphala* with their bioactivity (Sumantran, 2010). Additional studies of this nature would be useful.

Studies which perform qualitative and quantitative comparisons between Ayurvedic and modern quality control parameters are also very useful. For example, there is an interesting report which carefully compared the properties of the *Ghana* (*Ghanasara* extract) of *Guduchi* (*T. cordifolia*) produced by the classical method versus that produced by a Soxhlet apparatus (Umretia *et al.*, 2009). Interestingly, the study found that the content of water and alcohol soluble compounds in classical *Guduchi Ghana* was 50 per cent lower than that of *Ghana* produced by the Soxhlet method. However, the total solid content of classical *Guduchi Ghana* was almost twice that produced by the Soxhlet method. The authors attributed this to a greater proportion of starch retained in the classical *Guduchi Ghana* versus the *Guduchi Ghana* prepared by the Soxhlet method. The prolonged boiling step in the *Ghanasara* extraction process would have solubilized starch from *Guduchi*. Notably, this solubilization of starch would not have taken place if the other milder methods of aqueous extraction (infusions and decoctions) were used.

In summary, we need more research in order to understand the chemical and biochemical differences between extracts of source herbs produced by traditional versus modern methods. Such studies would provide a deeper understanding of the complex chemical changes taking place within the 'crude extracts'. Such insights could also potentially lead to the development of new protocols for producing standardized Ayurvedic crude drugs. One expects that these new protocols would more closely mimic traditional Ayurvedic methods and would therefore ensure better safety, tolerability, and efficacy of the Ayurvedic drug. Finally, the ICMR regulations clearly indicate that there is importance and value in retaining Ayurvedic 'crude drugs' extracted by the traditional methods. Therefore, this is an ideal time to re-examine traditional Ayurvedic pharmaceutics, and gain insights for designing new extraction methods for drugs derived from Ayurvedic source herbs.

Summary of Section 1

This section has defined and discussed the importance of Ayurvedic crude drugs, types of formulations, and general principles of Ayurvedic pharmaceutics. We then described the major types of Ayurvedic formulations, and their properties (Tables 2.1 and 2.2). The advanced nature of Ayurvedic pharmacology and methods for preparation of formulations were explained with 2 specific examples. First, we used an example to show how a single herbal drug prepared by different extraction methods, shows dramatically different therapeutic effects (*Guduchi* in Figure 2.1). Second, we used the example of *Mahayogaraj Guggulu* to explain how the vehicle/*Anupana* used along with the primary drug, can significantly influence the medicinal properties and target sites of the drug (Figure 2.2). A quick survey of the literature is heartening, because we find that modern research is investigating traditional extraction methods of preparing Ayurvedic formulations. Next, we discussed quality control parameters in terms of Ayurvedic pharmaceutics and modern science, and highlighted the need for comparative research on both types of quality control parameters. Throughout this section, we emphasized the Drug and Cosmetics rules (6th amendment) 2010, and ICMR rules (ICMR, 2006). These rules recognize the value of traditional methods of herbal processing and extraction.

The principles of Ayurvedic pharmaceutics explained in the first section of the chapter, lay the foundation for exploring issues such as Ayurvedic diagnostics, therapeutics, and factors which determine what drug is used for a given disease in a given patient. Therefore, the second section of this chapter describes Ayurvedic concepts and approaches for personalized medicine.

SECTION 2: Ayurveda and Personalized Medicine

Ayurvedic Drugs and Personalized Medicine

The number of well documented reports of adverse reaction to drugs, and 'non-responsiveness' to certain drugs has been increasing for the past decade. Therefore, personalized medicine has become the **'holy grail'** of today's Pharmaceutical industry. In practical terms, personalized medicine means that the overall health and disease status of the patient is considered, before prescribing treating for a diseased individual. Accordingly, one must design drugs specifically for individual patients or sub-types of patients. Although the concept of personalized medicine is new to Modern Medicine, it is a familiar concept to those practicing Ayurveda. Ayurveda has always recognized variations in overall health, strength, immunity, and physiology amongst individuals. Indeed, Ayurveda utilizes knowledge of such individual variations to choose appropriate drugs for treating patients in a holistic manner. This Ayurvedic knowledge stems from one of its most valuable and unique tenets: the *Tridosha* concept. According to the *Tridosha* concept, people can be classified into three main groups on the bases of their 'body constitution'. Below, we discuss the *Tridosha* concept in some detail, because it has lead Ayurvedic doctors to practice 'customized/personalized medicine' from the very beginning. *Doshas* are three biological elements which maintain health if they are in homeostasis. In a healthy individual, the three *Doshas, Dhatus* (tissues), and *Mala* (wastes) are in a state of balance. Any deviation from this balance creates a diseased state or *Vikriti*. A discussion of the *Tridosha* concept is given in the next paragraph.

The Three Doshas

Ayurveda believes that the 3 *Doshas* are biological manifestations of *Panchamahabhutas*, and play a central role in health and disease. The *Doshas* can be considered as physiological entities which regulate and control all functions of body and mind. Accordingly, Ayurveda believes that the 3 *Doshas* are primarily responsible for maintenance of overall health. Therefore, imbalance or vitiation of one or more *Doshas* produces disease.

The *Vata Dosha* is dry, light, swift, cool, rough and controls all activities, movements. Vitiation of *Vata Dosha* causes pain, increased catabolic rate, muscle wasting, dryness etc (Paradkar, 2000 B). The *Pitta Dosha* is penetrating, hot, liquid and responsible for all digestive, metabolic or transformation processes (like conversion of food into energy). Vitiation of Pitta causes burning, rash, bleeding, and inflammation (Paradkar, 2000 C). *Kapha Dosha* has slimy, heavy, slow, stable attributes which help in formation of body mass, shape, and form. The important properties of energy storage and biosynthesis are also associated with the *Kapha Dosha*. Vitiation of the Kapha *Dosha* causes increased secretions, dullness, lethargy, heaviness and excess quantity of tissues (Paradkar, 2000 D).

The recent global interest in personalized medicine has inspired Indian scientists to undertake new research on the *Tridosha* concept. Thus, there is a study which developed a new mathematical method to quantitate the levels of *Doshas* (Joshi, 2004). There are also studies to determine whether the three different *Doshas* are correlated with specific clinical and genetic markers (Patwardhan *et al*, 2008; Prasher *et al*, 2008). Pharmacogenetic and pharmacogenomic studies are explaining the role of liver genes and enzymes which control drug metabolism and drug tolerance (Zhou *et al.*, 2009); accordingly this information helps doctors to choose the optimum dose of a specific drug for a given individual. However, personalizing/customizing a drug's dose is only one aspect of personalized medicine. Since Ayurveda has been practicing personalized medicine for centuries, it has much to offer. Therefore, the next paragraph will explain concepts of *Tridosha* and *Prakriti* which lay the vital foundations for understanding Ayurvedic principles of diagnosis and therapeutics.

Dosha and Prakriti

In addition to the *Tridosha* concept, Ayurveda believes that the unique traits in each individual arise from a specific and permanent composition of *Doshas* at the time of conception. This *Dosha* pattern which is established at birth is termed *Prakriti*, which can be referred to as the individual's 'body constitution'. Thus, the dominance of one or more *Doshas* determines the *Prakriti* of an individual. For example, a person is said to have *Vata Prakriti*, if the *Vata Dosha* is dominant at birth; whereas another person could have a *Vata-Pitta Prakriti*, if both these *Doshas* were dominant at birth. The *Prakriti*/body constitution is a unique and specific trait for each person, and does not change during the individuals' lifespan. In other words, a person's *Prakriti* is considered to be 'permanent'. *Prakriti* can be of seven types: *Vata, Pitta, Kapha, Vata Pitta, Vata Kapha, Kapha Pitta* and *Sama* (all three *Doshas* in a balanced state).

Since, *Prakriti* is the root cause of individuality, Ayurveda believes that it plays a major role in all aspects of health and disease. As expected from the *Tridosha* concept, individuals with *Sama Prakriti* (all three *Doshas* in a state of balance) have good resistance and are usually not prone to diseases. Individuals whose *Prakriti* is associated with a single *Dosha* have better immunity and resistance than individuals whose *Prakriti* is associated with two *Doshas.* Interestingly, each of the 3 *Doshas* is associated with intrinsic differences in disease susceptibility. Thus, individuals with a *Vata Prakriti*, are more prone to disease than those with *Pitta Prakriti*, whereas individuals of *Kapha Prakriti* are less prone to disease than those with *Pitta Prakriti* (Acharya, 2000 C). Apart from susceptibility to disease, the Prakriti of an individual can also influence the severity of disease. Thus, a match between the *Doshas* of the *Prakriti* and the *Doshas* of the pathological factors; can increase the severity of disease. Therefore, individuals with *Vata, Pitta* or *Kapha Prakritis*, are expected to have severe diseases if the pathological factors themselves are of the *Vata, Pitta,* and *Kapha* types; respectively (Acharya, 2000 D). For example; heavy, cold, and oily causative factors, would cause severe 'Kapha disease' in an individual with *Kapha Prakriti*; whereas these same environmental factors would create milder disease in individuals with *Vata* or *Pitta Prakritis*.

In summary, the *Prakriti* of an individual affects susceptibility to disease and the severity of disease. Understanding of *Prakriti* is essential for Ayurvedic diagnosis and treatment of diseases. Knowledge of *Prakriti* also helps an Ayurvedic physician plan the preventive and curative management for a patient in accordance with the specific *Prakriti* of the individual, For example, a person with a *Pitta Prakriti*, would be advised to have a diet with 'cooling and 'light to digest' properties.

Ayurvedic Diagnostics and Therapeutics: Host and Drug Related Factors

According to Ayurveda, a given disease can originate and develop in different ways in different people. Thus, the modes and mechanisms of pathogenesis, the patient's response to the causative factors underlying the disease, and the severity of the disease; are patient specific. The intrinsic variations between patients (hosts), and variations in the pathogenesis process itself; cause heterogeneous manifestations of a given disease from patient to patient. In addition, the host's response to the ongoing pathogenic changes adds another complex layer of factors which affect the severity of illness. For successful diagnoses, the Ayurvedic physician considers various host dependent factors (Paradkar, 2000 E) and drug related factors (Paradkar, 2000 F). Below, we show a list of the Host factors and Properties of the drug as 2 separate columns.

Host Factors

- ☆ **Doshas** (Vata, Pitta, Kapha and combinations)
- ☆ **Dooshya** (Body tissues, systems, organs)
- ☆ **Desha** (Surroundings)
- ☆ **Bala** (Strength)
- ☆ **Kala** (Seasons/Diurnal variations)
- ☆ **Agni** (Digestive capacity)
- ☆ **Prakruti** (Constitution)
- ☆ **Age**
- ☆ **Satva** (Mental status and tolerance)
- ☆ **Acceptance** (Hypersensitivity)
- ☆ **Diet**
- ☆ **Diseases stage** (sub-types, patient conditions)

Properties of the Drug

- ☆ **Rasa** (Taste)
- ☆ **Vipaka** (Taste after digestion)
- ☆ **Veerya** (Potency/distinct attributes)
- ☆ **Gunas** (Attributes)
- ☆ **Prabhava** (Specific property of drug)
- ☆ **Sanyoga** (Synergetic/antagonist action)

Other Drug Related Factors: Formulation Type

- ☆ Presence of Anupana
- ☆ Processing methods
- ☆ Dose
- ☆ Mode of administration

Based on the host and drug properties explained in the above columns, the next two paragraphs discuss Ayurvedic Diagnostics and Therapeutics respectively.

Ayurvedic Diagnostics

The above columns describe host and drug related key factors which are considered in Ayurveda medicine. In terms of diagnoses, the analysis and classification of the *Dosha* abnormality is an essential step. For example, based on type of *Dosha* abnormality, the *Dosha* could be further sub-classified as *Kshaya* (depletion), *Vriddhi* (excess) or *Prakopa* (vitiation). Furthermore, an abnormal *Dosha* can interact with several organs, tissues, and body systems, and create additional types of *Dosha* abnormalities which can alter the body's physiology in a profound manner. Pathogenic factors also determine the nature and degree of vitiation of a particular *Dosha* and can also determine which body parts are affected. In this connection, we explain the concept of '*Dosha* dominance or vitiation'. Causative pathological factors bring about dominance or vitiation of a specific *Dosha* in the patient. The *Dosha* which gets vitiated, depends on the nature of pathological factors which interact with the *Prakriti* of the patient. Hence, an Ayurvedic physician must diagnose *Dosha* abnormalities and also identify the *Dosha* which has become vitiated or 'dominant' in terms of the patient's disease.

If we consider the various combinations of different degrees of *Dosha* abnormality and *Dosha* types, we get 63 possible types of *Dosha* abnormalities. Next, if we consider all permutations and combinations of the 63 possible *Dosha* abnormalities, with each of the host and drug factors shown above; we realize that Ayurvedic Diagnostics is extremely complex and sophisticated. Moreover, this complex system of diagnoses can identify the patient-patient variations in disease pathogenesis in great detail. Indeed, an Ayurvedic physician must note the detailed patient history to find probable causative factors for vitiation of a particular *Dosha*. Then, based on patient history and clinical examination; the physician identifies the treatment targets such as the specific *Doshas*, *Dhatus* (tissues), *Srotas* (systems) and the affected *Mala* (wastes). Using all these diagnostic parameters, the physician plans a treatment strategy which can reverse the disease pathogenesis, strengthen the affected tissues, and restore overall health.

Ayurvedic Therapeutics

In section 1 of this chapter, we explained Ayurvedic Therapeutics in the context of source herbs and different solvent extraction processes which are required and used. Other important factors such as bio-availability of the drug, drug stability, drug dose, and the time taken for the drug to exert its therapeutic effects; were also considered. In the next paragraph, we explain Ayurvedic Therapeutics in the context of patient's overall health, body constitution, and the drug's intrinsic properties.

Thus, we examine how Ayurvedic Therapeutics depends on the individual's host factors (including *Prakriti*), and the various types of *Dosha* abnormalities discussed above.

Having diagnosed the *Dosha* and *Samprapti* factors (like status of *Dhatu*, systems, digestive fire) in the patient, the Ayurvedic physician then considers the various properties and *Gunas* (attributes) of different drugs/types of formulations. The choice of drugs/types of formulations strongly depends on the *Dosha* and *Samprapti* factors (*Dhatu* abnormalities), because these factors help to determine which drugs are most suitable for the patient's disease, and whether these drugs can be tolerated by the patient (Paradkar, 2000 E). For example, individuals with *Vata Prakriti* and patients with '*Vata* dominant' diseases are usually treated with medicated oils. Similarly, persons with *Pitta Prakriti* or individuals with '*Pitta* dominant' disease, are advised *Ksheerapakas* (medicated milk) and *Himas* (cold infusions). Fermented extracts (*Asavas*) are often advised for individuals with *Kapha Prakriti* or individuals with '*Kapha* dominant' disease. Certain *Rasayanas* are also suitable for individuals with a specific *Prakriti*. For example, *Ksheerabala Taila* is a suitable *Rasayana* for individuals with *Vata Prakriti* (Panditrao, 1990 B). *Kushmandavaleha* is a suitable *Rasayana* for individuals with *Pitta Prakriti* (*Parashuramshastri*, 2005 F); and *Amrut Bhallataka* is indicated for individuals with *Kapha Prakriti* (Rajeshwardattashastri, 2005 D). Conversely, certain drugs are deemed unsuitable for individuals with a specific *Prakriti*. For example, *Bhallataka* (*Semecarpus anacardium*) is not used for patients with a *Pitta Prakriti* or for persons with '*Pitta* dominant' diseases. This is because both *Bhallataka* and *Pitta Dosha* possess similar, hot, and penetrating properties; and hence *Bhallataka* would be expected to further vitiate the *Pitta Dosha* of the individual (Bapalal, 1999 E).

In summary, the comprehensive nature of Ayurvedic Diagnostics and Therapeutics allows complete analysis of all the drug and host related factors. As a result, the Ayurvedic physician can generally 'predict' the patient's overall response to a given drug. This helps to 'narrow down' the drugs/types of formulations which are most likely to benefit the particular patient. This medical system of Ayurveda leads to 'personalized' patient care. Detailed examples of personalized treatment for different diseases will be discussed later in this section.

Generalized versus Personalized Medicine in Ayurveda

Before we discuss 'personalized' medicine with examples from Ayurveda, it is important to note that how and why the decision to use 'personalized' medicine must be carefully made. Thus, the choice between generalized versus personalized treatment depends on the disease, its stage, the symptoms and pathological processes associated with the disease, the nature and duration of treatment, and finally; the root causes for the disease. For example, immediate treatment of patients with fractures would be similar for all patients. Thus, immobilization of the joint with plaster, and management of pain/inflammation are common for all types patients with fractures. Since pain maybe a common feature in the different sub-types of Osteoarthritis, certain oils and hot fomentation can be used as external treatments for alleviation of pain in all these forms of Osteoarthritis. After addressing the acute symptoms of the fracture, the Ayurvedic physician would then select a 'customized *Rasayana*' for long-term bone healing. This customization of medicine is not done 'for the sake of customization'. Rather, customization is done only if it results in overall improvement of the patient's immediate symptoms and long-term health.

Having explained some of the basic concepts of Ayurvedic diagnostics and Therapeutics, we now give specific examples of diseases wherein Personalized Ayurvedic drugs are used. First, using the example of Osteoarthritis, we explain how different individuals suffering from this ailment may be treated with different Ayurvedic drugs. Personalized Ayurvedic drugs can also be used to treat a single disease as it progresses and changes within a single individual. Accordingly, in the second example we will explain how different (personalized) Ayurvedic drugs can be used to treat asthma as it progresses within a single individual.

Personalized Medicine in Ayurveda: The Examples of Arthritis

Modern medicine considers Osteoarthritis as a chronic and degenerative disease of the joints. In some cases, this disease has the additional complication of synovitis, inflammation, and gout. Ayurveda regards arthritis as *Sandhigata Vata* (Acharya, 2000 E). This term can be translated as 'joint disease associated with imbalance (vitiation) of *Vata dosha* and depleted *Dhatus*. Thus, the primary goal of Ayurvedic therapy is to rectify the vitiated *Vata dosha* in these patients. Treatment is comprehensive, and targets the total joint and its components (bone, cartilage, and muscle). However, Ayurveda considers Osteoarthritis arthritis in association with all other ailments within the patient. Accordingly, Ayurveda recognizes various sub-types of Osteoarthritis, and different personalized treatment modalities suitable for each sub-type are chosen (Table 2.3). Though every drug mentioned in Table 2.3 is indicated for *Sandhigata Vata*, its choice depends on the type of patient, stage of arthritis, and other specific problems associated with the arthritis in that particular patient.

Table 2.3 clearly shows that oral administration of some form of *Guggulu* is the primary drug used for arthritis. In fact, 6 different types of *Guggulu* can be used for 6 different sub-types of arthritis. These 6 sub-types include Joint stiffness, Osteoarthritis of Lumbo-sacral joint, Osteoarthritis due to trauma, Osteoarthritis in an obese patient, Osteoarthritis in a patient with Osteoporosis, and Chronic Rheumatoid arthritis. The type of *Guggulu* chosen depends on the sub-type of arthritis and associated ailments in the patient. For example, *Triphala Guggulu* is used to treat Osteoarthritis in an obese patient, because it can reduce the adipose tissue, alleviate the imbalance of *Vata*, and strengthen the joints (Paradkar, 2000, G). *Simhanad Guggulu* is useful for people suffering from arthritis of the lumbo-sacral region (Krishnanandaji, 1999 A), whereas *Trayodashanga Guggulu* is indicated for joint stiffness (Shastri, 2004). In addition, these 6 sub-types of arthritis require different supporting drugs and other treatments (enemas, massages, diet, and exercise). The type of formulation selected, also depends in part on its digestibility and the patients' tolerance. Thus, in the case of arthritis with inflammation and/or pain; oils and the drugs (which are heavy to digest), are avoided. Rather, decoctions and powders are used for these two sub-types of arthritis. In the case of arthritis with muscle wasting, anabolic effects are desired. Accordingly, *Ashvagandhadi Choorna* with *Chyavanaprasha* is prescribed for strengthening muscles (Panditrao, 1990 C; Acharya, 2000 F).

Table 2.3: Drugs used for different sub-types of Osteoarthritis

Disease Sub-type	*Therapeutic Targets (Dosha and Dhatu)*	*Main Drug (Oral)*	*Supportive Drug*	*Other Management*
Joint inflammation	Saama Rasa, Rakta Dushti, Asthi Kshaya, and Agnimandya	*Rasnadi Kvatha*	Shothahara Lepa (Local application)	Application of Vishagarbha Thailam
Joint Pain	Asthi Kshaya, Sandhi gata-Vata	*Ajamodadi Choorna*	Narayana Taila (Local application)	**Massage** with Balashvagandhadi Thailam
Joint stiffness	Vitiated Vata, Mamsagata Vata	*Trayodashang Guggulu*	Kaishora Guggulu (Oral)	**Massage** with Masha Tailam,
Arthritis of Lumbo-sacral joint	Kahpa Kshaya, Asthi Kshaya	***Simhanad Guggulu***	Trayodashang Guggulu (Oral)	Basti (oil **enema**) of Dhanvantar Thailam
Arthritis due to trauma	Rakta Dushti	***Lakshadi Guggulu***	Shothahara Lepa (Local application)	**Massage** with Prasarani Thailam
Obese patient	Kapha Vriddhi, Meda Kshaya	***Triphala Guggulu***	Arogyavardhini Vati (Oral)	Diet, exercise
Osteoporosis	Asthi Kshaya	***Panchatikta Ghrita Guggul***	Lakshadi Guggulu (Oral)	Guduchi Ksheerapaka Basti (retention **enema**)
Chronic Rheumatoid arthritis	Asthi Kshaya, Mamsa Kshaya, Shukra Kshaya	***Mahayogaraj Guggulu***	Dashamoolarishta (Oral) Shatagandha Vati	**Massage** with Sahacharadi Thailam
Muscle wasting	Mamsa Kshaya	*Ashvagandhadi Choorna*	Chyavanaprasha (Oral)	**Massage** with Bala Thailam, Pinda Sveda

Table 2.3 shows 9 different sub-types of Osteoarthritis for which different main drugs, supporting drugs, and other treatments can be given. 'Other management' may include enemas, diet, excercise, and application of oils with various types of fomentation.

Personalized Medicine in Ayurveda: The Example of Asthma

Another aspect of personalized treatment pertains to the well known fact that a disease evolves and progresses even within a single patient. Accordingly, an Ayurvedic physician has to use a sequence of different primary drugs and supportive treatments as the disease goes through different stages within the patient. In this context, we explain Ayurveda's approach with the example of asthma. Table 2.4 shows Ayurvedic treatment for a typical Asthma patient.

The first and most important goal of the treatment is to relieve breathlessness. For this purpose, combination of herbo-mineral drugs *Shvaskuthar Rasa* and *Rasasindur* (with *Pippalyasava* as *Anupana*) can be used (Rajeshwardattashastri, 2005 D; Panditrao, 1990 D) with oil massage and hot fomentation with steam inhalation (Acharya, 2000 G). Once the patient achieves normal breathing, treatment aimed at the overall respiratory system begins. If the patient history and clinical examination reveal that the patient suffers from vitiation of *Kapha Dosha,* then as shown in this example, the therapeutic emesis (*Vamana*) is planned (Acharya, 2000 H). After elimination of vitiated *Kapha* by the *Vamana* process, the physician attempts to improve digestion and metabolism of the patient by administering *Talisadi Choorna* (Acharya, 2000 I). Even when the patient does not show any asthmatic symptoms, continuous strengthening of the respiratory system is attempted, and achieved by administering specific *Rasayanas* such as *Chyavanaprasha* and *Vardhaman Pippali Rasayana* (using *Ksheerapaka* of *Pippali*) (Acharya, 2000 F; Acharya, 2000 J). The ideal end point of asthma management is to prevent recurrence. Therefore, other important and integral aspects of asthma management include non-pharmacological interventions such as *Vamana* (as therapeutic emesis), avoiding causes of asthma, diet-lifestyle modifications, and patient education.

In summary, Ayurvedic treatment targets each stage of asthma, ranging from broncho-dilatation to quality of life. Based on the patient's constitution, nature of the pathogenesis of asthma, and the stage of the disease; the Ayurvedic physician makes a detailed diagnoses of the asthma patient, and understands the unique requirements of that particular patient. Accordingly the physician chooses the optimum drugs, types of formulations, and plans the entire course of treatment. *In the example shown in* Table 2.4, *15 drugs are used in 7 different stages of Asthma.* However, more than 500 Ayurvedic drugs are indicated for Asthma (AyuSoft formulation data mining tool, 2006). The final and specific selection of drugs is done according to the disease progression of Asthma (through the 7 stages) in a particular patient. Therefore, the resulting Ayurvedic treatment is both dynamic and 'customized' to the patient's requirements.

The choice of drugs shown for Arthritis (Table 2.3) and Asthma (Table 2.4), represent a possible course of treatment which is used by some Ayurvedic physicians for treatment of '*Sandhigata Vata*' and Asthma respectively. However, due to variations between schools of thought, and traditional practices; physicians can use drugs other than those mentioned in Tables 2.3 and 2.4. Therefore, one would expect other courses of Ayurvedic treatment to be prescribed for patients with the different sub-types of Osteoarthritis and Asthma shown in Tables 2.3 and 2.4.

Table 2.4: Drugs for treating different stages of Asthma

Stage	*Symptoms*	*Objective of Treatment*	*Main Oral Drug (with Anupana)*	*Other Management*
Acute stage	Breathlessness	Bronco-dilatation and Control of respiratory secretions	***Shvaskuthar Rasa + Rasasindur (Pippalyasava)***	Oil massage, hot fomentation to chest Steam inhalation
Stable patient but Asthma persists	Cough with expectoration	Symptomatic relief, Control of respiratory secretions	***Lakshmivilas Rasa*** + ***Yavakshara*** (**with** Honey)	*Preparation for Vamana* Medicated Ghee consumed daily Oil massage and fomentation (4 -7 days)
Stable patient	Cough, Anorexia, Salivation Heaviness	**Elimination of** excess and vitiated ***Kapha***	***Madanaphala Kvatha***	***Vamana*** (Therapeutic Emesis)
Patient after *Vamana*	Loss of appetite	**Improve digestion and metabolism**	***Talisadi Choorna*** (with Honey + Ginger juice)	Restricted diet. Lifestyle advice. Avoid causative factors
Partial Relief	Cough	Symptomatic relief	Sitopaladi Churna (with Honey)	Follow diet and lifestyle advice
Complete Relief	Asymptomatic	*Strengthen respiratory system* (***Rasayana*** **Therapy**)	***Chyavanaprasha*** and Anu Taila	**Yoga** and breathing techniques ***Nasya*** (nasal administration of Anu Taila)
Preventive Measures	Asymptomatic	**Avoid recurrence.** Reverse lung damage. Prevent pathogenesis of Asthma. **Improve quality of life.**	***Vardhaman Pippali Rasayana*** (***Pippali Ksheerapaka***)	Continue diet, lifestyle modifications, Yoga ***Rasayana*** **Therapy**

Table 2.4 shows the seven different stages of Asthma which are treatable with different Main drugs + *Anupanas*. 'Other Management' includes *Vamana, Yoga, Rasayana Therapy*, and changes in diet and lifestyle.

Summary and Conclusions

This chapter has been inspired by the wish to bring out three important issues affecting Ayurveda today. First, the new rules pertaining to clinical use of Ayurvedic Formulations must be considered. Second, the trends in modern pre-clinical and clinical research on Ayurvedic Formulations and drugs should be examined. Third, the current requirements of mainstream medicine must also be considered. We deal with each issue separately.

First, the latest ICMR guidelines recognize the importance of traditional Ayurvedic Formulations, as they have been exempted from the rules which apply to new chemical entities. This renewed interest and perspective on traditional Ayurvedic Formulations is also reflected in publications on Ayurvedic Formulations as Crude Extracts. Thus, as stated earlier, there is *"No longer a crude view of the crude extract"* (Deocaris *et al.*, 2008). With these thoughts in mind, the first section of this chapter focused on Ayurvedic crude drugs and the most commonly used types of drug formulations (Tables 2.1 and 2.2). The versatile and sophisticated nature of Ayurvedic crude drugs was illustrated with specific examples. Thus, the example of *Guduchi* showed how a single herbal drug prepared by different extraction methods, has dramatically different therapeutic effects (Figure 2.1). The example of *Mahayogaraj Guggulu* explained how the vehicle/*Anupana* used along with the primary drug significantly influences the medicinal properties and target sites of the drug (Figure 2.2).

The second important issue concerns the trends in modern scientific research on Ayurvedic Formulations. Therefore, section 1 of the chapter cited studies wherein scientists are comparing different types of extraction methods for drugs derived from Ayurvedic source herbs. We also discussed quality control of drugs using parameters of Ayurveda and modern science, and highlighted the need for more research in this area. First, we require studies which identify new ways to link the HPLC profiles of the Ayurvedic herbal components with their bioactivities. Second, comparative research on herbal products using Ayurvedic and modern quality control parameters may identify crucial biochemical aspects of Ayurvedic drug preparation. Insights from such research could potentially lead to new protocols for producing standardized Ayurvedic crude drugs by protocols which more closely mimic traditional methods, and would therefore ensure better safety, tolerability, and efficacy of the Ayurvedic drug.

The third important issue concerns the current requirements of mainstream medicine. Personalized medicine is an area of intense interest and research by the medical community and pharmaceutical industry. Therefore, the second section of this chapter explains Ayurvedic Diagnostics and Therapeutics, and how these methods allow an Ayurvedic physician to 'predict' the patient's overall response to a given drug. This unique aspect of the Ayurvedic medical system leads to 'personalized' patient care. Accordingly, the remainder of the second section of gives specific examples of personalized medicine in Ayurveda.

The key points of the entire chapter are summarized in Table 2.5. This Table compares the different approaches and perspectives of Modern versus Ayurvedic

Table 2.5: Comparison of modern versus ayurvedic pharmaceutics

Process	*Modern Pharmaceutics and Pharmacology*	*Ayurveda Pharmaceutics and Pharmacology*
Objective of Extraction	Purification of Active chemical principle from source to obtain a drug AND Production of New Chemical Entity to produce a Novel Therapeutic drug	Reduces potential toxicity of some source herbs. Extraction Method ensures bioavailability and appropriate therapeutic effect of drug.
Use of Solvents and Extraction Methods	Water and Organic Solvents are used in different ratios. Percolation or Maceration with Solvent. Soxhlet apparatus maybe used to concentrate active components.	Aqueous extracts (decoctions, hot or cold infusion). Fermented Extracts. Milk/Ghee/Oil extract.
Type of Formulation Administered to Patient	*Constant* Based on data obtained from Pharmacokinetic and Pharmacodynamic studies of the drug in clinical trials.	*Variable and Personalized* Properties of drug, solvent and *Anupana* considered for proper targeting of drug. *Dosha* Abnormality and Tolerability of drug by patient are considered.
Therapeutic Targets	Targets Abnormal Gene Expression Targets Abnormal Cell Signaling pathways in Diseased Tissue.	Targets Causative factors of disease *Dosha, Dhatu, Srotas.* Aims for Therapeutic effects (Short and Long-term). Promotes Self-Healing and Regeneration of Tissues.

The Table 2.5 highlights key differences between Modern versus Ayurvedic Pharmaceutics.

Pharmaceutics. We begin with differences in Objectives of Extraction and Extraction Methods, that culminate in a major difference in the Types of Formulations which are administered to Patients. The approach taken by Modern Pharmaceutics leads to a single/constant type of drug for a given disease, whereas Ayurvedic Pharmaceutics is 'personalized'. Table 2.5 clearly shows that there are significant gaps between the 2 systems of medicine. These gaps can be bridged with new lines of research which re-examine traditional methods of preparation of Ayurvedic drugs in preclinical (*in vitro* and *in vivo*) research. Notably, the Ayurvedic approach of using appropriate vehicles and extraction methods for source herbs; in order to direct drugs towards specific targets, is worth investigating. Indeed, section 1 of this chapter also gives certain experimental approaches which could provide useful data on traditional versus modern methods herbal extraction, and quality control of herbal drugs. Moreover, the use of powerful modern analytical tools to probe the complexities of Ayurvedic Pharmaceutics, could allow Natural product based drug development to effectively integrate ancient wisdom and modern technology. Such interdisciplinary studies could facilitate better understanding of disease process and mechanism of action. Indeed, Ayurveda is not merely a source of 'raw materials' for potential drug candidates. Rather, some of the principles and practices of Ayurvedic Pharmaceutics we have discussed, could be explored for innovative drug development.

In section 2, the discussions on Osteoarthritis and Asthma are examples of personalized medicine which can be analyzed by modern science. Clinical research could replicate some of the clinical questions and conditions used by Ayurvedic physicians. For example, factors such as disease sub-types, overall physical health status of patient, and the strength and immunity of the patient could be included. Accordingly, clinical research could examine the importance of the *Samprapti* factors (status of tissues (*Dhatus*), body systems, and digestive fire), when administering drugs to patients. Due consideration of the *Samprapti* factors may lead to appropriate extraction methods of source herbs, which in turn could result in better efficacy and safety of drugs. Finally, it is important to realize that some of the goals of patient care in Ayurveda are absent from modern medicine. For example, modern drugs seek to rectify specific disease related abnormalities at the biochemical and genetic level, whereas Ayurvedic drugs aim to reverse disease symptoms, restore long-term health, and rejuvenate the entire body. This holistic approach of Auyurveda and other alternate forms of medicine, is precisely what attracts people all over the world. Accordingly, such a holistic approach could be considered by practioners of modern medicine.

We hope the information in this chapter will stimulate dialogue and interactions between Modern clinicians and Ayurvedic physicians. Such interactions may result in new insights and approaches for personalized medicine, which may well be the 'future of all medicine'.

Acknowledgements

We sincerely thank the Editor of this book Dr. V.K.Gupta, for inviting us to write this chapter. We also acknowledge Mr. Ranjit Puranik for sharing his views and providing information regarding crude extracts. We are very grateful to Dr. Sandip Ubale, Dr. Pramod Dhumal, and Dr. Neelima Tillu, for their useful suggestions.

GLOSSARY (Alphabetical)

Asthi Dhatu: Bone tissue.

***Dhatu*:** Body constituents (or tissues) with specific structure and functions. *Doshas* reside within the *Dhatus*

Dhatu Dushti: Vitiation of one or more Dhatus

Dhatu Kshaya: Depletion of Dhatus

Dhatu Vriddhi: Excessive growth of Dhatus

***Dosha*:** Three biological elements which maintain health if they are in homeostasis. Vitiation of *dosha* or *doshas* produces disease

Dosha Prakopa: The vitiation of one or more *Dosha*. Causative factors determine the nature and degree of Vitiation of a particular *Dosha*, and which body parts are affected.

Dravya: Any substance having attributes and functions.

***Guna*:** Attributes those are responsible for any pharmacological activity

Kapha Dosha: The *Dosha* which form body mass, shape and form. It is heavy, stable, smooth, unctuous, viscous, slimy and moist

Karma: Pharmacological activity

Majja Dhatu: The tissue enclosed by bone cavity, it includes brain.

Mamsa Dhatu: Muscle tissue that covers other body parts.

Marana: The process that converts metal/mineral into a *Bhasma* form. It includes specific methods of heating.

Meda Dhatu: Oily tissue responsible for lubrication with soft, slimy, smooth attributes.

Panchamahabhoota: Five primordial entities that constitute any substance.

Pitta Dosha: The *Dosha* responsible for all digestive, metabolic or transformation processes and conversion for one form to another. It is hot, penetrating, liquid.

Rakta Dhatu: Blood.

Rasa Dhatu: Dhatu in form of body fluid that nourishes the body and mind.

Rasa: The taste

Rasayana: *Rasayana* therapy aims at producing tissues of optimum quality and quantity.

Saama: Abnormal condition wherein *Dosha or Dhatu* are contaminated with undigested or semi-processed material. The condition can initiate inflammation and other pathogenic processes.

Samprapti: Events in Pathogenesis

Samskara: Processing methods

Sandhi: Joints (generally joints containing bones)

Shodhana: The procedure that converts a harmful substance into safe, absorbable form.

Shukra Dhatu: The entity responsible for reproduction and regeneration.

Srotasa: Channels (and micro-channels) which transmit signals and nutrients required for physiological processes to take place.

Vata Dosha: The *Dosha* which controls all activities, movements. It is dry, movable and imperceptible.

References

Acharya, Y.T. (2000 A). Charaka Samhita. Varanasi: Chowkhamba Surbharati Varanasi. (reprint).pp 81-83. (Charaka Samhita, Sootrasthana 13/14-15).

Acharya, Y.T. (2000 B). Charaka Samhita. Varanasi: Chowkhamba Surbharati Varanasi. (reprint), pp 150 (Charaka Samhita, Sootrasthana 28/27).

Acharya, Y.T. (2000 C). Charaka Samhita. Varanasi: Chowkhamba Surbharati Varanasi. (reprint).pp 256. (Charaka Samhita, Vimansthana 6/15).

Acharya, Y.T. (2000 D). Charaka Samhita. Varanasi: Chowkhamba Surbharati Varanasi. (reprint).pp 256. (Charaka Samhita, Vimansthana 6/16-18).

Acharya, Y.T. (2000 E). Charaka Samhita. Varanasi: Chowkhamba Surbharati Varanasi. (reprint).pp 618. (Charaka Samhita, Chikitsa Sthana 28/37).

Acharya, Y.T. (2000 F). Charaka Samhita. Varanasi: Chowkhamba Surbharati Varanasi. (reprint).pp 379. (Charaka Samhita, Chikitsa Sthana 1-1/63 - 73).

Acharya, Y.T. (2000 G). Charaka Samhita. Varanasi: Chowkhamba Surbharati Varanasi. (reprint).pp 536. (Charaka Samhita, Chikitsa Sthana 17/71).

Acharya, Y.T. (2000 H). Charaka Samhita. Varanasi: Chowkhamba Surbharati Varanasi. (reprint).pp 537. (Charaka Samhita, Chikitsa Sthana 17/89).

Acharya, Y.T. (2000 I). Charaka Samhita. Varanasi: Chowkhamba Surbharati Varanasi. (reprint).pp 466. (Charaka Samhita, Chikitsa Sthana 9/145 - 148).

Acharya, Y.T. (2000 J). Charaka Samhita. Varanasi: Chowkhamba Surbharati Varanasi. (reprint).pp 385. (Charaka Samhita, Chikitsa Sthana 1-3/36 - 40).

AyuSoft. (2006). A Decision Support System developed by Centre for Development of Advanced Computing, Pune (www.ayusoft.cdac.in).

Bapalal, Vaidya. (1999 A). Nighantu Adarsh Chaukhambha Bharati Academy Varanasi, 2nd edition, pp 638.

Bapalal, Vaidya. (1999 B). Nighantu Adarsh Chaukhambha Bharati Academy, Varanasi, 2nd edition, pp 22.

Bapalal, Vaidya. (1999 C). Nighantu Adarsh Chaukhambha Bharati Academy, Varanasi, 2nd edition, pp 456.

Bapalal, Vaidya. (1999). Nighantu Adarsh Chaukhambha Bharati Academy Varanasi, 2nd edition, Vol 1. pp 320.

Custer, C. Deocaris., Nashi, Widodo., Renu, Wadhwa., and Sunil, C. Kaul. (2008). Merger of Ayurveda and tissue culture-based functional genomics: inspirations from systems biology. *Journal of Translational Medicine*, **6**: 14

Funk, J.L., Frye, J.B., Oyarzo, J.N., and Timmermann, B.N. (2009). Comparative effects of two gingerol-containing *Zingiber officinale* extracts on experimental rheumatoid arthritis. *J Nat Prod.*,**72(3)**: 403-7.

Gogte, V. M. (2000 A). Ayurveda Pharmacology and Therapeutic uses of medicinal plants (Dravyagunavidnyan), Bharatiya Vidya Bhavan, Mumbai. pp 444.

Gogte, V. M. (2000 B). Ayurveda Pharmacology and Therapeutic uses of medicinal plants (Dravyagunavidnyan), Bharatiya Vidya Bhavan, Mumbai. pp 359.

ICMR. (2006). Ethical guidelines for biomedical research on human participants, Indian council of medical research, New Delhi. pp 51 – 55.

Jain, H., Parial, S.D., Jarald, E., Daud, A.S., and Ahmed, S. (2010). Extraction of ash by conventional extraction methods and evaluation of its anti-stress activity. *Int. J. Green Pharmacy*, **4(3)**: 183-185.

Joshi, R.R. (2004). A biostatistical approach to Ayurveda: quantifying the tridosha. *J Altern Complement Med.*, **10(5)**: 879-89.

Joshi, Y. G. (2003 A). Charaka Samhita, Vaidyamitra Prakshan, Pune, Vol 2, pp 221. (Chikitsa Sthana 8/103 – 104).

Joshi, Y. G. (2003 B). Charaka Samhita, Vaidyamitra Prakshan, Pune, Vol 2, pp 29. (Chikitsa Sthana 1 – 3/36-40).

Krishnanandaji. (1999 A). Rasatantrasara (Siddhaprayoga Sngraha), 14th reprint, A publication by Krishnagopal Charitable Trust, Kaleda. pp 643.

Laakmann, G., Dienel, A., and Kieser, M. (1998). Clinical significance of hyperforin for the efficacy of *Hypericum* extracts on depressive disorders of different severities. *Phytomedicine,* **5:** 435-442.

Lalithakumari, K., Krishnaraju, A.V., Sengupta, K., Subbaraju, G.V., and Chatterjee, A. (2006). Safety and toxicological evaluation of a novel, standardized 3-O-Acetyl-11-keto-beta-Boswellic Acid (AKBA)-enriched *Boswellia serrata* extract (5-Loxin(R)). *Toxicol Mech Methods.,* **16(4)**: 199-226.

Lieu, C.A., Kunselman, A.R., Manyam, B.V., Venkiteswaran, K., and Subramanian, T. A. (2010). Water extract of *Mucuna pruriens* provides long-term amelioration of parkinsonism with reduced risk for dyskinesias. *Parkinsonism Relat Disord.,* **6(7)**: 458-65.

Maity, P., Hasda, D., Bandhyopadhya, U., and Mishra, D. K. (2009). Biological activities of crude extracts and chemical constituents of Bael, (*Aegle marmelos*). *Ind J. Exp Biol.,* **47**: 849-861.

Mishra, A.K., Gupta, A., Gupta, V., Sannd, R., and Bansal, P. (2010). Asava and Aristha: an Ayurvedic medicine – an overview. *Int J of Pharm and Bio Archives,* **1(1)**: 24 –30.

Panditrao, D.V., Nair, S., and Arya, M. S. (1990 A). Sahasrayogam, Publication of Literature Research Unit, Central Council for Research in Ayurveda and Siddha, New Delhi. pp 215.

Panditrao, D.V., Nair, S., and Arya, M. S. (1990 B). Sahasrayogam, Publication of Literature Research Unit, Central Council for Research in Ayurveda and Siddha, New Delhi. pp 275.

Panditrao, D.V., Nair, S., and Arya, M. S. (1990 C). Sahasrayogam, Publication of Literature Research Unit, Central Council for Research in Ayurveda and Siddha, New Delhi. pp 173 – 174.

Panditrao, D.V., Nair, S., and Arya, M. S. (1990 D). Sahasrayogam, Publication of Literature Research Unit, Central Council for Research in Ayurveda and Siddha, New Delhi. pp 349.

Paradkar, Harishatri. (2000 A). Ashtang Hridayam, Chaukhambha Sanskrit Series Office, Varanasi. Reprint pp 25 (Ashtang Hridayam 1/125).

Paradkar, Harishatri. (2000 B). Ashtang Hridayam, Chaukhambha Sanskrit Series Office, Varanasi. Reprint pp 201. (Ashtang Hridayam 12/49-50).

Paradkar, Harishatri. (2000 C). Ashtang Hridayam, Chaukhambha Sanskrit Series Office, Varanasi. Reprint pp 201. (Ashtang Hridayam 12/51-52).

Paradkar, Harishatri. (2000 D). Ashtang Hridayam, Chaukhambha Sanskrit Series Office, Varanasi. Reprint pp 201. (Sootrasthana 12/52-53).

Paradkar, Harishatri. (2000 E). Ashtang Hridayam, Chaukhambha Sanskrit Series Office, Varanasi. Reprint pp 207. (Sootrasthana 12/67).

Paradkar, Harishatri. (2000 F). Ashtang Hridayam, Chaukhambha Sanskrit Series Office, Varanasi. Reprint pp 207. (Sootrasthana 9/1 – 28).

Paradkar, Harishatri. (2000 G). Ashtang Hridayam, Chaukhambha Sanskrit Series Office, Varanasi. Reprint pp 943. (Uttarasthana 40/48).

Parashuramshastri, Vidyasagar. (2005 A). Sharangdhar Samhita, Chaukhambha Orientalia, Varanasi. Sixth edition. pp 241.

Parashuramshastri, Vidyasagar. (2005 B). Sharangdhar Samhita, Chaukhambha Orientalia, Varanasi. Sixth edition. pp 137.

Parashuramshastri, Vidyasagar. (2005 C). Sharangdhar Samhita, Chaukhambha Orientalia, Varanasi. Sixth edition. pp 138.

Parashuramshastri, Vidyasagar. (2005 D). Sharangdhar Samhita, Chaukhambha Orientalia, Varanasi. Sixth edition. pp 173.

Parashuramshastri, Vidyasagar. (2005 E). Sharangdhar Samhita, Chaukhambha Orientalia, Varanasi. Sixth edition. pp 202.

Parashuramshastri, Vidyasagar. (2005 F). Sharangdhar Samhita, Chaukhambha Orientalia, Varanasi. Sixth edition. pp 209.

Parashuramshastri, Vidyasagar. (2005 G). Sharangdhar Samhita, Chaukhambha Orientalia, Varanasi. Sixth edition. pp 212.

Patwardhan, B., and Bodeker G. (2008). Ayurvedic genomics: establishing a genetic basis for mind-body typologies. *J Altern Complement Med.*, **14(5)**: 571-6.

Prasher, B., Negi, S., Aggarwal, S., Mandal, A.K., Sethi, T.P, Deshmukh, S.R., Purohit, S.G., Sengupta, S., Khanna, S., Mohammad, F., Garg, G., Brahmachari, S.K., and Mukerji M. (2008). Indian Genome Variation Consortium: Whole genome expression and biochemical correlates of extreme constitutional types defined in Ayurveda. *J Transl Med.*, **6**: 48.

Rajeshwardattashastri. (2005 A). Bhaishajya Ratnavali, Chaukhambha Sanskrit Sansthan, Varanasi, 18th Reprint. pp 11-14.

Rajeshwardattashastri. (2005 B). Bhaishajya Ratnavali, Chaukhambha Sanskrit Sansthan, Varanasi, 18th Reprint. 648.

Rajeshwardattashastri. (2005 C). Bhaishajya Ratnavali, Chaukhambha Sanskrit Sansthan, Varanasi, 18th Reprint. pp 601.

Rajeshwardattashastri. (2005 D). Bhaishajya Ratnavali, Chaukhambha Sanskrit Sansthan, Varanasi, 18th Reprint. pp 901. (54/197-206) and pp 466. (16/44-45).

Sakchai, Wittaya-areekul., and Kornkanok, Ingkaninan. (2007). Comparison of various extraction methods of *Bacopa monnieri*, Naresuan University Journal, **15(1)**: 29-34.

Sathe, Appashastri. (1991). Gharguti Audhadhe, Ayurved Bhavan, Mumbai. pp 112.

Savrikar, S.S., Dole, V., Ravishakar, and B., Shukla, V.J. (2010). A Comparative Pharmaceutical study of 3 samples of Guduchi Ghritha for adaptogenic activity against forced swimming induced gastric ulceration and hematological changes in albino rats. *Int J. Ayur Res.*, **1(2)**: 67-72.

Sharma, Dharmanand. (1996). Rasa Ratna Samuchchaya, Reprint, Motilal Banarasidas Publication, Delhi. (28/1). pp 459.

Sharma, P.V. (1982). Dhanwantari nighantu, Chaukhambha Orientalia, Varanasi, Guduchyadi Varga.

Shastri, Lakshmipati. (2004). Yogratnakar. Chaukhambha Sanskrit Bhavan, Varanasi. Purvardha (Vatvyadhi Chikitsa). pp 527-8.

Sumantran, V.N. (2010). Novel Approaches for the activity-based standardization of herbal drugs. Commentary: *Current Science*, **98(5)**: 610-611.

Swami S, Handa., Khanuja, S., Longo, G., and Dev Dutt, R. (Scientific Editors). (2008). Book: Extraction Technologies for Medicinal and Aromatic plants, International Centre for Science and High Technology Trieste.

Thorat, S., and Dahanukar, S.A. (1991). Can we dispense with Ayurvedic Samskaras?, Journal of Postgraduate Medicine, **37**: 3.

Umretia, B., Patguiri, B.J., Prajapati, P.K., and Ravishakar, B. (2009). A comparative pharmaceutical study of Guduchi Ghana prepared by kwatha and aqueous extract. AYU, 30 (1): 77-82.

Yang, N. S., Shyur, L.F., Chen, C.H., Wang, S.Y., and Tzeng, C.M. (2004). Medicinal herb extract and a single-compound drug confer similar complex pharmacogenomic activities in mcf-7 cells. *J Biomed Sci.*, **11**: 418-422.

Zhou, S.F., Liu, J.P., and Chowbay, B. (2009). Polymorphism of human cytochrome P450 enzymes and its clinical impact. Review in *Drug Metab Rev.*, 41(2): 89-295.

Traditional and Folk Herbal Medicine: *Pages* **53–66**
Recent Researches Vol. 1 (2012)
Editor: **V.K. Gupta**
Published by: **DAYA PUBLISHING HOUSE, NEW DELHI**

3

Bioactivities of *Duabanga grandiflora* for Human Skin Cells

Muneo Tsukiyama[1], Takio Sugita[1], Hiroko Kikuchi[1], Yoichi Yasuda[1], Masaki Arashima[1], Hidenobu Okumura[1], Sorasak Lhieochaiphant[2] and Yukihiro Shoyama[3*]

ABSTRACT

The aim of this investigation was to evaluate the bioactivities of Duabanga grandiflora (Sonneratiaceae), which has been used as a traditional Thai medicine on human skin. The leaf extract of D. grandiflora actively affected several human skin cells having bioactivities such as skin whitening, anti-aging and anti-inflammation. Interestingly it became clear that the extract stimulated the production of type III collagen. The crude extract of D. grandiflora was purified by activity guided fractionation by type III collagen production, and finally by HPLC to isolate an active compound determined as eugeniin by EI-mass, ^{13}C NMR, ^{1}H NMR and acidic hydrolysis. Eugeniin indicated strong dose dependent activity for type III collagen production, with this being the first example of stimulation activity for type III collagen production.

Keywords: *Duabanga grandiflora*, Sonneratiaceae, Eugeniin, Type III collagen, Anti-aging, Skincare, Cosmetic.

1 Noevir Co.Ltd., Kobe 650-8521, Japan.

2 The Faculty of Pharmacy, Chiang Mai University, Chiang Mai, Thailand, 50200.

3 Faculty of Pharmaceutical Science, Nagasaki International University, 2825-7 Huis Ten Bosch, Sasebo, Nagasaki 859-3298, Japan.

* *Corresponding author*: E-mail: shoyama@niu.ac.jp, Tel: +81-956-20-5653
Fax: +81-956-20-5622

Introduction

It has become well known that natural products like flavonoids (Sato *et al.*, 2007) and other phenolics (Pan *et al.*, 2001; Gitzingera *et al.*, 2009) are effective for skin care. In our ongoing investigation related to skin care, we examined the anti-skin tumor production activity in mice, which resulted in that crocin in saffron was an active component (Konoshima *et al.*, 1998). From this result we started to survey the active component contained in herbs for skin care. Many human cell lines related to skin care like skin whitening, anti-aging by cell viability and neutral lipid accumulation are now available. Moreover, type III collagen has attracted our attention because type I and III collagens are major components of the dermis and are organized into fibrils although type I to type VII collagens are now elucidated. Clear growth of fibrils has been observed during development of the body, and these fibrils are damaged by several skin diseases (Fleischmajer *et al.*, 1981). Recently it has been known that type III collagen is more widely distributed in baby skin rather than in adult skin by surveying them with an immunostaining method and/or enzyme-Linked immunosorbent assay (ELISA) using an anti-type III collagen monoclonal antibody (MAb) (Gay *et al.*, 1976; Sundarraj *et al.*, 1982; Sakai *et al.*, 1982; Varani *et al.*, 2006). Therefore, in order to evaluate the activity for anti-aging decline of skin, we decided to find out the active compound for stimulating type III collagen analyzed by ELISA using anti-type III collagen MAb.

In a preliminary assay, twenty three Thai plants were analyzed using the anti-peroxidized activity, resulting in strong activity in *Duabanga grandiflora*. *D. grandiflora* is an evergreen tree that grows up to be 30 meters high and is common by streams and along valleys in tropical evergreen forests in Thailand and southern China. The topography is mountainous with an elevation of more than 500 meters. The seeds have been used as a folk medicine for abdominal pain, food poisoning and peptic ulcer, and the leaves are traditionally supplied for the skin in some parts of northern Thailand. However, there is no evidence against such decoction.

In this review we assayed the biological activity of the crude extract of *D. grandiflora* using several human skin cell lines, and finally isolated an active polyphenol, eugeniin, as an activator of type III collagen.

Materials and Methods

Chemicals

Gallic acid, ellagic acid, N, O-bis(trimethylsilyl)acetoamide and BSA were purchased from Nacalai Tesque Ind., Ltd. (Kyoto, Japan). Eugeniin was obtained from Nagara Science Ind., Ltd (Gifu, Japan). Deuterium oxide and Methanol D_4 for NMR analysis were from EURISO-TOP (Andover, USA). All other chemicals were standard commercial products of analytical grade.

Cell Culture

Normal human fibroblasts (NHFs) were purchased from KURABO (Osaka, Japan) and maintained in a Dulbecco's modified Eagle medium (DMEM) containing 10 per cent fetal bovine serum (FBS; Intergen Co., Purchase, NY). Normal human preadipocytes (Cryo HprAd-SQ) were obtained from Cambrex Bio Science Walkersville, Inc. (Wakersville, MD) and maintained in a preadipocyte growth medium with 10 per cent FBS, 2 mM L-glutamine, 100 units/ml penicillin and 100 units/ml streptomycin. Preadipocytes were differentiated into adipocytes by culturing in a standard medium containing 10 μg/ml insulin, 1 μM dexamethasone, and 500 μM 3-isobutyl-1-methylxanthine (IBMX). Normal human epidermal melanocytes (NHEM) were purchased from KURABO (Osaka, Japan). The culture medium consists of Medium254 (M-254-500; KURABO, Osaka, Japan) and HMGS(S-002-5, KM-6350; KURABO, Osaka, Japan) supplemented with 10 per cent FBS, 3 μg/ml recombinant human basic fibroblast growth factor, 5 μg/ml insulin and 0.5 μM hydrocortisone. All cells were incubated in a humidified atmosphere containing 5 per cent CO_2 in air at 37 °C.

Cell Viability Assay

Cell viability was measured by a modified MTT assay using 3-(4, 5-dimethylthiazol-2-yl)-2, 5-diphenyltetrazolium bromide (MTT; Sigma Chemical Co., St. Louis, U.S.A.; Tada *et al.*, 1986). NHFs were plated in a 96-well culture plate containing DMEM medium supplemented with 1 per cent FBS as 2.0×10^4 cells per well. After 24 hr incubation the media were changed with a freshly prepared DMEM medium containing samples and cultured for 48 hr. The medium was removed and MTT reagent solution was added to adjust to 400 μg/ml, then incubated for 2 hr. Released formozan was extracted by 2-propanol and analyzed using a micro-plate reader at 550 nm and 650 nm, respectively. From the differences between the two measured values, the cell activity was evaluated in comparison to the control of a DMEM medium containing 1 per cent (negative control) and 5 per cent FBS (positive control). The value was shown as a relative intensity.

ELISA for Type III Collagen

NHFs were plated in a 96-well plate containing DMEM medium supplemented with 5 per cent FBS as 2.0×10^4 cells per well. After 24 hr incubation the media were changed with a freshly prepared DMEM medium supplemented with 0.5 per cent FBS containing samples and cultured for 48 hr. After 48 hr, the Type III collagen in the cell culture supernatant was measured by ELISA using mouse α collagen type III (SIGMA; C7805 Clone FH-7A), rat anti-mouse IgG_1-HRP (ZYMED Human; 04-6120) and standard human Collagen type III purified protein (CHEMICON; CC054).

Anti-peroxide Fatty Test (tert-Butylhydroperoxide Stress Experiments)

Cell viability against tert-butylhydroperoxide stress was measured by using modified Kim *et al.'s* method (Kim *et al.*, 2005). Human epidermal keratinocyte cell lines (HaCaT) were plated in a 96-well plate containing DMEM medium supplemented with 10 per cent FBS as 2.0×10^4 cells per well. After 24 hr incubation the media were changed with a freshly prepared DMEM medium containing samples and cultured for 24 hr. The cultured medium was changed with Hanks(+) solution containing tert-butylhydroperoxide (Sigma Chemical Co., St. Louis, U.S.A.) and incubated for 2 hr, then changed with PBS(-) solution containing 150 μg/ml of neutral red and incubated at 37 °C for 2 hr. The medium was removed and 50 per cent EtOH containing 1 per cent acetic acids was added.

Tyrosinase Assay Using Normal Human Epidermal Melanocytes

Tyrosinase activity in normal human epidermal melanocytes (NHEM) was assessed by measuring the rate of oxidation of L-3, 4-dihydroxyphenylalanine (L-DOPA) (Yokozawa *et al.*, 2007). NHEM (Kurabo Co. Ltd., Tokyo) were plated using medium 154S (Kurabo Co. Ltd., Tokyo) in a 96-well plate as 3.0×10^4 cells per well. The medium was removed after 24 hr culturing, and changed to a new medium 154S supplemented with individual concentration of samples. After 48 hr culturing the medium was removed and a phosphate buffer solution containing 1 per cent Triton-X (75 µl) was added. To the crude enzyme solution (50 µl), phosphate buffer containing 0.05 per cent L-DOPA was added and incubated at 37 °C for 2 hr. The concentration of dopamelanin was measured at 405 nm. The concentration of dopamelanin was calculated by the following equation. The concentration of dopamelanin released = {(405nm value incubated – 405 value before incubation) -2.1661/5.238}. The concentration of protein in an individual well was determined with BCA Protein Assay Kit (Pierce Co. Ltd., USA) and the specific activity was calculated.

Inhibition Test for Neutral Lipids Accumulation

Inhibitory activity for neutral lipids accumulation was investigated followed to and partly modified the published method (Omatsu-Kanbe *et al.*, 2006). Neutral lipids accumulated in the cytoplasmic space were stained and assayed using Oil Red O. For neutral lipid staining, cells cultured on coverslips were fixed with a 10 per cent formaldehyde neutral buffer solution (Nacalai Tesque Ind., Ltd., Kyoto, Japan) for 10 min at room temperature. The cells were then stained using the Oil Red O solution (Oil Red O stock solution: H_2O=3:2; Oil Red O stock solution is 0.6 per cent Oil Red O in isopropanol) for 10 min, and washed with 400 µl of distilled water. After washing with distilled water, each coverslip was mounted on a slide glass in an aqueous mounting medium (Mount-quick Aqueous; Daido Sangyo CO., LTD. JAPAN) and observed using bright field microscopy. Quantitation of neutral lipids was performed as described previously (Ramírez-Zacarías *et al.*, 1992). Cells grown and differentiated in 96-well culture plates were rinsed three times with PBS and fixed with a 10 per cent formaldehyde neutral buffer solution for 10 min at room temperature. After removing the formaldehyde solution, the cells in each well were stained with 50 µl of Oil Red O solution for 10 min and washed three times with 400 µl of distilled water. Dye trapped in intracellular lipid droplets was eluted with 100 µl of methanol per well and absorbance analysis was performed at 550 nm. Analysis was performed on 3 wells for each condition.

Plant Material and Preparation of Extracts

D. grandiflora leaves were collected in Chiang Mai, Thailand and dried at room temperature. A voucher specimen of the plant has been deposited in the Herbarium of Chiang Mai University. The dried powdered leaves (100 g) were extracted with 50 per cent MeOH twice at room temperature. MeOH was evaporated under reduced pressure and then dried by lyophilization to give extractives (19.1 g).

Separation of Active Compound 1

The crude extractives were separated using a Diaion HP-20 (Mitsubishi Kasei Co. Ltd, Tokyo) guided by stimulation of type III collagen production that used NHFs with water, 20, 40, 60, 80 and 99.5 per cent EtOH, successively. The most active fraction (40 per cent EtOH eluant) was further purified by HPLC equipped with Inertsil ODS-3 (20×150 mm; GL Science Co. Ltd.) monitored by UV absorption (210-400 nm) and gradually eluted with 0.2 per cent formic acid and MeOH mixture (70:30).

Compound 1, electron impact mass spectrometry (EIMS) m/z 937 $[M]^+$, 769 (M-170), 599 (M-2x170); ^1H NMR (CD_3OD:D_2O=1:1, 400 MHz) δ3.92(1H, d, J=12.8 Hz), 4.43(1H, m), 5.12(1H, t, J=9.8 Hz), 5.22(1H, m), 5.50(1H, t, J=8.7 Hz), 5.70(1H, t, J=9.6 Hz), 6.06(1H, d, J=7.8 Hz), 6.34(1H, s), 6.64(1H, s), 6.88(2H, s), 6.91(2H, s), 7.00(2H, s). ^{13}C NMR (CD_3OD:D_2O=1:1, 100 MHz) δ 64.2(CH_2), 71.4(CH), 72.5(CH), 73.4(CH), 74.0(CH), 94.1(CH), 108.3(CH), 109.0(CH), 110.9(CH×2), 111.0(CH×2), 111.2(CH×2), 116.3(C), 119.5(C), 120.0(C), 125.6(C), 126.1(C), 137.4(C), 137.5(C), 140.3(C), 140.4(C), 140.9(C), 145.1(C), 145.9(C), 146.0-146.3(C×10), 166.8(C=O), 167.5(C=O), 168.1(C=O), 169.6(C=O),170.3(C=O).

Hydrolysis of Compound 1 by Acid Solution

Compound 1 (20 mg) was treated with 1 mol/L HCl at 100 °C for 3 hr. The reaction solution was partitioned with EtOAc. The EtOAc extract was purified by HPLC described as above to give compounds 2 and 3. Compounds 2 and 3 were identified with the authentic samples of gallic acid and ellagic acid, respectively by EIMS and NMR analyses. The water portion was neutralized with $AgCO_3$ and dried under reduced pressure. The residue was treated with N, O-bis (trimethylsilyl) acetoamide and applied to GC-MS analysis (Hara *et al.*, 1987) confirming D-glucose.

Data Analysis

All of the values were expressed as mean ± standard deviation (S.D.). The statistically significant difference was examined by the Student's *t*-test.

Results

We analyzed 23 Thai plant species using several human skin cells. In the preliminary assay used anti-peroxidized fat activity we found that the leaves of *D. grandiflora* had stronger activity. Therefore, the crude extract of this plant was assayed using epidermal keratinocytes, melanocytes, human adipocytes and dermal fibroblast cell lines, individually.

Anti-peroxidized Fat Activity of *D. grandiflora* Leaf Extract

It became evident that the crude extract of *D. grandiflora* decreased skin damage dose-dependently by peroxidized fat using keratinocyte as indicated in Figure 3.1. Especially addition of 125 μg/ml of crude extract strongly decreased skin damage.

Inhibitory Activity for Tyrosinase

Skin whitening effect was tested for the inhibitory activity of tyrosinase using melanocytes (Yokozawa *et al.*, 2007). Figure 3.2 showed the inhibitory activity for

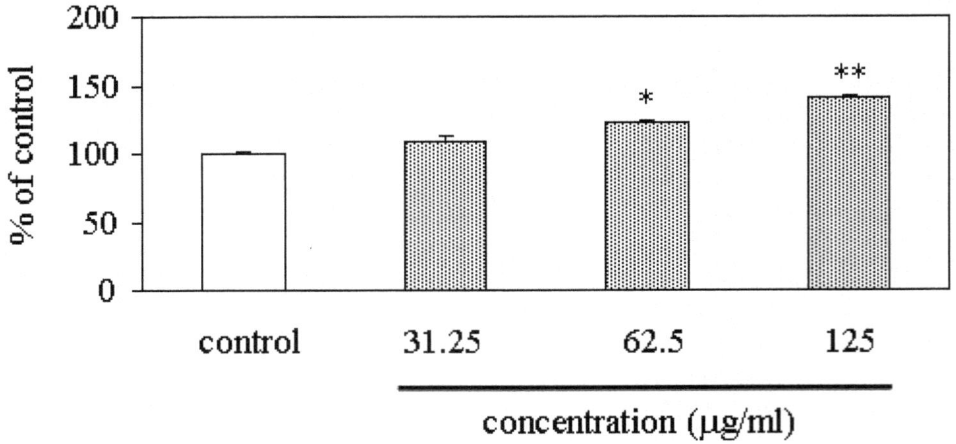

Figure 3.1: Cell viability against tert-butylhydroperoxide (t-BuOOH) stress.

The cells were cultured to sub-confluence then incubated for 24 h in the presence of the extract of *Duabanga grandiflor* (*D.grandiflor*), then assayed for viability of HaCaT cells treated with t-BuOOH, as detailed in Materials and Methods. Results are expressed as a percentage of control and each column represents the mean ± S.D. of three determinations. Asterisks indicate a significant difference compared with the control group, *p<0.05 and **p<0.01.

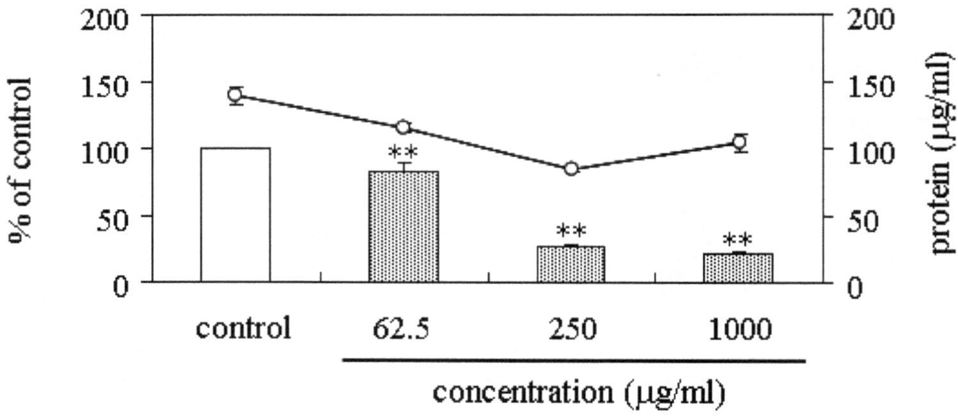

Figure 3.2: Effects of tyrosinase activity on normal human melanocytes.

Human tyrosinase activity in NHEM Cells after 24 h of treatment with or without the extract of *D.grandiflor*. Enzyme assay was performed as described in Materials and Methods. The bar graph shows the quantity of dopamelanin (per cent of control), and the dot graph shows the quantity of protein applied with each sample concentration. Data are expressed as a percentage of the control, and each column represents the mean ± S.D. of three determinations. Asterisks indicate a significant difference compared with the control group, **p<0.01.

tyrosinase using *D. grandiflora* resulting in strong inhibition of melanin preparation. When 1 mg/ml of the crude extract was added, the melanin concentration was reduced to approximately 15 per cent compared to the control. In this investigation the melanoma cells could propagate even at higher concentration of crude extract without cell death.

Neutral Lipid Accumulation

Neutral lipid accumulation assay using adipocytes (Omatsu-Kanbe *et al.*, 2006) is shown in Figure 3.3. In this case, a lower dose of the crude extract of *D. grandiflora* (31 μg/ml) inhibited the accumulation of fat. Approximately 40 per cent of fat accumulation was reduced by the addition of 125 μg/ml of crude extract compared to that of the control (Figure 3.3).

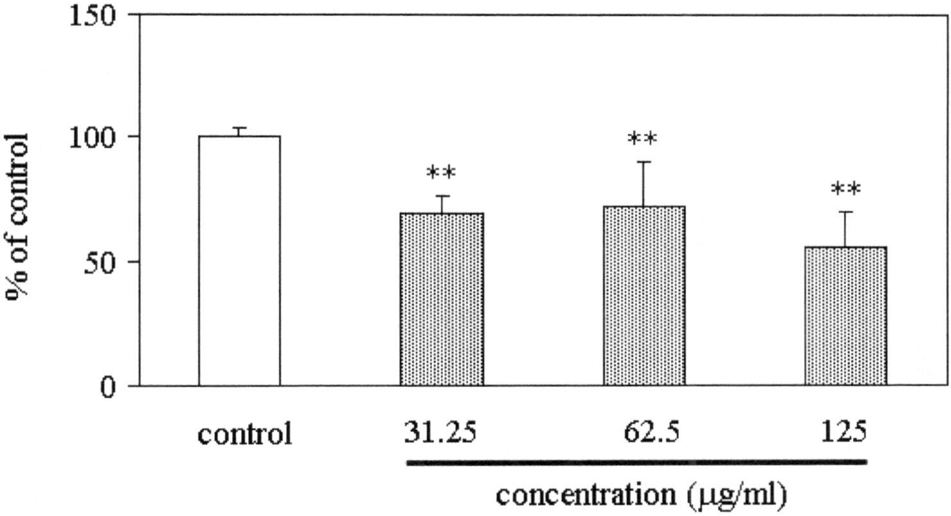

Figure 3.3: Effects of the extract of *D. grandiflor* in the process of neutral lipid accumulation on the adipocytes.

Preadipocytes were differentiated into adipocytes by culturing in standard medium containing insulin, dexamethasone, and IBMX. The adipocytes were cultured for 14 days in the presence of the extract of *D.grandiflor*, then assayed for neutral lipid accumulation levels, as detailed in Materials and Methods. Data are expressed as a percentage of the control, and each column represents the mean ± S.D. of three determinations. Asterisks indicate a significant difference compared with the control group, **$p<0.01$.

Cell Activation Activity

Figure 3.4 indicated cell activation using dermal fibroblasts. The activity increased dose-dependently against the concentration of the crude extract of *D. grandiflora*.

Stimulation of Type III Collagen

The increase of type III collagen was tested using dermal fibroblasts (Chan *et al.*, 1990). The addition of *D. grandiflora* crude extract clearly increased the production of type III collagen dose-dependently. Interestingly 100 μg/ml of *D. grandiflora* crude extract increased the 3.5 times production compared to that of the control as indicated in Figure 3.5. This evidence is very unique because there is no report of tests with plant samples. Therefore, we investigated further by the guided fractionation measuring type III collagen concentration using dermal fibroblasts to purify an active component contained in the crude extract of *D. grandiflora*.

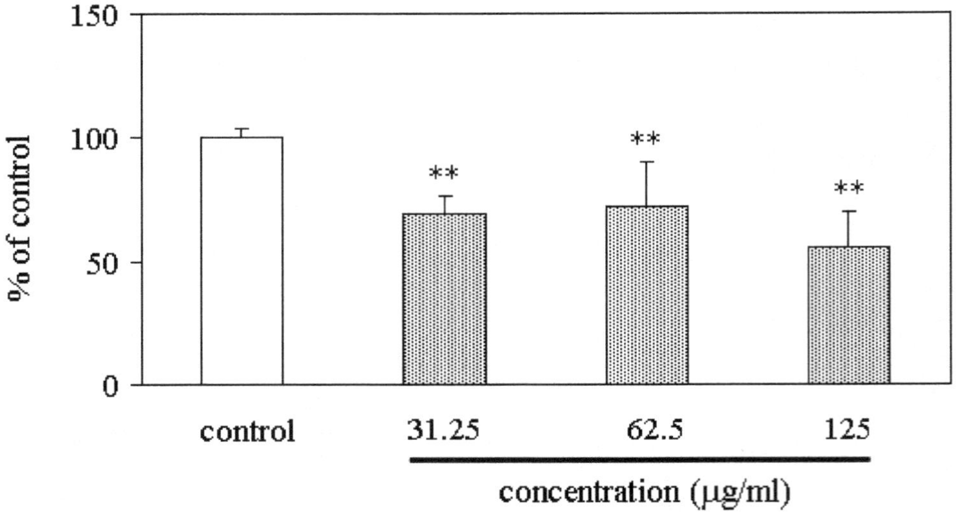

Figure 3.4: Effects of the extract of *D.grandiflor* on the viability of NHFs.

Cells were treated with various doses of the extract of *D.grandiflor* and were examined by MTT assay. Data are expressed as a percentage of the control, and each column represents the mean ± S.D. of three determinations. Asterisks indicate a significant difference compared with the control group, *p<0.05 or **p<0.01.

Purification of Stimulation Active Component for Type III Collagen

The crude extract of *D. grandiflora* was purified being guided by the increase of type III collagen on Diaion HP-20 column chromatography eluted by water and graduated from 20, 40, 60, 80 and 99.5 per cent EtOH, successively. Fractions eluted by 20, 40 and 60 per cent EtOH indicated higher activity. Among them the strongest activity was found in 40 per cent EtOH elution, then this fraction was assayed as described above and the result is shown in Figure 3.6. When compared to the crude extract, which had an effect at 25 μg/ml as indicated in Figure 3.5, 6.25 μg/ml of 40 per cent ethanol fraction clearly increased the production of type III collagen. This phenomenon easily suggested that 40 per cent ethanol fraction evidently contained the active components. This fraction was further purified by HPLC. Fourteen peaks appeared in this fraction as shown in Figure 3.7. Individual peaks were accumulated by repeated HPLC chromatography and tested by the same assay system as described above. Peak A, indicating the strongest stimulation activity of type III collagen was repeated purified by HPLC and finally compound 1 was isolated.

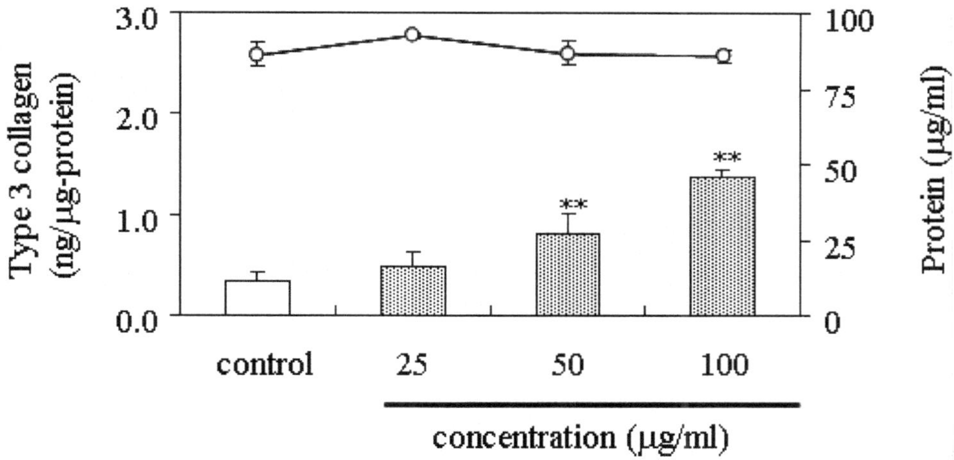

Figure 3.5: Effects of the extract of *D.grandiflor* on the synthesis of type III collagen in NHFs.

Type III collagen was assayed by ELISA as described in Materials and Methods. Bar graphs show the production of type III collagen, and dot graphs show the quantity of protein applied with each sample concentrations. Data are the mean ± S.D. of three determinations. Asterisks indicate a significant difference compared with the control group, **$p<0.01$.

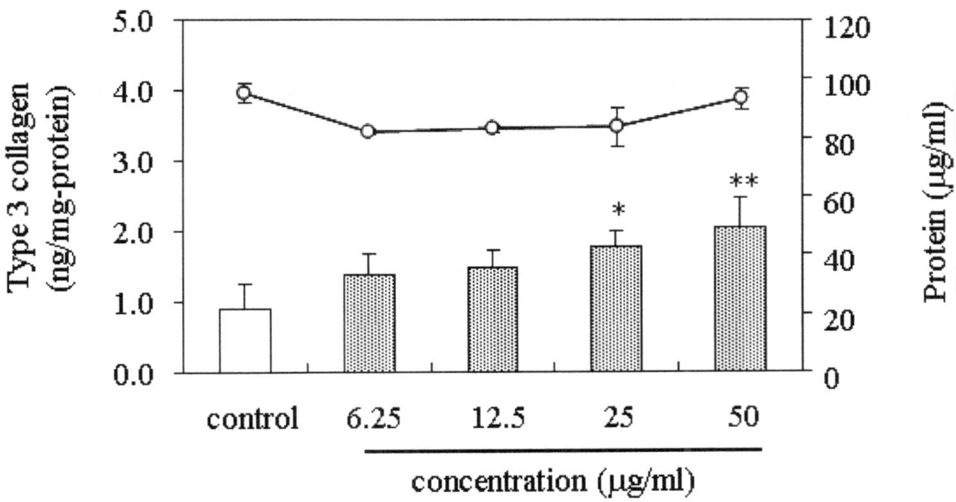

Figure 3.6: Effects of 40 per cent ethanol fraction from the crude extract of *D. grandiflor* on the synthesis of type III collagen in NHFs.

Type III collagen was assayed by ELISA as described in Materials and Methods. The experimental protocol and representation of the graphs are as in Figure 5. Data are the mean ± S.D. of three determinations. Asterisks indicate a significant difference compared with the control group, *$p<0.05$ or **$p<0.01$.

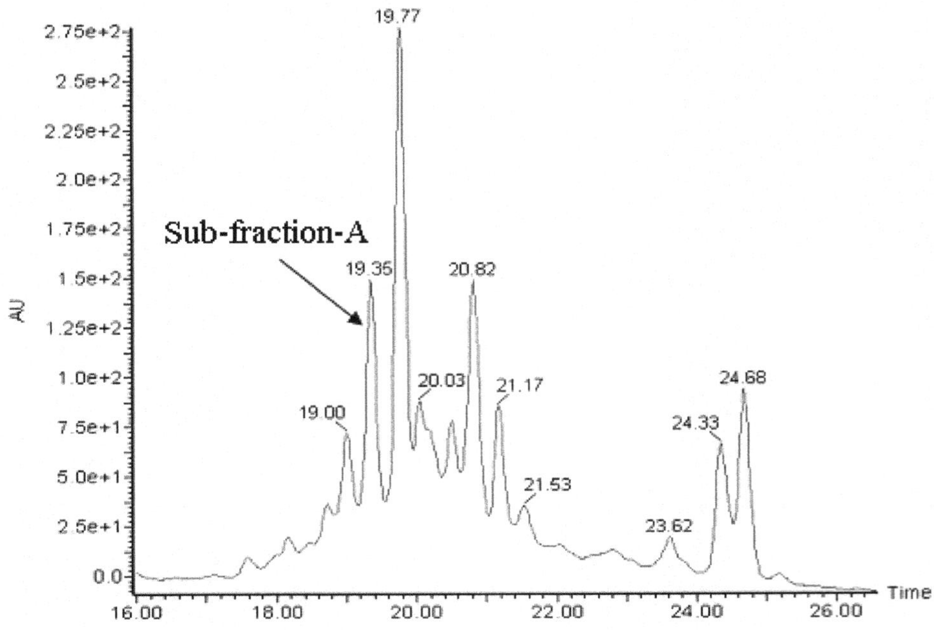

Figure 3.7: HPLC-UV chromatogram of 40 per cent ethanol fraction from the crude extract of *D. grandiflor.*

The 40 per cent ethanol fraction was subjected to reversed-phase HPLC on an Xterra Prep MSC18 column, and UV (210-400 nm) absorption of the eluate was continuously measured with a photodiode array detector. Each sub-fraction was collected in correspondence to the UV absorbing peaks. We isolated 14 UV-absorbing constituents from the 40 per cent ethanol fraction by LC-MS.

Determination of Type III Collagen Stimulating Active Compound

The EI-mass spectrum of compound 1 indicated a simple fragmentation as indicated in Figure 3.8-A showing ES+ 961 as a molecular ion coupled with sodium ion and ES- 937, which was an eliminated proton signal from the molecular ion. Therefore, the molecular ion was suggested to be 938 and then 170 and 338 fragment ions were eliminated to give 769^+ and 599^+, respectively. This fragmentation easily suggested that compound 1 contained gallic acid and ellagic acid in a molecule. ^{13}C NMR data also suggested that compound 1 possesses gallic acid, ellagic acid and sugar in a molecule.

Acidic hydrolysis of compound 1 gave compounds 2 and 3. The isolated compounds 2 and 3 were identified to be gallic acid and ellagic acid, respectively compared to the authentic samples by EI-mass and ^1H NMR analyses. The water portion was analyzed by GC-MS after being treated with N, O-bis (trimethylsilyl) acetoamide (Hara *et al.*, 1987) confirming D-glucose. From these results, compound 1

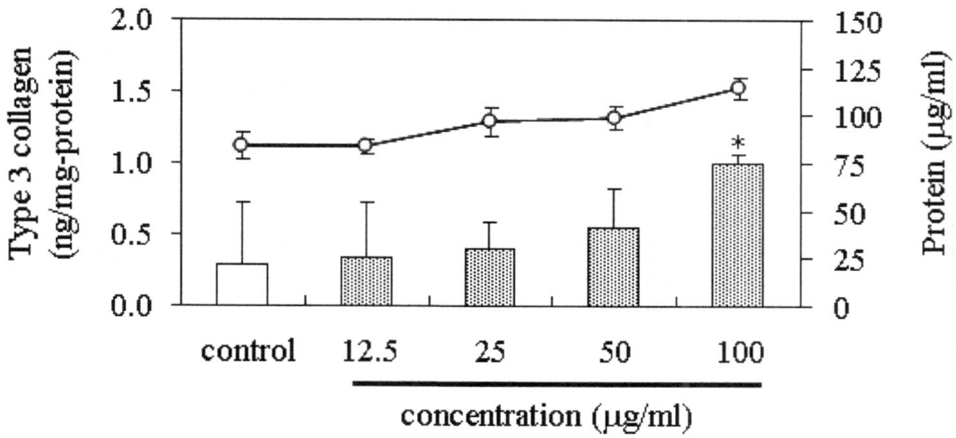

Figure 3.8: Effects of the Peak A subfraction from the 40 per cent ethanol fraction on the synthesis of type III collagen in NHFs.

Type III collagen was assayed by ELISA as described in Materials and Methods. The experimental protocol and representation of the graphs are as in Figure 5. Data are the mean ± S.D. of three determinations. Asterisks indicate a significant difference compared with the control group, **p<0.01.

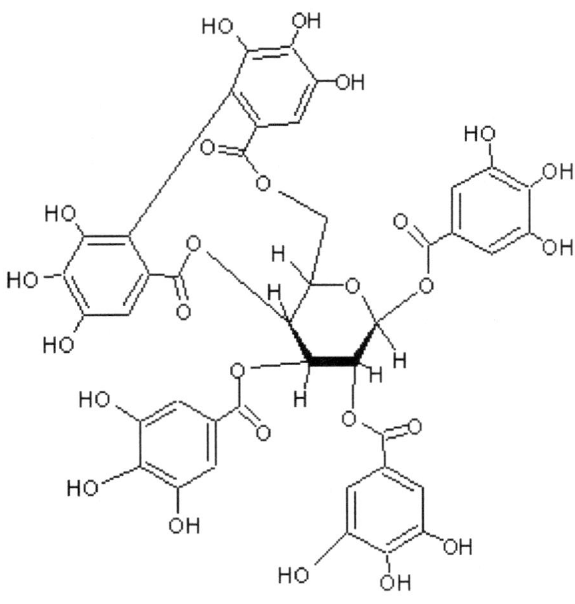

Figure 3.9: Chemical structure of eugeniin.

was clearly suggested to be eugeniin (Kurokawa *et al.*, 1998; Kurokawa *et al.*, 2001; Toda *et al.*, 2000). Finally compound 1 was identified with the authentic sample of eugeniin.

Discussion

In general, the fatty membrane covers the skin layer resulting in an anti-desiccant (Schmuth *et al.*, 2005). However, since the fat is damaged, peroxidized fat is produced, which gives a serious damage for skin when it is contacted to oxygen (Kim *et al.*, 2005). Growth of fibrils is found during development of organs and may be disturbed in some skin diseases (Lakos *et al.*, 2004). However, the mechanism of controlling this process is still obscure. Interstitial collagens are synthesized in the form of procollagens that possess additional extension amino- and carboxypropeptides (Fleischmajer *et al.*, 1981). Each precursor-specific peptide is removed by specific proteases (procollagen NH2-terminal and COOH-terminal proteases) presumably after release of procollagens from the cell (Fleischmajer *et al.*, 1981). On the other hand type III collagen is a major component of the dermis and organized into fibrils varying in diameter (Douglas *et al.*, 1987). However, it has not been established whether type I and III collagens are present in the same fibrils.

D. grandiflora, which is used as a folk medicine for anti-aging activities of skin in northern Thailand, was tested using several human skin cells in order to determine the recovery effect of *D. grandiflora* from skin damages.

From the extracts of twenty three Thai plant species we selected the extract of *D. grandiflora,* because the preliminary assay used by ordinal cell lines like MTT assay, anti-peroxide fatty test, tyrosinase assay, and especially type III collagen assay were used to evaluate that of *D. grandiflora* which was stronger than the other species. The present investigation was performed to compare several human skin cell lines. It became clear that the type III collagen is distributed more widely in baby skin than in adult skin by surveying them with the immunostaining method and/or ELISA using anti-type III collagen MAb (Gay *et al.*, 1976; Sundarraj *et al.*, 1982; Sakai *et al.*, 1982; Varani *et al.*, 2006). Therefore, we selected type III collagen distribution analyzed by ELISA using anti-type III collagen MAb as an assay system in order to evaluate the effect for anti-aging decline of skin. Since type III collagen increased dose-dependently by addition of crude extract, we started to purify the really active component in *D. grandiflora* by column chromatography packed with Diaion HP-20, resulting in an active fraction eluted by 40 per cent ethanol. Final purification was performed by HPLC to give a pure component, which was determined to be eugeniin, compared directly to the authentic sample by its spectroscopic data.

Eugeniin, a hydrolysable tannin, was isolated from *Syzgium aromaticum* as an α-glucosidase inhibitor (Toda *et al.*, 2000) and anti-herpesvirus compound (Kurokawa *et al.*, 1998; Kurokawa *et al.*, 2001). We confirmed the accumulating activity for type III collagen using a dermal fibroblast cell line, resulting in opening a new function of hydrolysable tannin. Fleischmajer *et al.* (1981) proposed that specific procollagen NH2-terminal proteases regulated the growth of type III fibrils. However, since no active compound against such protease has been found until now, this is the first example, although the activation mechanism is obscure. Moreover, it may be possible to supply the crude extract of *D. grandiflora* for the cosmetic field. We also found that the crude extract had inhibitory activities for phospholipase A2, superoxide dismutase (SOD) and MMP-1 (metalo matrix protease-1) (data not shown). These results suggest that *D. grandiflora* contains not only eugeniin, but also other active components having anti-inflammatory activity. Further investigation has been continuing. Those results will be reported elsewhere.

Acknowledgements

This work was supported by the program of Japan Society for the Promotion of Science, "Cooperative Research and Educational Center for Breeding and Standardization of Medicinal Plants and Drug Development between Japan and China".

References

Chan D, SR Lamande, WG Cole and JF Bateman (1990). Regulation of procollagen synthesis and processing during ascorbate-induced extracellular matrix accumulation *in vitro*. *Biochem. J.*, **269**: 175-181.

Fleischmajer R, R Timpl, L Tuderman, L Raisher, M Wiestner, JS Perlish and PN Graves (1981). Ultrastructural identification of extension aminopropeptides of type I and III collgesns in human skin. *Proc. Natl. Acad. Sci. USA*, **78**: 7360-7364.

Gay S, PK Müller, WN Meigel and K Kühn (1976). Polymorphie des Kollagens. Neue Aspekte für Struktur and Funktion des Bindegewebes. *Hautarzt.*, **27**: 196–205.

Gitzingera M, C Kemmera, M D El-Baba, W Webera, and M Fusseneggera (2009). Controlling transgene expression in subcutaneous implants using a skin lotion containing the apple metabolite phloretin. *Proc. Natl. Acad. Sci. USA*, **106**: 10638-43.

Hara, S., H. Okabe and K. Mihashi (1987). Gas-liquid chromatographic separation of aldose enantiomers as trimethylsilyl ethers of methyl 2-(polyhydroxyalkyl) thiazolidine-4(R)-carboxylates. *Chem. Pharm. Bull.*, **35**: 501-507.

Keene DR, LY Sakai, HP Bächinger and RE Burgeson (1987). Type III collagen can be present on banded collagen fibrils regardless of fibril diameter. *J. Cell Biol.*, **105**: 2393-2402.

Kim YS, H Zhuang, RC Koehler and S Doré (2005). Distinct protective mechanisms of HO-1 and HO-2 against hydroperoxide-induced cytotoxicity. *Free Radic. Biol. Med. J.*, **38**: 85-92.

Konoshima T, M Takasaki, H Tokuda, S Morimoto, H Tanaka, E Kawata, LJ Xuan, H Saito, M Sugiura, J Molnar and Y Shoyama (1998). Crocin and crocetin derivatives inhibit skin tumor promotion in mice. *Phytotherapy Res.*, **12**: 400-404.

Kurokawa M, T Hozumi, P Basnet, M Nakano, S Kadota, T Namba, T Kawana and K Shiraki. Purification and characterization of eugeniin as an anti-herpesvirus compound from *Geum japonicum* and *Syzygium aromaticum* (1998). *J. Pharmacol. Exp. Ther.*, **284**: 728-735.

Kurokawa M, T Hozumi, M Tsurita, S Kadota, T Namba and K Shiraki (2001). Biological characterization of eugeniin as an anti-herpes simplex virus type 1 compound *in vitro* and *in vivo*. *J. Pharmacol. Exp. Ther.*, **297**: 372-379.

Lakos G, S Takagawa, SJ Chen, AM Ferreira, G Han, K Masuda, KJ Wang, LA DiPietro and J Varga (2004). Targeted disruption of TGF-beta/Smad3 signaling modulates skin fibrosis in a mouse model of scleroderma. *Am. J. Pathol.*, **165**: 203-17.

Omatsu-Kanbe M, K Inoue, Y Fujii, T Yamamoto, T Isono, N Fujita and H Matsuura (2006). Effect of ATP on preadipocyte migration and adipocyte differentiation by activating P2Y receptors in 3T3-L1 cells. *Biochem. J.*, **393**: 171-180.

Phan TT, L Wang, P See, RJ Grayer, SY Chan and ST Lee (2001). Phenolic compounds of Chromolaena odorata protect cultured skin cells from oxidative damage: implication for cutaneous wound healing. *Biol Pharm Bull.*, **24**: 1373-9.

Ramírez-Zacarías J. L., F Castro-Munozledo. and W Kuri-Harcuch (1992). Quantitation of adipose conversion and triglycerides by staining intracytoplasmic lipids with Oil red O. *Histochemistry*, **97**: 493-497.

Sakai LY, E Engvall, DW Hollister and RE Burgeson (1982). Production and characterization of a monoclonal antibody to human Type IV collagen. *Am. J. Pathol.*, **108**: 310-318.

Sato T, A Takahashi, M Kojima, N Akimoto, M Yano and A Ito (2007). A citrus polymethoxy flavonoid, nobiletin inhibits sebum production and sebocyte proliferation, and augments sebum excretion in hamsters. *J Invest Dermatol.*, **127**: 2740-8.

Schmuth M, AM Ortegon, M Mao-Qiang, PM Elias, KR Feingold and A Stahl (2005). Differential expression of fatty acid transport proteins in epidermis and skin appendages. *J. Invest. Dermatol.*, **125**: 1174-1181.

Soeda S, T Ochiai, H Shimeno, H Saito, K Abe, H Tanaka and Y Shoyama (2007). Pharmacological activities of crocin in saffron. *J Nat Med.*, **61**: 102-111.

Sundarraj N and J Willson (1982). Monoclonal antibody to human basement membrane collagen type IV. *Immunology*, **47**: 133-140.

Varani J, MK Dame, L Rittie, SE Fligiel, S Kang, GJ Fisher and JJ Voorhees (2006). Decreased collagen production in chronologically aged skin: roles of age-dependent alteration in fibroblast function and defective mechanical stimulation. *Am. J. Pathol.*, **168**: 1861-1868.

Yokozawa T and YJ Kim (2007). Piceatannol inhibits melanogenesis by its antioxidative actions. *Biol. Pharm. Bull.*, **30**: 2007-2011.

Traditional and Folk Herbal Medicine: *Pages* **67–91**
Recent Researches Vol. 1 (2012)
Editor: **V.K. Gupta**
Published by: **DAYA PUBLISHING HOUSE, NEW DELHI**

4

Chromolaena odorata, A Traditional Medicinal Herb with Hemostatic and Wound Curative Properties: An Overview

Hataichanok Pandith[1] and Wandee Gritsanapan[1]*

ABSTRACT

Chromolaena odorata (L.) King and Robinson or Siam weed is a perennial scandent or semi-woody shrub of the family Asteraceae. It is native to Central and South America and spreads throughout the tropical and subtropical regions. C. odorata is used as a medicinal herb in many countries of Tropical Africa and Asia for variety of ailments such as fever, influenza, cold, cough, diabetes, inner hemorrhage, malaria, bleeding and wound. Many studies both in vitro and in vivo have demonstrated that C. odorata extracts can enhance hemostasis as well as heal wounds. In addition, it also possesses antioxidant, anti-inflammatory, antibacterial, antimalarial, and antituberculosis properties. The leaf and flower extracts of C. odorata contain many phytochemical constituents including volatile oils, phenols, tannin, saponins, terpenoids, alkaloids and flavonoids, which are the main component. Several flavonoids have been isolated from C. odorata extracts and are being investigated for the pharmacological and biological activities. For instant, 4′,5,6,7-tetramethoxy-flavone, the active compound for blood coagulation was isolated for the first time by preparative thin layer chromatography (PTLC). The major flavonoid, kaempferol-4′-methylether that could be used as a marker was isolated by reverse phase high performance liquid chromatography (RP-HPLC). It showed a good antioxidant activity tested by superoxide scavenging, DPPH radical scavenging and tyrosinase inhibition. Furthermore, naringenin-4′-methylether (isosakuranetin), which displayed anti-inflammatory, antioxidant and antistaphylococcol activities, was also isolated by RP-HPLC and could be used as a bioactive marker. Toxicity of

1 Department of Pharmacognosy, Faculty of Pharmacy, Mahidol University 447 Sri-Ayudthaya Road, Ratchatevi, Bangkok 10400, Thailand

* *Corresponding author*: E-mail: pywgs@mahidol.ac.th

aqueous and alcoholic extracts of C. odorata were studied on brine shrimps, cell lines, gills of fish and animal models. The alcoholic extract from the aerial part of C. odorata was found to be non-toxic to brine shrimps, normal kidney (Vero) and normal liver (WRL-68) cell lines although the ethanolic leaf extract showed acute toxicity of 96 hours on the gills of tilapia fish (Oreochromis niloticus Linn.) (LD_{50} 30.94 mg/l). Histological changes were found in dose dependent studies. For aqueous extract, only fibroblast cell damage was observed when cultured in 250 µg/ml of aqueous leaf extract, but there was no significant damage to endothelial cells and keratinocytes. Moreover, the aqueous lyophilized extract from fresh leaves of C. odorata did not irritate rabbit skin even at high dose (500 mg/ml) and did not show acute toxicity after oral administration (LD_{50}>5 g/kg) for one week. The ointment of aqueous leaf extract had been licensed for clinical use to treat soft tissue wounds and burns in Vietnam already. Hydro-gel containing C. odorata aqueous extract with hemostatic and antibacterial effects was developed in Thailand. Antiseptic and wound curative basic gel and cream formulations with pure flavonoids and essential oils isolated from C. odorata had been developed for skin care products in Malaysia. Since C. odorata is an effective hemostatic agent, it would be a beneficial choice of herb to be developed into skin care products. The potential products derived from C. odorata extracts would be a cost-effective and more readily available alternative to synthetic medicine.

Keywords: *Chromolaena odorata*, *Eupatorium odoratum*, Formulation, Preparation, Siam weed, Stop bleeding, Toxicity, Wound healing.

Introduction

Chromolaena odorata (L.) King and Robinson was formerly known as *Eupatorium odoratum* Linn. It is commonly called Siam weed. This plant is a perennial scandent or semi-woody shrub in the Asteraceae (Compositae) family. It is native to Central and South America and spreads throughout the tropical and subtropical areas of the world (Muniappan and Marutani, 1991). *C. odorata* has been used as a medicinal herb in many countries of Tropical Africa and Asia for variety of ailments such as fever, influenza, cold, cough, diabetes, bleeding, wounds and malaria (Pisutthanan *et al.*, 2005). Many studies both *in vitro* and *in vivo* have demonstrated that *C. odorata* extracts can enhance many pharmacological and biological activities. It can stimulate hemostasis (Akah, 1990; Wongkrajang *et al.*, 1990; Triratana *et al.*, 1991; Thongpraditchot *et al.*, 1994; Wongkrajang *et al.*, 1994; Phithanchort *et al.*, 2006; Khengraeng, 2007), wound healing, anti-inflammatory and antioxidant effects (Phan *et al.*, 1996; Phan *et al.*, 1998; Phan *et al.*, 2000; Phan *et al.*, 2001a; Phan *et al.*, 2001b; Phan *et al.*, 2001c). The plant extracts also show many biological activities *i.e.* antibacterial, antimalarial and antituberculosis (Suksamrarn *et al.*, 2004; Chomnawang *et al.*, 2005; Pisutthanan *et al.*, 2005; Ling *et al.*, 2006; Ling *et al.*, 2007c; Molina-Salinas *et al.*, 2007). Previous chemical studies revealed that extract of *C. odorata* contain many phytochemical constituents such as volatile oils, phenols, tannins, saponins, terpenoids, alkaloids and flavonoids, which are the main composition (Inya-Agha *et al.*, 1987; Umukoro and Ashorobi, 2006). Several flavonoids had been isolated from *C. odorata* extracts and were investigated for the pharmaceutical activities (Bose *et al.*, 1973; Naengchomnong, 1989; Suksamrarn *et al.*, 2004; Pisutthanan *et al.*, 2005; Ling *et al.*, 2007a; Ling *et al.*, 2007b; Ling *et al.*, 2007c). In

Thailand, 4′,5,6,7-tetramethoxy-flavone, the active compound for blood coagulation, was isolated from the leaves for the first time by preparative thin layer chromatography (Naengchomnong, 1989). The toxicity of aqueous and alcoholic extracts of *C. odorata* were studied with brine shrimps, cell lines, gills of fish and animal models (Thongpraditchot *et al.*, 1994; Phan *et al.*, 1998; Srisuk, 2000; Phan *et al.*, 2001a; Pisutthanan *et al.*, 2005). Although *C. odorata* has been used as herbal medicine in many countries for several decades and there are many reports of useful pharmacological and biological activities, there is only one product which has been licensed from *C. odorata* leaf extract. The ointment of aqueous leaf extract has been licensed for clinical use to treat soft tissue wounds and burns in Vietnam. Thailand and Malaysia also try to develop skin care products from this plant. However, comprehensive scientific evidences and clinical trials of products have not been studied yet. This review gives an overview of botanical aspects which include morphology, taxonomy, distribution, ethnopharmacology, chemical studies and its pharmacological activities as well as toxicity.

Botanical Aspects

Morphology (Figures 4.1 and 4.2)

Chromolaena odorata is an aromatic perennial scandent or semi-woody shrub, mostly 80-250 cm tall; covered in fine grey hairs; stem rounded, erect or sprawling to subscandent; petioles 5-20 mm; leaf blade (3-nerved) ovate to diamond shaped, 3-10 by 1-4 cm, acuminate, with yellow dots below, margin coarsely dentate to subentire; head (terminal or lateral) usually 5-50+ in corymbiform arrays; involucres cylindric, (7-)8-10 mm; phyllaries in 4-6(-8) series, apices of the inner appressed, rounded to truncated (sometimes slightly white-petaloid or expanded); $2n = 40, 60, 70$; corollas purplish to light blue to nearly white or slightly pinkish, flowering mostly in October-December, sometimes year round; fruit is one-seeded (Muniappan and Marutani, 1991; Nesom, 2006).

Taxonomy

Chromolaena odorata (L.) King and Robinson (1970), synonyms: *Eupatorium odoratum* L. (1759), *Eupatorium conyzoides* Vahl (1794) and *Osmia odorata* (L.) Schultz-Bip. (1866), is classified into Domain: *Eukaryota*, Kingdom: *Plantae*, Subkingdom: *Tracheobionta*, Division: *Magnoliophyta*, Class: *Magnoliopsida*, Subclass: *Asteridae*, Order: *Asterales*, Family: *Asteraceae*, Genus: *Chromolaena*, Species: *odorata* (L.) King and Robins. Common names are Siam weed, Agonoi, Baby Tea, Bitterbush, Butterfly-Weed, Christmasbush, Chromolaena, Crucita, Devilweed, Hagonoy, Huluhagonoi, Jack in the Bush, Ngesngesil, Paraffin-Weed, Triffidweed and Wisolmatenrehwei (Nesom, 2006).

Distribution

C. odorata is of neotropical origin, Central and South America. It is cold-sensitive and dies at temperatures below 0°C (Lu *et al.*, 2008). It was introduced to humid tropical Asia and Africa (India, Sri Lanka, China, Taiwan, Indonesia, Malaysia, Thailand, and also Australia) by early navigators in the mid 1800s and became a

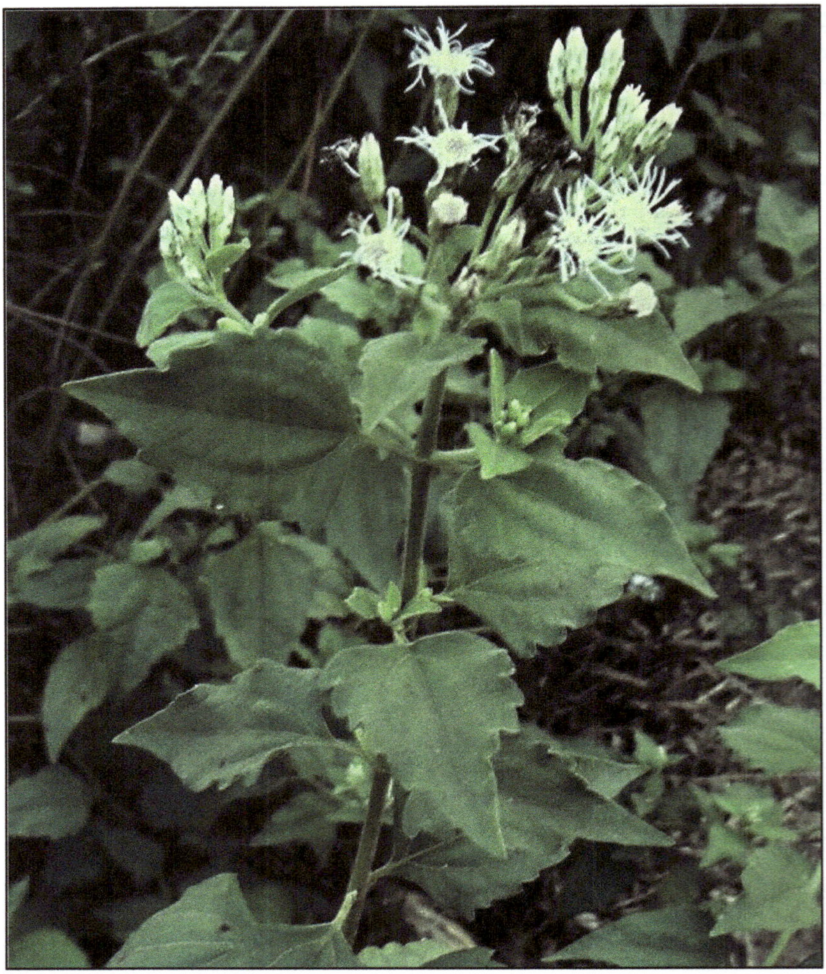

Figure 4.1: Photograph of *Chromolaena odorata*.

weed problem in West, Central and Southern Africa, South and Southeast Asia and Micronesia in the mid 1900s. Its invasion disturbed forests, vacant lands, wildlife reserves, riverbanks, pastures and plantation crops. The biotype of *Chromolaena* in South Africa was different from others that invaded the rest of the world. The origin of this biotype had been narrowed down to northern Caribbean region (Muniappan *et al.*, 2005; Simmonds, 1976). In India, *C. odorata* had successfully invaded in open areas in high rainfall regions, especially the Eastern and Western Himalayas and the Western Ghats. It competed heavily with indigenous plant species (Tripathi *et al.*, 2006). *C. odorata* had gradually replaced the primary forest regenerative species in the degraded forest site in Bangladesh (Islam *et al.*, 2001). In China, *C. odorata* was introduced in the early 20th century and became an invasive species. Since it was intolerant to cold weather, its distribution was restricted within Southern China, *e.g.*

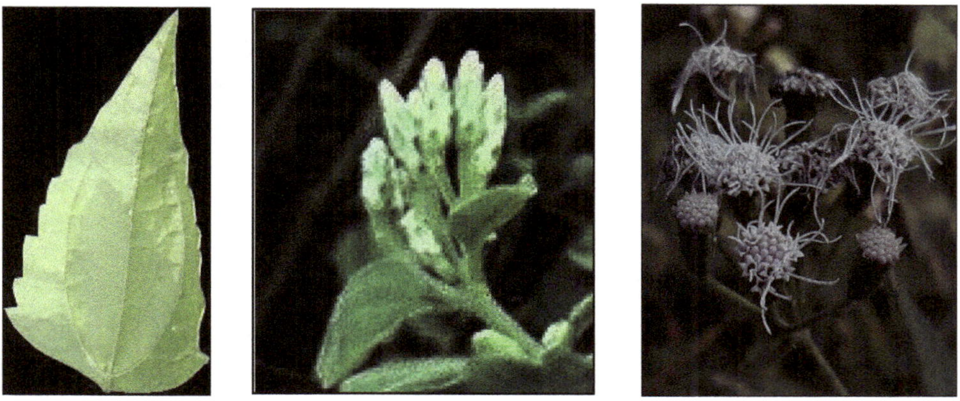

(A) (B) (C)

Figure 4.2: Parts of *C. odorata*; (A) leaf, (B) young flowers and (C) mature flowers.

Hainan province (Qin *et al.*, 2008), Honghe River Basin (Xu *et al.*, 2006), dry valleys of Yunnan and Sichuan provinces where it was relatively arid and hot with the maximum temperature of 40.4°C (Lu *et al.*, 2008). The distribution of this plant is shown in Figure 4.3.

Ethnopharmacology and Utilization

Although *C. odorata* has become a major weed of crops, plantations and pastures in many parts of the world, it has been used in folk medicine as a cure of various ailments including fever, influenza, cold, cough, diabetes, bleeding, wounds, internal hemorrhage and malaria (Inya-Agha *et al.*, 1987; Bamba *et al.*, 1993). Ethnopharmacological uses are shown in Table 4.1.

Fresh cut leaves of *C. odorata* are used as an antibacterial and hemostatic agent to stop bleeding in rural areas of Thailand, Vietnam, Malaysia, Nigeria and Ivory Coast (Wongkrajang *et al.*, 1990; Bamba *et al.*, 1993; Phan *et al.*, 2001a; Phan *et al.*, 2001b; Phan *et al.*, 2001c; Umukoro and Ashorobi, 2006; Ling *et al.*, 2007a; Ling *et al.*, 2007b; Ling *et al.*, 2007c). In Vietnam, the fresh leaves, aqueous extracts, or decoction have been used throughout the country to treat leech bites, soft tissue wounds, burnt wounds, skin infections, and infections in dento-alveolitis (Phan *et al.*, 2001a). The macerated leaves are usually applied to swollen portion of the body to relieve inflammation among the rural populace in the southern part of Nigeria (Umukoro and Ashorobi, 2006). In Malaysia, *C. odorata* is known as Pokok kapal terbang, Pokok Jerman, Rumput Jepun or Rumput Siam. Even a poultice of leaves is traditionally applied onto cuts or wounds to stop bleeding and promote healing. It is also applied topically as an antidote to the sting from the spine of the common sea catfish. An aqueous decoction of the roots is used as an antipyretic and analgesic remedy, while a leaf extract with salt is used as a gargle for sore throats and colds (Ling *et al.*, 2007b).

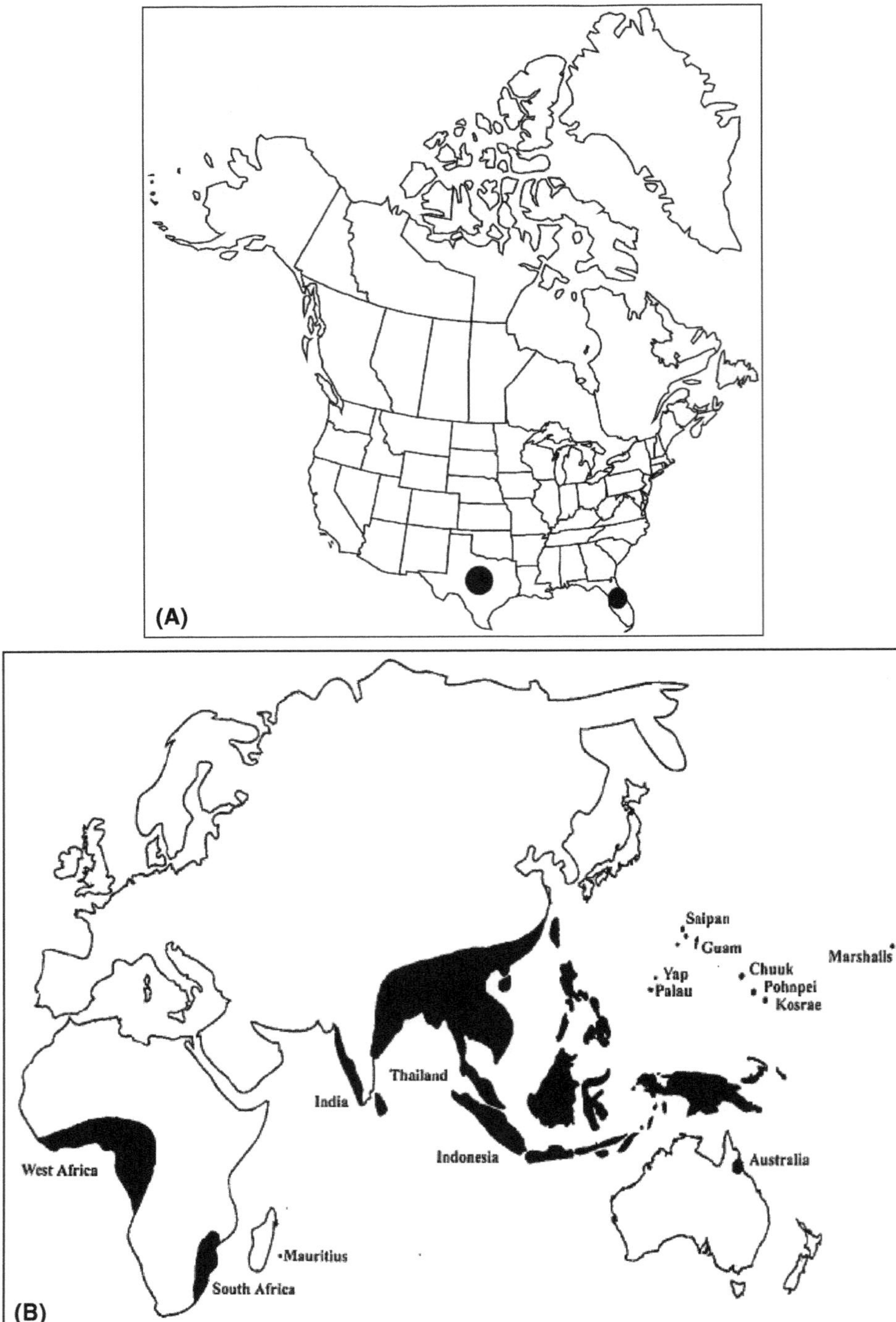

Figure 4.3: Distribution of *C. odorata* in many parts of the world shown in black, (A) Distribution in Central America (Nesom, 2006) (B) Distribution in Africa, Asia, Miconesia and Australia (Muniappan *et al.,* 2005).

Table 4.1: The ethnopharmacological uses of *C. odorata.*

Part Used	*Ethnopharmacological Activity*	*Reference*
Leaf	☆ stops bleeding (from leech bite)	Wongkrajang *et al.*, 1990;
	☆ heals soft tissue and burnt wound	Phan *et al.*, 2001b;
	☆ relieves diarrhea	Suksamrarn *et al.*, 2004;
	☆ relieves day or night blindness	Pisutthanan *et al.*, 2005;
	☆ relieves internal hemorrhage	Umukoro and Ashorobi, 2006;
	☆ antibacterial activity (relieves skin and dento-alveolitis infection)	Ling *et al.*, 2007b
	☆ anti-inflammatory activity	
	☆ antitussive activity	
	☆ antimalarial activity (used with *Psidium guajava* leaf and *Cymbopogon citratus*)	
	☆ antispasmodic effect	
	☆ antihypertensive effect	
	☆ antidote against the sting from spine of common sea catfish	
	☆ a gargle for sore throats, cough and cold	
	☆ diuretic	
	☆ astringent	
Aerial part	☆ relieves flatulence	Suksamrarn *et al.*, 2004
	☆ relieves engorged breast	
	☆ absorbs pus	
	☆ anti-inflammatory activity	
Whole plant	☆ treats tetanus	Suksamrarn *et al.*, 2004
Flower	☆ relieves apthous ulcer	
	☆ relieves cold and fever	
	☆ tonic and cardiotonic	Suksamrarn *et al.*, 2004
Root	☆ antipyretic and analgesic activities	Ling *et al.*, 2007b

C. odorata is also exploited as a mulch in plantation (Hulugalle *et al.*, 1987). It can be used for energy production, in particle board industry and fish poison in India (Talapatra *et al.*, 1974; Jagadeesh *et al.*, 1990; Kalita *et al.*, 1999).

Chemical Constituents

The extracts from *C. odorata* contain many phytochemical components such as volatile oils, phenols, tannins, saponins, terpenoids, alkaloids, and flavonoids, which are major components. Pyrrolizidine alkaloids are found in the roots and mature flower heads (Thongpraditchot *et al.*, 1994; Pisutthanan *et al.*, 2005; Umukoro and Ashorobi, 2006). In Nigeria, Inya-Agha and coworkers reported the phytochemical screening of aqueous extract, indicating the presence of tannins, phenols, and saponins and absence of anthraquinones. They were the first investigators of volatile oils in *C. odorata* fresh and dry leaves extracted by steam distillation, resulting in 0.40 per cent v/w yield and 22 volatile components. Comparing with authentic samples by GC-MS, the volatile oil contains 6 major components that was identified as α-pinene (19.32 per cent), cadinene (19.09 per cent), (+) camphor (15.46 per cent), limonene (10.22 per cent), β-caryophyllene (7.05 per cent) and cadinol isomer (6.36 per cent). Some minor constituents of the oil are citronellal, p-cymeme, and geraniol. The oils from both fresh and dry leaves illustrated similar antibacterial activity against *Staphylocoocus aureus* and *Escherichia coli*. They showed that drying temperature of leaves and daytime influence the oil content and its antimicrobial activity. The yield of oil and the antibacterial activity decreased with higher drying temperature. The oil content was the highest in the morning and lowest in the mid day but showed no qualitative difference in the composition (Inya-Agha *et al.*, 1987). The quantitative and qualitative levels of oil extracted from this plant were different when it was collected from different locations, such as that from Nigeria, Ivory Coast, and Thailand (Inya-Agha *et al.*, 1987; Bamba *et al.*, 1993; Pisutthanan *et al.*, 2005). Hydrodistillation of fresh plant from Ivory Coast and Thailand yielded 0.30 per cent and 0.20 per cent v/w oils, respectively (Bamba *et al.*, 1993; Pisutthanan *et al.*, 2005). The Ivorian oil was composed of 26 components, more than the oil analyzed in Nigeria with the absence of camphor and presence of geijerene and pregeijerene. Nevertheless α-pinene was still the major component (18.8 per cent). Other major components consisted of pregeijerene (14.3 per cent), β-pinene (10.5 per cent), germacrene-D (8.2 per cent) and β-caryophyllene (6.5 per cent) (Bamba *et al.*, 1993). The antibacterial activities of these essential oils were also reported. The Ivorian oil has no effect on *S. aureus* but exhibited notable activity against the Gram-negative species such as *E. coli*, *Klebsiella pneumonia*, and specifically against *Pseudomonas aeruginosa* (Bamba et *al.*, 1993). In contrast, *C. odorata* essential oils from Nigeria and Thailand demonstrated the greatest activity on gram-positive bacteria especially *S. aureus*. However, the chemical constituents of oils from these two locations were different (Inya-Agha *et al.*, 1987; Pisutthanan *et al.*, 2005).

Since flavonoids are the major components of *C. odorata*., many of them were isolated from *C. odorata* and investigated for the pharmaceutical activities (Bose *et al.*, 1973; Naengchomnong, 1989; Suksamrarn *et al.*, 2004; Pisutthanan *et al.*, 2005; Ling *et al.*, 2007a; Ling *et al.*, 2007b; Ling *et al.*, 2007c). Odoratin, a new chalcone and salvigenin, were isolated for the first time in 1973 and 1974, respectively (Bose *et al.*, 1973; Talapatra *et al.*, 1974). Odoratin was obtained from light petroleum (b.p. 60-80°C) extract by column chromatography over silica gel. The flavanone, isosakuranetin, was also isolated from acetone extract (Bose *et al.*, 1973). Odoratin was isolated from fresh aerial part including flower of *C. odorata*, extracted with chloroform by Soxhlet distillation. Moreover, two flavanones, isosakuranetin monomethyl ether and 4′,5-dihydroxy-3′,7-dimethoxyflavone, which had not been found in any member of *Eupatorium* genus, were obtained from this extract by column chromatography over silica gel. These three flavonoids did not have significant cytotoxicity to KB and P388 cell lines (Arene *et al.*, 1987). The rarely occurring flavones, salvigenin, was isolated from cold light petroleum (b.p. 60-80°C) extract of the whole plant by column chromatography over silica gel eluted with chloroform. Furthermore, the poorly yielded triterpene alcohols, lupeol and β-amyrin, were isolated by elution with benzene (Talapatra *et al.*, 1974). The active compound for blood coagulation, 4',5,6,7-Tetramethoxy-flavone, was isolated for the first time by preparative thin layer chromatography (Naengchomnong, 1989). The methanol leaf extract had antioxidant, anti-inflammatory and antistaphylococcol properties, and showed no cytotoxic effect on the selected normal cell lines (Ling *et al.*, 2006). Eleven flavonoids (1, kaempferol-3-*O*-rutinoside; 2, quercetin-3-*O*-rutinoside; 3, quercetin-4'methyl ether; 4, aromadendrin-4'-methyl ether; 5, taxifolin-7-methyl ether; 6, taxifolin-4'-methyl ether; 7, quercetin-7-methyl ether; 8, naringenin-4'-methyl ether; 9, kaempferol-4'-methyl ether; 10, eriodictyol-7,4'-dimethyl ether; 11, quercetin-7,4'-dimethyl ether) could be detected by RP-HPLC using photodiode array detection (DAD) as a detector (Ling *et al.*, 2007b). Flavonoid aglycones (flavonols, dihydroflavonols, flavonones) were presented as the major group of compounds. The major flavonoid, kaempferol-4'-methylether had a good antioxidant activity that could be used as a marker while naringenin-4'-methylether (isosakuranetin), which showed anti-inflammatory, antioxidant and antistaphylococcol activities, could be used as a bioactive marker (Ling *et al.*, 2007c). The 95 per cent methanolic extract provided six flavonoids including 3,5,3'-trihydroxy-7,4'-dimethoxyflavanone, 5,3'-dihydroxy-7,4'-dimethoxyflavanone, 5,7-dihydroxy-4'-methoxyflavanone, 3,5,4'-trihydroxy-7-methoxy flavanone, 5,7,3'-trihydroxy-5'-methoxyflavanone and 3,5,7-trihydroxy-4'-methoxy flavanone. The last three were first isolated from this plant (Pisutthanan *et al.*, 2005). The ethylacetate soluble portion of the 95 per cent ethanolic extract from the aerial part was isolated and purified by silica gel and sephadex LH-20 chromatography. Eight compounds were isolated and identified as kaempferol-3-methoxy, rhamnetin, tamarixetin, quercetin, kaempferol, apigenin, luteolin and dihydrokaempferide. This was the first time to report the isolation of kaempferol-3-methoxy, rhamnetin and apigenin from the genus *Eupatorium*, while dihydrokaempferide, which was a rare flavonoid, was obtained for the first time from the family Compositae (Yuan *et al.*, 2007).

From the *C. odorata*'s flowers, eight flavonoids were isolated and identified concluding four flavanones, *i.e.* isosakuranetin (5,7-dihydroxy-4'-methoxyflavanone), persicogenin (5,3'-dihydroxy-7,4'-dimethoxyflavanone), 5,6,7,4'-tetramethoxy-flavanone and 4'-hydroxy-5,6,7-trimethoxyflavanone; two chalcones, *i.e.* 2'-hydroxy-4,4',5',6'-tetramethoxy chalcone and 4,2'-dihydroxy-4',5',6'-trimethoxychalcone and two flavones, *i.e.* acacetin (5,7-dihydroxy-4'-methoxy flavone) and luteolin (5,7,3',4'-tetrahydroxyflavone) (Suksamrarn *et al.*, 2004).

The petroleum ether extract from the roots of *C. odorata* could isolate a novel triterpene, 3'-hydroxy-28-carboxyolean-12-ene/eupatoric acid along with seven known compounds poriferasterol, octadecane, butyrospermol acetate, bis(2-ethylhexyl) phthalate, chrysophanol, physcion and palmitic acid (Amatya and Tuladhar, 2005).

Table 4.2: Chemical constituents isolated from several parts of *C. odorata.*

Plant Part	*Chemical Constituent*	*Reference*
Leaf	**Volatile oils** (+) Camphor, β-Eudesmol, Hexenal, 3-Hexenol(*Z*), 1-Hexenol, 2-Hexenol(*E*), Lupeol, 4-Terpineol, T-cadinol, Elemol, α-Cadinol, Eupatol, γ-Eudesmol, β-Amyrin, Limonene, Myrcene, α-Pinene, β-Pinene, Sabinene, β-Caryophyllene, Ocimene(*E*), β-Phellandrene, α-Copaene, β-Elemene, Σ-Muurolene, α-Humulene, Geijerene, Pregeijerene, γ-Cadinene, δ-Cadinene, Germacrene-D, γ-Elemene, α-Elemene, *n*-Decane, Tridecane, Tetradecane, Hexadecane, Pentadecane, Undecane, Dodecane, d-,l-Eupatene	Bose *et al.,* 1973; Talapatra *et al.,* 1974; Arene *et al.,* 1987; Inya-Agha *et al.,* 1987; Naengchomnong, 1989; Bamba *et al.,* 1993; Pisutthanan *et al.,* 2005; Ling *et al.,* 2007b
	Flavonoids Odoratin (2'-hydroxy-4,4',5',6'-tetramethoxychalcone), Isosakuranetin (naringenin-4'-methyl ether/5,7-dihydro xy-4'- methoxyflavanone), Salvigenin, 4',5,6,7-Tetramethoxy-flavone, Kaempferol-4'-methyl ether, Kaempferol-3-*O*-rutinoside, Quercetin-3-*O*-rutinoside, Quercetin-4'methyl ether, Quercetin-7-methyl ether, Quercetin-7,4'-dimethyl ether, Aromadendrin-4'-methyl ether, Taxifolin-7-methyl ether, Taxifolin-4'-methyl ether, Eriodictyol-7,4'-dimethyl ether	
	Miscellaneous trihydric alcohol, anisic acids	
Flower	**Flavonoids** Isosakuranetin (naringenin-4'-methyl ether/5,7-dihydro xy-4'-methoxyflavanone), Persicogenin (5,3'-dihydroxy-7,4'-dimethoxyflavanone), 4,5,6,7'-Tetramethoxyflavanone, 4'-Hydroxy-5,6,7-trimethoxyflavanone, 2'-Hydroxy-4,4',5',6'- tetramethoxychalcone, 4,2'-Dihydroxy-4',5',6'-trimethoxychalcone, Acacetin (5,7-dihydroxy-4'-methoxyflavone) Luteolin (5,7,3',4'-tetrahydroxyflavone)	Suksamrarn *et al.,* 2004
Aerial part	**Flavonoids** Odoratin (2'-hydroxy-4,4',5',6'-tetramethoxy) chalcone, Isosakuranetin monomethyl ether, Quercetin, Kaempferol, Kaempferol-3-methoxy, Rhamnetin, Tamarixetin, Apigenin, Luteolin,	Arene *et al.,* 1987; Pisutthanan *et al.,* 2005; Yuan *et al.,* 2007

Contd...

Table 4.2–*Contd...*

Plant Part	*Chemical Constituent*	*Reference*
	Dihydrokaempferide, 4',5-Dihydroxy-3',7-dimethoxyflavone, 3,5,3' Ttrihroxy-7,4'-dimethoxyflavanone, 5,3'-Dihydroxy-7,4'-dimethoxyflavanone, 5,7-Dihydroxy-4'-methoxyflavanone, 3,5,4'-Trihydroxy-7- methoxyflavanone, 5,7,3'-Trihydroxy-5'- methoxyflavanone, 3,5,7-Trihydroxy-4'- methoxyflavanone	
Root	**Terpenoids** 3'-Hydroxy-28-carboxyolean-12-ene (eupatoric acid)	Amatya and Tuladhar, 2005
	Miscellaneous Poriferasterol, Octadecane, Butyrospermol acetate, Bis(2-ethylhexyl)phthalate, Chrysophanol, Physcion and Palmitic acid	

Pharmacological Activities

Hemostatic Activity

The fresh cut leaves of *C. odorata* is widely used in many countries, such as in rural area of Thailand, Vietnam, Malaysia, Nigeria and Ivory Coast, as a hemostatic agent to stop bleeding *e.g.* nosebleed (Akah, 1990; Wongkrajang *et al.*, 1990; Bamba *et al.*, 1993; Phan *et al.*, 2001a; Phan *et al.*, 2001b; Phan *et al.*, 2001c; Umukoro and Ashorobi, 2006; Ling *et al.*, 2007a; Ling *et al.*, 2007b; Ling *et al.*, 2007c). Hemostasis has been described as "a process by which the body spontaneously stops bleeding and maintains blood in the fluid state within the vascular compartment". Immediately after a blood vessel injury takes place, a number of hemostatic processes are activated (Rosado *et al.*, 2009). There are at least four major systems that are involved in this complex process, *i.e.* the vasculature system, the platelets, the fibrin-forming system, and the fibrin-lysing system. Hemostasis has also been further segregated into stages or steps. These are primary hemostasis, which is the interaction of the injured blood vessel and platelets; secondary hemostasis, which is referred to historically as the blood coagulation process; and tertiary hemostasis, which is the process of fibrinolysis (clot destruction). The first two stages are assessed by blood coagulation tests that evaluate platelets, circulating coagulation factors, and blood vessels. The vascular response consists of vasoconstriction that minimizes vessel diameter and slows bleeding. The instantaneous vascular contraction reduces blood flow from the vessel wound. The vascular response is accompanied by primary hemostasis involving the formation of a platelet plug. This process requires the interaction of platelets with components of the vessel wall, such as collagen, plasma proteins including thrombin, and certain autocrine factors released by platelets, such as ADP. In addition, vascular injury induces secondary hemostasis as a result of blood coagulation, a process that involves a complex cascade of coagulation factors, resulting in the transformation of fibrinogen into polymerized fibrin, which supports the clot. These events are followed by the growth of fibrous tissue within the vessel wall that initiates the repair process and the clot in fibrinolysis (DeMoranville, 2006).

Many studies underwent to investigate the effect of *C. odorata* on these hemostatic mechanisms. The lyophilized water extract from dry leaves induced the vascular constriction in the concentration-dependent fashion. It could contract both rat and guinea pig vasa deferentia and rabbit arterial strips while having no effect on isolated guinea pig ileum and rat stomach strip preparations. The contractions were halted by low concentrations of prazosin (alpha adrenoceptor antagonist). This suggested that the hemostatic action of *C. odorata* might be partly due to α-receptor mediated vasoconstriction (Akah, 1990). On the contrary, the lyophilized water extract from fresh leaves had no effect on the rabbit aortic strip tone. But the extract increased the vascular tension induced by submaximal dose of phenylephrine in dose-dependent manner (Thongpraditchot *et al.*, 1994). The water extract could also stimulate the constriction of pig coronary artery but the other extracts from organic solvents such as hexane, chloroform and ethanol could not induce the contraction (Narongsanti *et al.*, 1991).

The aqueous extract of dry leaves of *C. odorata* could increase hemostasis. The extract was used to stimulate hemostasis of whole blood containing disodium citrate as an anticoagulant. Several concentration of this extract *i.e.* 2.5 per cent, 5.0 per cent and 6.5 per cent w/v significantly increased blood viscosity with increasing concentration (Wisetwongsa and Taksintam, 1997). The clot formation was examined *in vitro* with coagulogram study, prothrombin time (PT), *viz.*, activated partial thromboplastin (APTT) and thrombin time (TT) but the results varied from different sources. The aqueous and 95 per cent ethanolic extracts from fresh and dry leaves of *C. odorata* activated the clot formation both in extrinsic and intrinsic pathway when detected by PT and APTT after administration in human and cow plasma with/without exogenous calcium (Phithanchort *et al.*, 2006; Khengraeng, 2007). In contrary, the study showed that lyophilized water extract from fresh leaves could clot human plasma if detected by PT without exogenous calcium but was undetectable by APTT. Therefore it could be concluded that *C. odorata* extract enhanced the hemostatic activity only extrinsicly (Wongkrajang *et al.*, 1990). Absolute ethanol extract from dry leaves of *C. odorata* was able to enhance the activity of clotting factors in the intrinsic pathway when detected by APTT (Rawiwong *et al.*, 1988; Triratana *et al.*, 1991). Moreover, many studies illustrated that *C. odorata* extracts could not enhance blood coagulation when detected by both PT and APTT (Soogarun *et al.*, 2005). Some extracts of *C. odorata* could accelerate the hemostatic activity by shortening the clotting plasma time without the presence of exogenous calcium. These evidences inferred that *C. odorata* leaves provide calcium to promote hemostatis. To prove this hypothesis, calcium was precipitated from *C. odorata* water extract to further examine the bleeding time and clotting factors. The extract with no calcium neither shorten bleeding time nor clotting time in PT test until the exogenous calcium was added (Wongkrajang *et al.*, 1994). Furthermore, the calcium content in water and 95 per cent ethanolic extract were inquired. The calcium content in water extract from dry leaves was ten times higher than in 95 per cent ethanolic extract, agreeing with coagulating activity that showed that the water extract promoted coagulation better than the 95 per cent ethanolic extract (Khengraeng, 2007). This result proved that calcium in *C. odorata* extract was a significant contributor to enhance blood coagulation. Another reported flavonoids isolated from the leaf extract was an active component for hemostatic activity. The active compound for blood coagulation, 4′,5,6,7-tetramethoxy-flavone, was isolated and identified for the first time by preparative thin layer chromatography (Naengchomnong, 1989). Triratana and coworkers reported that this active compound isolated from the 95 per cent ethanolic dry leaf extract accelerated blood coagulation through an intrinsic pathway detected by APTT test (Triratana *et al.*, 1991).

The effect of this plant to platelet aggregation was also studied. The lyophilized water extract from fresh leaves stimulates platelet aggregation by adding ADP as an aggregating agent. The high calcium content in leaves could explain the enhancement on platelet aggregation (Wongkrajang *et al.*, 1994). So far, there were no reports of active compound that promoted this pathway.

The *C. odorata* antibacterial hydro-gel that stopped bleeding was developed. The hydrogel containing aqueous extract stimulated clotting when tested with whole blood clotting time (WBCT) and it also inhibited *S. aureus*. This product will be further investigated in clinical trials for commercialization (Khengraeng, 2007).

Wound Healing Activity

The exploitation of local and herbal remedies to treat wounds are common and of paramount since it is inexpensive and easily available. *C. odorata* extract has been used as a traditional medicine to heal wounds in many countries such as Thailand, Vietnam, Malaysia and Nigeria for several decades. New formulations of old remedies were made into antiseptic, topical cosmetic products to treat wounds and were improved by adding isolated and active flavonoids and essential oils. However, the mechanisms involved in the wound healing process is unknown, and therefore many researches are going on currently to investigate the effects of these promising products. One formulated ointment from this plant extract is Eupolin, prepared from the aqueous extract of *C. odorata*'s leaves. It has been licensed for clinical use to treat soft tissue wounds and burns in Vietnam (Phan *et al.*, 2001a). Eupolin extract enhanced the wound healing process of fibroblast cell, endothelial cell, and keratinocyte. Phan and coworkers reported that it could inhibit the contraction of collagen lattices in normal human fibroblast, an *in vitro* model of wound contraction. The visualization of cells in the lattices indicated that 50 to 150 μg/ml of the extract exhibited living cells after incubation in a tetrazolium salt for 2 h. In contrast, all cells were killed in 300 or 400 μg/ml of the extract. These results showed the inhibitory effect of eupolin extract on collagen contraction. Therefore, its effect on wound contraction and scar should be clinically evaluated (Phan *et al.*, 1996). By using indirected immunofluorescence, the up-regulation effects of the eupolin extract on human keratinocyte production of the basement membrane zone components and extracellular matrix protein were demonstrated (Phan *et al.*, 2000). Those of cells were seeded in 24 or 96-well plates at different denesties. Different doses (0.1-300 μg/ml) of eupolin solution were added and cell growth was monitored by a colorimetric assay. Only keratinocyte was monitored cell migration by the closure of a denuded area scratched in a confluent monolayer. This demonstrated that eupolin extract increased all cells growth. The concentrations at 10 μg/ml and 100 μg/ml of eupolin extract enhanced fibroblast and endothelial cell growth. Low concentrations (0.1-5 μg/ml) of extract stimulated keratinocyte proliferation, caused migration at higher concentrations (5-60 μg/ml) and cell differentiation at the highest concentrations (50-300 μg/ml). Toxicity of the extract to fibroblast was observed at 250 μg/ml but there was no significant damage at this dose to endothelial cells and keratinocytes (Phan *et al.*, 1998; Phan *et al.*, 2001a). Furthermore, the extract was tested for endotoxin level using a standard Limulus amebocyte lysate assay. The eupolin extract at 10 μg/ml contained < 0.1 EU/ml endotoxin. This established that at concentration of the extract up to 5 μg/ml, which had the maximal stimulatory activity, the effect was not likely to be due to the presence of endotoxin (Phan *et al.*, 2001a). This meant that *C. odorata* aqueous extract stimulated wound healing process by inhibiting the contraction of collagen lattices and accelerating the fibroblast cell, endothelial cell, and keratinocyte formation.

Additionally, the clinical studies in animal models were investigated. The aqueous fresh leaf extract in different stages of maturity and dry leaf powder were studied *in vivo*. The treatment of extracts from young leaves, matured leaves, and all leaves, and dry powdered leaves were studied with regards to their effects on the castration wound in the piglets compared with the treatment of tingeriodine as a control. The extracts and the powder were sprayed to the castration wounds along with tingeriodine. The wounds treated with *C. odoratum* extracts had the mean healing time, the mean wound connection and oedema time less than that treated with tingeriodine. The all leaves extract displayed higher efficacy than mature and young leaves, respectively. Only the wound treated with dry leaves powder had the same mean of healing time as tingeriodine. It could be concluded that *C. odorata* aqueous leaf extract and dry powdered leaves could be used to treat the castration wound in piglet instead of tingeriodine, which had to be imported to Thailand (Sintaweewarakul *et al.*, 2008a; Sintaweewarakul *et al.*, 2008b). In addition, the aqueous leaf extract was screened for wound healing effect in rabbits. Complete filling and healing of wounds was observed by 11th–13th day after daily application of the extract to laboratory created wounds (Biswal *et al.*, 1997).

Antioxidant Activity

Oxidants are involved in a number of aspects of burnt injury and tissue repair. Oxygen free radicals contribute to further tissue damage in the events following skin injury and are known to impair healing process. Antioxidants, on the other hand, significantly prevent tissue damage and stimulate wound healing. There are few reports of antioxidant activity of *C. odorata*. The absolute ethanolic extract was used to investigate antioxidant effect in terms of wound healing and anti-inflammatory activities. Phan and coworkers investigated the effect of absolute ethanolic extract on human fibroblast and keratinocytes against oxidative damage in order to assess the potential role of the extract as an antioxidant (Phan *et al.*, 2001b). The antioxidant effects of both total ethanol and polyphenolic extracts from plant leaves on hydrogen peroxide and hypoxanthine-xanthine oxidase to human fibroblast and keratinocytes were studied. Cell viability was monitored by 3-(4,5-Dimethylthiazol-2-yl)-2,5-diphenyltetrazolium bromide (MTT) colorimetric assay. It was found that within the range of 50-800 μg/ml of the extract exhibited significant protection on cultured fibroblast and keratinocytes against hydrogen peroxide and hypoxanthine-xanthine oxidase. In fibroblast, the toxicity of hydrogen peroxide and hypoxanthine-xanthine oxidase on cells was dose-dependent. Total ethanolic extract (TEE) at 400 and 800 μg/ml showed maximum and consistent protective cellular effect on oxidant toxicity at low and high doses of oxidants. Regarding keratinocytes, a dose-dependent relationship of oxidant toxicity was only present with hydrogen peroxide but the protective action of the extract correlated with oxidant dosage. TEE at 400 and 800 μg/ml displayed dose-dependent effects with both low and high concentrations of oxidants. Polyphenolic extract exhibited a slight antioxidant effect. Protection of cells against destruction by inflammatory mediators may be one of the ways in which the extracts of this plant contribute to wound healing. They also found the active compounds that were fractionated and identified from crude extract using liquid chromatography coupled with UV spectrometry and mass spectrometry. The phenolic acids and complex mixture of lipophilic flavonoid aglycones were powerful antioxidants that protected cultured skin cells against oxidative damage (Phan *et al.*, 2001c). Corresponding with the report of Ling and coworkers, eleven flavonoids were studied for the antioxidant activities by superoxide scavenging, DPPH radical scavenging and tyrosinase inhibition. The results established that the major flavonoid, kaempferol-4'-methylether showed a good antioxidant activity while naringenin-4'-methylether (isosakuranetin) showed antioxidant, anti-inflammatory, and antistaphylococcol activities and it could be used as a bioactive marker (Ling *et al.*, 2007c). The antioxidant of this plant extract was also studied in term of anti-inflammatory effect. The ethanolic extract was also investigated *in vitro* on inflammation caused by *Propionibacterium acnes*, pus forming bacteria triggering an inflammation acne in terms of free radical scavenging and cytokine reducing properties.

Free radical scavenging activity of the extract from *C. odorata* was determined by DPPH scavenging and Nitroblue tetrazolium dye (NBT) reduction assay. The extract displayed moderate antioxidant effect, IC_{50} was 67.55 μg/ml and percent superoxide radical inhibition ratio was 44.89±1.38 per cent (Chomnawang *et al.*, 2007). These evidences established that *C. odorata* extract, having flavonoids as active components, was an competant antioxidant and anti-inflammatory agent and could promote the wound healing process. Furthermore, the antioxidant enzymes, peroxidase isoenzymes of *C. odorata* water extract from fresh leaves, were studied for their kinetics and thermal stability. These enzymes could be used as a low-cost time-temperature indicator strip, electrochemical biosensors for bioanalytical application, medical diagnostic and biotransformations (Rani and Abraham, 2006a; Rani and Abraham, 2006b). The thermal stresses on antioxidant enzymes of *C. odorata* were studied. Plants were grown in growth chamber with cold and heat treatments. *C. odorata* had less degree of membrane damage, as indicated by low monodehydroascorbate content, and the coordinated increase of the oxygen. Detoxifying enzymes were observed in heat-treated plants, but the antioxidant enzymes were unable to operate in cold stress. This indicated that *C. odorata* had a better oxygen radical scavenging activity in heat stress than in cold stress (Lu *et al.*, 2008).

Anti-inflammatory Activity

The anti-inflammatory effect of *C. odorata* extracts were already studied both *in vitro* and *in vivo* investigation. The 95 per cent ethanolic extract and all partitioned fractions (partitioned with water, dichloromethane, methanol, hexane and butanol) of the aerial and subterranean parts were investigated *in vitro*. Immortalized mouse PGH-1 and PGH-2 null cells were analyzed for prostaglandin E2 (PGE_2) concentrations by radioimmunoassay (RIA). The resulting radioactive counts were used to calculate the percent binding of ^3H-PGE_2 (percent inhibition against COX-1 and COX-2), which were then used for the estimation of PGE_2 concentrations from standard curves. Only butanol fraction from the aerial part showed anti-inflammatory that exhibited 48 per cent inhibition against COX-1 and 51 per cent inhibition against COX-2 at the concentration 10 μg/ml (Pisutthanan *et al.*, 2005). The ethanolic extract was also investigated *in vitro* in an inflammatory study caused by *Propionibacterium acnes*, pus forming bacteria triggering an inflammation acne, in terms of free radical scavenging as described before, and cytokine reducing properties. The cytokine reducing properties was determined by inhibiting pro-inflammatory cytokines (TNF-α) using ELISA when 5 and 50 μg/ml of the extract were added. Both concentrations of alcoholic extract also displayed high inhibitory activity against TNF-α production (>70 per cent inhibition) but the higher concentration could inhibit better (Chomnawang *et al.*, 2007).

The aqueous leaf extract of *C. odorata* was investigated *in vivo* on acute and sub-chronic inflammatory and membrane-stabilizing effects. The acute inflammation was studied in carrageenin-treated rats and the sub-chronic inflammation was induced by formaldehyde. The extract concentrations 100-400 mg/kg were given orally to rats. The circumference and percent inhibition oedema were calculated. The membrane stabilizing effect of 0.5 - 2.0 mg/ml extract was studied using its ability to reduce the levels of hemolysis of rat Red Blood Cells (RBC) exposed to hypotonic solution. The results showed that the extract possessed anti-inflammatory property, as it significantly reduced oedema formation induced by phlogistic agents in rats. At concentration ranged from 1.0-2.0 mg/ml, the extract offered significant protection of RBC against the hemolytic effect of hypotonic solution, and indicated the membrane-stabilizing activity, played a significance role in its inflammatory activity. Moreover, the extract was studied for acute toxicity by giving 0.10-2.0 g/kg extract orally to rats and then observing mortality and toxic symptoms within 24 h after administration. This result demonstrated that the extract has high safety profile as no death was observed. The behavioral changes that could be observed at toxic doses were sedation, ataxia and hyperventilation (Umukoro and Ashorobi, 2006).

Other Biological Activities

Many extracts of *C. odorata* were determined for biological activities such as antibacterial, antimalarial and antituberculosis activities.

Antibacterial Activity

The oil from both fresh and dry leaves collected from Nigeria and from both aerial and subterranean parts (including leaves, twigs, branches and stems) collected from Thailand were extracted by hydrodistillation. These extracts demonstrated the highest antibacterial activity against *Staphylocoocus aureus* and moderate activity against *Escherichia coli* (Inya-Agha *et al.*, 1987; Pisutthanan *et al.*, 2005). But the essential oil of this plant collected from Ivory Coast provided the different results. The Ivorian plant oil had no effect on *S. aureus* but it exhibited notable activity towards the gram-negative species such as *E. coli, K. pneumonia*, and specifically towards *P. aeruginosa* as described before (Bose *et al.*, 1973).

The aerial and subterranean parts of *C. odorata* from Thailand were separately extracted with 95 per cent ethanol and further partitioned with water, dichloromethane, methanol, hexane and butanol. These fractions were tested for bioactivity screening including antibacterial assays, brine shrimp lethality test, cytotoxicity assay, anti-inflammatory assay (as described before), antimalarial assay, antituberculosis assay and anti-herpes simplex virus assay. The determination of antibacterial activity was performed by disc diffusion method. Most extracts showed antibacterial activity against these two types of bacteria. The chloroform fraction from subterranean part showed the highest activity against both types of bacteria (diameter of inhibition zone is 10.5 and 7.5 mm, respectively). Only water extract did not show antibacterial activity against both bacteria (Pisutthanan *et al.*, 2005).

The methanol leaf extract of *C. odorata* possessed antistaphylococcol activity and the flavonoids were identified as main constituents (Ling *et al.*, 2006; Ling *et al.*, 2007c). The investigation to find out extraction parameters that gave the highest flavonoids content was undertaken. The 50 per cent ethanolic extract from extraction at 85°C for 30 min with sample to solvent ratio of 1:50 gave the highest flavonoid content. This extract and flavonoids were tested for antistaphylococcol activity against 2 strains of multi-drug susceptible and 3 strains of multi-drug resistant *S. aureus*. These extracts enhanced antistaphylococcol activity with isosakuranetin as active compound (Ling *et al.*, 2007c). *C. odorata* extract also demonstrated potent antibacterial activity against *P. acnes* and *S. epidermidis* with equal MIC of 0.625 mg/ml and MBC of 1.25 and > 5 mg/ml, respectively (Chomnawang *et al.*, 2005). Molina-Salinas and coworkers reported that methanol extract from the leaves of *C. odorata* had higher activity against *S. pneumonia* than the extracts from flowers and roots. The R1Sp strain showed MIC of these three extracts to be 250, 500 and inactive, respectively. While the R2Sp strain showed MICs at 250, 500 and 500, respectively (Molina-Salinas *et al.*, 2007).

Antimalarial Activity

The dry leaf powder of this plant was extracted with hexane, ether, chloroform, ethyl acetate, and ethanol. The results indicated that the ether and chloroform extracts were the most effective against *Plasmodium falciparum* with MICs 12.5 and 6.25 μg/ml, respectively. The fraction from ether extract promoted the highest antimalarial activity with MIC 1.56 μg/ml. Isosakuranetin was isolated from this fractionate but did not present the antimalarial activity against *Plasmodium falciparum* at the concentration of 5 μg/ml (Thakong, 1999). The subfractions from 95 per cent ethanolic extract from aerial and subterranean parts of *C. odorata* with many organic solvents were tested against *Plasmodium falciparum* (K1, multi-drug resistant strain)

(Pisutthanan *et al.*, 2005). Quantitative assessment of *in vitro* antimalarial activity was determined by microculture radioisotope techniques. Only dichloromethane fraction from subterranean part showed antimalarial activity with EC_{50} of 9.39 µg/ml (Pisutthanan *et al.*, 2005).

Antituberculosis Activity

The subfractions from of 95 per cent ethanolic extract from aerial and subterranean parts of *C. odorata* with many organic solvents were tested (Pisutthanan *et al.*, 2005). Anti-*Mycobacterium tuberculosis* was performed in 96-well microplates by adding Alamar Blue solution. Wells were observed for a color change from blue to pink. Visual MICs is defined as the lowest concentration of sample that prevents a color change. Only dichloromethane fraction from aerial and subterranean parts displayed moderate activity against *Mycobacterium tuberculosis* (MIC 100 and 200 µg/m, respectively) (Pisutthanan *et al.*, 2005). The persicogenin (5,3'-dihydroxy-7,4'-dimethoxyflavanone) isolated from the flowers of *C. odorata* exhibits moderate activity against *M. tuberculosis* with MIC value of 174.8 µM while 4'-hydroxy-5,6,7-trimethoxyflavanone, acacetin and luteolin exhibit weak activity with MICs 606.0, 704.2 and 699.3 µM, respectively (Suksamrarn *et al.*, 2004).

Anti-herpes Simplex Virus Activity

The subfractions from of 95 per cent ethanolic extract from aerial and subterranean parts of *C. odorata* with many organic solvents were tested. Anti-herpes simplex virus type 1 activity was tested against HSV-1 strain ATCC VR 260 by using colorimetric microtiter plate assay. The extracts were tested at non-cytotoxic concentrations that inhibited cell growth < 50 per cent. The extracts of which inhibited cell growth > 50 per cent were tested to determine the concentrations that inhibit viral activity by 50 per cent (IC_{50}). Dichloromethane fraction from the aerial part of *C. odorata* showed potent anti-HSV-1 activity with IC_{50} 1.74 µg/ml (Pisutthanan *et al.*, 2005).

Antigonorrhoeal Activity

The 50 per cent alcoholic extract from the leaves of *C. odorata* exhibited high activity against *Neisseria gonorrhoeae* with diameter of inhibition zone at approximately 11 mm (Caceres *et al.*, 1995).

Toxicity

C. odorata is used as a fish poison in the Himalayas, India. It is known as a fever plant in Puerto Rico and is toxic to arthropods (Arene *et al.*, 1987). *C. odorata* extracts were also investigated for their toxicities to brine shrimps, cell lines and animal models.

Brine Shrimp Lethality Test

The fractions from partition of 95 per cent ethanolic extract of aerial and subterranean parts of *C. odorata* with many organic solvents were tested for brine shrimp lethality (Pisutthanan *et al.*, 2005). The microwell cytotoxicity assay using *Artemia salina* (brine shimp) was employed. The 2 mg/ml of test samples were prepared and made serial dilutions in 96-well microplates before adding the nauplii (baby brine shrimps). The results revealed that all extracts did not show any toxicity to brine shrimps (LD_{50}>850 µg/ml) (Pisutthanan *et al.*, 2005). The root extract fractions were further tested for their cytotoxicity in the brine shrimp lethality test. Depending on the extraction methods and associating solvents (petroleum ether extract, hexane soluble and dichloromethane-soluble sub-fraction of solid residue left after ethanol extraction), the LC_{50} values showed significant or mild cytotoxicity. (Amatya and Tuladhar, 2005).

Cytotoxicity Assay

Toxicity of the aqueous extract from *C. odorata* to fibroblast cell was observed at concentration of 250 µg/ml but there was no significant damage at this dose to endothelial cells and keratinocytes (Phan *et al.*, 1998; Phan *et al.*, 2001a). The fractions from partition of 95 per cent ethanolic extract from aerial and subterranean parts of *C. odorata* with many organic solvents were also tested by cytotoxicity assay (Pisutthanan *et al.*, 2005). The cytotoxicity tests were performed on human epidermoid carcinoma of cavity, ATCC CCL-17 (KB), breast cancer cell (BC) and human small cell lung cancer cell line (NCI-H187) and were determined by colorimetric cytotoxicity assay to measure cell growth from cellular protein content. All extracts showed no cytotoxicity against KB, BC and NCI-H187 (IC_{50}>20 mg/ml) (Pisutthanan *et al.*, 2005). The 50 per cent ethanolic extract of *C. odorata* and its flavonoids extract did not have any cytotoxicity against Vero (normal kidney) and WRL-68 cell lines (Ling *et al.*, 2007c). Some flavonoids such as odoratin, isosakuranetin monomethyl ether and 4′,5-dihydroxy-3′,7-dimethoxyflavone, which could be isolated from fresh aerial part including flowers of *C. odorata* extracted with chloroform by Soxhlet. They were not cytotoxic to KB and P33 cell lines (Arene *et al.*, 1987). In contrast, some pure flavonoids that was isolated from the flowers of this plant showed cytotoxicity. The acacetin (5,7-dihydroxy-4'-methoxyflavone) exhibited moderate cytotoxicity against NCI-H187 cells with MIC 24.6 µM while luteolin (5,7,3',4'-tetrahydroxyflavone) exhibited moderate toxicity against NCI-H187 and weak toxicity against BC cells with MIC 19.2 and 38.4 µM, respectively (Suksamrarn *et al.*, 2004).

Animal Models

The ethanolic leaf extract showed acute toxicity of 96 hour on the gills of tilapia (*Oreochromis niloticus* Linn.) fish (LD_{50} 30.94 mg/l). The histological changes were found in dose dependent effect (Srisuk, 2000). The lyophilized extract from fresh leaves of *C. odorata* did not irritate the rabbit skin even at high dose (500 mg/ml) and did not show acute toxicity after a week of oral administration (LD_{50}>5 g/kg) (Thongpraditchot *et al.*, 1994).

Discussion

C. odorata is native to Central and South America, now its distribution is mainly in the tropical zone such as in Nigeria and Ivory Coast in Africa; India, Thailand, Vietnam and Malaysia in Asia; and also in Micronesia and Australia. It can tolerate high temperature, grows very quickly, and is easily distributed by wind. Therefore, it becomes a conservational and agricultural problem as an invasive plant in the forests, vacant lands, wildlife reserves, riverbanks, pastures and plantation crops of many countries. The mechanical and chemical means are employed to destroy this plant but it is costly. Therefore, taking advantage of its usefulness would be a better solution than finding the way to destroy it. There are many reports of ethnopharmacology and others usages of this plant, mainly as a hemostatic agent to stop bleeding and heal wound. Many of these ethnopharmacological uses have already been tested in both *in vitro* and *in vivo* biological studies. The active components were also reported. Essential oil shows antibacterial activity while flavonoids, which are mainly components, exhibit hemostatic activity, anti-inflammatory, antioxidant, and many other biological activities. Flavonoids obtained from flowers/aerial parts including flowers present the cytotoxicity. The flavonoids obtained from other parts except flowers do not show any cytotoxicity. These results indicate that the leaves and aerial parts of this plant that contain high amount of flavonoids are more preferable to be developed than flowers. Although there were some reports on the toxicity of *C. odorata* extracts in insects and fish, it is not toxic to higher animals even at high doses, confirming that *C. odorata* can be used as skin care products. Despite its many medical and pharmaceutical applications, only a few products are launched into the market. Not only would it be beneficial to farmers by helping them rid of weeds, the development from the product would be a more commercially available and safer alternative.

Acknowledgements

This review is a part of the project on Ph.D. **Sandwich Program** in The Strategic Scholarships Fellowships Frontier (CHE-PhD-SW) No. CHE510780, which is financial supported by Office of The Higher Education Commission. The authors would like to thank Mr. Panupon Khumsupan for helping in proofing the manuscript.

References

Akah, PA. (1990). Mechanism of hemostatic activity of *Eupatorium odoratum. International Journal of Crude Drug Research*, **28(4):** 253-256.

Amatya, S., and Tuladhar, S.M. (2005). Eupatoric acid: A novel triterpene from *Eupatorium odoratum* L. (Asteraceae). *Journal of Chemical Sciences*, **60:** 1006-1011.

Arene, E.O., Pettit, G.R., and Ode, R.H. (1987). The Isolation of Isosakuranetin methyl ether from *Eupatorium odoratum. Lloydia*, **41(1):** 68-70.

Bamba, D., Bessiere, J.M., Marion, C., Pelissier, Y., and Fouraste I. (1993). Essential oil of *Eupatorium odoratum. Planta Medica*, **59:** 184-185.

Biswal, P.R., Sardar, K.K., Parija, S.C., Mishra, P.R., and Misra, S.N. (1997). Wound healing effect of *Eupatorium odoratum* Linn. and Himax in rabbits. *Indian Journal of Indigenous Medicines*, **19:** 71-74.

Bose, B.K., Chakrabarti, P., Chakravarti, S., Dutta, S.P., and Barua, A.K. (1973). Flavonoid constituents of *Eupatorium odoratum. Phytochemistry*, **12:** 667-668.

Caceres, A., Menendez, H., Mendez, E., Cohobon, E., Samayoa, B.E., Jauregui, E., Peralta, E., and Carrillo, G. (1995). Antigonorrheal activity of plants used in Guatemala for the treatment of sexually transmitted diseases. *Journal of Ethnopharmacology*, **48:** 85-88.

Chomnawang, M.T., Surassmo, S., Nukoolkarn, V.S., and Gritsanaparn, W. (2005). Antimicrobial effects of Thai medicinal plants against acne-inducing bacteria. *Journal of Ethnopharmacology*, **101:** 330-333.

Chomnawang, M.T., Surassmo, S., Nukoolkarn, V.S., and Gritsanapan, W. (2007). Effect of *Garcinia mangostana* on inflammation caused by *Propionibacterium acnes*. *Fitoterapia*, **8(6):** 401-408.

DeMoranville, V.E. (2006). Coagulation tests. *The Gale Encyclopedia of Nursing and Allied Health. In:* Detroit Gale, Vol. 1, Ed. By Longe, J.L. pp.578-583.

Fifield, F.W., and Kealey, D. (1995). Principles and Practice of Analytical Chemistry. Blackie Hall, London.

Hulugalle, N.R., Lal, R., and Opara-Nadi, O.A. (1987). Management of plant residue for cassava (*Manihot esculenta*) production on an acid Ultisol in Southeastern Nigeria. *Field Crops Research*, **16:** 1-18.

Inya-Agha, S.I., Oguntimein, B.O., Sofowora, A., and Benjamin, T.V. (1987). Phytochemical and antibacterial studies on the essential oil of *Eupatorium odoratum*. *International Journal of Crude Drug Research*, **25:** 49-52.

Islam, K.R., Ahmed, M.R., Bhuiyan, M.K., and Badruddin, A. (2001). Deforestation effects on vegetative regeneration and soil quality in tropical semi-evergreen degraded and protected forests of Bangladesh. *Land degradation and development*, **12(1):** 45-56.

Jagadeesh, K.S., Geeta, G.S., and Reddy, T.K.R. (1990). Biogas production by anaerobic digestion of *Eupatorium odoratum* L. *Biological Wastes*, **33:** 67-70.

Kalita, D., Ghosh, S.R. and Saikia, C.N. (1999). Medium density particle board from weeds. *Journal of Scientific and Industrial Research*, **58**: 705-710.

Khengraeng, U. (2007). Development of *Eupatorium odoratum* leaf extract hydro-gel for prevention of bleeding and anti-microbacteria : commercialized purpose. A special project of bachelor degree (Medical Technology), Faculty of Associated Medical Sciences, Chiangmai University, Chiangmai, Thailand.

Ling, S.K., Mazura, M.P., Khoo, M.G.H., Vimala, S., Ong, B.K., Mastura, M., Nor Azah M.A., Salbiah, M., Siti Asha, A.B., and Rasadah, M.A. (2006). Chemical constituents and therapeutic potential of the leaf extract from *Chromolaena odorata* (L.) King and Robinson. *In:* Highlights of FRIM's IRPA Projects 2005, Ed. By Nik Zanariah, N.M., Sarifah, K.A., and Nor Azman, H., Kuala Lumpur, Malaysia, pp. 108–114

Ling, S.K., Mazura, M.P., and Salbiah, M. (2007a). Platelet-activating factor (PAF) receptor-binding antagonist activity of the methanol extracts and isolated flavonoids from *Chromolaena odorata* (L.) King and Robinson. *Biological and Pharmaceutical Bulletin*, **30:** 1150–1152.

Ling, S.K., Abdull Rashih, A., Salbiah, M., Siti Asha, A.B., Mazura, M.P., Khoo M.G.H., Vimala, S., Ong, B.K., Mastura, M., and Nor Azah, M.A. (2007b). Extraction and simultaneous detection of flavonoids in the leaves of *Chromolaena odorata* by

RP-HPLC with DAD. *In*: Highlights of FRIM's IRPA Projects 2006, Ed. By Nik Zanariah, N.M., Sarifah, K.A. and Nor Azman, H. Kuala Lumpur, Malaysia, pp.32–37.

Ling, S.K., Nor Azah, M.A., Mastura, M., Khoo, M.G.H., Saidatul Husni, S., Salbiah, M., Abdull Rashih, A., Vimala, S., Ong, B.K., Siti Asha, A.B., and Nuraini, A.M. (2007c). Standardisation and formulation of *Chromolaena odorata* for herbal preparation. *In*: Highlights of FRIM's IRPA Projects 2007, Ed. By Nik Zanariah NM, Sarifah KA and Nor Azman H. Kuala Lumpur, Malaysia, pp.176–182.

Lu P., Sang, W.G., and Ma, K.P. (2008). Differential responses of the activities of antioxidant enzymes to thermal stresses between two invasive *Eupatorium* species in China. *Journal of Integrative Plant Biology*, **50(4):** 393-401.

Molina-Salinas, G.M., Prez-Lpez, A., Becerril-Montes, P., Salazar-Aranda, R., Said-Fernndez, S., and Torres, N.W. (2007). Evaluation of the flora of Northern Mexico for *in vitro* antimicrobial and antituberculosis activity. *Journal of Ethnopharmacology*, **109:** 435–441.

Muniappan, R., and Marutani, M. (1991). Distribution and control of *Chromolaena odorata* (Asteraceae). *Micronesica*, **3:** 103-107.

Muniappan, R., Reddy, G.V.P., and Lai, P.Y. (2005). Distribution and biological control of *Chromolaena odorata*. *Invasive Plants: Ecological and Agricultural Aspects*, 223-233.

Naengchomnong, W. (1989). Chemical constituents of *Jatropha curcas* (Euphorbiaceae) and *Eupatorium odoratum* (Compositae). Ph.D. thesis (Organic Chemistry), Faculty of Graduate Studies, Mahidol University, Bangkok, Thailand.

Narongsanti, A., Reangdenchai, R., and Unjitti, S. (1991). The usage of Eupatorium odoratum L. in pharmaceutical practice. A special project of bachelor degree (Pharmacy), Faculty of Pharmacy, Chulalongkorn University, Bangkok, Thailand.

Nesom, L.G. (2006). *Chromolaena odorata*. *In:* Flora of North America, Vol. 21, Ed. By Boufford D.E., Freeman C.C., Brouillet L., Gandhi K.N., Elisens W.J., Gillespie L.J., Hartman R.L. *et al.*, Oxford University press, USA, pp.544-545.

Phan, T.T., Hughes, M.A, Cherry, G.W., Le, T.T., and Pham, H.M. (1996). An aqueous extract of leaves of *Chromolaena odorata* (formerly *Eupatorium odoratum*) (Eupolin) inhibits hydrated collagen lattice contraction by normal human dermal fibroblasts. *Journal of Alternative and Complementary Medicine*, **2(3):** 335-343.

Phan, T.T., Hughes, M.A., and Cherry, G.W. (1998). Enhanced proliferation of fibroblasts and endothelial cells treated with an extract of the leaves of *Chromolaena odorata* (Eupolin), an herbal remedy for treating wounds. *Plastic and Reconstructive Surgery*, **101(3):** 756-765.

Phan. T.T., Allen, J., Hughes, M.A., Cherry, G.W., and Wojnarowska, F. (2000). Upregulation of adhesion complex proteins and fibronectin by human keratinocytes treated with an aqueous extract from the leaves of *Chromolaena odorata* (Eupolin). *European Journal of Dermatology*, **10(7):** 522-527.

Phan, T.T., Hughes, M.A., and Cherry, G.W. (2001a). Effect of an aqueous extract from the leaves of *Chromolaena odorata* (Eupolin) on the proliferation of human keratinocytes and on their migration in an *in vitro* model of reepithelialization. *Wound Repair and Regeneration*, **9:** 305-313.

Phan, T.T., Patrick, S., Teik, L.S., and Yung, C.S. (2001b). Antioxidant effects of the extracts from the leaves of *Chromolaena odorata* on human dermal fibroblast and epidermal keratinocytes against hydrogen peroxide and hypoxanthine-xanthine oxidase induced damage. *Burns*, **27:** 305-313.

Phan. T.T., Wang, L., See, P., Grayer, R.J., Chan, S.Y., and Lee, S.T. (2001c). Phenolic compounds of *Chromolaena odorata* protect cultured skin cells from oxidative damage: Implication for on cutaneous wound healing. *Biological and Pharmaceutical Bulletin*, **24(12):** 1373-1379.

Phithanchort, C., Jannoi, S., Tositarat, T., Wongpayapkul, L., and Pornprasert, S. (2006). Effect of *Eupatorium odoratum* leaf extract on hemostasis. *Journal of the Medical Association of Thailand*, **34:** 1711-1720.

Pithayanukul, P., Nithitanakool, S., and Bavovada, R. (2009). Hepatoprotective Potential of Extracts from Seeds of *Areca catechu* and Nutgalls of *Quercus infectoria*. *Molecules*, **14:** 4987-5000.

Pisutthanan, N., Liawruangrath, S., Bremner, J.B., and Liawruangrath, B. (2005). Chemical Constituents and Biological Activities of *Chromolaena odorata*. *Chiang Mai Journal of Science*, **32(2):** 139-148

Qin, W.H., Wang, Z., Xu, W.G., and Jiang, M.K. (2008). Investigation and analysis on alien invasive plants in three national nature reserves in Hainan Province. *Journal of Plant Resources and Environment*, **17:** 44-49.

Rani, D.N., and Abraham, T.E. (2006a). Kinetics study of a purified anionic peroxidase isolated from *Eupatorium odoratum* and its novel application as time temperature indicator for food materials. *Journal of Food Engineering*, **128:** 594-600.

Rani, D.N., and Abraham T.E. (2006b). Kinetics and Thermal stability of two peroxidase isozymes from *Eupatorium odoratum*. *Applied Biochemistry and Biotechnology*, **128:** 215-226.

Rawiwong, R., Triratana, T., Thebtaranonth, Y., and Cheawsin P. (1988). The testing of herb's extraction in intrinsic pathway of hemostatic mechanism. Research support from Mahidol University, Bangkok.

Recio, M.C., Ginet, R.M., Uribun, L., Manez, S., Cerda, M., De La Fuente, J.R., and Rios, J.L. (2000). *In vivo* activity of pseudoguaianolide sesquiterpene lactones in acute and chronic inflammation. *Life Sciences*, **16:** 2509-2518.

Rosado, J.A., Redondo, P.C., and Granados, M.P. (2009). Textbook of Hemostasis and Blood Coagulation. Research Signpost, India.

Simmonds, F.J. (1976). Some recent puzzles in biological control. *Entomophaga*, **21:** 327-332.

Sintaweewarakul, C., Sintaweewarakul, V., and Harnthavee, M. (2008a). The use of the *Eupatorium odoratum* Linn. leaves crude extract to treat the castration wound in piglet. The proceeding of 46th Kasetsart University Annual Conference, January 29 - February 1, Bangkok, Thailand, 115-121.

Sintaweewarakul, C., Sintaweewarakul, V., and Taai K. (2008b). Using *Eupatorium odoratum* Linn. leaves powder to treat the castration wound in piglet. Rajamangala University of Technology Conference, August 27-29, Trang, Thailand, 131-138.

Soogarun, S., Wiwanitkit, V., and Suwansaksri, J. (2005). Effect of *Chromolaena odorata* on prothrombin time and activated partial thromboplastin time. *Hema,* **8(1):** 149.

Srisuk, P. (2000). Effect of *Chromolaena odorata* Linn., Leaves extract on Gills of Tilapia *Oreochromis niloticu* Linn. A special project of bachelor degree (Biology), Faculty of Sciences, Chulalongkorn University, Bangkok, Thailand.

Suksamrarn, A., Chotipong, A., Suavansri, T., Boongird, S., Timsuksai, P., Vimuttipong, S., and Chuaynugul, A. (2004). Antimycobacterial Activity and Cytotoxicity of Flavonoids from the Flowers of *Chromolaena odorata. Archives of Pharmacal Research,* **27(5):** 507-511.

Talapatra, S.K., Bhar, D.S., and Talapatra, B. (1974). Flavonoid and terpenoid constituents of *Eupatorium odoratum. Phytochemistry,* **13:** 284-285.

Thakong, K. (1999). A study on the antimalarial constituents and chemical composition of *Eupatorium odoratum* (L.). A master degree thesis (Pharmaceutical chemistry and phytochemistry), Faculty of Graduate Studies, Mahidol University, Bangkok, Thailand.

Thongpraditchot, S., Suvitayavat, W., and Temsiririrkkul, R. (1994). Effect of *Eupatprium odoratum* Linn. on vascular tone and primary irritation test. *The Mahidol University Journal of Pharmaceutical Sciences,* **21(2):** 44-49.

Tripathi, R.S., Kushwaha, S.P.S., and Yadav, A.S. (2006). Ecology of three invasive species of *Eupatorium*: A review. *International Journal of Ecology and Environmental Sciences,* **32:** 301-326.

Triratana, T., Suwannuraks, R., and Naengchomnong, W. (1991). Effect of *Eupatorium odoratum* on blood coagulation. *Journal of the Medical Association of Thailand,* **74(5):** 283-287.

Umukoro, S., and Ashorobi, R.B. (2006). Evaluation of the anti-inflammatory and membrane-stabilizing effects of *Eupatorium odoratum. The International Journal of Pharmaceutics,* **2(5):** 509-512.

Wisetwongsa, P., and Taksintam, R. (1997). A study of the effect of *Eupatorium odoratum* Linn. on blood viscosity during coagulation process. A special project (Pharmacy), Faculty of Pharmacy, Mahidol University, Bangkok, Thailand.

Wongkrajang, Y., Muagklum, S., Peungvicha, P., Jaiarj, P., and Opartkiattikul, N. (1990). *Eupatorium odoratum* linn : An enhancer of hemostasis. *Mahidol University Journal of Pharmaceutical Sciences,* **17:** 9-13.

Wongkrajang, Y., Thongpraditchote, S., Nakornchai, S., Chuakul, W., Muangklum, K., and Jaiarj, P. (1994). Hemostatic activities of *Eupatoriujm odoratum* Linn.: Calcium removal extract. *Mahidol University Journal of Pharmaceutical Sciences*, **21**: 143-148.

Xu, C., Dong, X., and Lu, S. (2006). Invasive plants in Honghe River Basin of Yunnan Province, China. *Chinese Journal of Ecology*, **25:** 194-200.

Yuan, J.Q., Yang, J.S., and Miao, J.H. (2007). Studies on flavonoids of *Eupatorium odoratum* L. *Journal of Chinese medicinal materials*, **30:** 657-660.

Traditional and Folk Herbal Medicine: *Pages* **93–132**
Recent Researches Vol. 1 (2012)
Editor: **V.K. Gupta**
Published by: **DAYA PUBLISHING HOUSE, NEW DELHI**

5

Chemistry, Pharmacology and HPTLC Standardisation of *Asparagus racemosus* (Shatavari): An Important Ayurvevic Herb

Musarat Amina[1], V.K. Sharma[1], N.K. Satti[1], Prabhu Dutt[1], V.K. Gupta[2], B.D. Gupta[1] and K.A. Suri[1]*

ABSTRACT

Man's existence on this earth has been made possible only because of the vital role by the plant kingdom in sustaining his life. The three important necessities of life, food, clothing and shelter as well as most of the other useful products are supplied to a greater extent by the plant kingdom. The nature has provided a complete storehouse of remedies to cure all ailments of mankind. Since the dawn of civilization, in addition to food crops, man cultivated herbs for his medicinal needs. The knowledge of drugs has accumulated over thousands of years as a result of man's inquisitive nature, so that today we possess many effective means of ensuring health care (Newman et al., 2000). Asparagus racemosus Willd. (Asparagaceae) is an important medicinal plant of tropical and subtropical India. Its medicinal usage has been reported in the Indian and British Pharmacopoeias and in traditional system of medicine such as Ayurveda, Unani and Siddha. Asparagus racemosus is mainly known for its phytoesterogenic properties. The aim of this article is to review the literature on Asparagus racemosus, up to date. This

1 Natural Product Chemistry Division,
2 Pharmacology Division,
Indian Institute of Integrative Medicine, Council of Scientific and Industrial Research, Canal Road, Jammu-Tawi – 180 001, India.

* *Corresponding author*: E-mail: kasuri_rrl@yahoo.com; Phone: +91 0191 2569001; Fax: +91 0191 2569019.

review summarizes the chemistry, proton and carbon spectral data, biological activities, pharmacological applications and clinical studies of Asparagus racemosus in an attempt to providing a direction for further research.

Keywords: *Asparagus racemosus*, Asparagaceae, Chemical constituents, Sapogenins, Steroidal saponins, Spectral data, Pharmacology, HPTLC standardization.

Introduction

Asparagus racemosus Willd. is commonly mentioned as a Rasayana in the Ayurveda and is mentioned as Shatavari which means "she who has hundred husbands" referring to the rejuvenative effect of the herb on female reproductive organs. A study of ancient classical Ayurvedic literature claimed several therapeutic attributes for the roots of *Asparagus racemosus* and has been specially recommended in case of threatened abortions and as a galactogogue (Dev, 1999). The plant belongs to the genus *Asparagus* of family Asparagace. Genus *Asparagus* has been recently moved from the subfamily Asparagae in the family Liliaceae to a newly created family Asparagaceae (Sharma and Lal, 2003). The name is taken from the Greek/ Persian "Asparag" referring to tender 'short' or stalk that can be consumed. It is known to have been in cultivation since 200 BC. Because of the prickly nature of the foliage, the fronds have been used in Greece to stuff up the holes of mice and rats. An old Chinese custom was to save the selected roots for family and friends in the belief that it would increase feelings of compassion and love. *Asparagus* is highly regarded herb worldwide and is considered to be of medicinal importance because of the presence of steroidal saponins and sapogenins in various parts of plant (Oketah-Rabah, 1998). The name "Asparagus" was given by Theophrastus. *Asparagus racemosus* has been used in Ayurveda for various conditions (Aswal *et al.*, 1996). Though its main use has been as a glactogogue to increase the milk secretion (Joglekar *et al.*, 1967), it is also used as general tonic (Tewari and Pandey, 1993) and as an aphrodisiac (Singh, 1998). *Asparagus racemosus* is recommended in Ayurvedic texts for prevention and treatment of gastric ulcers, dyspepsia and as a galactogogue. It has been used successfully by some Ayurvedic practitioners for nervous disorder, inflammation, liver diseases and certain infectious diseases (Sharma and Lal, 2003). Recent chemical analyses have now thrown some light on its efficacy. We give here references to a number of recent analyses of its multiple medicinal properties.

Plant Profile

Taxonomical Classification

Kingdom: Plantae (plants); Phylum: Anthophyta; Class: Monocotyledons; Order: Asparagales; Family: Asparagaceae; Genus: *Asparagus*; Species: *racemosus.*

Common Names

Satawar or satmuli (in Hindi); Satavari (Sanskrit); Shatamuli (in Bengali); Shatavari or Shatmuli (in Marathi); Satawari (in Gujarati); Toala-gaddalu or Pilligaddalu (in Telegu); Shimaishadavari or Inli-chedi (in Tamil); Chatavali (in Malayalam); Majjigegadde or Aheruballi (in Kanada); Kairuwa (in Kumaon); Narbodh or Satmooli (in Madhya Pardesh); Norkanto or satawar (in Rajasthan) catavari, Shaqaqule Hindi, Songga Langit.

Possible Synonyms

Asparagus rigidulus, Asparagus schoberioides, Protasparagus racemosus.

Distribution

The plant species, *Asparagus racemosus,* is chiefly distributed at low altitudes in shade and in tropical climates throughout India, Asia, Australia and Africa. It is native to Australia through western Africa to southern Asia (Anonymous, Wealth of India, 1985). It grows wild and is also cultivated throughout the tropical and subtropical parts of India.

Botanical Description

Asparagus racemosus is an ascending, spinous, much branched, perennial woody climber growing to 1-2 m in height. It contains adventitious root system with rootstock bearing numerous fusiform, succulent tuberous roots 30-100 cm long and 1-2 cm thick. Stem woody, whitish grey or brown armed with strong, straight or recurved spines 5-13 mm long cladodes more or less acicular, 2-6 nate, falcate finely acuminate. Leaves are uniform, small and reduced to sub-erect or sub-spines like pine needles. Flowers are white, fragrant, small, profuse in simple or branched racemes up to 7 cm long; berries globose scarlet, trilobed, 4-6 mm in diameter, 1-seeded red. The plant is rather variable, three varieties are generally recognized although not distinguished in trade (Anonymous, Wealth of India, 1985).

1. Var. *racemosus* with slightly compressed, acuminate falcate cladodes and longer, decurved spines common in partial shade in plains and across upper ghats in Hassan district (Karnataka) in India
2. Var. *javanicus* Baker syn. *Asparagopsis javanica* Miq. has very slender, flexuous racemes with terminal tufts of cladodes and narrower, liner-oblong perianth, distributed in the Deccan peninsula, Karnataka, Saurashtra (Gujarat) and Bhopal (Madhya Pradesh) in India.
3. Var. *subacerosa* Bakers distributed in Sikkim at an altitudes of 300-1200 m. The plant is very common in the upper Gangetic plains and the Bihar plateau, flowering after rains when it becomes conspicuous by its masses of white fragrant flowers. The white tuberous roots are collected for food and constitute a kind of safed musali used medicinally. It is often cultivated in gardens near Dehradun and is ornamental when in flower. Often the whole plant is seen in flower bearing no cladodes.

Chemistry

The Chemical constituents of *Asparagus racemosus* have been studied to some extent. The compounds so far reported include flavonoids, oligosaccharides, amino acids, sulphur containing acids and steroidal Saponins. The major active constituents of *Asparagus racemosus* are steroidal saponins. Phytochemical investigation of root has resulted in the isolation of 3-O-{[α-L-rhamnopyranosyl(1→2)][β-D-glucopyranosyl(1→4)]-β-D-glucopyranosyl}-26-O-(β-D-glucopyranosyl)-(25S)-5β-furostan-3β,22α,26-triol **(1)** (Shatavarins-I), Shatavarin-II (no reported structure), 3-O-{[α –L-rhamnopyranosyl(1→2)][β-D-glucopyranosyl(1→4)]-β-D-glucopyranosyl}-(25S)-5β-spirostan-3β-ol **(2)** (shatavarin-IV), glycoside-AR-4(incomplete structure elucidation) (Ravi Kumar *et al.*, 1987; Joshi and Sukh, 1988; Jadhav and Bhutani, 2006), 4,6-dihydroxy-2-O-(2′-hydroxyisobutyl) benzaldehyde **(3)** (Paiwal *et al.*, 1991), Asaragamine A **(4)** (Sekine *et al.* 1994), 9,10-dihydro-1,5-dimethoxy-8-methyl-2,7-phenanthrenediol **(5)** (racemosol) (Sekine *et al.*, 1997), 8-methoxy-5,6,4′-trihydroxyisoflavone-7-O-β-D-glucopyranoside **(6)** (Saxena and Chourasia, 2001), 3-O-{[α-L-rhamnopyranosyl (1→2)][α-L-rhamnopyranosyl-(1→4)-O-β-D-glucopyranosyl}-(25S)-spirostan-3β-ol **(7)** (immmunoside) (Handa *et al.*, 2003), racemofuran **(8)** (Wiboonpun *et al.*, 2004), 3-O-{[α-L-rhamnopyranosyl(1→2)][β-D-glucopyranosyl(1→4)]-β-D-glucopyranosyl}-(25S)-5β-spirostan-3β-ol **(9)** (Shatavarin-V) (Hayes *et al.*, 2006), 3-O-{[β-D-glucopyranosyl (1→2)]- [α-L-rhamnopyranosyl (1→4)]- β-Dglucopyranosyl}-(22S,25S)-5 β-spirostan-3 β-ol **(10)** (Shatavarin-VI), 3-O-{[β-D-glucopyranosyl (1→2)]- [α-L-rhamnopyranosyl(1→4)]-β-D-glucopyranosyl}-5β-spirost-25(27)-en-3β-ol **(11)** (Shatavarin-VII), 3-O-{[β-D-glucopyranosyl(1→2)]-[α-L-arabinopyranosyl(1→4)]-[β-D-glucopyranosyl (1→6)]-β-D-glucopyranosyl}-(25S)-5-β-spirostan-3β-ol **(12)** (Shatavarin-VIII), 3-O-{[β-D-glucopyranosyl(1→2)][β-D-glucopyranosyl(1→4)]-β-D-glucopyranosyl}-(25S)-5β-spirostan-3β-ol **(13)** (Shatavarin-IX), 3-O-{[α-L-rhamnopyranosyl(1→2)][β-D-6-acetyl-glucopyranosyl (1→4)]-β-D-glucopyranosyl}-(25S)-5β-spirostan-3-β-ol **(14)** (Shatavarin- X) (Hayes *et al.*, 2008) from this plant. Also sarsasapogenin and kaempferol have been isolated from the woody portion of tuberous roots from *Asparagus racemosus*. The leaves of the plant have resulted in the isolation of diosgenin and quercetin. Chemical examination of flowers and fruits of *Asparagus racemosus* have resulted in the isolation of compounds such as quercetin, rutin (2.5 per cent dry basis) hyperoside, glycoside of quercitin, rutin, hyperoside, sitosterol, stigmasterol, sarsasapogenin, cyanidin-3-galactoside, cyanidin-3-glucorhamnoside (Anonymous, Wealth of India, 1987; Sharma *et al.*, 1981; Thomsen, 2002). Three steroidal saponins, (25S)-5β-spirostan-3 β -ol-3-O-{β-D-glucopyranosyl (1→6)-[α-L-rhamnopyranosyl (1→6)-β-D-glucopyranosyl (1→4)]-β-D-glucopyranoside **(15)** (racemoside-A), (25S)-5β-spirostan-3βol-3-O-α-L-rhamnopyranosyl-(1→6)-β-D-glucopyranosyl-(1→6)-β-D-glucopyranoside **(16)** (racemoside- B), (25S)-5β-spirostan-3β-ol-3-O-{α-L-rhamnopyranosyl-(1→6)-[α-L-rhamnopyranosyl-(1→4)]-β-D-glucopyranoside **(17)** (racemoside-C) were also present in the fruits of *Asparagus racemosus* (Mandal *et al.*, 2006). Apart from this the plant contains 3-O-β-D-glucopyranosyl-D-mannopyranose(a disaccharide), starch, tannin, cetanoate, mucilage, crude fibre and inorganic matter (Landge *et al.*, 1970; Ahmad *et al.*, 1991).

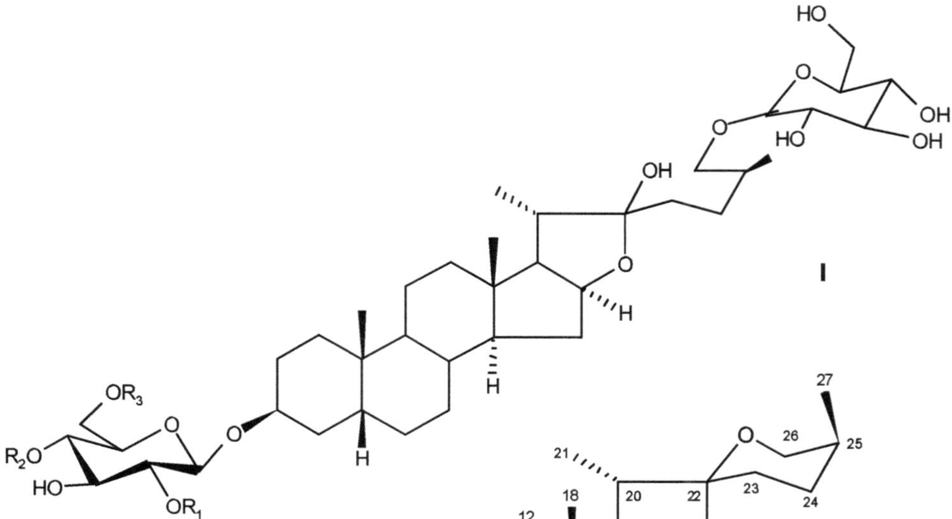

I

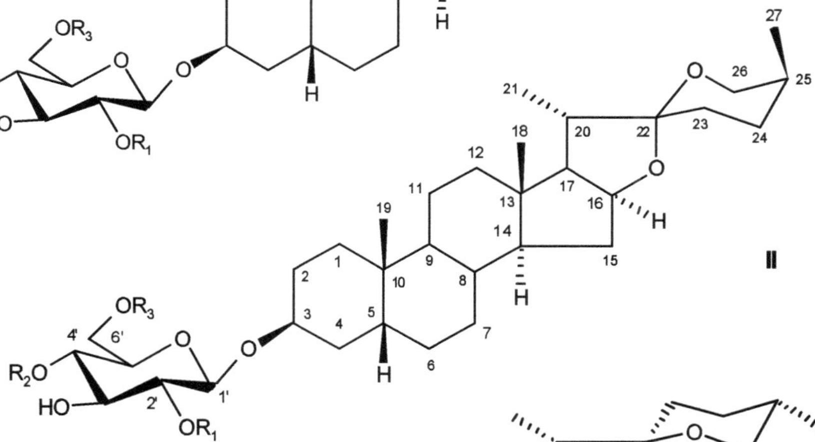

II

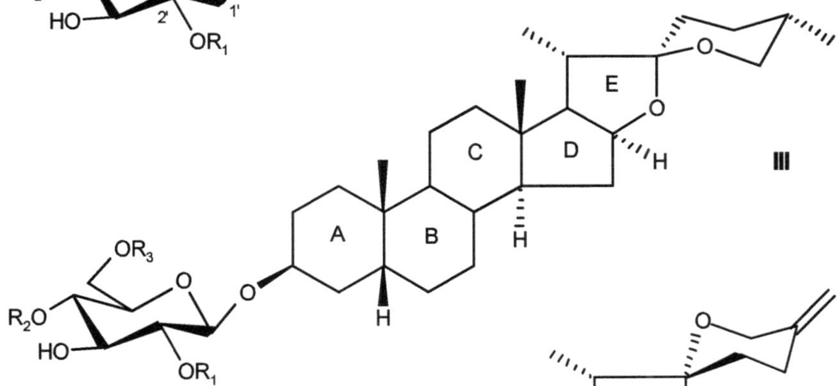

III

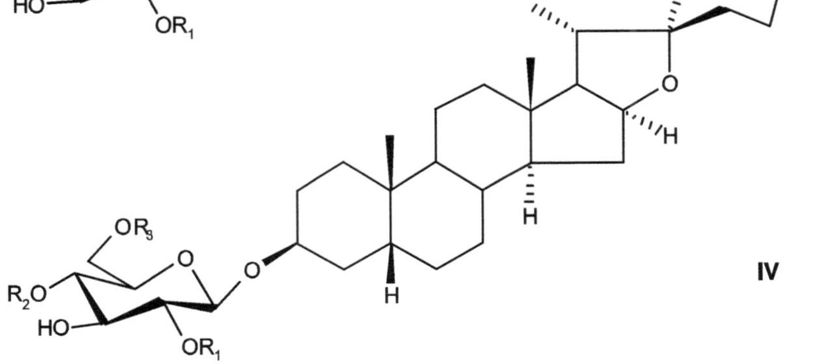

IV

Compound No.	*Skeleton Type*	*R-1*	*R-2*	*R-3*	*Name*
1	I	β-D-Glc	α-L-Rha	H	Shatavarin-I
2	II	β-D-Glc	α-L-Rha	H	Shatavarin-IV
7	II	α-L-Rha	α-L-Rha	H	Immunoside
9	II	α-L-Rha	β-D-Glc	H	Shatavarin-V
10	III	β-D-Glc	α-L-Rha	H	Shatavarin-VI
11	IV	β-D-Glc	α-L-Rha	H	Shatavarin-VII
12	II	β-D-Glc	α-L-Ara	β-D-Glc	Shatavarin-VIII
13	II	β-D-Glc	β-D-Glc	H	Shatavarin-IX
14	II	α-L-Rha	β-D-Glc (6-OAc)	H	Shatavarin-X

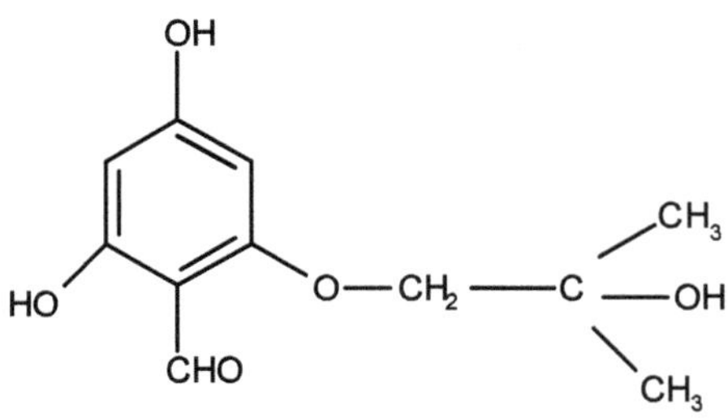

3

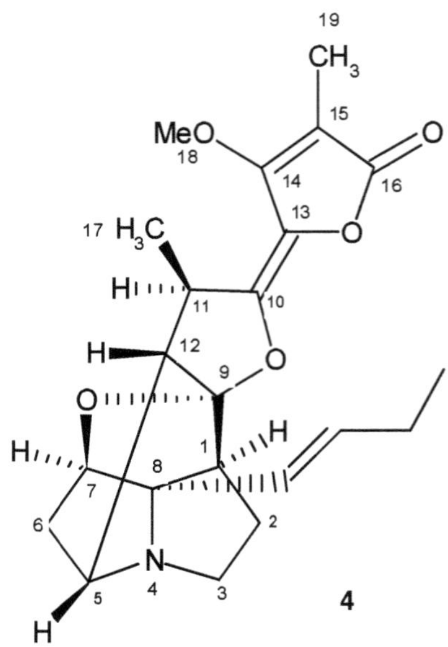

4

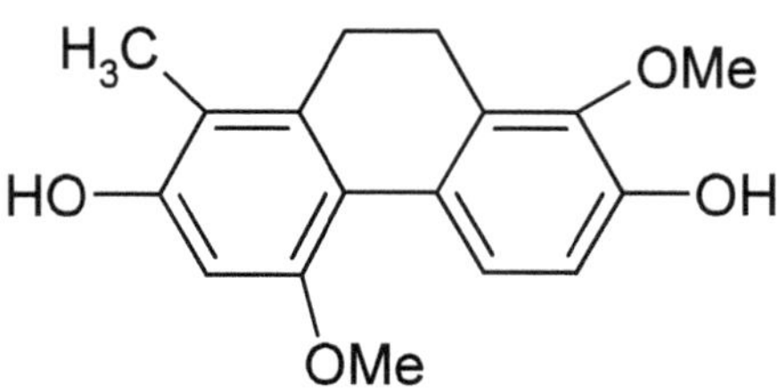

5

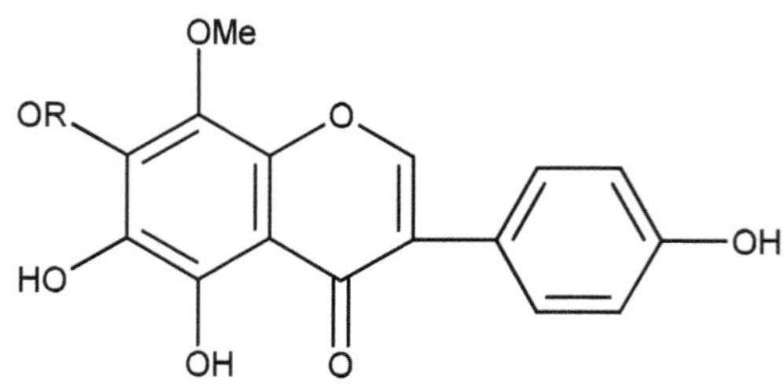

R = β-D-glucosyl

6

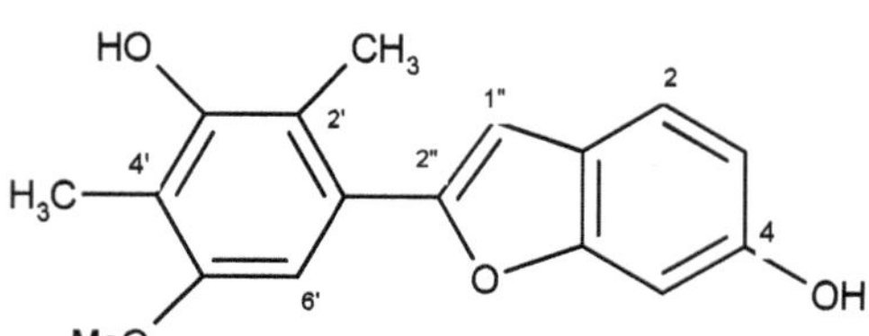

8

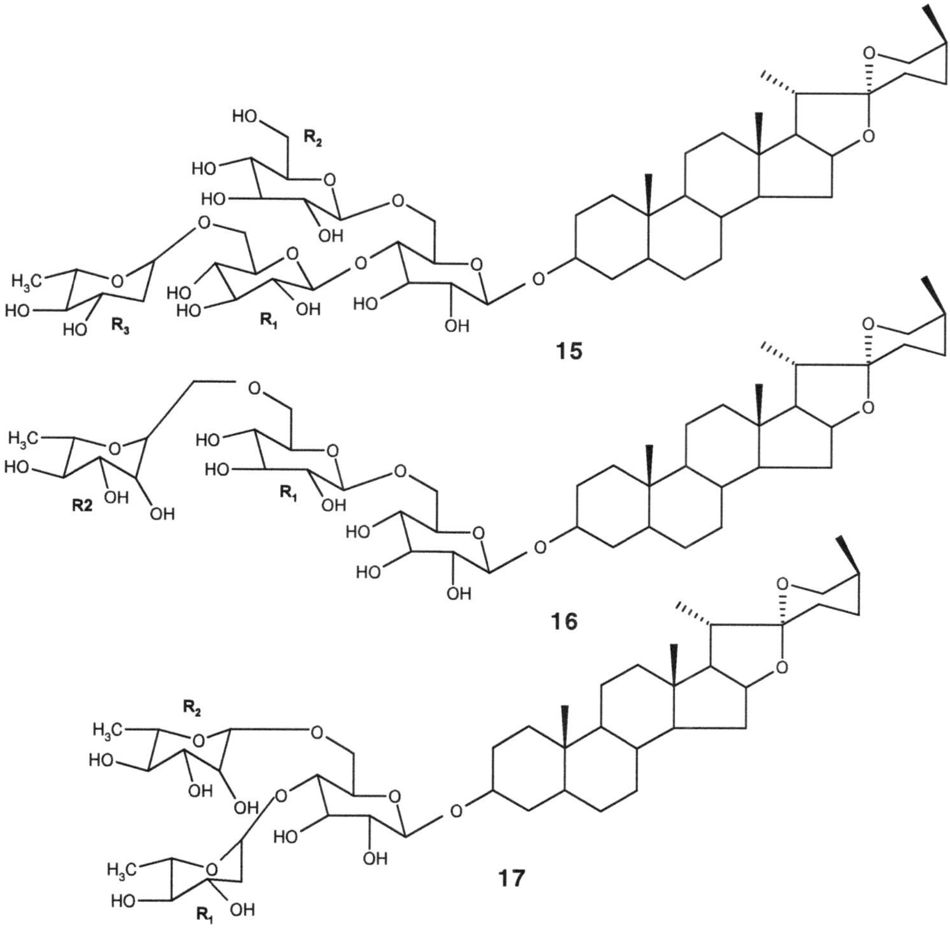

Figure 5.1: Compounds isolated from *Asparagus racemosus.*

Spectral Studies

We have tried to sum up the ^{1}H and ^{13}C spectral data of the above described compounds. ^{1}H NMR data are given in Table 5.1 and ^{13}C NMR data are given in Table 5.2.

Medicinal Uses

Almost all parts of this plant are used by the Indian traditional system of medicine (Ayurved and Unani) for the treatment of various ailments in human beings. Tuber, leaves and fruits are used in gonorrhea, piles, diabetes, rheumatism, cough, diarrhoea, dysentery, gastric troubles and headache, also for increasing lactation.

In Ayurveda and Siddha medicine system the plant is used for treating *madhura rasam, madhura vipakam, seeta-veeryam,* polyuria, chronic fevers, *soma rogam,* white

Table 5.1: ^{1}H NMR spectral data of saponins (algycone Part) (δ Values).

Hydrogens	*Shatavarins I (1) Pyridine-d_5*	*Shatavarins IV (2) Pyridine-d_5*	*Immunoside (7) Pyridine-d_5*	*Shatavarins (V) (9) Pyridine-d_5*
H-1	1.75–1.83 m	0.94 m 1.24 m	–	1.82 m 1.48 m
H-2	1.14 m H-a 1.78 m H-b	1.45 m 2.05 ddd (5.9, 7.6, 12.6)	–	1.48 m H-a 1.90 m H-b
H-3	4.24 m	4.26 m	3.84 m	4.24 m
H-4	1.43 m H-a 1.78 m H-b	1.82 m 1.82 m	–	1.83 m H-a 1.83 m H-b
H-5	2.16 m	2.21 m	–	2.14 m
H-6	1.47 m H-a 1.82 m H-b	1.53 m 1.91 m	–	1.53 m H-a 1.91 m H-b
H-7	1.20 m H-a 1.90 m H-b	1.17 m 1.79 m	–	0.97 m H-a 1.28 m H-b
H-8	1.46 m	1.48 m	–	1.50–1.56 m
H-9	1.24 m	1.28 m	–	1.21–1.31 m
H-11	1.20 m H-a 1.30 m H-b	1.20 m 1.32 m	–	1.21 m H-a 1.32 m H-b
H-12	1.08, m H-a 1.71 m H-b	1.10 m 1.71 m	–	1.08 m H-a 1.68 m H-b
H-14	1.04 m	1.06 m	–	1.06 m
H-15	1.40 ddd (6.1, 6.1, 12.8) 2.00 m	1.45 m 1.98 m	–	1.35–1.41 m H-a 2.01 m H-b
H-16	4.97 dt (7.1, 7.1)	4.63 dt (7.2, 7.2)	4.40 q (7.1)	4.56 m
H-17	1.95 m	1.88 dd (6.7, 8.5)	–	1.81 m
H-18	0.86 s	0.85 s	0.77 s	0.80 s
H-19	0.94 s	0.97 s	0.98 s	1.06 s
H-20	2.22 dq (6.8, 6.8)	1.96 dq (6.8, 6.8)	–	1.90 m
H-21	1.31 d (6.8)	1.19 d (6.9)	1.13 d H-a (6.9) 1.01d H-b (6.0)	1.14 d (7.0)
H-23	1.95 m H-a 2.06 m H-b	1.47 m 1.84 m	–	1.88 m H-a 1.42 m H-b
H-24	1.66 m H-a 2.06 m H-b	1.40 m 2.17 dt (4.4, 12.7)	–	1.33 m H-a 2.12 m H-b
H-25	1.91 m	1.61 m	–	1.57 m
H-26	3.46 m H-a 4.08 m H-b	3.43 br d (10.6) 4.11 dd (2.7, 11.2)	3.39 bd H-a (11.0) 3.58 dd H-b (11.0,2.0)	3.35 br d H-a (12.0) 4.06 m H-b
H-27	1.01 d (6.8)	1.10 d (7.1)	1.09 d (7.2)	1.06 d (6.8)

Contd...

Table 5.1–*Contd...*^{1}H NMR spectral data of saponins (sugar Part) (δ Values).

Hydrogens	*Shatavarins I (1) Pyridine-d_5*	*Shatavarins IV (2) Pyridine-d_5*	*Immunoside (7) Pyridine-d_5*	*Shatavarins (V) (9) Pyridine-d_5*
	3-O-α-D-Glucose	*3-O-α-D-Glucose*	*3-O-β–D-Glucose*	*3-O-β-D-Glucose*
H-1'	4.85 d (7.8)	4.87 d (7.6)	4.87 d (7.7)	4.80 d (7.8)
H-2'	4.28 dd (8.0, 9.0)	4.30 dd (7.6, 9.0)	4.28	4.25 dd (7.8, 9.3)
H-3'	4.26 dd (9.0, 9.0)	4.26 dd (9.0, 9.0)	4.23	4.20 m
H-4'	4.48 dd (9.1, 9.1)	4.34 dd (9.0, 9.0)	4.40	4.20 m
H-5'	3.59 ddd (2.3, 3.1, 9.2)	3.61 ddd (2.1, 3.6, 9.6)	3.68	3.81 ddd (3.1, 3.8, 12.0)
H-6'	4.08 dd (3.1, 12.0) 4.22 brd (12.0)	4.04 dd (3.6, 12.2) 4.19 dd (2.1,12.2)	4.09 H-a 4.24 H-b	4.44 dd (3.1, 12.0) 4.50 dd (3.9, 12.0)
	2'-O-β-D-Glucose	*2'-O-β-D-Glucose*	*2'-O-α-L-Rhamnose*	*2'-O-α-L-Rhamnose*
H-1"	5.46 d (7.8)	5.48 d (7.8)	6.40 brs	6.36 d (1.4)
H-2"	4.08 dd (8.0, 9.2)	4.07 dd (8.4, 8.4)	4.86	4.71 dd (1.4, 3.3)
H-3"	4.26 dd (9.2, 9.2)	4.27 dd (8.8, 8.8)	4.68	4.53 dd (3.3, 9.3)
H-4"	4.34 dd (9.2, 9.2)	4.23 dd (8.8, 8.8)	4.36	4.31 dd (9.4, 9.4)
H-5"	3.97 ddd (3.2, 4.5, 9.2)	3.99 ddd (2.9, 5.5, 8.8)	4.97	4.79 dq (6.2, 9.4)
H-6"	4.49 dd H-a (4.5, 11.8) 4.58 dd H-b (3.2, 11.8)	4.43 dd H-a (5.8, 12.0) 4.61 dd H-b (2.9, 12.0)	1.72	1.72 d (6.2)
	4'-O-α-Rhamnose	*4'-O-α-L-Rhamnose*	*4'-O-α-L-Rhamnose*	*4'-O-β-D-Glucose*
H-1"'	5.92 brs	5.78 d (1.4)	5.73 brs	5.12 d (7.6)
H-2"'	4.76 dd (1.5, 3.3)	4.68 dd (1.4, 3.2)	4.70	4.05 dd (7.6, 9.1)
H-3"'	4.53 dd (3.3, 9.3)	4.56 dd (3.2, 9.4)	4.62	4.20 dd (9.3, 9.3)

Contd...

Table 5.1–*Contd...*

Hydrogens	*Shatavarins I* (1) Pyridine-d_5	*Shatavarins IV* (2) Pyridine-d_5	*Immunoside* (7) Pyridine-d_5	*Shatavarins (V)* (9) Pyridine-d_5
	4'-O-α-Rhamnose	*4'-O-α-L-Rhamnose*	*4'-O-α-L-Rhamnose*	*4'-O-β-D-Glucose*
H-4'''	4.32 dd (9.3, 9.3)	4.36 dd (9.4, 9.4)	4.34	4.27 dd (9.2, 9.2)
H-5'''	4.99 dq (6.3, 9.3)	4.91 dq (6.1, 9.3)	4.95	3.95 ddd (2.5, 4.9, 9.2)
H-6'''	1.67 d (6.3)	1.67 d (6.2)	1.59	4.31 dd (5.0, 12.2) 4.43 dd (2.5, 12.2)
	26-O-β-D-Glucose			
H-1''''	4.81 d (7.8)			
H-2''''	4.03 dd (7.8, 9.0)			
H-3''''	4.25 m			
H-4''''	4.25 m			
H-5''''	3.95 ddd (2.7, 5.1, 9.3)			
H-6''''	4.40 dd H-a (5.2,11.8) 4.56 brd H-b (2.7, 11.8)			

discharge, internal heat and as tonic. Generally the root is employed in diarrhoea as well as in chronic colic and dysentery problems. Root boiled with some bland oil, is applied in various skin diseases. Root is boiled in milk and the milk is administered to relieve bilious dyspepsia and diarrhoea and to promote appetite; root is also used in rheumatism. The root also finds use as a fertility enhancer, as an antidiabetic, in fever, in haemostatic, in whooping cough, to check excessive bleeding during menstruation, in urinary troubles, impotency, leucorrhoea, burning micturition, post partum complaints, body ache and in infection of umbilicus. Bark is used in amoebic dysentery. The twigs are used in wound healing, dysentery, bacterial, viral, plastic diseases and digestive disturbances. Tubers are candied and taken as a sweetmeat. Fresh root juice is given with honey as a demulcent. Boiled leaves smeared with ghee are applied to boils, smallpox, etc., in order to prevent their confluence. Juice of this drug taken with milk is useful in gonorrhea. The plant finds use in about 64 ayurvedic formulations (Nishritha and Sexena, 2007) which include traditional formulations such as 'Shatavari kalpa', 'Phalaghrita', 'Vishnu taila', 'Shatavari guda' etc. (Unnikrishnan, 1998).

Table 5.1–*Contd...*^{1}H NMR spectral data of saponins (algycone Part) (δ Values).

Hydrogens	Shatavarin VI (10) Pyridine-d_5	Shatavarin VII (11) Pyridine-d_5	Shatavarin VIII (12) Pyridine-d_5	Shatavarin IX (13) Pyridine-d_5	Shatavarin X (14) Pyridine-d_5
H-1	1.43 m H-a 1.78 m H-b	1.42 m H-a 1.78 m H-b	1.46 m H-a 1.83 m H-b	1.45 m H-a 1.81 m H-b	1.50 m H-a 1.84 m H-b
H-2	1.21 m H-a 1.82 m H-b	1.14 m H-a 1.75 m H-b	1.47 m H-a 1.82 m H-b	1.48 m H-a 1.85 m H-b	1.51 m H-a 1.85 m H-b
H-3	4.32 m	4.28 m	4.22 m	4.22 m	4.22 m
H-4	1.73 m H-a 1.79 m H-b	1.77 m H-a 1.77 m H-b	1.75 m H-a 1.75 m H-b	1.77 m H-a 1.77 m H-b	1.82 m H-a 1.825 m H-b
H-5	2.18 m	2.16 brs	2.22 m	2.17 m	2.16 m
H-6	1.47 m H-a 1.83 m H-b	1.46 m H-a 1.84 m H-b	1.18 m H-a 1.84 m H-b	1.15 m H-a 1.78 m H-b	1.35 m H-a 1.98 m H-b
H-7	0.95 m H-a 1.22 m H-b	0.91 m H-a 1.20 m H-b	0.92 m H-a 1.23 m H-b	0.92 m H-a 1.22 m H-b	0.98 m H-a 1.28 m H-b
H-8	1.45 m	1.48 m	1.47 m	1.47 m	1.54 m
H-9	1.27 m	1.25 m	1.25 m	1.25 m	1.31 m
H-11	1.19 m H-a 1.31 m H-b	1.16 m H-a 1.28 m H-b	1.18 m H-a 1.31 m H-b	1.18 m H-a 1.31 m H-b	1.22 m H-a 1.32 m H-b
H-12	1.05 m H-a 1.63 m H-b	1.05 m H-a 1.66 m H-b	1.08 m H-a 1.66 m H-b	1.06 m H-a 1.67 m H-b	1.08 m H-a 1.68 m H-b
H-14	0.96 m	1.08 m	1.05 m	1.03 m	1.08 m
H-15	1.44 m H-a 2.02 m H-b	1.43 m H-a 2.03 m H-b	1.40 m H-a 2.09 m H-b	1.40 m H-a 2.00 m H-b	1.38 m H-a 2.01 m H-b
H-16	4.37 m	4.63 dt (6.8, 7.6)	4.58 m	4.57 m	4.56 m
H-17	1.63 dd (6.5, 8.7)	1.86 dd (6.8, 7.6)	1.81 m	1.82 m	1.81 m
H-18	0.94 s	0.84 s	0.81 s	0.81 s	0.81 s
H-19	0.94 s	0.97 s	0.98 s	0.96 s	1.06 s
H-20	2.32 dq (6.5, 6.5)	1.99 dq (6.8, 7.0)	1.91 m	1.91 m	1.90 m
H-21	1.00 d (7.2)	1.08 d (7.0)	1.12 d (7.0)	1.14 d (7.0)	1.14 d (7.0)
H-23	1.46 m H-a 1.67 dbrt H-b (2.8,12.0)	1.82 m H-a 1.82 m H-b	1.45 m H-a 1.90 m H-b	1.44 m H-a 1.90 m H-b	1.42 m H-a 1.88 m H-b
H-24	1.55 m H-a 1.55 m H-b	2.24 brd H-a (12.0) 2.70 td H-b (5.4, 12.0)	1.35 m H-a 2.12 m H-b	1.35 m H-a 2.13 m H-b	1.34 m H-a 2.13 m H-b
H-25	1.63 m	–	1.59 m	1.57 m	1.56 m
H-26	3.56 dd H-a (10.5, 11.0) 3.70 rdd H-b (5.6, 11.0)	4.06 d H-a (12.0) 4.49 d H-b (12.0)	3.36 brd H-a (11.0) 4.07 brdd H-b (2.2, 11.0)	3.36 brd H-a (11.6) 4.07 bdd H-b (2.9, 11.6)	3.36 brd H-a (11.6) 4.06 bdd H-b (2.6, 11.6)
H-27	0.68 d (6.4)	4.80 brs H-a 4.82 brs H-b	1.07 d (6.7)	1.07 d (7.2)	1.06 d (7.2)

Contd...

Table 5.1–*Contd*...^{1}H NMR spectral data of saponins (sugar Part) (δ Values).

Hydrogens	*Shatavarin VI (10) Pyridine-d_5*	*Shatavarin VII (11) Pyridine-d_5*	*Shatavarin VIII (12) Pyridine-d_5*	*Shatavarin IX (13) Pyridine-d_5*	*Shatavarin X (14) Pyridine-d_5*
	3-O-β-D-Glucose	*3-O-β-D-Glucose*	*3-O-β-D-Glucose*	*3-O-β-D-Glucose*	*3-O-β-D-Glucose*
H-1'	4.85 d (8.2)	4.88 d (8.0)	4.75 d (8.2)	4.84 d (8.0)	4.82 d (7.8)
H-2'	4.20 dd (8.8, 9.0)	4.27 dd (9.2, 8.0)	4.11 dd (8.2, 9.2)	4.26 dd (9.2, 8.0)	4.24 m
H-3'	4.23 dd (9.1, 9.1)	4.28 dd (9.2, 9.2)	4.16 dd (9.2, 9.2)	4.29 dd (9.2, 9.2)	4.22 m
H-4'	4.45 dd (9.1, 9.1)	4.38 dd (9.1, 9.1)	4.34 dd (9.2, 9.2)	4.23 dd (9.1, 9.1)	4.15 dd (9.4, 9.4)
H-5'	3.60 brd (9.2)	3.62 brd (9.0)	3.75 brd (9.2)	3.80 brd (9.0)	3.84 ddd (3.3, 3.3, 9.4)
H-6'	4.04 dd H-a (5.7, 12.0) 4.73 brd H-b (12.0)	4.07 dd H-a (3.6, 12.0) 4.23 brd H-b (12.0)	4.60 dd H-a (1.3, 12.0) 4.73 brd H-b (2.8, 12.0)	4.42-4.51 m	4.47 m
	2'-O-β-D-Glucose	*2'-O-β-D-Glucose*	*2'-O-β-D-Glucose*	*2'-O-β-D-Glucose*	*4'-O-β-D-Glucose*
H-1"	5.43 d (7.6)	5.45 d (7.8)	5.36 d (8.0)	5.40 d (8.0)	5.06 d (7.9)
H-2"	4.07 dd (7.6, 8.6)	4.07 dd (7.8, 8.6)	3.98 dd (8.2, 9.0)	4.02 dd (8.0, 8.8)	4.00 dd (8.0, 9.3)
H-3"	4.24 dd (8.8, 8.6)	4.25 dd (8.9, 8.9)	4.23 dd (9.1, 9.1)	4.24 dd (8.8, 9.0)	4.17 dd (9.3, 9.3)
H-4"	4.31 dd (8.8, 8.8)	4.27 dd (8.8, 8.8)	4.22 dd (9.2, 9.0)	4.29 dd (9.0, 9.0)	3.93 dd (9.3, 9.3)
H-5"	3.94 ddd (3.4, 4.9, 9.2)	3.97 ddd (2.7, 5.1, 9.2)	3.95 ddd (3.1, 4.9, 9.2)	3.95 ddd (3.3, 5.0, 9.0)	4.10 ddd (2.2, 8.0, 9.3)
H-6"	4.48 dd H-a (4.9, 12.0) 4.56 brd (12.0)	4.45 dd H-a (5.2, 12.0) 4.58 dd (2.7,12.0)	4.44 dd H-a (5.0, 12.0) 4.57 brd (3.1,12.0)	4.47 brd H-a (12.0) 4.56 brd H-b (12.0)	4.57 dd (8.0, 12.0) 4.90 dd (2.2, 12.0)
6-o-Ac					171.6 (Me)
	4'-O-α-L-Rhamnose	*4'-O-α-L-Rhamnose*	*4'-O-α-L-Rhamnose*	*4'-O-β-D-Glucose*	*2'-O-α-L Rhamnose*
H-1'''	5.90 brs	5.82 brs	5.29 d (8.1)	5.14 d (8.3)	6.32 brd (1.1)
H-2'''	4.66 brs	4.67 brs	4.43 dd (8.2, 8.3)	4.05 dd (8.3, 8.7)	4.72 brs
H-3'''	4.53 dd (2.9, 9.2)	4.55 dd (3.8, 9.1)	4.22 m	4.19 m	

Contd...

Table 5.1–*Contd...*

Hydrogens	*Shatavarin VI (10) Pyridine-d_5*	*Shatavarin VII (11) Pyridine-d_5*	*Shatavarin VIII (12) Pyridine-d_5*	*Shatavarin IX (13) Pyridine-d_5*	*Shatavarin X (14) Pyridine-d_5*
	4'-O-α-L-Rhamnose	*4'-O-α-L-Rhamnose*	*4'-O-α-L-Rhamnose*	*4'-O-β-D-Glucose*	*2'-O-α-L Rhamnose*
H-4'''	4.31 dd (9.2, 9.2)	4.35 dd (9.2, 9.2)	4.22 m	4.21 m	4.52 dd (3.2, 9.6)
H-5'''	4.97 dq (5.8, 9.2)	4.93 dq (6.2, 9.2)	4.22 m	3.98 m	4.30 dd (9.6, 9.6)
H-6'''	1.66 d (5.8)	1.66 d (6.2)	–	4.28 m 4.49 brd (12.0)	4.78 dq (6.3, 9.6) 1.73 d (6.3)
			6'-O-β-D-Glucose		
H-1''''			5.27 d (8.0)		
H-2''''			4.04 dd (8.0, 8.7)		
H-3''''			4.22 dd		
H-4''''			4.22 m (8.8, 8.8)		
H-5''''			4.24 dd (9.8, 9.6)		
H-6''''			4.35 dd (5.2, 12.0) 4.49 dd (2.8, 12.0)		

Pharmacology

Anti-diarrhoeal Potential

Diarrhoea has long been recognized as one of the most important health problems in the developing countries (Snyder *et al.*, 1982). Worldwide distribution of diarrhoea accounts for more than 5-8 million deaths each year in infants and small children of age less than 5 year. According to WHO estimation for the year 1998, there were about 7.1 million deaths due to diarrhoea (Park, 2000). Secretory diarrhoea is the most dangerous symptom of gastrointestinal problems (Fontaine, 1988) and is associated with excessive defecation and stool outputs, the stools being of abnormally loose consistency (Aranda-Michel and Gianella, 1999).

The ethanol and aqueous extract of *Asparagus racemosus* exhibited significant anti-diarrhoeal activity against castor oil induced diarrhoea in rats. The extracts had a similar activity as loperamide, when tested at 200 and 250 mg/kg and statistically

Table 5.1–*Contd...*^{1}H NMR spectral data of saponins (algycone Part) (δ Values).

Hydrogens	*Racemoside A (15) Pyridine-d_5*	*Racemoside B (16) Pyridine-d_5*	*Racemoside C (17) Pyridine-d_5*
H-1	1.78 dd H-a (3.6, 11.4) 1.46 H-b	1.78 dd (3.6, 11.4) 1.46	1.78 dd (3.6, 11.4) 1.46
H-2	1.92 H-a 1.46 H-b	1.92 H-a 1.46 H-b	1.92 H-a 1.46 H-b
H-3	4.28	4.2	4.2
H-4	1.75 H-a 1.85 H-b	1.75 1.85	1.75 1.85
H-5	2.15	2.15	2.15
H-6	1.77 H-a 1.14 H-b	1.77 H-a 1.14 H-b	1.77 H-a 1.14 H-b
H-7	1.19 H-a 0.91 brd H-b (10.8)	1.19 0.91 brd (10.8)	1.19 0.91 brd (10.8)
H-8	1.46	1.46	1.46
H-9	1.24	1.24	1.24
H-10	–		
H-11	1.31 H-a 1.14 H-b	1.31 H-a 1.14 H-b	1.31 H-a 1.14 H-b
H-12	1.66 H-a 1.04 H-b	1.66 H-a 1.04 H-b	1.66 H-a 1.04 H-b
H-14	1.04	1.04	1.04
H-15	2.01 ddd H-a (4.8, 7.2, 11.4) 1.46 H-b	2.01 ddd H-a (4.8, 7.2, 11.4) 1.46 H-b	2.01 ddd H-a (4.8, 7.2, 11.4) 1.46 H-b
H-16	4.35	4.35	4.35
H-17	1.80 dd (6.6, 8.4)	1.80 dd (6.6, 8.4)	1.80 dd (6.6, 8.4)
H-18	0.82 s	0.82 s	0.82 s
H-19	0.96 s	0.96 s	0.96 s
H-20	1.83	1.83	1.83
H-21	1.17 d (6.6)	1.17 d (6.6)	1.17 d (6.6)
H-23	1.89 brd H-a (9.6) 1.24 H-b	1.89 brd H-a (9.6) 1.24 H-b	1.89 brd H-a (9.6) 1.24 H-b
H-24	2.15 H-a 1.37 H-b	2.15 H-a 1.37 H-b	2.15 H-a 1.37 H-b
H-25	1.66	1.66	1.66
H-26	4.10 H-a 3.38 d H-b (10.8)	4.10 H-a 3.38 d H-b (10.8)	4.10 H-a 3.38 d H-b (10.8)
H-27	1.09 d (6.6)	1.09 d (6.6)	1.09 d (6.6)

Contd...

Table 5.1–*Contd...*¹H NMR spectral data of saponins (sugar Part) (δ Values).

Hydrogens	*Racemoside A (15) Pyridine-d_5*	*Racemoside B (16) Pyridine-d_5*	*Racemoside C (17) Pyridine-d_5*
	3-O-α-D-Glucose	*3-O-α-D-Glucose*	*3-O-β-D-Glucose*
H-1'	4.82 d (7.2)	4.85 d (7.8)	4.82 (7.8)
H-2'	4.26	4.07 t (9)	4.0
H-3'	4.08t (9.0)	4.23	4.20
H-4'	4.35	4.25	4.15
H-5'	3.85 dd (5.4, 8.4)	3.99 dd (4.8, 8.4)	3.87 dd (3.6,7.2)
H-6'	4.23 dd 6H-a (1.8, 1.2) 3.64 d 6H-b (11.4)	4.27b 6H-a 3.66 d 6H-b (10.8)	4.36 H-a 3.98 dd H-b (4.8, 11.4)
	4'-O-β-D-Glucose	*6'-O-β-D-Glucose*	*4'-O-α-L-Rhamnose*
H-1"	4.78 d (7.8)	4.84 d (7.8)	5.40 brs
H-2"	4.43 t (8.4)	4.49	4.64 d (3)
H-3"	4.03	4.07 t (9)	4.52 dd (3, 9)
H-4"	4.18	4.18 t (8.4)	4.26
H-5"	4.59	4.13	4.36
H-6"	4.62 dd (2.4, 11.4) 4.54	4.75 d (11.4) 4.35 dd (4.2, 11.4)	1.68 d (6.4)
	6'-O-β-D-Glucose	*6"-O-α-L-Rhamnose*	*6'-O-α-L-Rhamnose*
H-1"'	5.47 d (7.8)	5.63 brs	5.67 brs
H-2"'	4.11	4.69 brs	4.68 d (2.4)
H-3"'	4.33	4.57 t (8.4)	4.57dd (3,9)
H-4"'	4.36	4.29 t (9)	4.36

Contd...

significant reduction in the frequency of defecation and the wetness of the faecal droppings when compared to untreated control rats. Ethanol and aqueous extracts of *Asparagus racemosus* possess significant anti-diarrhoeal activity due to inhibitory effect both on gastrointestinal propulsion and fluid secretion (Venkatesan *et al.*, 2005).

Table 5.1–*Contd...*

Hydrogens	*Asparagamine A (4) $CDCl_3$*	*Racemosol (5) $CDCl_3$*	*8-methoxy-5,6,4'-trihydroxyisoflavone-7-O-β-D-glucopyranoside (6) $CDCl_3$*	*Racemofuran (8) $CDCl_3$*
H-1	2.79 d (5.9)	–	–	–
H-2	1.82 m H-a 1.73 m H-b	5.66 brs (OH)	8.2 s	7.14 d (6.1)
H-3	3.03 m H-a 2.93 mH-b	6.83 d (9.0)	–	6.64 dd (6.1, 2.2)
H-4	–	7.85 d (9.0)	–	4.84 s (OH)
H-5	3.45 brs	–	–	7.15 d (2.2)
H-6	1.89d H-a (12.2) 1.74 m H-b	6.39 s	–	–
H-7	4.14 brs	5.00 brs (OH)	–	–
H-8	4.08	–	3.8 s (-OMe)	–
H-9	–	2.68 m (2H)	–	–
H-10	–	2.74 m (2H)	–	–
H-11	3.04 m	3.76 s (3H)	–	–
H-12	1.77 m	3.80 s (3H)	–	–
H-13	–	2.16 s (3H)	–	–
H-14	–	–	–	–
H-15	–	–	–	–
H-16	–	–	–	–
H-17	1.32 d (6.4)	–	–	–
H-18	4.08 s	–	–	–
H-19	2.01 s	–	–	–
H-20	5.44 d (15.5)	–	–	–
H-21	5.71 dt (15.5, 6.4)	–	–	–
H-22	2.01 m	–	–	–
H-23	0.93 t (7.5)	–	–	–
H-1'	–	–	–	–
H-2'	–	–	7.43 d (8.6)	2.39s ($-CH_3$)

Contd...

Table 5.1–*Contd...*

Hydrogens	*Asparagamine A (4) $CDCl_3$*	*Racemosol (5) $CDCl_3$*	*8-methoxy-5,6,4'-trihydroxyisoflavone-7-O-β-D-glucopyranoside (6) $CDCl_3$*	*Racemofuran (8) $CDCl_3$*
H-3'	–	–	7.2 d (8.6)	5.18 s (-OH)
H-4'	–	–	2.18 s (3H)	2.18 s ($-CH_3$)
H-5'	–	–	7.2 d (8.6)	3.87 s (-OMe)
H-6'	–	–	7.43 d (8.6)	6.91 s
H-1"	–		–	6.88 s
Sugar protons	–	–	4.80-5.42 m	

Antioxidant Properties

The deleterious effects of ionizing radiation, especially those having low linear energy transfer (LET), in biological systems are mainly mediated through the generation of oxygen-derived intermediates in the form of free radicals and excited states (Von Sonntag, 1987). Cellular membranes are among the most important targets of radiation damage (Giusti *et al.*, 1998). Due to the presence of polyunsaturated fatty acids, membranes are highly susceptible for oxidative damage induced by reactive oxygen species (ROS), generated during radiation (Rice-Evans and Burdon, 1993). Adverse alterations in biomembranes can directly lead to cytotoxicity and/or indirectly to genotoxicity (Esterbauer, 1996). Hence compounds capable of protecting cellular membranes against ionizing radiation in particular and free radicals in general will have potential benefits as radioprotectors, antioxidants and antimutagens (Stavric, 1994; Odin, 1997). There is sufficient evidence to suggest that adequate antioxidant defense by vitamin E and the other antioxidants can provide protection from the high levels of free radicals generated (Packer and Ong., 1998). Currently, great interest centres on the possible protective value of a wide variety of plant-derived antioxidant compounds, particularly those from fruits and vegetables, against radiation damage (Stavric, 1994; Maulik *et al.*, 1997; Zhang *et al.*, 1997). Jayashree *et al.* (2000) found antioxidant effects of crude extract and a purified aqueous fraction of *Asparagus racemosus* against membrane damage induced by the free radicals generated during γ-radiation examined in rat liver mitochondria. γ-Radiation in the dose range of 75–900 Gy, induced lipid peroxidation as assessed by the formation of thiobarbituric acid reactive substances (TBARS) and lipid hydroperoxides (LOOH). Using an effective dose of 450 Gy, antioxidant effects of *Asparagus racemosus* extract were studied against oxidative damage in terms of protection against lipid peroxidation, protein oxidation, depletion of protein thiols and the levels of the antioxidant enzyme, superoxide dismutase. An active fraction consisting of

Table 5.2: ^{13}C NMR spectral data of (aglycone part) saponins.

Carbons	*Shatavarins I (1) Pyridine-d_5*	*Shatavarins IV (2) Pyridine-d_5*	*Immunoside (7) Pyridine-d_5*	*Shatavarins (V) (8) Pyridine-d_5*
C-1	30.7	27.1	31.13	31.0
C-2	26.7	26.5	27.71	26.9
C-3	75.3	75.4	63.03	76.1
C-4	30.8	30.7	32.35	30.9
C-5	36.7	36.8	37.32	37.2
C-6	27.0	26.8	26.60	26.8
C-7	26.8	26.8	27.0	26.9
C-8	35.5	35.3	35.81	35.6
C-9	40.2	40.3	40.60	40.4
C-10	35.3	35.6	35.49	35.3
C-11	21.2	21.2	21.40	21.2
C-12	40.4	40.4	32.35	40.4
C-13	41.2	40.9	41.18	40.9
C-14	56.4	56.5	56.86	56.6
C-15	32.4	32.2	40.69	32.2
C-16	81.3	81.4	81.59	81.4
C-17	64.0	63.0	60.79	63.0
C-18	16.7	16.6	16.94	16.6
C-19	24.0	24.0	24.05	23.4
C-20	40.7	42.5	42.70	42.5
C-21	16.5	15.0	15.12	14.9
C-22	110.7	109.8	109.85	109.7
C-23	37.2	30.9	27.0	26.4
C-24	28.4	26.3	26.39	26.3
C-25	34.5	27.6	27.0	27.5
C-26	75.4	65.2	65.42	65.1
C-27	17.5	16.3	16.58	16.3

Contd...

polysaccharides was effective even at a low concentration of 10 µg/ml. Both the crude extract as well as the polysaccharides fraction significantly inhibited lipid peroxidation and protein oxidation. The antioxidant effect of polysaccharides fraction was more pronounced against lipid peroxidation, as assessed by TBARS formation, while that of the crude extract was more effective in inhibiting protein oxidation. Both the crude extract and polysaccharides fraction also partly protect against radiation-induced loss of protein thiols and inactivation of superoxide dismutase. The inhibitory effects of these active principles, at the concentration of 10 µg/ml, are

Table 5.2–*Contd...*^{13}C NMR spectral data of (Sugar part) saponins.

Hydrogens	*Shatavarins I (1) Pyridine-d_5*	*Shatavarins IV (2) Pyridine-d_5*	*Immunoside (7) Pyridine-d_5*	*Shatavarins (V) (8) Pyridine-d_5*
	3-O-α-D-Glucose	*3-O-α-D-Glucose*	*3-O-β-D-Glucose*	*3-O-β-D-Glucose*
C-1'	101.9	101.9	101.60	102.0
C -2'	82.7	81.4	79.46	76.6
C -3'	76.4	76.5	78.30	78.2
C -4'	77.2	77.4	77.20	82.3
C -5'	77.1	77.1	76.79	76.2
C -6'	61.2	61.3	61.65	62.1
	2'-O-β-D-Glucose	*2'-O-β-D-Glucose*	*2'-O-α-L-Rhamnose*	*2'-O-α-L-Rhamnose*
C -1"	105.6	105.6	102.23	101.4
C -2"	77.1	76.5	72.22	72.4
C -3"	77.9	78.0	72.60	72.7
C -4"	71.8	71.9	73.81	74.1
C -5"	78.6	78.6	69.52	69.4
C -6"	62.9	63.0	18.60	18.8
	4'-O-α-Rhamnose	*4'-O-α-L-Rhamnose*	*4'-O-α-L-Rhamnose*	*4'-O-β-D-Glucose*
C -1'''	102.4	102.5	103.01	105.3
C -2'''	72.6	72.6	72.43	75.0
C -3'''	72.8	72.8	72.70	78.3
C -4'''	74.0	74.0	73.97	71.3
C -5'''	70.3	70.3	70.57	78.5
C -6'''	18.5	18.6	18.80	62.1
	26-O-β-D-Glucose			
C -1''''	105.2			
C -2''''	75.3			
C -3''''	78.6			
C -4''''	71.7			
C -5''''	78.6			
C -6''''	62.8			

Contd...

Table 5.2–*Contd...* ^{13}C NMR spectral data of (aglycone part) saponins.

Carbons	*Shatavarin VI (10) Pyridine-d_5*	*Shatavarin VII (11) Pyridine-d_5*	*Shatavarin VIII (12) Pyridine-d_5*	*Shatavarin IX (13) Pyridine-d_5*	*Shatavarin X (14) Pyridine-d_5*
C-1	30.9	31.1	30.9	31.2	31.1
C-2	27.0	27.2	27.1	27.0	26.9
C-3	75.4	75.6	75.1	75.6	76.0
C-4	30.8	30.8	30.5	31.0	31.1
C-5	36.8	36.8	36.5	36.8	37.3
C-6	26.8	26.9	26.9	27.1	26.8
C-7	26.9	27.0	26.8	26.9	26.9
C-8	35.3	35.8	35.6	35.7	35.7
C-9	40.4	40.4	40.4	40.3	40.4
C-10	35.3	35.6	35.3	35.4	35.4
C-11	21.2	21.4	21.2	21.4	21.2
C-12	40.8	40.5	40.3	40.5	40.4
C-13	41.5	41.3	40.9	41.1	41.0
C-14	55.7	55.7	56.5	56.7	56.7
C-15	33.2	32.4	33.2	32.3	32.2
C-16	81.0	81.9	81.4	81.5	81.4
C-17	62.8	63.2	63.0	63.1	63.0
C-18	17.0	16.9	16.6	16.7	16.9
C-19	24.0	24.2	24.1	24.1	23.9
C-20	42.2	42.2	42.5	42.5	42.4
C-21	16.8	15.2	14.9	15.0	14.9
C-22	110.6	109.8	109.8	109.9	109.8
C-23	28.3	33.5	26.5	26.6	26.5
C-24	28.2	29.2	26.2	26.3	26.2
C-25	30.8	144.6	27.6	27.7	27.7
C-26	69.6	65.3	65.2	65.2	65.2
C-27	17.4	109.1	16.3	16.3	16.3

Contd...

comparable to that of the established antioxidants glutathione and ascorbic acid. Jayashree *et al.* (2000) showed that the crude as well as polysaccharide fraction, as well as inhibiting lipid peroxidation and protein oxidation, also protected superoxide dismutase and protein thiols from inactivation. The latter two components significantly contribute to the cellular antioxidant defense systems. The restoration of levels of these defenses also can partly explain the observed protection against lipid peroxidation and protein oxidation. The established antioxidants glutathione and vitamin C require 20–30 times more concentration for effective inhibition of TBARS formation comparable to that by *Asparagus racemosus* extracts.

Table 5.2–*Contd...*^{13}C NMR spectral data of saponins (Sugar Part) (δ Values).

Hydrogens	*Shatavarin VI (10) Pyridine-d_5*	*Shatavarin VII (11) Pyridine-d_5*	*Shatavarin VIII (12) Pyridine-d_5*	*Shatavarin IX (13) Pyridine-d_5*	*Shatavarin X (14) Pyridine-d_5*
	3-O-β-D-Glucose	*3-O-β-D-Glucose*	*3-O-β-D-Glucose*	*3-O-β-D-Glucose*	*3-O-β-D-Glucose*
C-1'	102.5	101.7	101.3	101.7	102.2
C-2'	82.9	82.1	80.6	81.4	76.8
C-3'	76.5	76.6	76.2	76.6	78.1
C-4'	77.4	77.6	79.6	81.4	83.6
C-5'	77.1	77.0	75.0	76.3	76.1
C-6'	61.4	61.3	68.3	62.0	62.2
	2'-O-β-D-Glucose	*2'-O-β-D-Glucose*	*2'-O-β-D-Glucose*	*2'-O-β-D-Glucose*	*4'-O-β-D-Glucose*
C-1''	105.7	105.4	105.2	105.5	105.7
C-2''	77.1	76.6	79.6	77.1	75.0
C-3''	78.0	77.7	78.4	77.9	78.1
C-4''	71.9	71.7	71.7	72.1	72.1
C-5''	78.6	78.5	78.7	78.6	75.2
C-6''	63.0	62.9	63.3	63.0	64.8
6-o-Ac					171.6 (C=O) 20.8 (Me)
	4'-O-α-L-Rhamnose	*4'-O-α-L-Rhamnose*	*4'-O-α-L-Arabinose*	*4'-O-β-D-Glucose*	*2'-O-α-L-Rhamnose*
C-1'''	101.9	102.5	105.2	105.0	101.9
C-2'''	72.6	72.4	72.6	75.1	72.4
C-3'''	72.8	72.6	74.7	78.5	74.2
C-4'''	74.0	73.9	72.3	71.6	74.2
C-5'''	70.3	70.3	67.8	78.5	69.7
C-6'''	18.5	18.5	–	62.3	18.9
			6'-O-β-D-Glucose		
C-1''''			105.1		
C-2''''			75.3		
C-3''''			78.0		
C-4''''			69.9		
C-5''''			78.5		
C-6''''			62.8		

Contd...

Table 5.2–*Contd...* ^{13}C **NMR spectral data of (aglycone part) saponins.**

Carbons	*Racemoside A (15) Pyridine-d_5*	*Racemoside A (16) Pyridine-d_5*	*Racemoside A (17) Pyridine-d_5*
C-1	31.3	31.2	31.2
C-2	26.7	26.7	26.8
C-3	76.2	75.6	75.4
C-4	31.0	32.5	30.4
C-5	37.1	37.4	37.4
C-6	27.3	27.3	27.3
C-7	27.1	27.1	27.1
C-8	35.6	35.6	35.9
C-9	40.6	40.6	40.7
C-10	35.9	35.9	35.6
C-11	21.5	21.5	21.5
C-12	40.6	40.7	40.6
C-13	41.2	41.2	41.2
C-14	56.8	56.8	56.8
C-15	32.5	32.5	32.5
C-16	81.9	81.7	81.7
C-17	63.3	63.3	63.3
C-18	16.9	16.9	17.0
C-19	24.3	24.2	24.2
C-20	42.8	42.8	42.8
C-21	16.6	16.6	16.7
C-22	110.0	110.1	110.1
C-23	27.0	27.1	27.2
C-24	26.5	26.5	26.6
C-25	27.9	27.9	27.9
C-26	65.4	65.4	65.4
C-27	15.3	15.2	15.3

Contd...

Gastrointestinal Effects

The powdered dried root of *Asparagus racemosus* is used in Ayurveda for dyspepsia. Oral administration of powdered dried root of *Asparagus racemosus* has been found to promote gastric emptying in healthy volunteers. Its action is reported to be comparable with that of the synthetic dopamine antagonist metoclopromide. In this study, 2 g powdered root of *Asparagus racemosus* was compared to a standard treatment of metoclopramide (10 mg tablet) in eight normal healthy male volunteers, and the gastric emptying halftime was observed. There was no statistically significant

Table 5.2–*Contd...*^{13}C NMR spectral data of (sugar part) saponins.

Carbons	Racemoside A (15) Pyridine-d_5	Racemoside B (16) Pyridine-d_5	Racemoside C (17) Pyridine-d_5
	3-O-α-D-Glucose	3-O-α-D-Glucose	3-O-β-D-Glucose
C-1'	102.2	103.5	103.6
C-2'	76.4	75.0	75.8
C-3'	80.5	76.9	77.1
C-4'	81.7	74.5	79.8
C-5'	75.2	75.5	75.9
C-6'	68.1	68.2	67.5
	4'-O-β-D-Glucose	6'-O-β-D-Glucose	4'-O-α-L-Rhamnose
C-1"	106.0	106.2	102.4
C-2"	72.6	72.6	72.7
C-3"	74.9	75.2	73.1
C-4"	69.9	69.9	74.4
C-5"	81.7	81.2	70.2
C-6"	63.3	67.6	19.1
	6'-O-β-D-Glucose	6"-O-α-L-Rhamnose	6'-O-α-L-Rhamnose
C-1"'	106.0	102.7	103.3
C-2"'	77.4	72.8	73.00
C-3"'	78.3	73.2	73.1
C-4"'	72.1	75.5	74.2
C-5"'	79.2	70.2	70.9
C-6"'	67.1	19.1	18.9
	6"'-O-α-Rhamnose		
C-1""	102.5		
C-2""	72.8		
C-3""	73.1		
C-4""	74.4		
C-5""	70.2		
C-6""	19.1		

Contd...

difference between the actions of *Asparagus racemosus* and metoclopramide. They hypothesized that shatavari might be a mild dopamine agonist (Dalvi *et al.*, 1990). *Asparagus racemosus* has also been mentioned for the treatment of ulcerative disorders

Table 5.2–*Contd...*^{13}C NMR spectral data of other compounds from *Asparagus racemosus.*

Carbons	*Asparagamine A (4) $CDCl_3$*	*Racemosol (5) $CDCl_3$*	*8-methoxy-5,6,4'-trihydroxyisoflavone-7-O-β-D-glucopyranoside (6) $CDCl_3$*	*Racemofuran (8) $CDCl_3$*
C-1	55.1	143.2	–	118.4
C-2	26.8	146.6	152.6	104.3
C-3	47.9	112.4	124.8	107.9
C-4	–	125.0 C-4 126.4 C-4a 116.8 C-4b	176.2	149.0
C-5	60.7	155.4	162.2	124.8
C-6	32.8	97.9	154.2	156.1
C-7	80.6	152.8	122.2	–
C-8	82.9	112.6 C-8 139.3 C-8a	114.7	–
C-9	112.7	25.7	154.3	–
C-10	148.3	22.1 C-10 130.6 C-10a	124.6	–
C-11	34.5	61.4	–	–
C-12	47.5	55.5	–	–
C-13	127.7	11.2	–	–
C-14	162.8	–	–	–
C-15	98.3	–	–	–
C-16	169.7	–	–	–
C-17	18.2	–	–	–
C-18	58.8	–	–	–
C-19	9.0	–	–	–
C-20	126.4	–	–	–
C-21	133.2	–	–	–
C-22	25.2	–	–	–
C-23	13.4	–	–	–
C-1'	–	–	112.6	128.4
C-2'	–	–	112.8	112.4 113.1, (2'- CH_3)
C-3'	–	–	115.8	153.0
C-4'	–	–	145.4	114.5 8.5, (4' -CH_3)
C-5'	–	–	98.1	156.0 55.8, (5'-O CH_3)
C-6'	–	–	152.4	103.3
C-1"	–	–	98.8	101.7
C-2"	–	–	72.6	154.9
C-3"	–	–	77.2	–
C-4"	–	–	68.6	–
C-5"	–	–	76.8	–
C-6"	–	–	66.2	–

of stomach and Parinama Sula, a clinical entity akin to the duodenal ulcer diseases. The juice of fresh root of *Asparagus racemosus* has been shown to have definite curative effect in patients of duodenal ulcers (Kishore *et al.* 1980). The ulcer protective effect of methanolic extract of fresh roots of *Asparagus racemosus* 25-100 mg/kg given orally, twice daily for 5 days, was studied on different gastroduodenal ulcer models. *Asparagus racemosus* 50 mg/kg, twice daily, orally (total saponins 0.9 per cent) showed significant protection against acute gastric ulcers induced by cold restraint stress (CRS), pyloric ligation, aspirin plus pyloric ligation, and duodenal ulcers induced by cysteamine. *Asparagus racemosus* in the above dose also significantly healed chronic gastric ulcers induced by acetic acid after 10 days treatment. However, *Asparagus racemosus* was ineffective against aspirin- and ethanol-induced gastric ulcers. Further, gastric juice and mucosal studies showed that *Asparagus racemosus* significantly increased the mucosal defensive factors like mucus secretion, cellular mucus, life span of cells and also possessed significant anti-oxidant effect, but had little or no effect on offensive factors like acid and pepsin (Sairam *et al.*, 2003). *Asparagus racemosus* along with *Terminalia chebula* are reported to protect gastric mucosa against pentagastrin and carbachol induced ulcers, by significantly reducing both severity of ulceration and ulcer index. Decreased volume and increased pH of the secretions in drug treated rats suggest a reduced responsiveness of the gastric parietal cells to secretogogues and narcotizing agents (Dahanukar *et al.*, 1983). Cytoprotective effect has been suggested to be due to increased output of mucus.

Singh *et al.* (1986) showed that Shatavari promptly and persistently relieves the pain and burning sensation as well as other dyspeptic symptoms due to duodenal ulcer. Since Shatavari did not have antacid and anti-secretory properties, the observed mild reduction in acid secretion may be due to some changes in gastric mucosa. Shatavari has been suggested to heal the ulcers by potentiating defensive factors and many hypotheses have been put forward for its possible mechanism.

1. It may prolong the life span of mucosal cells, increase the secretion and viscosity of mucus and strengthen the mucosal barrier and thus reduces H^+ ion back diffusion.
2. Shatavari may form a complex with mucus of other substances at the base of ulcer which may protect the ulcer from the corrosive and proteolytic effects of acid-pepsin.
3. It may have cytoprotective action like that of prostaglandins. In addition to antiulcerogenic activity of *Asparagus racemosus* in clinical trials, De *et al.* (1997) demonstrated similar effects of fresh root juice of *Asparagus racemosus* in rats, using cold stress and pyloric-ligation induced gastric ulcer. In contrast to previous report (Singh *et al.*, 1986) these workers suggested a reduction in acid and pepsin contents (aggressive factors) and increase in mucin-bicarbonate secretions and life span of the mucosal cells (defensive factors). Anti-ulcerogenic effect is suggested to be due to the regulation of the above two factors (Goeletal, 1991).
4. *Asparagus racemosus* has been shown to cause contraction of smooth muscles of rabbit's duodenum, guinea pig's ileum and rat's fundal strip without

affecting peristaltic movement. These actions were found to be similar to that of acetylcholine and were blocked by atropine, suggesting a cholinergic mechanism of action (Jetmalani *et al.*, 1967). However, no effect was observed on isolated rectus abdominus. Hence, the gastro duodenal ulcer protecting and gastric ulcer healing effect of *Asparagus racemosus* may be due its predominant effect on the mucosal defensive factors rather than offensive factors.

Galactogogue Effect

Extract of Shatavari has been shown to increase both the weight of mammary lobuloaveolar tissue and the milk yield. The root extract of *Asparagus racemosus* is prescribed in Ayurveda to increase milk secretion during lactation (Nadkarni, 1954). Joklekar *et al.* (1967) observed that *Asparagus racemosus* in combination with other herbal substance in the form of 'Ricalex' tablets (Aphali Pharmaceutical Ltd; 40 mg concentrated root extract per tablet) has been shown to increase milk production in female suffering from deficient milk secretion. Systemic administration of aqueous extract of *Asparagus racemosus* roots increased the weight of mammary glands in post-partum, oestrogen-primed rats, the uterine weight in the oestrogen-primed group and maintained milk secretion. This effect may be due to the action of released corticoids and prolactini Oral administration of alcoholic extract of *Asparagus racemosus* rhizome (30mg/100 g body weight, daily for 15 days) to adult pregnant female albino rats had an oestrogenic effect on the female mammary glands and genital organs (Pandey *et al.*, 2005).

Asparagus racemosus in the form of a commercial preparation, 'Lactare' (TTK, Pharmaceuticals; Chennai) is reported to enhance milk output in women complaining of scanty breast milk on 5th day after delivery (Sholapurkar, 1962). A significant increase in milk yield has also been observed in guinea pigs and goats after feeding 'Lactare' which also increased growth of the mammary glands, alveolar tissues and acini in guinea Pigs (Narendranath *et al.*, 1986). Patel *et al.* (1969) have shown galactogogue effect of roots of *Asparagus racemosus* in buffaloes. However, Sharma *et al.* (1996) conducted randomized controlled trials to evaluate the lactogogue effect of *Asparagus racemosus* in lactational deficiency among women who had delivered at term without complications. Each 100g dose of medicine contained 15g *Aspargus racemosus* root extract. After 4-week treatment, they did not observe any increase in prolactin levels in females complaining of secondary lactational failure, suggesting that *Asparagus racemosus* has no lactogenic effect.

Effects on Uterus

Inspite of cholinergic activity of *Asparagus racemosus* on guinea Pig's ileum, ethyl acetate and acetone extracts of the root of *Asparagus racemosus* blocked spontaneous motility of the virgin rat's uterus (Jetmalani *et al.*, 1967). These extracts also inhibited contraction, induced by spasmogen like acetylcholine, barium chloride and 5-hydroxytryptamine whereas alcoholic extract was found to produce a specific block of pitocin induced contractions. On the other hand petroleum ether as well as ether extracts of the powdered roots did not produce any uterine activity. It indicates the presence of some particular substances in the alcoholic extract which specifically blocks pitocin sensitive receptors though not other receptors in the uterus, confirming that shatavari can be used as uterine sedative. 'U-3107' a herbal preparation (formulated by the Himalayia Drug Co., Banglore) is used to treat various menstrual disoders and threatened abortion. Administration of 'U-3107' in normal rats increased wet and dry uterine weights and also resulted in marked increase in osterogen levels with no change in progesterone levels as compared to control. The primary changes in uterine tissues are controlled by oestrogen and progesterone. The oestrogenic effect in this case was observed only in the presence of functional ovaries indicating that the formulation *per se* dose not possess any oestrogenic activity. The effect is only evident in cases where the ovaries are functional. The rats from both controlled and treated group showed normal oestrous cycle (Mitra *et al.*, 1999). Further, a glycoside, Shatavarin-I, isolated from the root of *Asparagus racemosus* has been found to be responsible for the competitive block of oxytocin-induced contraction of rat, guinea pig and rabbit's uteri, *in vitro* as well as *in vivo* (Gaitonde and jetmalani, 1969). Nevrekar *et al.* (2002) observed that administration of *Asparagus racemosus* in the form of 'EveCare' capsules is effective in the treatment of women suffering from of dysfunctional uterine bleeding (DUB). Seventy women in the age group of 20-45 years with DUB were included in this study. They found 63 women out of seventy had achieved a regularized menstrual cycle. Another study conducted by Swarup and Umadevi (1998) found that, a group of 40 patients suffering from dysmenorrhoea and pre-menstrual syndrome (PMS) were symptom free after treatment with 'EveCare'. Dhaliwal (2003) patented that a drug prepared from *Asparagus racemosus* was found to be effective in the treatment of PSM in human females who experience adverse system.

Adaptogenic Effects

Asparagus racemosus is one of the six rasayana plants from Ayurveda, which were studied for their adaptogenic potential. The whole, aqueous, standardized extracts of selected plants (*Tinospora cordifolia, Asparagus racemosus, Emblica officinalis, Withania somnifera, Piper longum* and *Terminalia chebula*) were administered orally to experimental animals, in a dose extrapolated from the human dose, after which they were exposed to a variety of biological, physical and chemical stressors. The plant extracts were found to protect against the stressors, as measured by markers of stress responses and objective parameters for stress manifestations. Using a model of cisplatin induced alterations in gastrointestinal motility; the ability of the plants to exert a normalizing effect, irrespective of direction of pathological change was tested. All the plants reversed the effects of cisplatin on gastric emptying, while *Asparagus racemosus* also normalized cisplatin-induced intestinal hypermotility. All the plant drugs were found to be safe in both acute and subacute toxicity studies. Studies on the mechanisms of action of the plants revealed that they all produced immunostimulation (Rege *et al.*, 1999). EuMil, a polyherbal formulation consisting of standardised extracts of *Withania somnifera* (L) Dunal, *Ocimum sanctum* L, *Asparagus racemosus* Willd. and *Emblica officinalis* Gaertn., is used as an anti-stress agent to attenuate the various aspects of stress related disorders. Anti-stress activity of EuMil were evaluated by measuring the rat brain monoamine neurotransmitter levels and tribulin activity. EuMil (100 mg/kg, p.o., 14 days) treatment normalized the perturbed regional NA, DA, 5HT concentrations, induced by chronic stress. EuMil also significantly attenuated the stress-induced increase in the rat brain tribulin activity. The amelioration of chronic stress-induced neurochemical perturbations by EuMil explains the neurochemical mechanisms underlying the observed putative anti-stress activity of the product. Latter Bhattacharya *et al.* (2004) undertook a study to investigate the adaptogenic activity of 'Siotone' (a herbal formulation consisting of *Withania somnifera, Ocimum sanctum, Asparagus racemosus, Tribulus terristris* and shilajit) against chronic unpredictable, but mild, foot shock stress induced perturbations in the behaviour (depression), glucose metabolism, suppressed male sexual behaviour, immunosuppression and cognitive dysfunction in albino rats. The stress indices for evaluation were gastric ulceration, adrenal gland and spleen weights, ascorbic acid and corticosterone concentration of adrenal cortex and plasma corticosterone levels. Powdered roots of Panax ginseng, a reputed 'rasayana' herb was used as the standard adaptogenic agent for comparison. The results showed that 'Siotone' had significant ($p<0.05$) adaptogenic activity in that it was able to reverse chronic stress-induced biochemical, physiological and behavioural perturbations and was qualitatively comparable to Panax ginseng. Both the 'Siotone' and 'Eumil'were found to devoid of any toxic effect against the acute and subacute toxicity studies and had identical effects on similar types of stresses, it may be because both the formulations contain *Withania somnifera, Ocimum sanctum* and *Asparagus racemosus.*

Dhingra (2007) undertook a study to investigate the effect of *Asparagus racemosus,* on depression in mice using tail suppression test (TST) and forced swim test (FST). Oral administration of methanolic extract (50, 100 and 200 mg/kg p.o) for 14 successive days decreased immobility periods significantly in a dose-dependent manner in both the TST and FST, indicating significant anti-depressant activity.The efficacies of the extract were found to be comparable to fluoxetine and imipramine in both TST and FST. Methanolic extract (200 mg/kg) did not show any significant effect on locomotor effect in mice. Pretreatment of animals with sulpiride (50 mg/kg) or parzosin (62.5mg/kg) or p-CPA (100 mg/kg) or baclofen (10 mg/kg) significantly blocked the decrease of immobility time elicited by the extract. Methanolic extract administrated for 14 successive days to mice significantly decrease MAO-A and MAO-B activities levels as compare to control. Therefore anti-depressant-like activity of *Asparagus racemosus* was probably by inhibiting MAO-A and MAO-B; through interaction with adrenergic, dopaminergic, serotonergic and GABAergic system.

'Shatavari mandur'(root extract of *Asparagus racemosus* as major ingredient) an Ayurvedic herbo-formulation was investigated for its efficacy in the treatment of cold restrain stress-induced gastric ulcer in rats. It showed significant ($p<0.05$) protection against pyloric ligation induced gastric ulcer due to the involvement of mucosal defensive factors (like mucosal secretion) as against the involvement of offensive factors(such as acid and pepsin secretion) during the treatment of ulcers with the allopathic drug Ranitidine (Datta *et al.,* 2002).

Immunomological Activities

Immune system from a state of immune deficiency to one of more normal function would likely to have a significant impact on disease and the patient it affects. Such agents would not be a cure, but it would control the manifestation and course of disease (Gottlieb *et al.*, 1987). Ayurveda (with particular reference to plants) may play an important role in modern health care, particularly where satisfactory treatment is not available. *Asparagus racemosus* is believed to be useful in strengthening the immune system of an individual (Patwardhan *et al.*, 1990). The immunoadjuvant potential of *Asparagus racemosus* was evaluated in experimental animals immunized with diphtheria, tetanus, pertussis (DTP) vaccine. Oral administration of root extract of *A. racemosus* (100 mg/kg body weight) has resulted in significant increase (P = 0.0052) in antibody titers to *Bordtella pertussis* as compared to untreated (control) animals. Reduced mortality accompanied with overall improved health status observed in treated animals after intra-cerebral challenge of *B. pertussis* indicated development of protective immune response (Gautam *et al.*, 2004). *Asparagus racemosus* extract was evaluated for neuroendocrine immune modulating effect. It prevents stress-induced increase in plasma cortisol along with an activation of peritoneal macrophages and inhibition in gastric vascular damage (Dahanukar and Thatte, 1988). A significant decrease was observed in the intraperitoneal adhesion scores induced by caecal rubbing in animals receiving *Asparagus racemosus* providing a novel approach for the prevention and management of post-operative adhesions (Rege *et al.*, 1989). The protective effects of *Asparagus racemosus* and *Tinospora cordifolia* against myelosuppression induced by single doses of CP have been previously reported. Leukopenia produced by CP was prevented, to varying degrees suggesting potent immunostimulatory activity (Thatte and Dahanukar,1988). The crude saponins obtained from asparagus shoots (edible part of asparagus) were found to have antitumor activity. It inhibited the growth of human leukemia HL-60 cells in culture and macromolecular synthesis in a dose and time-dependent manner (Shao *et al.*, 1996). Intra-abdominal sepsis are major causes of mortality following trauma and bowel surgery. Immunomodulating property of *Asparagus racemosus* has been shown to protect the rat and mice against experimental induced abdominal sepsis. Oral administration of decoction of powdered root of *Asparagus racemosus* has been reported to produce leucocytosis and predominant neutrophilia along with enhanced phagocytic activity of the macrophages and polymorphs. Percentage mortality of *Asparagus racemosus* treated animals was found to be significantly reduced while survival rate was comparable to that of the group treated with a combination of metronidazole and gentamicin (Dahanukar *et al.*, 1986; Thatte *et al.*, 1987). Since *Asparagus racemosus* is reported to be devoid of antibacterial action, so protection offered by *Asparagus racemosus* against sepsis by altering function of macrophages, indicates its possible immunomodulatory property. Dhuley (1997) has reported the revival of macrophage chemotaxis and interleukin-I (IL-I) and tumor necrosis factor a (TNFα) production by the oral treatment of *Asparagus racemosus* root extract in ochratoxin A treated mice. Alcoholic extract has been found to enhance both, humoral and cell mediated immunity of albino mice injected with sheep red blood cells as particulate antigen (Muruganadan *et al.*, 2000). The role of *Asparagus racemosus* as an immunoadjuvant in traditional therapy is well documented and therefore it can be applied the toxic side effects of synthetic chemotherapeutic drugs without compromising its anti-tumour activity. Intrestingly, in Ayurvedic medicine, AIDS is thought to be a disease of decreased 'ojas', defined as the essential energy of the body.

Shatavari is said to aid in the formation of 'ojas' and has been used in immune therapy (Canadian AIDS treatment Information Exchange, 2005). It is in the situations like these that the function of *Asparagus racemosus* as an immunoadjuvant can be scrutinized for use in adjuvant therapy in the management of HIV.

Cardiovascular Effects

A herbo-mineral formulation 'Abana' (10 mg *Asparagus racemosus* extract per tablet) has shown significant hypocholesterolaemic effects in rats. It was found that the use of drug significantly lower (37-45 per cent) of the total cholesterol, phospholipids and triglyceride levels as against control and therefore demonstrated a potential for use as cardio-protective agent (Khanna *et al.*, 1991). Since 'Abana' is a polyherbal formulation, further research needs to be conducted on the exact role that the *Asparagus racemosus* component plays in the hypolipidaemic action.

Alcoholic extract of the root of *Asparagus racemosus* has been reported to produce positive ionotropic and chronotropic effect on frog's heart with lower doses and cardiac arrest with higher doses. The extract was found to produce hypotension in cats which was blocked by atropine, indicating cholinergic mechanism of action. The extract also produced congestion and complete stasis of blood flow in mesentric vessels of mice and rat. Slight increase in the bleeding time and no effect on clotting time was observed on I.V. administration of the extract in rabbits (Roy *et al.*, 1971).

Another study conducted by Visavadiya and Narasimhacharya (2005) found that powered root supplements of *Asparagus racemosus* decreased lipid peroxidation and caused a dose-dependent reduction in lipid profiles. The total lipids, total cholesterol and triglycerides in plasma low-density lipoprotein (LDL) and very low-density lipoprotein(VLDL) cholesterol decreased by more than 30 per cent.

Neuroprotective Effect

Parihar and Hemnani (2004) investigated the potential of extract of *Asparagus racemosus* against kainic acid (KA)-induced hippocampal and striatal neuronal damage. Intra-hippocampal and intra-striatal injections of kainic acid (KA; 0.25 microg in a volume of 0.5 microl) to ketamine and xylazine (200 and 2 mg/kg b.w. respectively) anesthetized mice has resulted in the production of excitotoxic lesions in the brain. After KA injection, impairment of hippocampus and striatal regions of brain was obsevered accompanied by increase in lipid peroxidation, increase in protein carbonyl content, decline in glutathione peroxidase (GPx) activity and reduced glutathione (GSH) content. The *Asparagus racemosus* extract supplemented mice displayed an improvement in GPx activity and GSH content and reduction in membranal lipid peroxidation and protein carbonyl. They concluded that the minimizing effect of *Asparagus racemosus* extract on oxidative damage in addition to the elevation of GPx activity and GSH content could eventually result in protective effect on the KA-induced excitotoxicity.

Neither stimulant nor depressant action of 'Lactare' on central nervous system has been reported in albino mice (Narendranath *et al.*, 1986). Shatavari did not produce catalepsy in experimental rats even with massive oral doses suggesting that its action may be outside the blood-brain barrier, similar to that of metoclopromide (Davi *et al.*, 1990).

A polyherbal preparation 'EuMil' combination of standardized extracts of *Asparagus racemosus, Withania somnifera, Ocimum sanctum* and *Emblica officinalis* was evaluated for its anti-stress activity in rats. Chronic electroshocks stress for 14 days was found to increase the rat brain tribulin activity and decrease the monoamine neurotransmitter levels. 'EuMil' treatment normalized the perturbed nor-adrenalin, dopamine and 5-hydroxytryptamine concentrations and also attenuated the tribulin activity (Bhattacharya *et al.*, 2002).

'Mentat' a polyherbal psychotropic preparation containg *Asparagus racemosus* has been found to be effective in the treatment of alcohol abstinence induced withdrawal symptoms such as tremors, convulsions, hallucinations and anxiety in ethanol administrated rats due to its anticonvulsant and anxiogenic actions. However, it is likely that these are the only reasons for its de-addiction potential and can be examined further. Keeping in mind the encouraging leads and the limited data regarding the use of *Asparagus racemosus* in treating neurological disorders, more studies need to be conducted to fully exploit the potential of shatavari in the area.

Antioxidant Activity

The antioxidant effects of crude extract and a purified aqueous fraction of *Asparagus racemosus* against membrane damage induced by the free radicals generated during γ-radiation were examined in rat liver mitochondria. γ-Radiation, in the dose range of 75–900 Gy induced lipid peroxidation as assessed by the formation of thiobarbituric acid reactive substances (TBARS) and lipid hydroperoxides (LOOH). The crude and purified aqueous extract of *Asparagus racemosus* (effective dose of 450 Gy) were studied against oxidative damage in terms of protection against lipid peroxidation, protein oxidation, depletion of protein thiols and the levels of the antioxidant enzyme, superoxide dismutase. An active fraction consisting of polysaccharides (termed as P3) was effective even at a low concentration of 10 μg/ml. Both the crude extract as well as the P3 fraction significantly inhibited lipid peroxidation and protein oxidation. The antioxidant effect of P3 fraction was more pronounced against lipid peroxidation, as assessed by TBARS formation, while that of the crude extract was more effective in inhibiting protein oxidation. Both the crude extract and P3 fraction also partly protects against radiation-induced loss of protein thiols and inactivation of superoxide dismutase. The inhibitory effects of these active principles, at the concentration of 10 μg/ml, are comparable to that of the established antioxidants glutathione and ascorbic acid. The results indicate that extracts from *Asparagus racemosus* have potent antioxidant properties *in vitro* in mitochondrial membranes of rat liver (Kamat *et al.*, 2000).

Anti-lithic Effect

Shatavari has been shown to reduce urinary stone formation in rats. An ethanol extract of shatavari was evaluated for its inhibitory potential on lithiasis (stone formation), induced by oral administration of 0.75 per cent ethylene glycolated water to adult male albino Wistar rats for 28 days. The ionic chemistry of urine was altered by ethylene glycol, which elevated the urinary concentration of calcium, oxalate and phosphate ions, thereby leading to renal stone formation. The ethanol extract, however, significantly ($p < 0.05$) reduced the elevated level of these ions in urine. Furthermore, shatavari elevated the urinary concentration of magnesium, which is considered to be one of the inhibitors of crystallization. The high serum creatinine level observed in the ethylene glycol-treated rats was also reduced following treatment with the shatavari extract. The histopathological findings also showed signs of improvement after treatment with the extract (Christina *et al.*, 2005).

Antihepatotoxic Activity

Alcoholic extract of root of *Asparagus racemosus* has been shown to significantly reduce the enhanced levels of alanine transaminase, aspartate transaminase and alkaline phosphatase in CCl_4-induced hepatic damage in rats (Muruganadan *et al.*, 2000), indicating antihepatotoxic potential of *Asparagus racemosus.*

Antibactarial Activity

Different concentrations (50, 100, 150 mcg/mL) of the methanol extract of the roots of *Asparagus racemosus* showed considerable *in vitro* antibacterial efficacy against *Escherichia coli, Shigella dysenteriae, Shigella sonnei, Shigella flexneri, Vibrio cholerae, Salmonella typhi, Salmonella typhimurium, Pseudomonas putida, Bacillus subtilis* and *Staphylococcus aureus.* The effects produced by the methanol extract were compared with chloramphenicol (Mandal *et al.*, 2000a). The antimicrobial activity may be due to 9, 10-Dihydrophenanthrene (Boger *et al.*, 1985). The antibacterial effect of *Asparagus racemosus* may also be playing a secondary action with respect to other functions of the plant as well and therefore needs to be studied in greater details.

Antineoplastic Activity

Asparagus racemosus is well known for its phytoestrogenic properties and use as a hormone modulator (Mayo, 1998). Rao (1981) demonstrated the inhibitory action of *Asparagus racemosus* on DMBA-induced mammary carcinogenesis in rats. Chloroform/methanol (1:1) extract of fresh root of *A. racemosus* has been reported to reduce the tumor incidence in female rats treated with DMBA (7,12 dimethyl benz (a) anthracene). This action is suggested to be mediated by virtue of mammotropic and/or lactogenic influence of *Asparagus racemosus* on normal as well as estrogen- primed animals, which renders the mammary epithelium refractory to the carcinogen.

Antitussive Activity

In a isolated study, methanol extract of *Asparagus racemosus* root (200 and 400 mg/kg, p.o.) showed significant antitussive activity on sulfur dioxide-induced cough in mice, the cough inhibition (40.0 and 58.5 per cent, respectively) being comparable to that of 10-20 mg/kg of codeine phosphate (36.0 and 55.4 per cent, respectively) (Mandel *et al.*, 2000b).

Antidiabetic Effects

Asparagus racemosus root has previously been reported to reduce blood glucose in rats and rabbits. The ethanol extract and each of the hexane, chloroformand ethyl acetate partition fractions concentration-dependently stimulated insulin secretion in isolated perfused rat pancreas, isolated rat islet cells and clonal b-cells. The stimulatory effects of the ethanol extract, hexane, chloroform and ethyl acetate partition fractions were potentiated by glucose, 3-isobutyl-1-methyl xanthine IBMX, tolbutamide and depolarizing concentration of KCl. Inhibition of A. racemosus-induced insulin release was observed with diazoxide and verapamil. Ethanol extract and five fractions increased intracellular Ca_2C, consistent with the observed abolition of insulin secretory effects under Ca_2C-free conditions. These findings reveal that constituents of *Asparagus racemosus* root extracts have wide-ranging stimulatory effects on physiological insulinotropic pathways (Hannan *et al.*, 2007)

Effects on Respiratory System

Alcoholic extract of root of *Asparagus racemosus* at higher doses are reported to cause dilatory effect on bronchial musculature of guinea pigs but failed to antagonise the histamine induced broncho-constriction. The extract has also been reported to produce depression of respiration in cat (Roy *et al.*, 1971).

Toxic Effects

In Ayurveda, *Asparagus racemosus* has been described as absolutely safe for long term use, even during pregnancy and lactation. Systemic administration of higher doses of all the extracts did not produce any abnormality in behaviour pattern of mice and rat (Jetmalani *et al.*, 1967) LD[50] of the product lactare has not been assessed since it did not produce mortality even up to the oral dosages of 64 g/kg.

Teratogenicity

Methanolic extract of shatavari roots (1000 mg/kg/day for 60 days) showed teratological effects such as increased resorption of foetuses and gross malformations *e.g.* swelling in legs and intrauterine growth retardation with a small placental size in Charles Foster rats. Pups born to a mother exposed to shatavari roots for the full duration of gestation showed evidence of higher rate of resorption and therefore smaller litter size. The live pup showed significant decrease in body weight and length and delay of various developmental parameters when compared to respective control groups (Goel *et al.*, 2006). The dosage used would be equivalent to around 70 g or 140 ml of a 1:2 liquid extract per day for a human adult. This is more than 16 times the recommended dosage. Another study examining the teatrogeniciy of a formulation containing shatavari did not find any signs of teatrogenicity with the dosage used. Rats received 3 g/kg of the product once a day orally from day 0 of gestation till day 21 of lactation or just during gestation. No adverse effects were observed on pregnancy and growth of the foetus. The daily dose of shatavari was about 4 g (similar to the recommended dosage of shatavari 1:2 extract). However, shatavari should nevertheless, be used with caution in pregnancy.

Conclusions

Asparagus racemosus, commonly known as Shatavari has significant medicinal properties and forms the essential component of established Ayurvedic drugs. Traditionally this plant is used as a reproductive tonic, anthelmintic, for treating gonorrhea, piles, diabetes, rheumatism, cough, diarrhea, dysentery, gastric troubles and headache and also for increasing lactation. Due to its traditional usage and the fact that it contains steroidal saponins, shatavari is fast becoming a popular alternative to false unicorn root. Shatavari is used by Ayurvedic doctors for the prevention and treatment of gastric ulcers, dyspepsia and as a galactogogue. *Asparagus racemosus* has also been used successfully by some Ayurvedic practitioners for nervous disorders, inflammation, liver diseases and certain infectious diseases. However, few human studies justifying these uses of the root extract of *Asparagus racemosus* is available so far. Since most drugs containg Shatavari available in the market are in the form of polyherbal formulations, it is difficult to attribute medicinal action solely due to *Asparagus racemosus* component of drug. The pharmacological studies conducted on *Asparagus racemosus* indicate immense potential of the this plant in the treatment of various conditions although considerable work has to be done to exploit the biological activity and medicinal applications of this plant, countless possibilities for investigation remain in relatively newer areas of its function. The Western world has now to accept these traditional treatments after analyzing the chemistry of this plant. Here the most important thing to notice is that we have many traditional herbs and therapies which can cure many incurable diseases, but these traditional medicines are vanishing very rapidly since we do not have any serious program for promoting these medicines.

References

Ahmad, S. and Jain, P.C. (1991). Chemical examination of Shatavari (*Asparagus racemosus*). *Bulletin of Medical and Ethnobotanical Research*, **12**: 157–160.

Aswal, B.S., Goel, A.K., Kulshrestha, D.K., Melhotra, B.N. and Patnaik, G.K.(1996). Screening of Indian plants for biological activity.Part XV. *Indian journal of experimental Biology*, **34**: 444-467.

Aranda-Michel, J. and Gianella, R.A. (1999). Acute diarrhoea: A practical review. *American journal of Medicine*, **106**: 670-676.

Anonymous (1987). The Wealth of India, Raw materials. Publication and Information Directorate, CSIR, N. Delhi, pp. 468–472.

Boger, D.L., Mitscher, L. A., Mullican, M. D., Drake, S. D. and Kitos, P. (1985). Antimicrobial and cytotoxic properties of 9,10- dihydrophenanthrenes: structure-activity studies on juncusol. *Journal Medicine Chemistry*, **28**: 1543-1547.

Bhattacharya, A., Murugandam, A.V., Kumar, V. and Bhattacharya, S.K. (2002). Effect of polyherbal formulation, EuMil, on neurochemical perturbations induced by chronic stress. *Indian Journal of Experimental Biology*, **40** : 1161–1163.

Bhattacharya, S.K., Bhattacharya, A. and Chakrabarti, A. (2004). Adaptogenic activity of Siotone, a polyherbal formulation of Ayurvedic rasayanas. *Indian Journal of Experimental Biology*, **38**: 119–128.

Canadian AIDS Treatment Information Exchange.(2005). A Practical Guide to Herbal Therapies for people living with HIV. <http: //www.catie.ca/herbe.nsf/TOC/701292B81D86D0118525697A00767262?OpenDocument>.

Christina, A.J., Ashok, K., Packialakshmi, M., Tobin, G.C., Preethi, J. and Murugesh, N.(2005). Antilithiatic effect of *Asparagus racemosus* Willd on ethylene glycol-induced lithiasis in male albino Wistar rats. *Methods Find Exp Clin Pharmacol.*, **27**: 633-638.

Dahanukar, S.A., Date, S.G. and Karandikar, S.M. (1983).Cytoprotective effect of *Terminalia chebula* and *Asparagus racemosus* on gastric mucosa. *Indian Drugs*, **21**: 442-445.

Dahanukar, S., Thatte, U., Pai, N., Mose, P.B. and Karandikar, S.M. (1986). Protective effect of *Asparagus racemosus* against induced abdominal sepsis. *Indian Drugs*, **24**: 125-128.

Dalvi, S.S., Nadkarni, P.M. and Gupta, K.C. (1990). Effect of *Asparagus racemosus* (Shatavari) on gastric emptying time in normal healthy volunteers. *Journal of Postgraduate Medicine*, **36**: 91–94.

De, B., Maiti, R.N., Joshi, V.K., Agrawal, V.K. and Goel, R.K. (1997). Effect of some Sitavirya drugs on gastric secretion and ulceration. *Indian Journal of Experimental Biology*, **35**: 1084-1087.

Dhuley, J.N. (1997). Effect of some Indian herbs on macrophage functions inochratoxin A treated mice. *Journal of Ethnopharmacology*, **58**: 15–20.

Dev, S. (1999). Ancient-Modern concordance in Ayurvedic plants. *Environ Health Perspect*, **107**: 783-789.

Datta, G.K., Sairam, K., Priyambada, S., Debnath, P.K. and Goel, R.K. (2002). Antiulcerogenic activity of Satavari mandur–an ayurvedic herbo-mineral preparation. *Indian Journal of Experimental Biology*, **40**: 1173–1177.

Dhaliwal, K.S. (2003). Method and composition for treatment of premenstrual syndrome in women. US Patent number 698662.

Dhingra, Dinesh. and Kumar, V. (2007). Pharmacological evaluation for antidepressant-like activity of *Asparagus racemosus* Willd. In mice. *Pharmacologyonline*, **3**: 133-152.

Esterbauer, H.(1996). Estimation of peroxidative damage: a critical review. *Pathologie Biologie Paris*, **44** : 25–28.

Fontaine, O. (1988). Diarrhoea and treatment. *Lancet*, **28**: 1234-1235.

Gaitonde, B.B. and Jetmalani, M.H. (1969). Antioxycytocic action of saponin isolated from *Asparagus racemosus* Willd. (Shatavari) on uterine muscle. *Archives Internationales de Pharmacodynamie et de Therapie*, **179**: 121–129.

Gottlieb, A.A., Gottlieb, M.S. and Scholes, V.E. (1987). Reconstitution of immune function in AIDS/ARC. *Concepts in Immunopathology*, **4**: 261–274.

Goel, R.K. and Bhattacharya, S.K. (1991) Gastroduodenal mucosal defense and mucosal protective agents. *Indian Journal of Experimental Biology*, **29**: 701-714.

Giusti, A.M., Raimondi, M., Ravagnan, G., Sapora, O. and Parasassi, T. (1998). Human cell membrane oxidative damage induced by single and fractionated doses of ionizing radiation: a fluorescence spectroscopic study. *International Journal of Radiation Biology,* **74** : 595–605.

Gautam, M., Diwanay, S., Gairola, S., Shinde, Y., Patki, P. and Patwardhan, B.(2004). Immunoadjuvant potential of *Asparagus racemosus* aqueous extract in experimental system. *Journal of Ethnopharmacology,* **91**: 251–255.

Goel, R.K., Prabha, T., Kumar, M.M., Dorababu, M., Prakash. and Singh, G. (2006). Teratogenicity of *Asparagus racemosus* Willd. root, a herbal medicine. *Indian Journal of Experimental Biology,* **44**: 570-573.

Handa, S.S., Suri, O.P., Gupta, V.N., Suri, K.A., Satti, N.K., Bhardwaj,V., Bedi, K.L., Khajuria, A., Kaul, A., Parikh, K., Kulhe, P. and Krishnamurthy, R. (2005). Isolation of immunoside(3-O-a-L-rhamnopyranosyl-(1!2)-a-L-rhamnopyranosyl-(1→4)-3-b-D-glucopyranosyl-25(S)-spirostan-3-b-ol) and its use as immunomodulator. *PCT Int. Appl.,* 16.

Hannan, J. M. A., Marenah, L., Ali, Liaquat., Rokeya, B., Flatt, P. R. and Wahab, Y. H. A. (2007) Insulin secretory actions of extracts of *Asparagus racemosus* root inperfused pancreas, isolated islets and clonal pancreatic *b*-cells. *Journal of Endocrinology,* **192**: 159–168.

Jetmalani, M.H., Sabins, P.B. and Gaitonde, B.B. (1967). A study on the pharmacology of various extracts of Shatavari- *Asparagus racemosus* (Willd). *J Res Ind Med.,* **2**: 1-10.

Joglekar, G.V., Ahuja, R.H. and Balwani, J. H. (1967). Galactogogue effect of *Asparagus racemosus*. Preliminary communication. *Indian Medical Journal,* **61**: 165.

Joglekar, G.V., Ahuja, R.H. and Balwani, J.H. (1967). Galactogogue effect of *Asparagus racemosus*. *Indian Medical Journal,* **61**: 165.

Joshi, J. and Sukh, D.(1988). Chemistry of Ayurvedic crude drugs: Part VIII-Shatavari-2: structure elucidation of bioactive shatavarin-I and other glycosides. *Indian Journal Chemistry,* **27B**: 12–16.

Jayashree, P. K., Krutin, K. B., Thomas P. A. D. and Venkatachalam, S. R.(2000). Antioxidant properties of *Asparagus racemosus* against damage induced by γ-radiation in rat liver mitochondria. *Journal of Ethnopharmacology,* **71** : 425-435.

Jadhav, A.N. and Bhutani, K.K. (2006). Steroidal saponins from the roots of *Asparagus adscendens* Roxb and *Asparagus racemosus* Willd. *Indian Journal Chemistry,* **45B (6)** : 1515–1524.

Kishore, P., Pandey, P.N., Pandey, S.N. and Dash, S. (1980) Treatment of duodenal ulcer with *Asparagus racemosus* Linn. *J Res Indian Med Yog Homeo.,* **15** : 91-94.

Khanna, A.K., Chander, R. and Kapoor, N.K. (1991). Hypolipidaemic activity of *Abana* in rats. *Fitoterapia* (Italy), **62**: 271–275.

Kamat, J.P., Boloor, K.K., Devasagayam, T.P.A. and Venkatachalam, S.R.(2000). Antioxidant properties of *Asparagus racemosus* against damage induced by radiation in rat liver mitochondria. *Journal of Ethnopharmacology,* **71**: 425–435.

Landge, A. B. and Bose, J. L. (1970). Structure of disaccharide from *Asparagus racemosus* (Shatamuli). *Indian Journal of Chemistry*, **8** : 588-589.

Mandal S.C., Nandy, A., Pal, M. and Saha, B. P. (2000). Evaluation of antibacterial activity of *Asparagus racemosus* willd. root. *Phytother Research*, **14**: 118-119.

Mandal, S.C., Kumar, C.K.A., Lakshmi, M., Sinha, S., Murugesan, T., Saha, B.P. and Pal, M. (2000b). Antitussive effect of *Asparagus racemosus* root against sulphur dioxide-induced cough in mice. *Fitoterapia*, **71**: 686–689.

Maulik, G., Maulik, N., Bhandari, V., Kagan, V.E., Pakrashi, S. and Das, D.K. (1997). Evaluation of antioxidant effectiveness of a few herbal plants. *Free Radical Research*, **27**: 221–228.

Mayo, J.L. (1998). Black cohosh and chasteberry: herbs valued by women for centuries. *Clinical Nutrition Insights*, **6**: 1–4.

Mitra, S.K., Gopumadhavan, S., Venkataranganna, M.V., Sarma, D.N.K., Anturlikar,

Muruganadan, S., Garg, H., Lal, J., Chandra, S. and Kumar, D. (2000) Studies on the immunostimulant and antihepatotoxic activities of *Asparagus racemosus* root extract. *J Med Arom Pl Sci.*, **22**: 49-52.

Mitra, S.K.,Gopumadhavan,S., Venkataranganna, M.V., Sarma, D.N.K. and Anturlikar, S.D. (1999). Uterine tonic activity of herbal preparation in rats. *Indian Journal of Pharmacology*, **31**: 200–203.

Mandal, D., Banerjee, S., Mondal, N. B., Chakravarty, A. K. and Sahu, N. P. (2006). Steroidal saponins from the fruits of *Asparagus racemosus*. *Phytochemistry*, **67** : 1316–1321.

Nadkarni, A.K. (1954). Indian Materia Medica. Bombay: Popular Book Depot. **I** : 153-155.

Narendranath, K.A., Mahalingam, S., Anuradha, V. and Rao, I.S. (1986). Effect of herbal galactogogue (Lactare) a pharmacological and clinical observation. *Med Surg.*, **26**: 19-22.

Newman, D. J., Cragg, G. M. and Snader, K.M. (2000). The influence of natural products upon drug discovery. *Natural Product Reports*, 17-125.

Nevrekar, P., Bai, N. and Khanna, S. (2002). Eve Care capsules in DUB. *Obstetrics and Gynaecology Communications*, **3** : 51–53.

Odin, A.P. (1997). Vitamins as antimutagens: advantages and some possible mechanisms of antimutagenic action. *Mutation Research*, 386.

Oketah-Rabah, H.A. (1998). Phytochemical constituents of the genus *Asparagus* and their biological activites. *Hamdard*, **41**: 33-43.

Patwardhan, B. K., Kalbag, D., Patki, P.S. and Nagsampagi, A. (1990). Search of immunomodulatory agents: a review. *Indian Drugs*, **28**: 56–63.

Paiwal, M.K., Siddiqui, I.R., Singh, J. and Tiwari, H.P. (1991). Chemical examination of shatavari (*Asparagus racemosus*). *Journal of Indian Chemical Society*, **68**: 427-428.

Packer, L., Ong, A.S.H. (1998). *Biological Oxidants and Antioxidants: Molecular Mechanisms and Health Effects*, AOCS Press, Champaign, Illinois.

Park, K.(2000). Park's Text book of Preventive and Social Medicine. Banarsidas Bharat Publishers, Jabalpur, 122-175.

Parihar, M.S. and Hemnani, T. (2004). Experimental excitotoxicity provokes oxidative damage in mice brain and attenuation by extract of *Asparagus racemosus*. *Journal of Neural Transmission*, **111**: 1–12.

Pandey, S. K., Sahay, A R., Pandey, S. and Tripathi,Y. B.(2005). Effect of *Asparagus racemosus* rhizome (Shatavari) on mammary gland and genital organs of pregnant rat. *Phytotherapy Research*, **19**(1): 721-24.

Patricia, Y. H., Aisyah, H. J., Reg L., Kerry, P.,William, K. and James, J. De V.(2006). Asparinins, asparosides, curillins, curillosides and shavatarins: structural clarification with the isolation of shatavarin V, a new steroidal saponin from the root of *Asparagus racemosus*. *Tetrahedron Letters*, **47**: 8683–8687.

Patricia, Y. H., Aisyah, H. J., Reg, L., Kerry, P.,William, K. and James J. D. V. (2006). Structural revision of shatavarins I and IV, the major components from the roots of *Asparagus racemosus*. *Tetrahedron Letters*, **47**: 6965–6969.

Patricia, Y. H., Aisyah, H. J., Reg L., Kerry, P.,William, K. and James, J. De V. (2008). Steroidal saponins from the roots of *Asparagus racemosus*. *Phytochemistry*, **69**: 796–804.

Roy, R.N., Bhagwager, S., Chavan, S.R. and Dutta, N.K. (1971). Preliminary pharmacological studies on extracts of root of *Asparagus racemosus* (Satavari), Willd, N.O. Lilliaceae. *J Res Ind Med.*, **6**: 132-138.

Rao. (1981). Inhibitory action of *Asparagus racemosus* on DMBA-induced mammary carcinogenesis in rats. *International Journal of Cancer*, **28**: 607– 610.

Ravikumar, P. R., Soman, R., Chetty, G. L., Pandey, R. C. and Sukh, D. (1987). Chemistry of Ayurvedic crude drugs: Part VI (Shatavari-I): structure of shatavari-IV. *Indian Journal Chemistry*, **26B**: 1012–1017.

Rege, N.N., Nazareth, H.M., Isaac, A.A., Karadikar, S.M. and Dahanukar, S.A. (1989). Immunotherapeutic modulation of intraperitoneal adhesions by *Asparagus racemosus*. *Journal of Postgraduate Medicine*, **35**: 199–203.

Rice-Evans, C. and Burdon, R. (1993). Free radical-lipid interaction and their pathological consequences. *Progress in Lipid Research*, **32** : 71–110.

Rege, N.N., Thatte, U.M. and Dahanukar, S.A. (1999). Adaptogenic properties of six rasayana herbs used in Ayurvedic medicine. *Phytotherapy Research*, **13** : 275–291.

Snyder, J.D. and Merson, M.H. (1982). The magnitude of the global problem of acute diarrhoea disease: A review of active surveillance of data. *Bull WHO*, **60**: 605-613.

Sholapurkar, M.L. (1986). Lactare-for improving lactation. *Indian Practitioner*, **39**: 1023-1026.

Singh, K.P. and Singh, R.H. (1986). Clinical trial on Satavari (*Asparagus racemosus* Willd.) in duodenal ulcer disease. *J Res Ay Sid.*, **7**: 91-100.

Sekine, T., Fukasawa, N., Kashiwagi, Y., Ruangrungsi, N. and Murakoshi, I. (1994). Structure of asparagamine A, a novel polycyclic alkaloid from *Asparagus racemosus*. *Chemical and Pharmaceutical Bulletin*, **42**: 1360.

Stavric, B. (1994). Antimutagens and anticarcinogens in foods (Review). *Food and Chemical Toxicology*, **32**: 79–90.

Sharma, S., Ramji, S., Kumari, S. and Bapna, J.S.(1996). Randomized controlled trial of *Asparagus racemosus* (Shatavari) as a lactogogue in lactational inadequacy. *Indian Pediatr.*, **33**: 675-677.

Sekine, T., Fukasawa, N., Murakoshi, I. and Ruangrungsi, N. (1997). A 9,10-dihydrophenanthrene from *Asparagus racemosus*. *Phytochemistry*, **44**: 763–764.

Sharma, S.C., Sati, O.P. and Chand, R. (1981). Constituents of the fruits of *Asparagus racemosus* Willd. *Pharmazie*, **36** : 709.

Singh, G. and Mukherjee, P. (1998). Herbal aphrodisiacs : A review. *Indian Drugs*, **35**: 175-182.

Swarup, A. and Umadevi, K. (1998). Evaluation of EveCare® in the treatment of dysmenorrhoea and premenstrual syndrome. *Obstetrics Gynaecology Today*, **6** : 369–372.

Saxena,V.K. and Chourasia, S. (2001). A new isoflavone from the roots of *Asparagus racemosus*. *Fitoterapia*, **72**: 307–309.

Sairam, K., Priyambada, S., Aryya, N. C. and Goel, R. K. (2003). Gastroduodenal ulcer protective activity of *Asparagus racemosus*: an experimental, biochemical and histological study. *Journal of Ethanopharmacology*, **86** : 1-10.

Sharma, B.D. and Lal, H. (2003). *Asparagus racemosus*-An update. *Indian journal of Medical Sciences*, **57:** 408-414.

Shao, Y., Chin, C-K., Ho, C-T., Ma, W., Garrison, S.A. and Huang, M.T. (1996) Antitumour activity of the crude saponins obtained from asparagus. *Cancer Letters*, **104**: 31–36.

Thatte, U.M., Chhabria, S.N., Karandikar, S.M. and Dahanukar, S.A. (1987). Protective effects of Indian medicinal plants against cyclophosphamide neutropenia. *Journal of Postgraduate Medicine*, **33**: 185–188.

Thatte U.M. And Dahanukar S.A. (1988). Comparative study of immunomodulating activity of Indian medicinal plants, lithium carbonate and glucan. *Methods Find Exp Clin Pharmacol.*, **10**: 639-44.

Tewari, L.C. and Pandey, G. (1993). Satavari, *Asparagus racemosus* Willd. An important herbal tonic and therapeutic agent. *Sachitra Ayurved*, **46**: 53-54.

Thomsen, M. (2002). Shatavari–*Asparagus racemosus*. www.Phytomeicine.com

Unnikrishnan, P.M. (1998). Shatavari–*Asparagus racemosus* Willd. *Amruth*, 18–20.

Von Sonntag, C. (1987). The Chemical Basis of Radiation Biology. *Taylor and Francis*, London.

Venkatesan, N., Thiyagarajan, V., Narayanan, S., Arul,Arokya., Raja, S., Sengodan, G. V. K., Rajarajan, Thandavarayan. and Perianayagam, J. B. (2005). Anti – Diarrhoeal potential of *Asparagus racemosus* Wild root extracts in laboratory animals. *Journal of Pharma Pharmaceutical Sciences*, **8(1)**: 39-46.

Visavadiya, N.P. and Narasimhacharya, R.L.(2005). Hypolipidemic and antioxidant activities in *Asparagus racemosus* in hypercholesteremic rats. *Indian Journal of Pharmacology*, **37**: 376–380.

Wiboonpun,N., Phuwapraisirisan, P. and Tip-pyang, S. (2004). Identification of antioxidant compound from *Asparagus racemosus*. *Phytotherapy Research*, **18**: 771–773.

Zhang, C., Zheng, S., Zhang, Y., Luo, C. and Guo, C. (1997). The protective effective of polysaccharide and C-phycocyanin from *Spirulina platensis* on acute radiation injury in mice. *Acta Nutrimenta Sinica*, **18**: 327–331.

Traditional and Folk Herbal Medicine: *Pages* **133–154**
Recent Researches Vol. 1 (2012)
Editor: **V.K. Gupta**
Published by: **DAYA PUBLISHING HOUSE, NEW DELHI**

6

Ethnobotanical, Phytochemical, Pharmacological and Conservational Studies on *Glycyrrhiza glabra*: An Overview in Pakistan

Zabta Khan Shinwari[1*], Abdul Latif Khan[2,3] and In-Jung Lee[2]

ABSTRACT

Pakistan has more than 6000 species of higher plants. At least 12 per cent of the flora of Pakistan is used as traditional medicine. Glycyrrhiza or Liquorice has 30 species all over the world of which three species G. triphylla, G. uralensis, G. glabra are in Pakistan. It is a plant of dry temperate region. Local names of the plants are Mulathi, Mulei, Miti kthi, Mayoon, Khowaga Zela/Khushwalgye. Glycyrrhiza may be collected as whole plant or only specific parts used are collected. Root is the most extensively utilized part but its other products such as sap and extracts are also used. Local communities with indigenous knowledge relay on the plant for its nutritive value and as, emollient, for asthma, bronchitis, laxative and as an expectorant. Analyzing its active constituents, glycyrrhizin, a triterpenoid compound, accounts for the sweet taste of licorice root. Various other compounds belong mainly to terpenoids (sesquiterpene lactones and dimmers, diterpenes and triterpenoids) and flavonoids. Bioactive chemical constituents from G. glabra have presented a wide array of pharmacological potentials ranging from anti-inflammatory, anti-oxidative, hepatoprotective and anticancer activities. Local communities play crucial role in collection, processing and value addition of the

1 Professor and Chairman, Department of Biotechnology, Quaid-e-Azam University, Islamabad, Pakistan.

2 School of Applied Biosciences, College of Agriculture and Life Sciences, Kyungpook National University, Daegu Republic of Korea.

3 Kohat University of Science and Technology, Kohat, Pakistan.

* *Corresponding author*: E-mail: shinwari2002@yahoo.com

Glycyrrhiza. Almost 65 per cent of the product is lost during processing. To replenish the loss of natural stock of the Glycyrrhiza, cultivation on marginal and agricultural lands has been adopted.

Keywords: Ethnobotanical, Phytochemical, Pharmacological, Conservational, *Glycyrrhiza glabra*, Traditional medicine.

Introduction

Glycyrrhiza or Liquorice is famous for its sweet taste and wider use in herbal medicine in traditional Chinese medicines (TCM), Kampo medicines and Ayurvedic medicines (Zaidi, 1999; Shinwari *et al.*, 2006). It is described as grandfather of herbs (Ody, 2000). The roots of the plants are used by different communities in parts of Asia and Europe. It is distributed in various parts of Pakistan. Worldwide it has more than 30 species while in Pakistan it has three important species indigenous to various areas.

Pakistan exhibits a variety of geophysical and phytoecological conditions and has a rich resource base of plants (Stewart, 1972), particularly medicinal plants. More than 6,000 species of higher plants and 4000, species of fungi are recognized. At least 1000 medicinal plant species are reported from Pakistan, of which 500 species are actively used in health care practices and 350 are traded for billions of rupees in the national and international markets (Latif and Shinwari, 2006; Hussain *et al.*, 2009; Hussain *et al.*, 2010; Shinwari, 2010; Adnan *et al.*, 2010). Most of the work done on medicinal plants of Pakistan generally covers the northern uplands of temperate and alpine forests of Kashmir, Hazara, Malakand and Northern areas (Amin, 1961; Chaudri, 1958; Chaudri, 1961; Ikram and Hussain, 1978; Zaman and Khan, 1972). It is disappointing to note that no baseline data exists on the medicinal flora of the dry tropical and lower sub tropical forests of Pakistan, nor does the literature tell us anything regarding the conservation status of species in these forests. Various reviews of literatures on availability of economically important medicinal plants from different areas of Pakistan (Awan, 1978; Khan, 1951; Khan, 1985; Khan *et al.*, 2009; Ahmad, 1999; Ahmad *et al.*, 2002; Ahmad *et al.*, 2004; Shinwari *et al.*, 2006; Shinwari *et al.*, 2010) shows that the medicinal flora of the different fragile ecosystems needs intensive site-specific phyto-sociological assessment to acquire an in-depth knowledge of the threats and baseline data regarding the conservation status of species in the area.

Pakistan has four phytogeographical regions, the uniregionals, consisting of Irano-Turanian (46 per cent), Sino-Himalayan (10 per cent), Saharo-Sindian (9.5 per cent) and Indian element (4.5 per cent). Despite the biggest territory of Saharo-Sindian region, the diversity of uniregional elements is lowest in this region. In Pakistan, 70 per cent species are uni-regional and about 30 per cent of the species are bi- and pluri-regional (Figure 6.1) (Ali and Qaiser, 1986). Most of the medicinal plants, available in the market or supplied directly to the pharmaceuticals industries, are extracted from the natural sources (Chaudari, 1958; Amin, 1961; Chaudari, 1961, Zaman *et al.*, 1972; Zaman and Khan, 1972).

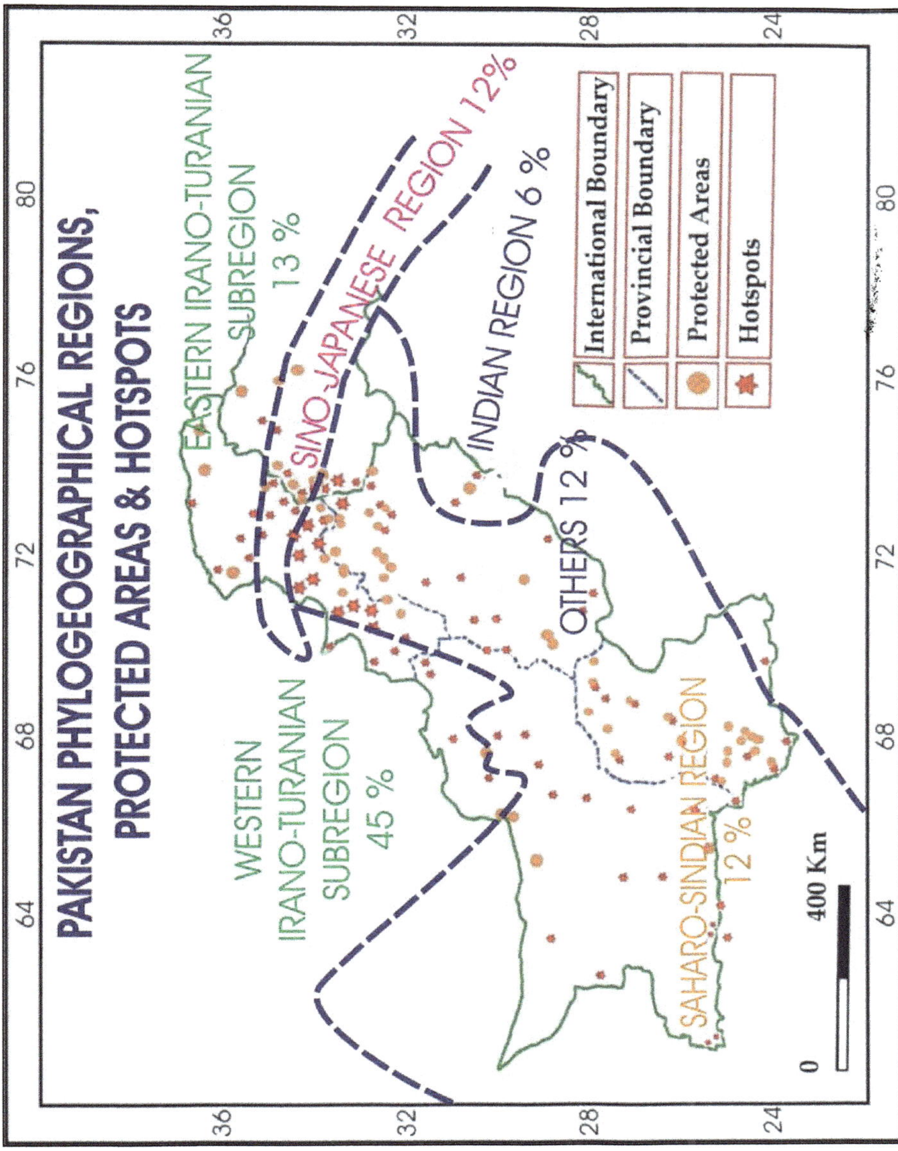

Figure 6.1: Floristic regions of Pakistan.

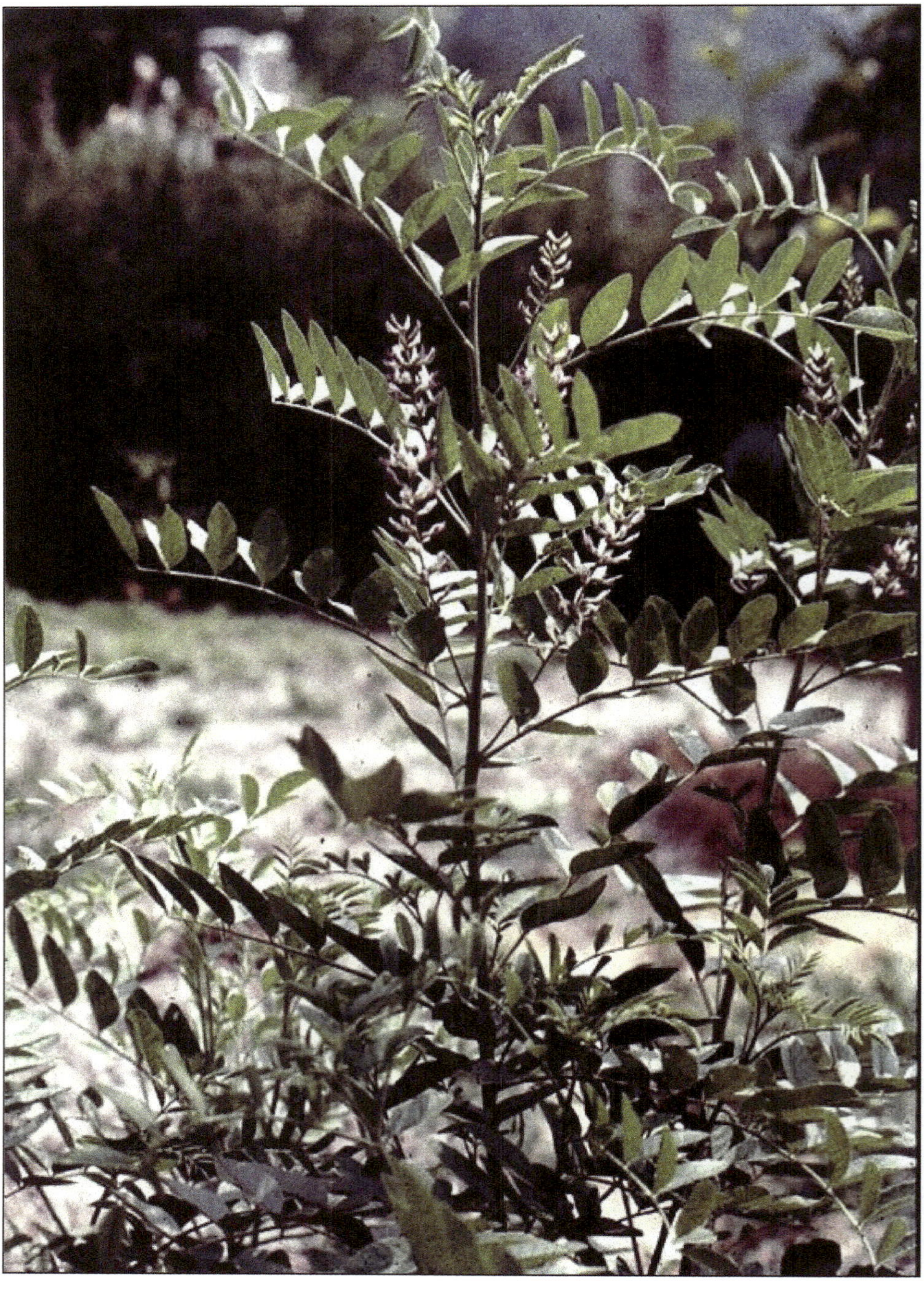

Figure 6.2: *Glycyrrhiza glabra* growing in wild.

Results reported in this chapter, unless otherwise referred to, are supported by our observations, analyses and decades long experience with the natural resource conservation of the area. Identification of plants was carried out with the help of the Flora of Pakistan (Stewart, 1972; Nasir and Ali, 1995; Ali and Qaiser, 2002). Field information regarding plant use is confirmed with Awan (1978) and Baquar (1989).

Botanical Description

It grows up to 120 cm high with a spread of 90 cm. Its bluish-green leaves are divided into oval leaflets (Figure 6.2). In late summer, it produces small, pale, purple-blue and white flowers on erect stems. The licorice shrub is a member of the pea family and grows in subtropical climates in rich soil. It has oval leaflets, white to purplish flower clusters, and flat pods. Below ground, the licorice plant has an extensive root system with a main taproot and numerous runners. The main taproot, which is harvested for medicinal use, is soft, fibrous, and has a bright yellow interior (Ahmad, 1999). A genus with about 30 species; represented by three species in Pakistan.

Taxonomy of *Glycyrrhiza*

Kingdom	:	Plantae
Division	:	Magnoliophyta
Class	:	Magnoliopsida
Order	:	Fabales
Family	:	Fabaceae (alt. Leguminosae) also placed in Papilionaceae
Sub Family	:	Faboideae
Tribe	:	Galegeae
Genus	:	*Glycyrrhiza*
Species	:	*G. triphylla, G. uralensis, G. glabra*
Habitat	:	Hardy plant growing well in deep, rich sandy soil
Common Names	:	Lacrisse, sweet licorice, Liquorice root, Reglisse (French), Lacrisse (German), Regolizia (Italian)
Local Names	:	Mulati (Urdu and Hindi), Mulei (Punjabi), Miti kthi (Sindhi), Mayoon (Chitrali), Khwaga Zela/Khwaga waly (Pashto)

Ethnobotanical Uses

Glycyrrhiza glabra is one of the most widely used species of *Glycyrrhiza*. It was introduced to England around the 15th century by Dominican friars who brought it to the Yorkshire Dales, where it became famous as an ingredient in Pontefract cakes. Liquorice is a top selling herb with many uses. Apart from its medicinal applications, it is employed as a flavoring agent in sweets and tobacco, and as a foaming agent in fire extinguishers and beer. It is used traditionally to relieve coughs and sore throats, and against gastric inflammation. One of the main active ingredients is glycyrrhizin, which has a cortisone-like effect and additionally, is 50 times sweeter then sucrose. Glycyrrhizin raises prostaglandin levels locally, increasing mucous secretion and promoting proliferation of cells in the stomach. Due to its cortisone-like action, it should not be used continuously. Sustained doses could, however, lead to high blood pressure, water retention or oedema. Liquorice should be avoided during pregnancy and by those with high blood pressure, inflammation or cirrhosis of the liver or potassium deficiency. In Unani medicines, *G. glabra* is classified as hot and dry. According to Indian Materia Medica, the herbs are classified by the four natures of hot, cold, dry, or moist and then by degrees, which can range from 1 to 4. Common foods and very mild herbs are usually classified as being in the first degree; the frequently used herbs of relatively mild nature (such as those listed above) are usually classed in the second degree, strong-acting herbs are classed in the third degree, and toxic herbs reach the fourth degree (Zaidi, 1999).

One of the important products prepared from *Glycyrrhiza* is *Joshanda* – a product of Hamdard and Qarshi Industries Pvt. Ltd Pakistan. It is a reputed remedy to relieve common cold and influenza. It comprises seven active ingredients, all of herbal origin [*viz., Althaea officinalis* (*Khatmi*), *Cordia latifolia* (*Sapistan*), *Glycyrrhiza glabra* (*Mulathi*), *Malva sylvestris* (*Khubazi*), *Onosma bracteatum* (*Gaozaban*), *Viola odorata* (*Banafsha*) and *Zizyphus jujuba* (*Unnab*)] (Shinawari *et al.*, 2006).

Glycyrrhiza is collected as whole or particular part wise. The main product is root but its other products such as sap and extracts are also used. Local communities with indigenous knowledge relay on the plant for nutritive, emollient, asthma, bronchitis, laxative and expectorant. *Glycyrrhiza* being identified to be salt tolerant, due to this property it is cultivated on various soils which are affected by salinity. Due to its importance, it was declared as medicinal plant of the year 2001. In this regard, the Postal Service of Pakistan (PSP) has issued a postal ticket (Figure 6.3).

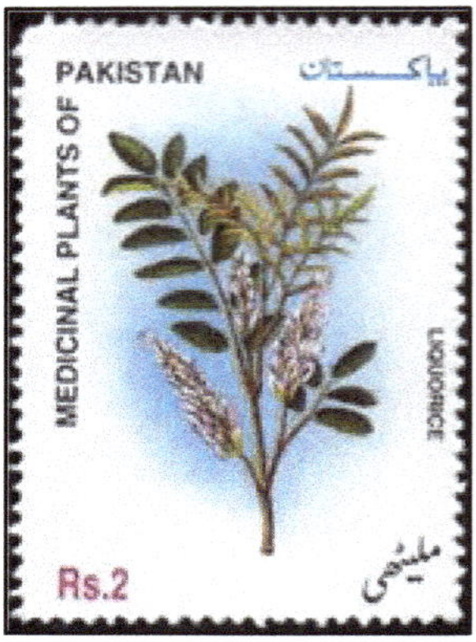

Figure 6.3: Postal ticket issued by PSP.

Bioactive Phytochemicals

A number of components have been isolated from licorice, including a water-soluble, biologically active complex that accounts for 40-50 percent of total dry material weight. This complex is composed of triterpene saponins, flavonoids, polysaccharides, pectins, simple sugars, amino acids, mineral salts, and various other substances (Zaidi, 1999). Glycyrrhizin, a triterpenoid compound, accounts for the sweet taste of licorice root (Figure 6.4). This compound represents a mixture of potassium-calcium-magnesium salts of glycyrrhizic acid that varies within a 2-25 percent range. Among the natural saponins, glycyrrhizic acid is a molecule composed of a hydrophilic part, two molecules of glucuronic acid, and a hydrophobic fragment, glycyrrhetic acid. The yellow color of licorice is due to the flavonoids and chalcones content of the plant, which includes liquiritin, liquiritigenin, rhamnoliquiritin, neoliquiritin, chalcones isoliquiritin, isoliquiritigenin, neoisoliquiritin, licuraside, glabrolide and licoflavonol (Figure 6.5; Hayashi *et al.*, 1998; Hatano *et al.*, 2000; Williamson, 2003; Cong and Lin, 2007). The isoflavones glabridin and hispaglabridins A and B have significant antioxidant activity and both glabridin and glabrene possess estrogen-like activity (Fukai *et al.*, 2005). The herb contains chalcones along with known glycyrrhizin, glycyrrhetinic acid, flavonoids and iso-flavonoids (Figure 6.5; Hoffmann,

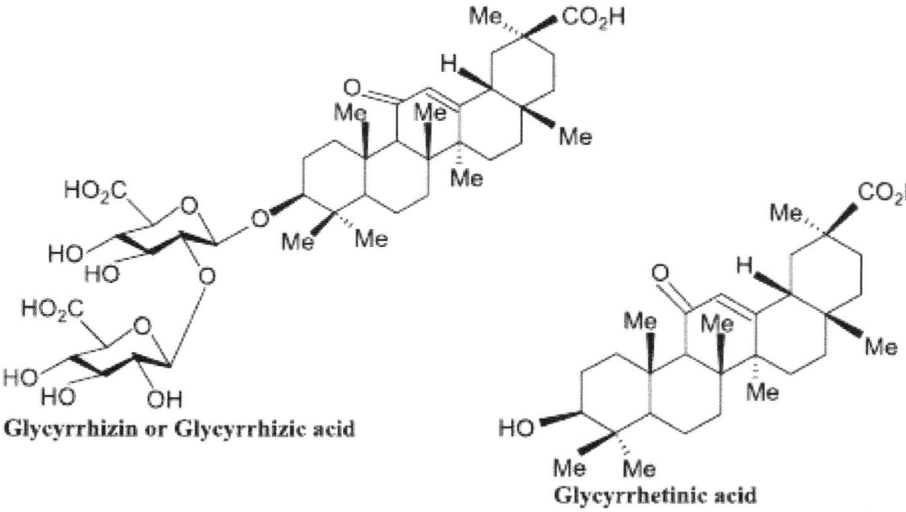

Figure 6.4: Glycyrrhizin and glycyrrhetinic acid isolated from *G. glabra.*

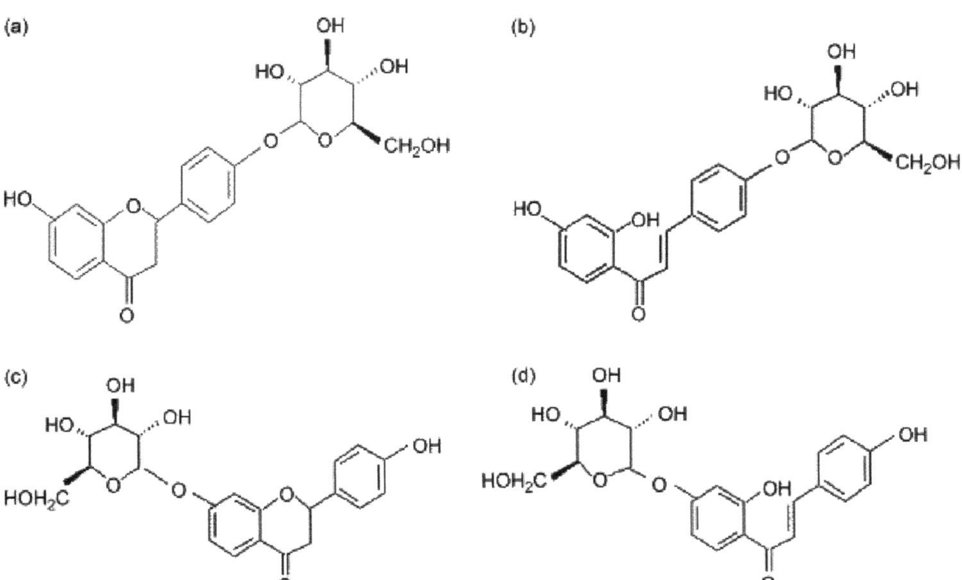

Figure 6.5: The chemical structure of flavonoid liquiritin and its isomers. (a) Liquiritin, (b) isoliquiritin, (c) neoliquiritin, (d) neoisoliquiritin (Kinoshita *et al.*, 2005).

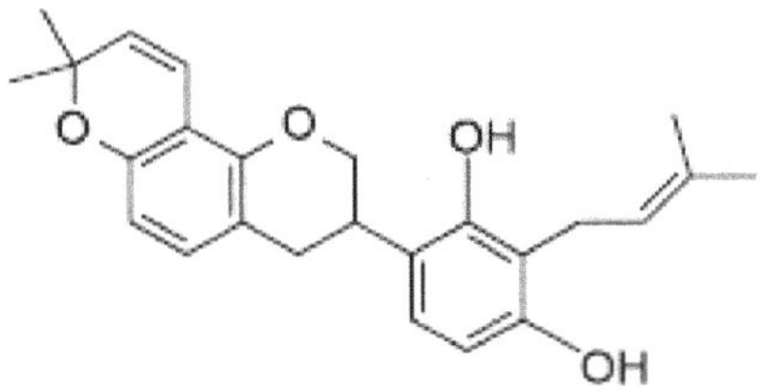

Hispaglabridin A

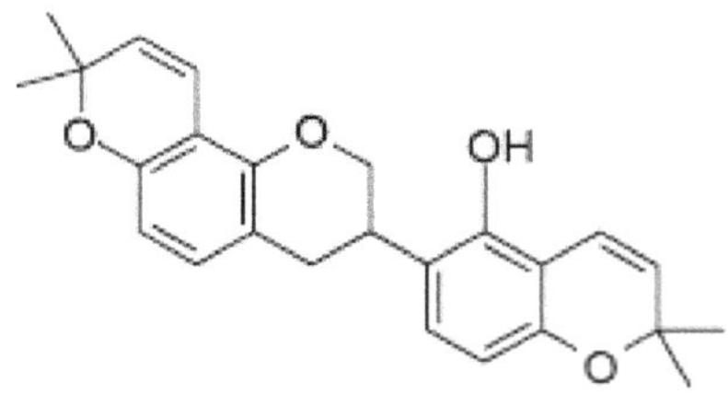

Hispaglabridin B

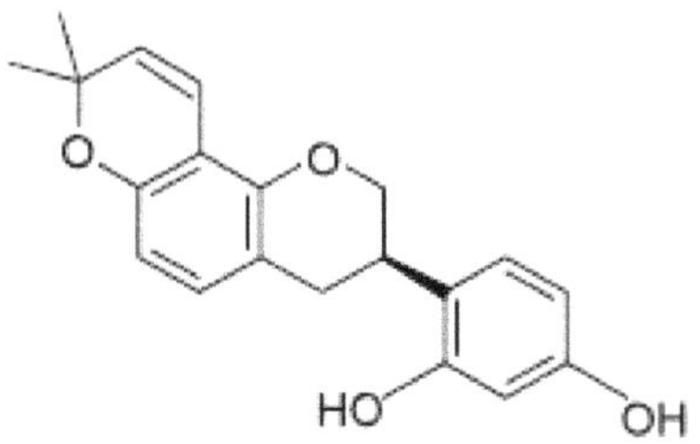

Glabridin

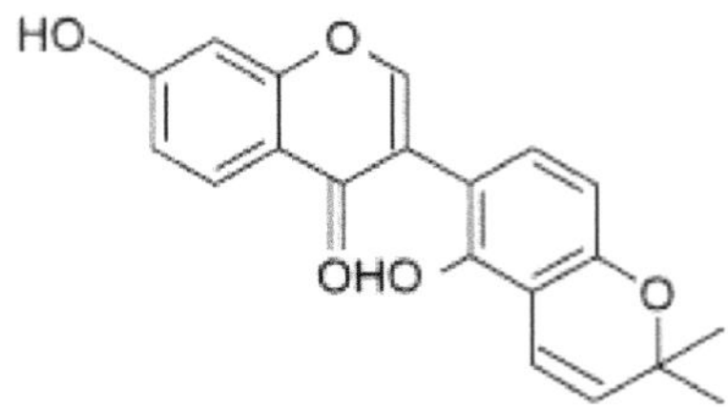

Glabrone

Figure 6.6: Chemical constituents isolated from *G. glabra.*

1990). Besides that, more recently a constituent, 5-hydroxy-8- methoxyl-flavone-7-O-beta-D-glucuronide, 5, 8-dihydroxy-flavone-7- O-beta-D-glucuronide, glychionide A, and glychionide B were isolated from the ethylacetate fraction of roots of *G. glabra* (Li *et al.,* 2005). Other class of bioactive chemical constituents is shown in Figure 6.6. Similarly some dervatives of dehydrostilbene were also characterized from *G. glabra* (Biondi *et al.* 2005).

Pharmacological Potential

Bioactive chemical constituents from *G. glabra* have presented a wide array of pharmacological potentials ranging from anti-inflammatory, anti-oxidative, hepatoprotective to anticancer activities. Inflammation is a process that elicits in response to microbial infections, mechanical injuries and burns. The process incorporate changes in destruction of tissue via inflammatory mediators such as prostaglandins (PGs) leukotriene platelet activating factors induced by phospholiphase A2, cyclooxygenase (COXs) and lipoxygenases and blood flow and increase in vascular permeability (Hinz and Brune, 2002; Wiart, 2007). iNOS and COX-2 expression, and NF-κB activation have been used as biomarkers for the screening of anti-inflammatory activity.

One of the major secondary metabolite from *G. glabra* is Glycyrrhetinic acid. During an *in vitro* study, it did not inhibit either cyclooxygenase 1 or 2 catalysed prostaglandin biosynthesis (Perera *et al.,* 2001). However, more recently, derivatives of glycyrrhetinic acid were inhibitive against interleukin-1b (IL-1b)-induced prostaglandin E2 (PGE2) production in normal human dermal fibroblasts (NHDF) (Tsukahara *et al.,* 2005).

In *G. glabra*, β-glycyhrritinic acid, a major secondary metabolite of glycyrrhizin, was observed for wide array of anti-inflammatory effects while analyzing it on different animals (Gumpricht *et al.*, 2005). It has been suggested that β-glycyhrritinic acid first suppress glucocorticoid metabolism and second classical complement pathway activation. In glucocorticoid metabolism the effect is activated to causes accumulation of glucocorticoids having anti-inflammatory properties. Oral or co-administration of β-glycyhrritinic acid or glycyrrhizin confirmed this result (Schleimer, 1991; Kroes *et al.*, 1997). However, if it is co-medicated with hydrocortisone while treating inflammatory lung disease, it will be much useful (Schleimer, 1991).

Glycyrrhizin on the other hand effected the production of reactive oxygen species (ROS) by neutrophills which are found potent mediator during tissue inflammation in *in vitro* studies (Wang and Nixon, 2001). Glyderinine and its derivatives from *G. glabra* exhibited anti-inflammatory effects while it was also that it can decrease myocardial inflammatory edema in experimental myocardial damage (Zakirov *et al.*, 1999; Tokiwa *et al.*, 2004).

Analyzing the hepatoprotective effect of glycyrrheitinic acid and glycyrrhizin, it was observed that the former is a better drug than the later one (Nose *et al.*, 1994). Furthermore, in a hepatocyte model of cholestatic liver injury, glycyrrhizin exhibited pro-apoptotic properties, whereas glycyrrheitinic acid is a potent inhibitor of bile acidinduced apoptosis and necrosis (Gumpricht *et al.*, 2005). This observation is in keeping with the protective effects of glycyrrhetinic acid against the carbon tetrachloride-induced hepatotoxicity and retrorsine-induced liver damage, respectively, in mice and rats (Jeong *et al.*, 2002).

In antioxidant capacity of isoflavone and their derivatives isolated from *G. glabra* such as glabridin protected mitochondrial functions from oxidative stresses (Haraguchi *et al.*, 2000). Hispaglabridin A, showed a potent antioxidative activity against peroxidation induced by Fe-ascorbate (Haraguchi, 2001).

Dhingra *et al.* (2004) and Dhingra and Sharma (2006) reported that the *G. glabra* can enhance memory and can work as antidepressant while assessing it on *in vitro* mice.

Glycyrrhiza in Pakistan

Ecology

The plant grows mostly in warm climate however it cannot endure severe freezing, but in cool weather interferes to cause the formation of useful juices and renders a woody structure of the plant. It has been found that climate favorable to the orange production is also suitable to grow liquorice (Zaidi, 1999). Owing to the depth to which the root penetrates and its ready propagation from detached pieces, the plant is a most persistent weed in cultivated grounds where it is indigenous and exceedingly difficult of extirpation. It is very healthy and robust and very little confronted to disease attacks, while at the same time successfully occupying the ground to compete with other plant species. For this reason, the continued natural supply of *Glycyrrhiza* spp. may be assured, though it is liable to suffer from severe reduction via over-collection (Zaidi, 1999).

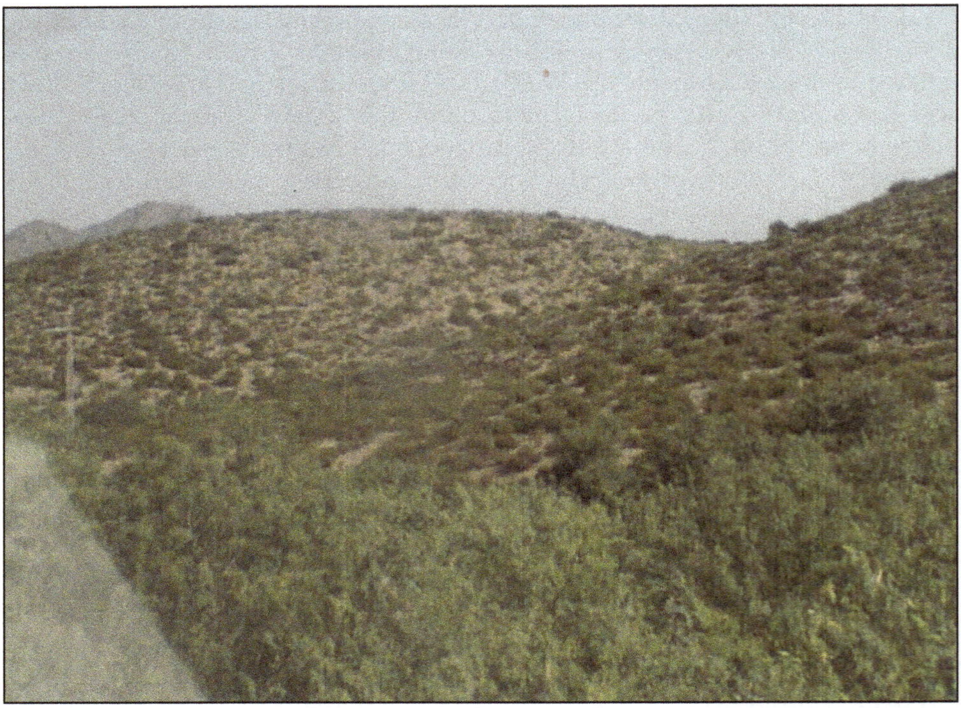

Figure 6.7: *G. glabra* growing at federally administrative tribal areas (FATA).

Distribution

1. ***Glycyrrhiza triphylla***
   Baluchistan: District Quetta, Ziarat (6000 ft), Loralai, Qila Saifullah

   NWFP: Kurram

2. ***Glycyrrhiza uralensis***
   Northern Areas: Gilgit, Bala Tangi, Wakhem (10000 ft)

   NWFP: Chitral

   Baluchistan: Harnai, Shahrig, upper Zhob, Pashin, Chaman, Mach hills

3. ***a) Glycyrrhiza glabra***
   Northern Areas: Gahkoch, Shagar, Askoli, Kwardo Hunza, Balatit, (8000 ft)

   NWFP: Chitral, Drosh, (4500 ft.), Kurram Valley

   Baluchistan: Harnai, Shahrig, upper Zhob, Pashin, Chaman, Mach hills

   b) Glycyrrhiza glandulifera
   NWFP: Kurram

If we analyze the research work done on the species from 1970 to 1996, it is revealed that more than 217 research publications in various forms have been published on different aspect of *Glycyrrhiza* (Figure 6.9).

Trade of *Glycyrrhiza*

There are approximately 400 manufacturers of herbal marketable products in Pakistan. Although there are 125 registered firms or industrial units with Pakistan Tibbi Pharmaceutical Manufacturing Association (PTPMA), but 77 are available on the paid roll call. The phytopharmaceutical industrial set up can be classified either

Figure 6.8: Occurrence and distribution of the three *Glycyrrhiza* species in various regions of Pakistan (images drawn in ArcView 3.1, USA).

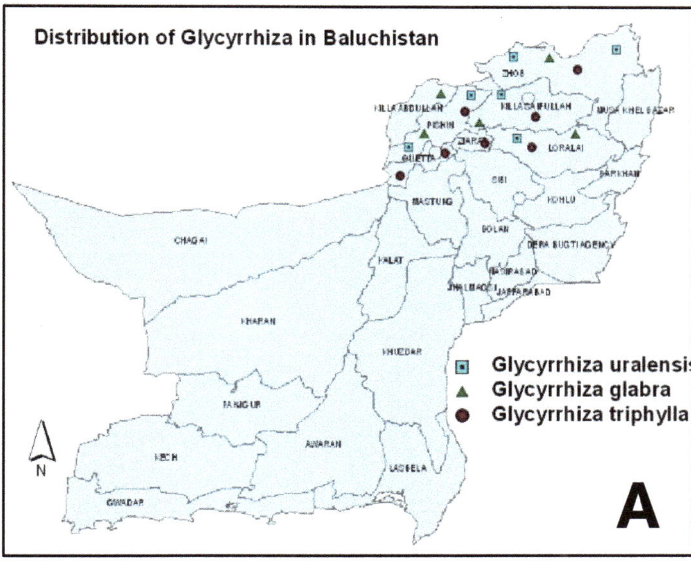

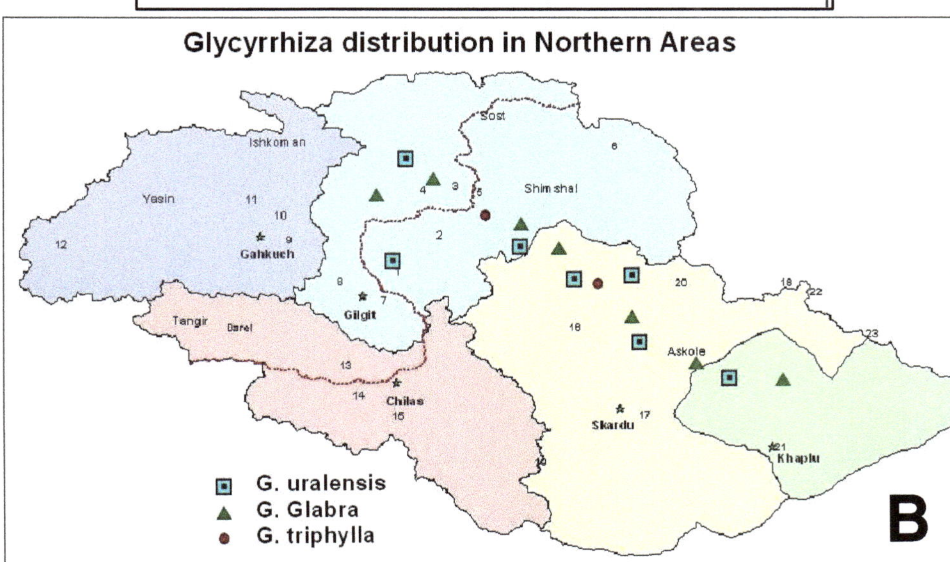

Contd...

Figure 6.8–*Contd...*

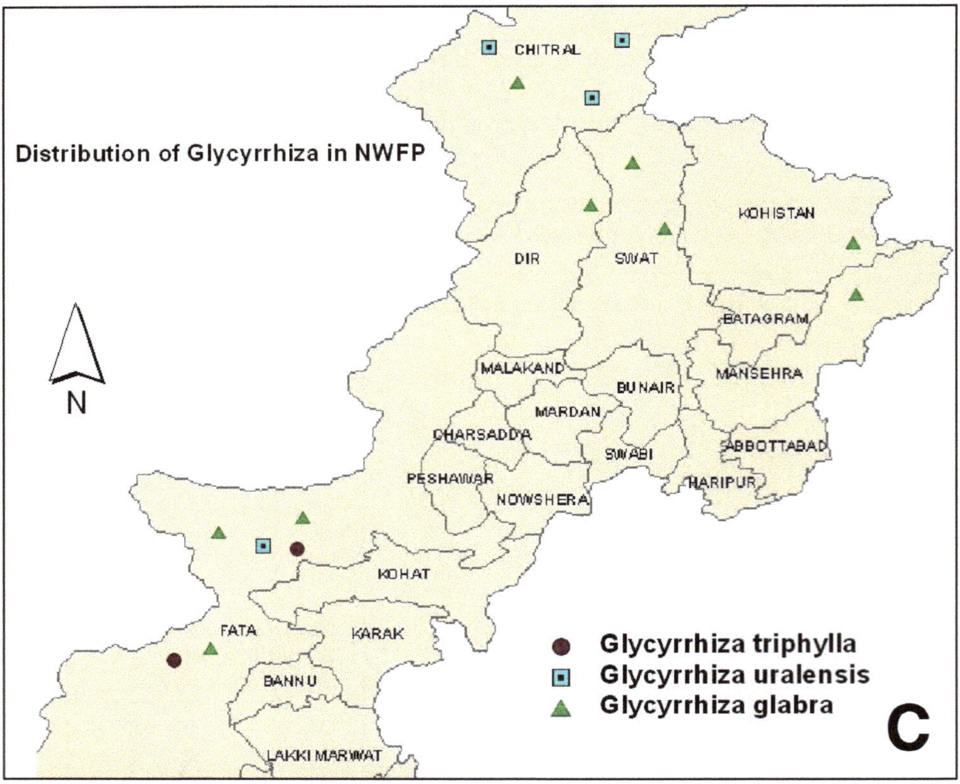

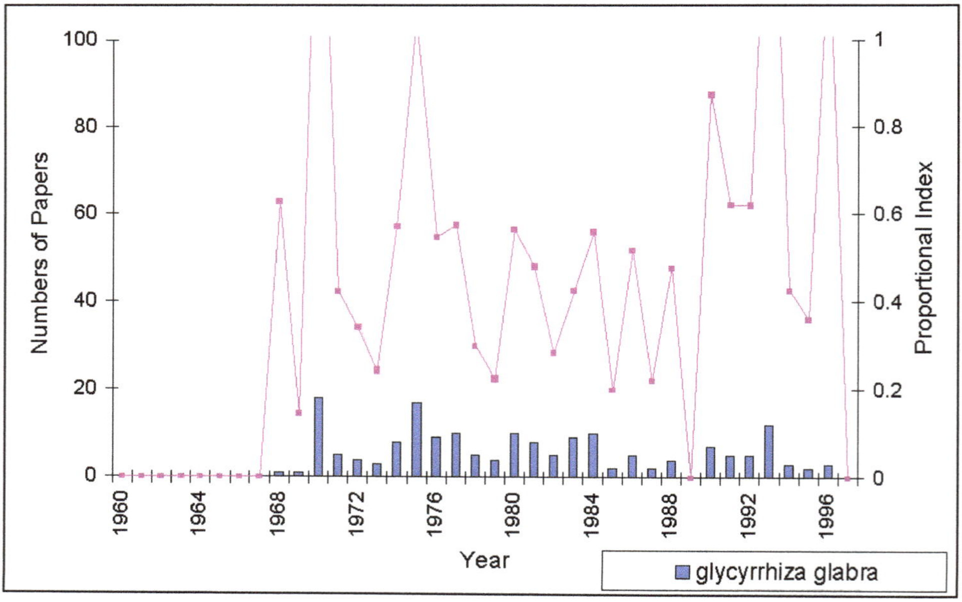

Figure 6.9: Number of publications on *Glycyrrhiza* spp. (*Source*: Agricola database 1970-1996).

as industrial or Dawakhana category (Hussain *et al.*, 2009; Hussain *et al.*, 2010; Shinwari, 2010). There are approximately 1300 or more raw material herbal suppliers (traders, shops, wholesalers), which serve the market as well as the industry for various herbal products and raw materials including *Glycyrrhiza* spp. (Gill, 2004).

The herbal product market in Pakistan is sizable but smaller interms of revenue *i.e.* 5.0 billion rupees as compared to 35 billion rupees of pharmaceuticals. It is a regular feature in Pakistan that 20-40 per cent average pharmacy and food stores have a specified area of trade mark herbal products (Hussain *et al.*, 2010; Shinwari, 2010). Approximately 300-400 different kinds of herbal products and 40-50 kinds of phytopharmaceutical are distributed in Pakistan. Domestic competition between Pakistani herbal manufacturing organizations is apparently invisible and as such, only 15-20 leading herbal manufacturers have out numbered the rest and these have captured the bulk of the consumer market (Shinwari *et al.*, 2006).

Collection at Local Level and Marketing

Local communities play crucial role in collection, processing and value addition of the *Glycyrrhiza*. The local collectors sell the products both in fresh and dry form. The price varies then accordingly. Since little knowledge is available with the local communities to adopt modern approaches for collection, drying, and packing, therefore, major chunk of the plant is wasted at local level. The marketing chain of the product is complex but the margin for profit for the local collector is much less than the actual national or international trading value. The product of *Glycyrrhiza* traded in the country passes through various chains, which are mainly four phases. These are (*i*) local collector level, (*ii*) middlemen level, (*iii*) local market levels and (*iv*) exporting level (Latif, 2004; Ahmad *et al.*, 2004).

1. Product is extracted by local communities, which processed them through indigenous methods and sell the home based finished product to middlemen.
2. Middlemen are normally the local shopkeeper or local traders. Shopkeepers buy the product and sell on local traders in the main market. Local people also sell products on local traders (Bey Pari), who visit regularly to the village or town for collection.
3. Local market level, are the main city/town markets, in which the local trader or middlemen sell the further processed product.
4. From the local market the product goes to exporting markets. After further processing, it is exported. Main exporting countries are Germany, Middle Eastern countries, Japan, India, UK, Switzerland, France etc.

During the whole process of marketing, more than 60-65 per cent is product is lost during the way to the main markets (Figure 6.10).

Prices

The trading price varies at various levels. In case of various medicinal plants the difference is even more than double while comparing the local and national/ international level. The current reported price of *Glycyrrhiza* at Akbari Mandi is Rs.

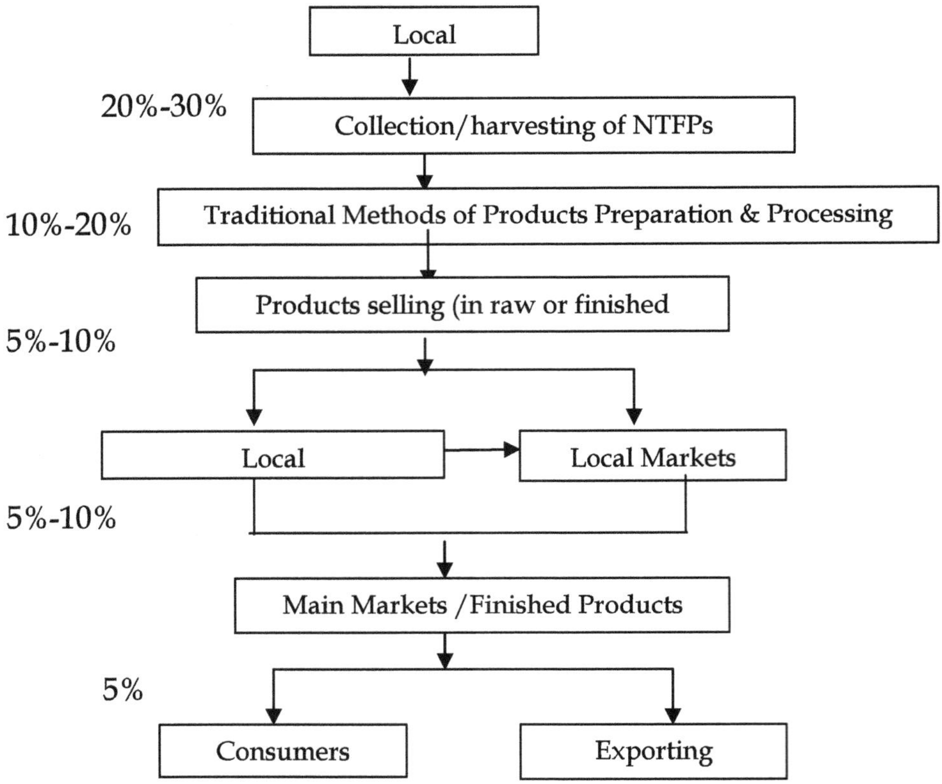

Figure 6.10: Market chain and loss of *Glycyrrhiza.*

130 per kg (dry), while the price of same product in fresh form is Rs. 15-20 per kg at Jodia Bazar Karachi (Zaidi, 1999). At local collector level the price is not more than Rs. 50 – 60 per kg in dry form while the fresh form has a price of Rs. 10 – 12 per kg. In 1990 the price of the same product was Rs. 30 – 35 per kg (Rizvi and Aftab, 2004).

Quantity Collected and Trade

It is estimated that more than 45 per cent of the product is used at local herbal industries. The remaining is exported to various countries like France, China, Gulf States, and Italy etc. The year wise data presented by the Export Promotion Bureau of Pakistan in 2003 is given in Table 6.1.

In 2002, the production of Liquorice roots were less 380,000 kg compared to the quantity of 1999 (554,000 kg) (Anwar and Masood, 2001). The value of the product is more due the price difference. As the price difference was doubled, therefore, it earned more revenue than previous years (Figure 6.11).

Besides the Liquorice roots, other products like Liquorice sap and extracts are also exported to different countries. In 1999 – 2000, total production was 60,000 kg and it has earned revenue of Rs. 4,454,000. While in 2000 – 01, total production was 100,000 kg that earned Rs. 7,334,000 (EPB, 2006).

Similarly the whole plant is also traded and used in various quantities. In 1999 – 2000, total production of *Glycyrrhiza* was 379,000 kg with a value of Rs. 6,227,000. In 2000 – 01, total production was 521,000 kg with a value of Rs. 24,592,000 (EPB, 2009).

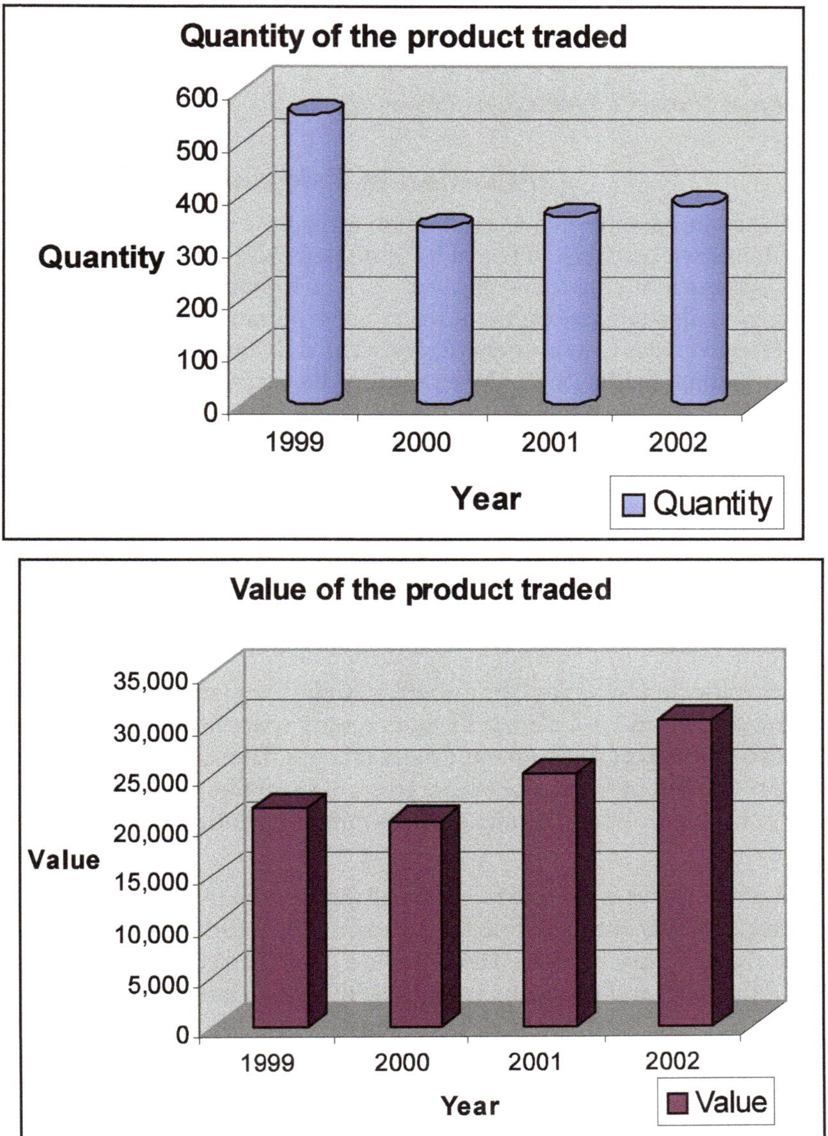

Figure 6.11: Quantity and value of liquorice roots.

Table 6.1: Year wise production and valuation of liquorice roots.

Year	*Quantity***	*Value**
1999	554	21,773
2000	341	20,460
2001	359	25,130
2002	380	30,400
Average	408	24,440

*Value in 000s rupees; **Quantity in 000s kg

Source: Muslim Trading Co. Karachi, EPB Pakistan

Cultivation in Pakistan

In Pakistan, experiments on cultivation were quite successful. The experiments were first initiated by Pakistan Forest Institute Peshawar. Latter on it was adopted by various institutes of respective climatic region. Kohat University of Science and Technology is also one among them which have initiated its cultivation in Kurram Valley. However, it is now successfully cultivated in the province of Punjab, NWFP and Baluchistan (Zaidi, 1999). On account of the depth to which the root strikes when the plant has room to flourish, the soil should have a good staple of mould 60 or 90 cm in depth and be manured if necessary.

The planting season is either October, or February and March; the latter is preferred. The plants are procured from old plantations, being waste from the harvesting process, consisting of those side roots or runners which have eyes or buds, cut into sections about 15 cm long. They are dibbled in, in rows 120 to 130 cm apart, about 10 cm underneath the surface and about 45 cm apart in the rows. In the autumn, the ground is dressed with farmyard manure, about 40 tons to the acre.

During the first two years the growth is slight, the plants not rising above 30 cm in the first season, and in Calabria the intervening space is generally utilized for the production of potatoes, cabbages and similar crops. The soil being heavily fertilized for the production of Liquorice, these crops are usually very luxuriant. After the second year, the growing Liquorice plants cover the entire soil to the exclusion of other growth.

Not until the end of the third season will the roots be ready to take up for use, but harvesting generally occurs only in the autumn of the fourth year. The soil is carefully removed from the space between the rows to a depth of 60 or 90 cm as required, thus exposing the roots and rhizomes at the side, the whole being then removed bodily. The earth from the next space is then removed and thrown into the trench thus formed and these operations are repeated continuously (Zaidi, 1999; Shinwari *et al.*, 2006).

Every portion of the subterranean part of the plant is carefully saved, the drug consisting of both runners and roots, the former constituting the major part. The roots are properly washed, trimmed and sorted, and either sold in their entire state or cut into shorter lengths and dried, in the latter case the cortical layer being sometimes removed by scraping. The older or 'hard' runners are sorted out and sold separately; the young, called 'soft,' are reserved for propagation.

The average yield per hectare is from 10 to 12 tons. The same ground yields a crop every three or four years, the fourth-year growth being the best. That of the third year and earlier is deficient in sweet substances, but immediately after the fourth year the texture begins to take on a tough, coarse and woody character. It is desirable also to collect the roots of those plants which have never borne fruit since that process exhausts the sweet substance of the sap. *Glycyrrhiza* being identified to be salt tolerant. Due to this property it is cultivated on various soils which are affected by salinity.

Future Prospects

The present trends in herbal medicine trade and production indicates that these fields are gaining significant importance throughout the world. It is expected that more and more importance will be given to the green medicines. There is a great economic potential of medicinal and aromatic plants in the near future. For example, the present export prices of' glycyrrhizic acid and glycyrhetinic acid are $240 and $500 per kg respectively. The demand of these two drugs in the European countries is estimated to be over 2500 kg per year. To produce these quantities of active ingredients, 300 tons of *Glycyrrhiza* roots will be required. Although there are some review reports available on *Glycyrrhiza* spp. (Asl and Hosseinzadeh, 2008) however, there is still need to actively work on various aspects of the plant. Following are few future guidelines for research and development of liquorice/*Glycyrrhiza* spp.:

- ☆ To meet the demand of the growing pharmaceutical industries of Pakistan and replenish the lost natural stock of the plant, its cultivation is deemed necessary. Cultivation at marginal lands is an alternative for the economic benefits of the local people. Not only, that marginal land be occupied thus reducing the soil erosion like problems. Now being known to be salt tolerant, the experiments of cultivation on such soils are appreciated by the government as well as the former. Although large scale cultivation is still to be adopted but work is in progress.
- ☆ *Glycyrrhiza* spp. alone or in combination with other plant species has widely been used for various diseases in traditional medicines. Indigenous uses of the plant created curiosity among scientists to isolate responsible bioactive constituents. Pursuing this, perspective metabolites were isolated from the plant. Further work is needed to identify and characterize bioassay guided active secondary metabolites belonging to various classes like sesquiterpenes, diterpenes, and saponins not only from the roots (being potentially used) but from other parts of the plant. Total synthesis of such active secondary metabolites and their metabolic pathway might be elaborated.
- ☆ Some of the active constituents like glycyrrheitinic acid and glycyrrhizin are assessed in their role against various malfunction while upto some extent clinical trials of some of the important health problems like inflammations and cancer have been researched however, still broader

clinical trials of other bioactive constituents and their effect against other health disease have to be elaborated. More work is needed to address the issues pertaining to pharmacokinetics and drug interactions.

- Studies should be planned so as to explore the allelopathic behavior of the plant as suggested by Khan *et al.* (2009).
- Studies explaining the ecological interactions with other species, wider distribution among various countries, production, trade and economic perspectives, and ethnobotany should be undertaken.
- Additional studies should be done to evaluate the genetic resources of the plant for variation in growth, morphology and yield-related characteristics which, in turn, can be used to identify high-yielding populations suitable for agronomical and plant breeding programs.
- Ecological role of *Glycyrrhiza* spp. In terms of tolerance to abiotic and biotic stresses might be evaluated especially when it is cultivated at large scale.
- Role of various plant growth promoting hormones and their effect on biosynthesis of plant's biologically active constituents may also needs further assessment.
- Similarly proximate and essential nuterients analysis of *Glycyrrhiza* spp.will be important as revealed by (Hussain *et al.*, 2009; 2010)

Acknowledgements

I am grateful to Mr. Javed Iqbal and Mr. Naveed Alam for their co-operation in correcting the paper.

Reference

Adnan, M., Hussain, J., Shah, M.T., Shinwari, Z.K., Farmanullah, Bahader, A., Khan, N., Khan, A.L., and Watanabe, T. (2010). Proximate and nutrient composition of medicinal plants of humid and sub-humid regions in Northwest Pakistan. *Journal of Medicinal Plants Research,* **4(4):** 339 - 345.

Ahmad, H. (1999). Medicinal Plants of Swat. WWF-Pakistan. Pp29.

Ahmad, H., A. Ahmad and M. M. Jan. (2002). The medicinal plants of Salt Range. *Journal of Biol. Sces.,* **2:** 175-177.

Ahmad, H., G.R. Bhatti and A. Latif (2004). Medicinal plants of Thar Pakistan. *Zonus Aridas,* ISSN 1013-445X.

Ali, S. I. and M. Qaiser (1986). A phytogeographical analysis of the phanerogames of Pakistan and Kashmir. *Proc.Royal.Soc.Edinburg,* **89B**: 89-101.

Ali, S. I. and M. Qasir (1995-2002). Flora of Pakistan. Department of Botany, University of Karachi.

Amin, M. (1961). A note on the plants and the pharmaceutical industry in Pakistan. *J.Sc. Ind. Res.,* **4:** 217-218.

Anwar, R. and S. Masood. (2001). Status of herbs and other economic plants in Pakistan. *In*: Proc. of the Meeting PGRI. *Medicinal Plants of Pakistan.* (Eds.): R. Anwar, N. Haq and S. Masood. PASTIC Press, Islamabad, 49-53 pp.

Asl, M.N., Hosseinzadeh, H. (2008). Review of pharmacological effects of *Glycyrrhiza* sp. and its bioactive compounds. *Phytotherapy Research*, **22 (6)**: 709–724.

Awan, H. M. H. (1978). Kitabul Mufarradat : Almarouf Khawasul Adwiya (Urdu). Sheikh Ghulam Hussain and Sons, Lahore.

Baquar, S. R. (1989). Medicinal and poisonous plants of Pakistan.

Biondi DM, Rocco C, Ruberto G. (2005). Dihydrostilbene derivatives from *Glycyrrhiza glabra* leaves. *J Nat Prod.*, **68:** 1099–1102.

Chaudari, I.I. (1961). Distribution of some important medicinal plants of Pakistan. *Pak. J. Sc. Ind. Res.* **4:** 207-211.

Chaudari, I.I. (1958). Medicinal plants resources of Pakistan and their development. *Pak. J. Sci.*, **10:** 34-56

Cong, J., Lin B. (2007). Separation of Liquiritin by simulated moving bed chromatography. *Journal of Chromatography A*, **1145:** 190–194.

Dhingra D, Parle M, Kularni SK. (2004). Memory enhancing activity of *Glycyrrhiza glabra* in mice. *J Ethnopharmacol.*, **91:** 361– 365.

Dhingra D, Sharma D. 2006. Antidepressant-like activity of *Glycyrrhiza glabra* L. in mouse models of immobility tests. *Progress Neuro-Psychopharmacol Bio Psychiat.*, **30:** 449–454.

Export Promotion Bureau of Pakistan (2006).

Export Promotion Bureau of Pakistan (2009).

Fukai T, Cai B-S, Maruno K, Miyakawa Y, Konishi M, Nomura T.(2005). An isoprenylated flavanone from *Glycyrrhiza glabra* and rec-assay of licorice phenols. *Phytochemistry*, **49:** 2005–2013.

Gill, A. (2004). Introduction of Medicinal Herbs and Spices as Crop in Pakistan. Pro. H. Ahmad, and A.A. Khan. WWF-Pakistan.

Gumpricht E, Dahl R, Devereaux MW, Sokol RJ. (2005). Licorice compounds glycyrrhizin and 18β-glycyrrhetinic acid are potent modulators of bile acid-induced cytotoxicity in rat hepatocytes. *J Biol Chem.*, **280:** 10556–10563.

Haraguchi H. (2001). Antioxidative plant constituents. In *Bioactive Compounds from Natural Sources: Isolation, Characterization and Biological Properties*, Tringali C (ed.). Taylor and Francis: New York, 348–352.

Haraguchi H, Yoshida N, Ishikawa H, Tamura Y, Mizutani K, Kinoshita T. (2000). Protection of mitochondrial functions against oxidative stresses by isoflavans from *Glycyrrhiza glabra*. *J Pharm Pharmacol.*, **52:** 219–223.

Hatano T, Aga Y, Shintani Y, Ito H, Okuda T, Yoshida T. (2000). Minor flavonoids from licorice. *Phytochemistry*, **55:** 959–963.

Hayashi H, Hiraoka N, Ikeshiro Y, Yamamoto H, Yoshikawa T. 1998. Seasonal variation of glycyrrhizin and isoliquiritigenin glycosides in the root of *Glycyrrhiza glabra L. Biol Pharm Bull.*, **21:** 987–989.

Hinz, B.; Brune, K. (2002). Cyclooxygenase – 2 - 10 years later. *J. Pharm. Experim. Therap.*, **300:** 367–375.

Hoffmann, D. (1990). The New Holistic Herbal, Second Edition, Element, Shaftesbury.

Hussain J., Bahader, A., Farmanullah, Rehman, N., Khan, A.L., Wasiullah, Shinwari, Z.K. (2010). Proximate and nutrient analysis of the locally manufactured herbal medicines and its raw material. *Journal of American Science*, **6(5):** 91-96

Hussain, J., AL Khan, N, Rehman, M. Hamayun, Z.K., Shinwari, Wasiullah, IJ. Lee. (2009). Assessment of herbal products and their composite medicinal plants through proximate and micronutrients analysis. *Journal of Medicinal Plant Research*, **3(12):** 1072-1077.

Hussain, J., Khan, AL., Rehman, N., Zainullah, Hussain, S.T., Khan, F., Shinwari, Z.K. (2009). Proximate and nutrient analysis of selected medicinal plant species of Pakistan. *Pakistan Journal of Nutrition*, **8 (1):** 620-624.

Ikram, M. and S. F. Hussain (1978). Compendium of medicinal plants. Pak. Council Sc. Ind. Research, Peshawar.

Jeong HG, You HJ, Park SJ *et al.* (2002). Hepatoprotective effects of 18β-glycyrrhetinic acid on carbon tetrachloride-induced liver injury: inhibition of cytochrome P450 2E1 expression. *Pharm Res.*, **46:** 221–227.

Khan, A. H. (1951). The Medicinal plants, their past and present (with special reference to the work being done in Pakistan. *The Pak. J. Forest*, **1 (4):** 353-367.

Khan, A. A. (1985). Survey of crude drug (herbal) markets in Pakistan.

Biological Sciences Research Division. Bulletin No.7, PFI, Peshawar.

Khan, A.L., J. Hussain, M. Hamayun, Z.K. Shinwari, H. Khan, Y.H. Kang, SM Kang, I.J. Lee. (2009). Inorganic profile and allelopathic effect of endemic *Inula koelzii* from Himalaya Pakistan. *Pakistan Journal of Botany*, **41 (5):** 2517-2527.

Khan, AL., M. Hamayun, J. Hussain, S.A. Gilani, Khan, H., A. Kikuchi, K.N Watanabe, Eung Ho Jung and In-Jung Lee. (2009). Assessment of allelopathic potential of selected medicinal plants of Pakistan. *African Journal of Biotechnology*, **8 (6):** 1024-1029.

Kinoshita T, Tamura Y, Mizutani K. (2005). The isolation and structure elucidation of minor isoflavonoids from licorice of *Glycyrrhiza glabra* origin. *Chem Pharm Bull.*, **53:** 847– 849.

Kroes BH, Beukelman CJ, Van Den Berg AJJ, Wolbink GJ, Van Dijk H, Labadie RP. (1997). Inhibition of human complement by beta-glycyrrhetinic acid. *Immunology*, **90:** 115–120.

Latif, A. 2004. A review on Non Timber Forest Products of Pakistan. WWF-Pakistan.

Latif, A. and Z.K. Shinwari (2006). NTFPs can help in poverty alleviation of Pakistan. *Journal of Science and Technology*. **24(1):** 38-46. http: / /www.fao.org/forestry/ site/nwfp/en/

Li JR, Wang YQ, Deng ZZ. (2005). Two new compounds from *Glycyrrhiza glabra. J Asian Nat Prod Res.*, **7:** 677–680.

Nasir, E. and S. I. Ali (1971-1995). Flora of Pakistan. Department of Botany, University of Karachi.

Nose M, Ito M, Kamimura K, Shimizu M, Ogihara Y. (1994). A comparison of the antihepatotoxic activity between glycyrrhizin and glycyrrhetinic acid. *Planta Med.*, **60:** 136–139.

Ody P. (2000). *The Complete Guide Medicinal Herbal.* Dorling Kindersley: London, 75.

Perera P, Ringbom T, Huss U, Vasange M, Bohlin L. (2001). Search for natural products which affect cyclooxygenase-2. In *Bioactive Compounds from Natural Sources: Isolation, Characterization and Biological Properties,* Tringali C (ed.). Taylor and Francis: New York, 458, 462.

Rizvi, M. A., Aftab, S. (2004). Trade of Medicinal Plants of Pakistan. Pro. H. Ahmad, and A.A. Khan. WWF-Pakistan.

Schleimer RP. (1991). Potential regulation of inflammation in the lung by local metabolism of hydrocortisone. *Am J Respir Cell Mol Biol.*, **4:** 166 –173.

Shinwari, Z.K., Mehboob, R. Takashi, W. Takadumi, Y. (2006). A pictorial guide to medicinal plants of Pakistan.

Shinwari, Z.K. (2010.) Medicinal plants research in Pakistan. *Journal of Medicinal Plants Research*, **4(3):** 161-176.

Stewart, R. R. (1972). An annotated catalogue of the vascular plants of West Pakistan. Fakhri Printing Press, Karachi.

Tokiwa T, Harada K, Matsumura T, Tukiyama T. (2004). Oriental medicinal herb, *Periploca sepium*, extract inhibits growth and IL-6 production of human synovial fibroblast-like cells. *Pharm Bull.*, **27:** 1691–1693.

Tsukahara M, Nishino T, Furuhashi I, Inoue H, Sato T, Matsumoto H. (2005). Synthesis and inhibitory effect of novel glycyrrhetinic acid derivatives on IL-1b-induced prostaglandin E2 production in normal human dermal fibroblasts. *Chem Pharm Bull.*, **53:** 1103–1110.

Wang ZY, Nixon DW. (2001). Licorice and cancer. *Nutr Cancer*, **39:** 1–11.

Wiart, C. (2007). *Ethnophamacology of Medicinal Plants Asia and the Pacific*; Hunama Press: Totowa, NJ, USA; pp. 31–33.

Williamson EM. (2003). Liquorice. In *Potter's Cyclopedia of Herbal Medicines.* C W Daniels: Saffron Walden, UK, 269– 271.

Zaidi, S. H. (1999). *Glycyrrhiza glabra* (Mulethi). A dollar earning medicinal plants of upper Chitral. *Newslet. Pak. Forest Inst. Peshawar*, **2:** 2.

Zakirov NU, Aizimov MI, Kurmukov AG. (1999). The cardioprotective action of 18-dehydroglycyrrhetic acid in experimental myocardial damage. *Eksp Klin Farmakol.*, **62:** 19–21.

Zaman, M.B., A.A. Khan and A. Ahmad (1972). Contribution to the knowledge of medicinal plants. PFI, Peshawar.

Zaman, M.B. and M.S. Khan (1972). List of medicinal and other flowering plants of Dir. PFI, Peshawar.

Traditional and Folk Herbal Medicine: *Pages* **155–173**
Recent Researches Vol. 1 (2012)
Editor: **V.K. Gupta**
Published by: **DAYA PUBLISHING HOUSE, NEW DELHI**

7

Pharmacotherapeutic Potential of Medicinal Plants against Cardiometabolic Diseases

Suman Bala Sharma[1*], Seema Garg[1], Shipra Gupta[1], Reenu Singh Tanwar[1], Santosh Kumar Shukla[1], Richa Gupta[1] and Shreesh Ojha[2]

ABSTRACT

Coronary artery disease (CAD) in association with diabetes is the leading cause of morbidity and mortality world over. The current therapy for CAD and or diabetes has only symptomatic and requires a long term life long management approach. Following the understanding of these disease processes, there is an immense interest in drug development for cardio metabolic diseases. But, the synthetic agents discovered, developed and which are in use suffer from the drawback of adverse effects on long term use. In parallel, the modern drugs of these segments are quite expensive, hard to access for everyone and may involve safety issues. Hence, evaluating agents which might offer a preventive as well as therapeutic approach are desirable in context of high rise of cardiovascular and metabolic diseases. Medicinal plants with reputed traditional use in cardiovascular diseases might provide valuable drugs and/or lead molecules. Viewed in this reference, certain medicinal plants like Terminalia arjuna, Commiphora mukul, Emblica officinalis, Allium sativum, Ocimum sanctum, Cassia auriculata, Eugenia jambolana and Inula racemosa were evaluated in animal models of diabetes, myocardial infarction, atherosclerosis and ischemic reperfusion injury. The present review focuses on the preclinical and clinical evidences of cardioprotective effect of these medicinal

1 Department of Biochemistry, University College of Medical Sciences and GTB Hospital, University of Delhi, Delhi – 110 095, India.

2 College of Medicine, University of Arizona, Tucson 85719, Arizona, USA.

* *Corresponding author*: E-mail: drsbs08@hotmail.com, shreeshojha@gmail.com; Tel.: +91-011-22582972; Fax: +91-11-22590495.

plants and their active constituents. On the basis of studies including ours it has been suggested that the hypolipidemic, antioxidant, antihyperglycemic and free radical scavenging properties contribute to the cardioprotective of these plants and should further evaluated for safety, efficacy and clinical use.

Keywords: Antioxidants, Atherosclerosis, Cardioprotection, Coronary Artery Disease, Herbal drugs, Hyperlipidemia, Diabetes mellitus, Myocardial infarction, Medicinal plants.

Introduction

Heart disease is a hardship that impairs the structure or function of the heart, such as atherosclerotic and hypertensive diseases, congenital heart disease, rheumatic heart disease, and other conditions. Among several types of heart diseases, coronary artery disease (CAD), the most common manifestation of heart diseases occurs when the arteries that supply blood to the heart become hardened and narrowed from fatty plaque buildup on the artery walls, a process called atherosclerosis. According to WHO predictions, India will have 100 million or 60 per cent of the total world's population as heart patient in near future parallel to high rise in developed countries (Gupta and Gupta, 2009). This figure puts forth a big challenge for developed as well as developing countries like us. Recent epidemiological studies have also established a direct relationship between serum cholesterol and coronary artery disease. High levels of total cholesterol and more importantly LDL-cholesterol are well acknowledged to be major coronary risk factors. Changes in life style, high stress environment, high fat diet, high serum lipid levels, hypertension and increased environmental pollution are several reasons which culminating into excessive formation of free radicals that lead atherosclerosis and other cardiovascular diseases. In parallel, several studies have also demonstrated that an increase in HDL cholesterol is associated with a decrease in coronary risk. Despite of the availability of modern drugs for symptomatic cure, there is a scarcity of primary preventive agent in healthy or vulnerable population. WHO in a report mentioned that about 80 per cent of the global population still relies on botanical drugs and simultaneously several herbal medicines have advanced to clinical use in modern time (Ernst, 2003; Sirtori *et al.*, 2009). A growing body of literature exists for several herbal drugs and potential medicinal plants in the prevention of coronary artery disease and hypertension (Chen *et al.*, 2009). Certain indigenous drug preparations with naturally available medicinal herbs have been in use for many decades for the treatment of hyperlipidaemia, diabetes and cardiovascular diseases (Fugh, 2000; Visioli *et al.*, 2000; Shukla *et al.*, 2010). Many of these herbal medicines are finding their way onto the world market as alternatives to prescribed drugs currently available to treat various disorders including cardiovascular diseases (Wattanapitayakul *et al.*, 2005). Recently, there has been renewed interest in medicinal herbs and food products derived from medicinal herbs that have been found to have certain preventive measures in the treatment of cardiovascular diseases (Chen *et al.*, 2008; Tripathi, 2009; Shukla *et al.*, 2010). The consumption of diets rich in plant foods are associated with reduced risk of cardiovascular diseases (Visioli *et al.*, 2000; Chen *et al.*, 2009). For atherosclerotic coronary vascular disease, in particular, hyperlipidemia is a major risk factor which has shown to culminate in high mortality. While this association has been observed in several epidemiological studies, a better understanding of the processes involved in atherosclerosis has stimulated the search for new drugs, which could prevent from CAD and protect against the ischemic damages of myocardium. The herbal products, medicinal herbs or dietary supplements have been widely recognized as naturally occurring antioxidants and protect against the deleterious consequences of oxidative stresses an underlying cause of atherosclerosis, diabetes and ischemic heart diseases (Miller, 1998; Dwivedi and Aggarwal, 2009). With a goal to reduce the morbidity and mortality and achieve maximal reduction from atherosclerosis risk, preventive strategies are aimed at reducing the elevated levels of circulatory lipids and correcting the levels of lipoproteins. Present review focused on the preventive properties of medicinal plants or their active constituents against the ischemic heart disease based on evidences from our laboratory and others experimental studies (Table 7.1). Considering the conventional use of herbs and traditional literature, we have chosen medicinal plants and evaluated their cardioprotective potential with an aim to provide a phytotherapeutic approach in CAD management as well as identifying lead compounds for the new drug discovery and development (Shukla *et al.*, 2010).

Table 7.1: Active phytochemicals which may be responsible for cardioprotection.

Plant	*Active constituent*
Terminalia arjuna	Arjunic acid, arjungenin, arjunolic acid and saponins
Commiphora mukul	E and Z guggulsterone
Emblica officinalis	Emblicanin-A, emblicanin-B
Allium sativum	Allyl propyl disulphide, diallyl disulphide
Ocimum sanctum	Eugenol
Eugenia jambolana	Flavoniods, gallic acid, ellagic acid, glycosides, triteripenoids, saponins, β-sitosterol, succinamic acid derivative (US Patent: US 6428825B2; Indian Process Patent: 188759; Indian Product Patent: 230753)
Cassia auriculata	Coumarin derivatives
Inula racemosa	Sesquipertenes, inuline, alantolactone, isoalantolactone

Plants Exhibited Cardioprotective Properties

Terminalia arjuna

Terminalia arjuna popularly known as "Arjuna" belongs to family Combretaceae. The bark of *T. arjuna* has been mentioned in ancient Indian medicinal literature to have beneficial effects in heart diseases (Dwivedi *et al.*, 1989). The active principles identified in the *T. arjuna* bark include triterpenoid (arjunic acid, arjungenin, arjunolic acid) and saponins (Row *et al.*, 1970; Ahmad *et al.*, 1983; Tripathi *et al.*, 1992). Ancient physicians used the powdered tree bark of arjuna for relief from "Hridshool" (angina).

Evidence of Efficacy

In several preclinical and convincing numbers of the clinical studies it has shown to an important medicinal plant for cardioprotection in modern medicine (Dwivedi,

2007; Maulik and Katiyar, 2009). Available evidences demonstrate its usefulness in ischemic heart diseases alone as well as adjuvant along with conventional cardiovascular drugs.

Preclinical Studies

T. arjuna acts as a hypolipidemic agent and decreases oxidative stress in cardiovascular diseases (Kumar *et al.*, 2009). In hyperlipidemic rabbits, oral administration of *T. arjuna* extract (100 mg-500 mg/kg, body wt) resulted in dose related decrease in total cholesterol and LDL-C (Ram *et al.*, 1997). Dried pulverized bark of *T. arjuna* when administered orally to Wistar albino rats (120-150 g) in two doses (500 and 750 mg/kg) in 2 per cent carboxymethyl cellulose, 6 days per week for 12 weeks caused a significant increase in myocardial TBARS as well as endogenous antioxidants (Gauthman *et al.*, 2001). This suggests that chronic administration of *T. arjuna* imparts better oxidant stress bearing capacity to rat's heart (Kumar *et al.*, 2009). Arjunolic acid (15 mg/kg, body wt, i.p.) decreases cardiac enzymes and restores electrocardiographic changes to normal in rats with myocardial necrosis induced by isoproterenol (Sumitra *et al.*, 2001). Moreover in isoproterenol-induced myocardial injury, treatment with alcoholic extract of *T. arjuna* (6.75 mg/kg body wt) resulted significant decrease in myocardial TBARS and increased endogenous antioxidants (Karthikeyan *et al.*, 2003). Co-treatment with butanolic extract of *T. arjuna* bark in rats shown to improve cardiac antioxidants and decrease lipid peroxidation following doxorubicin (20 mg/kg; i.p.)-induced cardiotoxicity (Singh *et al.*, 2008). Its cardioprotective effect shown to be further supported by decreased CK-MB and reduced degree of histopathological perturbations and myocardial salvage in rats (Singh *et al.*, 2008). Our laboratory studies also demonstrated the cardioprotective potential of hydroalcoholic extract of *T. arjuna* in animal model of myocardial infarction. Hydroalcoholic extract of *T. arjuna* ameliorated myocardial infarction and ischemic-reperfusion injury of myocardium in a dose dependent manner (Gauthaman *et al.*, 2001). Studies including ours clearly demonstrate that *T. arjuna* has definitive cardioprotective activity owing to antioxidant, free radical scavenging, anti-inflammatory and anti-apoptotic properties (Shukla *et al.*, 2009).

Clinical Studies

In several clinical studies, *T. arjuna* has been found effective in chest pain and other cardiovascular ailments (Maulik and Katiyar, 2009). *T. arjuna* bark extract (500 mg 8 hourly) given to patients with stable angina with provocable ischemia on treadmill exercise led to improvement in clinical and treadmill exercise parameters as compared to placebo (Bharani *et al.*, 2002). Moreover, it has been also observed that *T. arjuna* (1-2 gm/day) administered in three divided doses to patients exerted hypolipidemic effects evidenced by improved lipid profile, increased NO end products and left ventricular ejection fraction along with decreased Lp(a). No toxicity has been reported with dosage of 500 mg administered four times daily for 24 months. Based on several clinical studies, use of *T. arjuna* as an adjuvant to modern drugs has been suggested in vulnerable heart patients (Dwivedi, 2007).

Commiphora mukul

Commiphora mukul is commonly known as Guggul and belongs to family Burseracae. In Ayurvedic medicines, Guggul is a reputed herb in Ayurvedic medicine for over 2500 years (Singh *et al.*, 1994). It is very much appreciated herbal medicine from Ayurvedic medicine to modern medicine. It has been used to reduce joint pain and as a gargle in tooth disease. It is also used as an antiseptic, emmenagogue, expectorant, astringent and demulcent. It is quite popular as a dietary supplement in western world for its cholesterol-lowering, weight loss and arthritis treatment. It has also shown to increase thermogenesis by stimulating the thyroid hormones. In the recent decade, it has been well acknowledged to lower total and LDL cholesterol and triglycerides and inhibit platelet aggregation (Singh *et al.*, 2007). Recently, its active component, guggulsterone has been found to act as an antagonist for the nuclear receptor farnesoid X receptor (FXR) and bile acid receptor (BAR) that underlie its lipid-lowering property (Deng, 2007).

Evidence for Efficacy

In preclinical and clinical studies it has shown to be effective in reducing lipid levels and aid into cardioprotection (Deng, 2007).

Preclinical Studies

Guggul has been shown to be hypolipidemic in numerous experimental studies (Deng, 2007). *C. mukul* extract and synthetic guggulsterone (E and Z isomers) has also been shown to possess cardioprotective activity in isoproterenol-induced myocardial necrosis in rats (Chander *et al.*, 2003; Ojha *et al.*, 2008). *C. mukul* and its cholesterol-lowering component guggulsterone, effectively inhibit LDL oxidation (Wang *et al.*, 2004).

Clinical Studies

Numerous clinical trials have shown the hypolipidemic property of *C. mukul* (Deng, 2007; Singh *et al.*, 2007). Guggul (12-16 gm/day) administered for 12 weeks in 25 hyperlipidemic patients decreases total cholesterol by 35 per cent, triglyceride by 32.7 per cent and body weight by 1.4 kg/month. Administration of guggul powder (8 gm/day) to ischemic heart disease patients for 12 weeks improves pericardial pain (in 35 per cent patients), decreases body weight (1 kg/month), total cholesterol (by 27 per cent) and total triglyceride (by 36 per cent) (Singh *et al.*, 1994). Guggulsterone activates multiple nuclear receptors and induce CYP3A gene expression through pregnane receptor thereby lowered lipid levels (Wang *et al.*, 2004). It also maintains cholesterol and bile acid balance by its action as an antagonist of Farsenoid-X receptor and by up regulating bile salt export pump in liver. Moreover, it inhibits the activation of NF-κB which is an important regulator of inflammation. This suggests that guggalsterone has hypolipidemic as well as anti inflammatory properties (Brobst *et al.*, 2004). In clinical trials of guggulipid, generally 90 per cent or greater compliance has been reported with infrequent side effects of headache and mild nausea. Therefore, concern has been raised about the bioavailability of concomitantly administered cardiovascular drugs. Its use in hyperlipidemic patient as an adjuvant with other cardiovascular drugs requires further explorative studies. However, available studies demonstrate its hypolipidemic, antioxidant, anti-ischemic and free radical scavenging property and suggest its strong cardioprotective potential as a pharmaco-therapeutic agent (Sing *et al.*, 1994; Deng, 2007).

Emblica officinalis

Emblica officinalis Gaertn. commonly known as Amla, is a member of family, Euphorbiaceae. It widely grows in tropical and subtropical parts of China, India, Indonesia and Malaysia. All parts of the plant are mentioned to be useful for medicinal purposes and fruits are generally used for dietary purpose. The fresh or the dry fruit is used in traditional medicines for the treatment of diarrhea, jaundice and inflammatory conditions. Available studies have demonstrated its potent antioxidant, adaptogenic (Rege *et al.*, 1999) and cardioprotective properties. The fruits extract are commonly used in polyherbal and herbomineral formulations of Ayurvedic, Unani and Siddha system of medicine (Bafna and Balraman, 2005). The active principles mainly include tannoids, emblicanin-A (37 per cent) and emblicanin-B (33 per cent) and shown to exhibit protective in myocarditis (Wang *et al.*, 2009).

Evidence of Efficacy

Traditional literature and preclinical studies demonstrate the cardioprotective effect of *E. officinalis* owing to their potent antioxidant and free radical scavenging activity (Anila and Vijayalakshmi, 2002; Rajak *et al.*, 2004; Mohan *et al.*, 2006). In a small clinical study it has reported to reduce hypercholesteremia and suggested to improve lipid profile and reduce hyperlipidaemia as well as hyperglycemia (Jacob *et al.*, 1988). Though, randomized clinical studies are lacking to demonstrate its clinical usefulness in cardiometabolic conditions.

Preclinical Studies

It is observed that in hyperlipidemic rabbits the administration of amla juice for 60 days decreases total cholesterol by 82 per cent, triglycerides by 66 per cent, phospholipids by 77 per cent and LDL-C by 90 per cent. There was significant decrease in tissue lipids and aortic plaques were also regressed (Mathur *et al.*, 1996). The tannoid constituents of *E. officinalis* (50 mg/kg body wt) have been found to exert potent antioxidant activity both *in-vivo* and *in-vitro* against ischemia-reperfusion induced oxidative stress in rat heart (Bhattacharya *et al.*, 2002). Oral administration of *E. officinalis* in rats resulted in reduced serum TBARS and ox LDL levels (Kim *et al.*, 2005). Moreover, administration of flavonoids obtained from *E. officinalis* reported to significantly reduce lipid levels and lipid peroxidation in serum as well as in tissues. Cholesterol levels were reduced due to decrease in the activity of hepatic HMG CoA reductase and increased activity of endogenous antioxifants (Anila and Vijayalakshmi, 2002). Furthermore, a decrease in iNOS and Bax, a proapoptotic protein and an increase in bcl-2, an antiapoptotic protein in rats administered an extract of amla (10 mg/kg body weight) are strongly suggestive of its hypolipidemic effect and its beneficial role in ageing and obesity.

Clinical Studies

Fruits of *E. officinalis* supplemented for 38 days both in normal and hypercholesterolemic subjects have been shown to significantly decrease total cholesterol levels (Yokozawa *et al.*, 2007). In clinical observations, the fruits were found to be devoid of any toxicity at the therapeutic doses.

Allium sativum

Allium sativum is commonly known as "Garlic" and belongs to the family of Liliaceae. It is a common spice as well as flavoring agent and used since ancient times. Even nowadays garlic is well popular in many cultures throughout the world. Mainly, the bulbs are used as medicinal agents and dietary purposes. It is commercially available as garlic oil, garlic powder, and pills and popularly used for certain medicinal purposes, including lowering blood pressure and improving lipid profile. In experimental studies it has shown to possesses antithrombotic, hypolipidaemic and hypoglycemic agents (Ali *et al.*, 2000; Banerjee and Maulik, 2002). Its constituents include enzymes, alliinase and sulfur-containing compounds, including alliin, and compounds allicin produced enzymatically from alliin. Among several constituents, allyl propyl disulphide and diallyl disulphide are the major active principles (Banerjee *et al.*, 2003).

Evidence of Efficacy

It is one of the best-researched, best-selling herbal remedies and is also commonly used as a food and a spice. Traditionally, it has been employed to treat infections, wounds, diarrhea, rheumatism, heart disease, diabetes, and many other disorders. Evidences from preclinical and clinical studies clearly demonstrate the usefulness of garlic in hyperlipidemia and their therapeutic and preventive properties in ischemic heart diseases (Banerjee and Maulik, 2002; Thomson *et al.*, 2006; Gorinstein *et al.*, 2007).

Preclinical Studies

A study on the effect of intake of *A. sativum* (1 per cent added to standard chow for 10 weeks) on susceptibility to ventricular arrhythmias under ischemia and reperfusion investigated in isolated rat heart perfused with modified Kreb-Henseleit solution resulted in a significant decrease in incidence of ventricular tachycardia and fibrillation after ligation of the descending branch of left coronary artery for 20 min following *A. sativum* pretreatment. Rat pretreated with diallyl disulphide (80 mg/kg body wt.) for 5 weeks and garlic extract showed preventive effects on cardiac enzyme and lipids in isoproterenol-induced myocardial ischemia (Banerjee *et al.*, 2003; Sangeetha and Darlin, 2006). Long term dietary supplementation of fresh garlic extracts (rats) exerts a lipid lowering effects partially through reducing intestinal MTP gene (microsomal triglyceride transfer protein) expression, thus suppressing the assembly and secretion of chylomicron from intestine to the blood circulation (Lin *et al.*, 2002). Raw garlic (500 mg/kg) showed profound effects in reducing the glucose, cholesterol and triglycerides levels when administered orally for 4 weeks, whereas boiled garlic had little effect in controlling these parameters (Mukherjee *et al.*, 2009). Anti-atherosclerotic properties of *A. sativum* are mainly attributed to allicin. The allicin component affects atherosclerosis not only by acting as an anti-oxidant but also by other mechanism such as lipoprotein modification and inhibition of LDL uptake and degradation by macrophages (Gonen *et al.*, 2005; Gorinstein *et al.*, 2007).

Clinical Studies

Garlic derived diallyl disulphide is most effective for inhibition of HMG-CoA reductase, a regulatory enzyme of cholesterol biosynthesis (Kumar *et al.*, 1991). High content of G-glutamyl peptides are responsible for garlic's ability to inhibit angiotensin-converting enzyme (ACE). Clinical and experimental studies indicate that allyl disulphide is most likely to be responsible for the inhibition of cholesterol synthesis by *A. sativum* and inhibition mediated at sterol 4-alpha methyl oxidase (Singh and Porter, 2006). Allicor (garlic preparation) when administered to patients of acute myocardial infarction/CAD for 12 months resulted in decrease in LDL-C by 32.9 mg/dl in males 27.3 mg/dl in females (Sobenin *et al.*, 2007). Similarly, a decrease in lipid levels and an increase in antioxidant levels were observed in 70 patients (30-60 years) of primary arterial hypertension when they were administered garlic preparation (6 capsules per day for 30 days) along with standard antihypertensive therapy (Duda *et al.*, 2008). Therefore, garlic is advocated for its use as a potent vasorelaxant, hypolipidemic, antioxidant as well as potent antiatherogenic agent. The reported adverse effects are nausea, weight loss and gastro intestinal discomfort.

Ocimum sanctum

Ocimum sanctum Linn popularly known as 'Tulsi' in Hindi and 'Holy basil' in English is an indigenous plant sacred to Hindus (Satyavati *et al.*, 1984). The leaves and seeds of the plant have been claimed to be effective in wide spectrum of diseases (Mondal *et al.*, 2009). In traditional medicine including Ayurveda, it is acclaimed to possess therapeutic properties in treating infections, cold, cough, skin and liver diseases and as an antidote for snakebite (Gupta *et al.*, 2002). Indian Materia Medica describes the use of extract of *O. sanctum* leaves in a variety of disorders like bronchitis, rheumatism and pyrexia (Singh *et al.*, 2009). Several recent investigations have indicated that *O. sanctum* possesses significant anti-inflammatory, antioxidant and immunomodulatory properties (Singh *et al.*, 1996; Gupta *et al.*, 2002; Mediratta *et al.*, 2002). Its leaves are reported to contain 0.7 per cent volatile oil comprising of about 71 per cent eugenol and 20 per cent methyl eugenol. It has been also reported to contain several additional components such as carvacrol, sesquiterpine hydrocarbon caryophyllene, apigenin, luteolin, apigenin-7-O-glucuronide, orientin, molludistin and ursolic acid (Gupta *et al.*, 2002). Likewise the active principles of *O. sanctum* comprising of phenols and flavones have been shown to have significant antioxidant and anti-inflammatory activity both *in vivo* and *in vitro* (Singh *et al.*, 2002).

Evidence of Efficacy

The plant has a pungent bitter taste, stomachic and antipyretic properties and traditionally used for heart diseases (Gupta *et al.*, 2002). Evidences available from preclinical studies demonstrate its usefulness in several oxidative stress associated pathogenic conditions owing to antioxidant and adaptogenic activity (Rai *et al.*, 1997; Mohanty *et al.*, 2006). However, clinical data demonstrating its therapeutic effect against heart diseases are lacking but the experimental observations are encouraging for the cardioprotective efficacy of this medicinal plant (Sharma *et al.*, 2001).

Preclinical Studies

O. sanctum significantly decreases total cholesterol, triglycerides, LDL-C and faecal sterol contents of normal rabbits. Aqueous extract of *O. sanctum* showed significant liver and aortic tissue protection by inhibiting lipid peroxidation in hypercholesterolemic rabbits (Sarkar *et al.*, 1994; Kelm *et al.*, 2000). Wistar rats of either sex were fed with fresh leaves homogenate of *O. sanctum* by oral gavage in two different doses (50 mg/kg body wt. and 100 mg/kg body wt.) daily for 30 days, followed by isoproterenol administration (85 mg/kg, body wt.) subcutaneously for two doses at 24 h intervals in both control and *O. sanctum* fed rat to induce myocardial necrosis. The result indicate that long term feeding of *O. sanctum* offered significant protection against isoproterenol-induced myocardial necrosis and coronary artery ligated ischemia-reperfusion injury of myocardium through enhancement of endogenous antioxidant and adaptogenic and antiapoptotic activity (Sood *et al.*, 2005; Mohanty *et al.*, 2006). Administration of *O. sanctum* seed oil (0.8 g/kg body wt/day) for four weeks in cholesterol fed (100 mg/kg, body wt/day) rabbits significantly decreased serum cholesterol, triglycerides and LDL-cholesterol as compared to untreated cholesterol fed groups. There was significant fall in atherogenic index in *O. sanctum* seed oil treated group (Gupta *et al.*, 2006).

Eugenia jambolana

It is commonly known as 'Jamun' in Hindi and 'Black plum or Black berry' in English is a member of family Myrtaceae. The seeds and pulp part is used for medicinal purposes. It is a large tree found in all forests over the greater part of India from Sub-Himalayan tract to extreme south. Several studies have demonstrated the hypoglycemic properties of jamun fruit pulp and seed in animals. *Eugenia jambolana* seeds contain several biological active constituents such a flavonoids, gallic acid, ellagic acid, glycosides, triteripenoids and saponins. *E. jambolana* thus has hypoglycemic, anti-inflammatory, neuropsycho-pharmacological, anti-bacterial, anti-HIV, stomachic, astringent, diuretic and anti-diarrheal effects (Bhatia and Bajaj, 1975). The decoction and extract of dried stem bark is has been believed to be useful as a cardiotonic and in diabetes. Ethanolic extract of seeds not only exhibited hypoglycemic activity but also improves lipid profile and decreases the activity of key enzymes of cholesterol biosynthesis; HMG-CoA reductase, a rate liming enzyme (Prince *et al.*, 1998).

Evidence of Efficacy

The plant has been demonstrated to be useful in diabetes mellitus and heart diseases. Evidences available preclinical studies demonstrate its potential in reducing morbidity and mortality from heart diseases under the diabetic conditions (Sharma *et al.*, 2003). Clinical studies have indicated its potential usefulness in diabetic cardiomyopathy and suggested its therapeutic properties in hyperglycemia.

Preclinical Studies

Active principles possessing anti-hyperglycemic and hypolipidemic activity has been demonstrated in our laboratory in extensive manner. The characterization of these active components have been performed in fruit pulp (US Patent No. 6426826;

Aug 2002; Indian Patent No. 188759; May 2003) and seeds (patent filed, awaited). Biochemical parameters including markers of endothelial dysfunction as well as histopathological studies have revealed anti atherosclerotic properties of the compound along with a significant improvement in macro and microvascular complications on treatment with it in animals. On the basis of studies performed, it can be concluded that this compound, which is relatively non-toxic, and has potential to reverse dyslipidemia as well as associated complications of diabetes through the anti-atherosclerotic action. Based on our study findings, the use of *E. jambolana* as add on therapy may be optional in patients of diabetes who are on risk of developing cardiovascular disease.

Cassia auriculata

Cassia auriculata is popularly known as 'Tarwar' in Hindi and 'Tenner's cassia' or 'Mature tea tree' in English and belongs to family Caesalpiniaceae. It is an Indo-Malaysian species native to India and Sri Lanka and widespread to Southeast Asia. In India, it is particularly common in south India. It is a fast growing and profusely branched, evergreen shrub, characterized by the presence of leafy stipules. The roots of *C. auriculata* are traditionally used in Indian folk medicine for the treatment of diabetes mellitus. The roots are reported to contain flavonoids, polysaccharides, tannins, and saponins, among other components which may contribute to its diverse uses in folklore medicine (Rao *et al.*, 2000).

In Ayurvedic medicine, *C. auriculata* is widely known as 'Kalpa drug' which contains five parts of the shrub (roots, leaves, flowers, unripe fruits and bark) which are taken in equal quantity, dried and powdered to give "Avarai Panchaga Choornam" for the control of sugar levels and reduction of symptoms like polyurea and polydipsia in diabetics. It is one of the constituents of many antidiabetic polyherbal formulations such as Diamed, Hyponnoid, Dianex and Diasulin. It is also used in the treatment of rheumatism, eye diseases, skin diseases and urinary disorders (Pari *et al.*, 2001; Babu and Staneley-Mainzen, 2004; Mutalik *et al.*, 2005; Sarvanan and Pari, 2005).

Evidence of Efficacy

The plant has been traditionally known to be useful in diabetes and in experimental studies in our laboratory found to exhibit potent anti-hyperglycemic and hypolipidemic activity in both mild and severe diabetes comparable to acarbose (Abesundara *et al.*, 2004; Gupta *et al.*, 2009a, 2009b and 2009c). However, the plant is unexplored for its clinical usefulness despite of its ancient use and experimental data.

Preclinical Studies

Pari and Latha (2002) have shown the anti-hyperlipidemic activity of aqueous extract of *C. auriculata* flowers in streptozotocin-induced diabetic rats. Kalaivani *et al.* (2008) has reported the antioxidant effect of ethanolic extract of *C. auriculata* flowers in alloxan-induced diabetic rats. The leaves of *C. auriculata* have also been reported to posses anti-hyperlipidemic and antioxidant activity. Kumar *et al.* (2002) has reported hypolipidemic effect of aqueous extract of *C. auriculata* leaves in rats with alcoholic liver injury, where as Kalaivani *et al.* (2008) has shown antioxidant potential of ethanolic extract in experimental diabetes. We have also demonstrated lipid lowering potential of aqueous leaf extract in STZ-induced diabetic rats (Gupta *et al.*, 2009a and 2009b). The favourable effect of *C. auriculata* on lipid profile and oxidative stress strongly supports the notion that dietary supplementation with *C. auriculata* flowers and leaves may lead to reduction in the risk of developing heart diseases.

The further studies conducted in our lab revealed cardioprotective role of *C. auriculata* aqueous leaf extract in alloxan-induced and streptozotocin-induced diabetic rats (Surana *et al.*, 2008; Gupta *et al.*, 2009c; 2009d). Though hypolipidemic effect was more pronounced on serum triglycerides as compared to LDL and total cholesterol, a remarkable improvement was observed in different markers of endothelial dysfunction. Histopathological examination of heart of *C. auriculata* treated rats also showed the reversal of fatty changes towards normal. A thorough review of the literature did not reveal any information on clinical trials conducted with *C. auriculata.* The studies available did not show any adverse effect on gross behavior of the animals receiving high doses of *C. auriculata* extract. The results of sub-acute toxicity study conducted in our lab suggest that the plant seems to be devoid of toxic effects based on biochemical and histopathological studies on liver and kidney of treated animals.

Inula racemosa

Inula racemosa is commonly available bioactive representative of order inuleae of family compositeae which possess numerous medicinal properties including hypolipidemic and antiischemic (Mangathayaru *et al.*, 2009; Ojha *et al.*, 2010). It is used as a rasayana in ayurveda. In Ayurveda, roots are regarded as a diuretic, stimulant, anthelmentic, anti-inflammatory, digestant, bronchodilator, cardiotonic and aphrodisiac (Tripathi *et al.*, 1988). Sesquiterpenes are the key constituent's which accounts for the bioactivity of this herb (Su *et al.*, 2001). It is acknowledged to be a beta blocker of herbal origin. The herb has been demonstrated to possess antiadrenergic properties (Lokhande *et al.*, 2006).

Evidence of Efficacy

Numerous experimental and clinical studies have demonstrated the cardioprotective activity of *I. racemoa* used as a crude drug extract and in poly herbal formulation, lipistat (Seth *et al.*, 1998; Tripathi, 2009; Ojha *et al.*, 2010).

Preclinical Studies

The root powder of *I. racemoa* has been demonstrated to decrease plasma glucose and insulin levels following adrenaline-induced hyperglycemia in rats and exert negative inotropic and negative chronotropic effects in frog heart (Tripathi *et al.*, 1988). In another study, pretreatment with *Inula racemosa* exhibited significant cardioprotective activity by restoring the level of endogenous GOT, LDH, CPK, cAMP and energy substrates cortisol, pyruvate, lactate and glucose levels in isoproterenol-induced myocardial infarction in rats (Patel *et al.*, 1982). In rats, an herbal formulation consisting of inula in combination with guggul and arjuna at three doses 225, 350, 450 mg/kg/day for 60 days showed a significant reduction in isoproterenol-induced myocardial ischemia by restoration of the myocardial energy levels and preventing accumulation of lactate (Seth *et al.*, 1998).

Conclusions

Several of the medicinal plants have been found effective in animal models of hyperlipidemia, diabetes, atherosclerosis and other cardiovascular diseases (Tripathi, 2009). Some of them have been further studied in clinical studies too and found to confer beneficial effects in cardiovascular diseases. In preclinical studies improved hemodynamics, restored endogenous antioxidant network, prevention of rise in triglycerides level, total cholesterol, and antihyperglycemic effect along with improved histopathological characteristics assures the cardioprotective potential of medicinal plants. Our observation strongly supports the hypothesis that myocardial injury and its deleterious consequences can be prevented by intervention of medicinal plants rich in antioxidants (Mahmood *et al.*, 2010). The medicinal plants are used widely since ancient time as a traditional medicine and in contemporary herbal medicines for the treatment of cardiovascular diseases therefore, following the reverse pharmacological approaches of drug evaluation, more studies are needed to determine their efficacy and safety in clinical settings. In view of the safety, efficacy, traditional acceptability and clinical evaluations for other indications, well-controlled prospective clinical trials of the plants need to be contemplated to establish efficacy in the treatment of ischemic heart diseases.

References

Abesundara, K.J., Matsui, T. and Matsumoto, K. (2004). Alpha-Glucosidase inhibitory activity of some Sri Lanka plant extracts, one of which, *Cassia auriculata*, exerts a strong antihyperglycemic effect in rats comparable to the therapeutic drug acarbose. *Journal of Agricultural and Food Chemistry*, **52**: 2541-2545.

Ahmad, M.U., Mullah, K.B. and Ulla, J.K. (1983). Terminoic acid, a new trihydroxytriterpene carboxylic acid from bark of *Terminalia arjuna*. *Indian Journal of Chemistry*, **22:** 738-740.

Ali, M., Thomson, M. and Afzal, M. (2000). Garlic and onions: their effect on eicosanoid metabolism and its clinical relevance. *Prostaglandins Leukotrienes and Essential Fatty Acids*, **62**: 55-73.

Anila, L. and Vijayalakshmi, N.R. (2002). Flavonoids from *Emblica officinalis* and *Mangifera indica*-effectiveness for dyslipidemia. *Journal of Ethnopharmacology*, **79**: 81-87.

Babu, P.S. and Stanley Mainzen, P. (2004). Antihyperglycaemic and antioxidant effect of hyponidd, an ayurvedic herbomineral formulation in streptozotocin-induced diabetic rats. *Journal of Pharmacy and Pharmacology*, **56**: 1435-1442.

Bafna, P.A. and Balaraman, R. (2005). Antioxidant activity of DHC-1, an herbal formulation, in experimentally-induced cardiac and renal damage. *Phytotherapy Research*, **19**: 216-221.

Banerjee, S.K. and Maulik, S.K. (2002). Effect of garlic on cardiovascular disorders: a review. *Nutrition Journal*, **19**: 1-4.

Banerjee, S.K., Sood, S., Dinda, A.K., Das, T.K. and Maulik, S.K. (2003). Chronic oral administration of raw garlic protects against isoproterenol-induced myocardial necrosis in rat. *Comparative Biochemistry Physiology - Toxicology and Pharmacology,* **136**: 377-386.

Bharani, A., Ganguli, A., Mathur, L.K. Jamra, Y. and Raman, P.G. (2002). Efficacy of *Terminalia arjuna* in chronic stable angina: a double blind placebo, controlled crossover study. Comparing *Terminalia arjuna* with isosorbide mononitrate. *Indian Heart Journal,* **54**: 170-175.

Bhatia, I.S. and Bajaj, K.L. (1975). Chemical constituents of the seeds and bark of *Syzygium cumini. Planta Medica,* **28**: 346-352.

Bhattacharya, S.K., Bhattacharya, A., Sairam, K. and Ghosal, S. (2002). Effect of bioactive tannoid principles of *Emblica officinalis* on ischemia-reperfusion induced oxidative stress in rat heart. *Phytomedicine,* **9**: 171-174.

Brobst, D.E., Dingh, X. and Katrina, L. (2004). Guggulsterone activates multiple nuclear receptors and induce CYOP3A gene expression through the pregnane X receptor. *Journal of Pharmacology and Experimental Therapeutics,* **310**: 528-534.

Chander, R., Rizvi, F., Khanna, A.K. and Pratap, R. (2003). Cardioprotective activity of synthetic guggulsterone (E and Z isomers) in isoproterenol induced myocardial ischemia in rats: a comparative study. *Indian Journal of Clinical Biochemistry,* **18**: 71-79.

Chen, Z.Y, Peng, C., Jiao, R., Wong, Y.M., Yang, N. and Huang, Y. (2009). Anti-hypertensive nutraceuticals and functional foods. *Journal of Agriculture and Food Chemistry,* **57**: 4485-4499.

Chen, Z.Y., Jiao, R. and Ma, K.Y. (2008). Cholesterol-lowering nutraceuticals and functional foods. *Journal of Agricultural and Food Chemistry,* **56**: 8761-8773.

Deng, R. (2007). Therapeutic effects of guggul and its constituent guggulsterone: cardiovascular benefits. *Cardiovascular Drug Reviews,* **25**: 375-390.

Duda, G., Suliburska, J. and Pupek-Musialik, D. (2008). Effects of short-term garlic supplementation on lipid metabolism and antioxidant status in hypertensive adults. *Pharmacological Reports,* **60**: 163-170.

Dwivedi, S. (2007). *Terminalia arjuna* Wight and Arn - a useful drug for cardiovascular disorders. *Journal of Ethnopharmacology,* **114**: 114-129.

Dwivedi, S. and Aggarwal, A. (2009). Indigenous drugs in ischemic heart disease in patients with diabetes. *Journal of Alternative and Complementary Medicine,* **15**: 1215-1221.

Dwivedi, S., Chansouria, J.P.N., Somani, P.N. and Udupa, K.N. (1989). Effect of *Terminalia arjuna* on ischemic heart disease. *Alternative Medicine,* **43**: 115-122.

Ernst, E. (2003). Cardiovascular adverse effects of herbal medicines: a systematic review of the recent literature. *Canadian Journal of Cardiology,* **19**: 818-827.

Fugh, B.A. (2000). Herbs and dietary supplements in the prevention and treatment of cardiovascular disease. *Preventive Cardiology,* **3**: 24-32.

Gauthaman, K., Maulik, M., Kumari, R., Manchanda, S.C., Dinda, A.K. and Maulik, S.K. (2001). Effect of chronic treatment with bark of *Terminalia arjuna*: a study on the isolated ischemic-reperfused heart. *Journal of Ethnopharmacology*, **75**: 197-201.

Gonen, A., Harats, D. and Rabinkov, A. (2005). The antiatherogenic effect of allicin: possible mode of action. *Pathobiology*, **72**: 325-334.

Gorinstein, S., Jastrzebski, Z., Namiesnik, J., Leontowicz, H., Leontowicz, M. and Trakhtenberg, S. (2007). The atherosclerotic heart disease and protecting properties of garlic: contemporary data. *Molecular Nutrition and Food Research*, **51**: 1365-1381.

Gupta, S., Mediratta, P.K., Singh, S., Sharma, K.K. and Shukla, R. (2006). Antidiabetic antihypercholesterolemic and antioxidant effect of *Ocimum sanctum* (Linn) seed oil. *Indian Journal of Experimental Biology*, **44**: 300-304.

Gupta, S., Sharma, S.B. and Prabhu, K.M. (2009a). Ameliorative effect of *Cassia auriculata* L. leaf extract on glycemic control and atherogenic lipid status in alloxan-induced diabetic rabbits. *Indian Journal of Experimental Biology*, **47**: 974-980.

Gupta, S., Sharma, S.B., Bansal, S.K. and Prabhu, K.M. (2009b). Antihyperglycemic and hypolipidemic activity of aqueous extract of *Cassia auriculata* L. leaves in experimental diabetes. *Journal of Ethnopharmacology*, **123**: 499-503.

Gupta, S., Sharma, S.B., Bansal, S.K. and Prabhu, K.M. (2009c). Anti-hyperglycemic and hypolipidemic activity of aqueous extract of *Cassia auriculata* leaves in experimental diabetes. *Journal of Ethnopharmacology*, **123**: 499-503.

Gupta, S., Sharma, S.B., Prabhu, K.M. and Bansal, S.K. (2009d). Protective role of *Cassia auriculata* leaf extract on hyperglycemia-induced oxidative stress and its safety evaluation. *Indian Journal of Biochemistry Biophysics*, **46**: 371-377.

Gupta, S., Sharma, S.B., Singh, U.R., Bansal, S.K. and Prabhu, K.M. (2010). Elucidation of mechanism of action of *Cassia auriculata* leaf extract for its antidiabetic activity in streptozotocin-induced diabetic rats. *Journal of Medicinal Food*, **13**: 528-534.

Gupta, S.K., Prakash, J. and Srivastava, S. (2002). Validation of traditional claim of Tulsi, *Ocimum sanctum* Linn as a medicinal plant. *Indian Journal of Experimental Biology*, **40**, 765-773.

Jacob, A., Pandey, M., Kapoor, S. and Saroja, R. (1988). Effect of Indian gooseberry (Amla) on serum cholesterol level in man aged 35-55 years. *European Journal of Clinical Nutrition*, **42**: 939-944.

Kalaivani, A., Umamaheswari, A., Vinayagam, A. and Kalaivani, K. (2008). Antihyperglycemic and antioxidant properties of *Cassia auriculata* leaves and flowers on alloxan-induced diabetic rats. *Pharmacology online*, **81**: 204-209.

Karthikeyan, K., Bai, B.R., Gauthaman, K., Sathish, K.S. and Devaraj, S.N. (2003). Cardioprotective effect of the alcoholic extract of *Terminalia arjuna* bark in an *in vivo* model of myocardial ischemic reperfusion injury. *Life Science*, **73**: 2727-2731.

Kelm, M.A., Nair, M.G., Strasburg, G.M. and De Witt, D.L. (2000). Antioxidant and cyclooxygenase inhibitory phenolic compounds from *Ocimum sanctum* Linn. *Phytomedicine,* **7**: 7-13.

Kim, H.J., Yokozawa, T., Kim, H.Y., Tohda, C., Rao, T.P. and Juneja, L.R. (2005). Influence of amla (*Emblica officinalis*) on hypercholesterolemia and lipid peroxidation in cholesterol fed rats. *Journal of Nutrition and Science Vitaminology,* **51**: 413-418.

Kumar, R.S., Ponmozhi, M., Viswanathan, P. and Nalini, N. (2002). Effect of *Cassia auriculata* leaf extract on lipids in rats with alcoholic liver injury. *Asia Pacific Journal of Clinical Nutrition,* **11**: 157-163.

Kumar, R.V., Banerji, A., Kurup, K. and Ramasarma, T. (1991). The nature of inhibition of 3-hydroxy-3-methylglutaryl CoA reductase by garlic derived diallyl disulfide. *Biohemica Biophysica Acta,* **24**: 219-225.

Kumar, S., Enjamoori, R., Jaiswal, A., Ray, R., Seth, S. and Maulik, S.K. (2009). Catecholamine-induced myocardial fibrosis and oxidative stress is attenuated by *Terminalia arjuna* (Roxb). *Journal of Pharmacy and Pharmacology,* **61**: 1529-1536.

Lin, M.C., Wang, E.J., Lee, C., Chin, K.T., Liv, D., Chiv, J.F. and Kung, H.F. (2002). Garlic inhibits microsomal triglycoside transfer protein gene expression in human liver and intestinal cell lines and in rat intestine. *Journal of Nutrition,* **132**: 1165-1168.

Lokhande, P.D., Dhaware, B.S., Jagdale, S.C., Chabukswar, A.R. and Mulkalwar, S.A. (2006). Cardiac activity of isolated constituents of *Inula racemosa. Journal of Herbal Pharmacotherapy,* **6**: 81-88.

Mahmood, Z.A., Sualeh, M., Mahmood, S.B. and Karim, M.A. (2010). Herbal treatment for cardiovascular disease the evidence based therapy. *Pakistan Journal of Pharmaceutical Sciences,* **23**: 119-124.

Mangathayaru, K., Kuruvilla, S., Balakrishna, K. and Venkhatesh, J. (2009). Modulatory effect of *Inula racemosa* Hook. f. (Asteraceae) on experimental atherosclerosis in guinea-pigs. *Journal of Pharmacy and Pharmacology,* **61**: 1111-1118.

Mathur, R., Sharma, A., Dixit, V.P. and Varma, M. (1996). Hypolipidaemic effect of fruit juice of *Emblica officinalis* in cholesterol fed rabbits. *Journal of Ethnopharmacology,* **50**: 61-68.

Maulik, S.K. and Katiyar, C.K. (2010). *Terminalia arjuna* in cardiovascular diseases: making the transition from traditional to modern medicine in India. *Current Pharmaceutical Biotechnology,* **11**: 855-860.

Mediratta, P.K., Sharma, K.K. and Singh, S. (2002). Evaluation of immunomodulatory potential of *Ocimum sanctum* seed oil and its possible mechanism of action. *Journal of Ethnopharmacology,* **80**: 15-20.

Miller, A.L. (1998). Botanical influences on cardiovascular disease. *Alternative Medicine Reviews,* **3**: 422-431.

Mohan, I.K., Kumar, K.V., Naidu, M.U., Khan, M. and Sundaram, C. (2006). Protective effect of CardiPro against doxorubicin-induced cardiotoxicity in mice. *Phytomedicine*, **13**: 222-229.

Mohanty, I., Arya, D.S. and Gupta, S.K. (2006). Effect of *Curcuma longa* and *Ocimum sanctum* on myocardial apoptosis in experimentally induced myocardial ischemic-reperfusion injury. *BMC Complementary and Alternative Medicine*, **19**: 3-8.

Mondal, S., Mirdha, B.R. and Mahapatra, S.C. (2009). The science behind sacredness of Tulsi (*Ocimum sanctum* Linn.). *Indian Journal of Physiology and Pharmacology*, **53**: 291-306.

Mukherjee, S., Lekli, I., Goswami, S. and Das, D.K. (2009). Freshly crushed garlic is a superior cardioprotective agent than processed garlic. *Journal of Agriculture and Food Chemistry*, **57**: 7137-7144.

Mutalik, S., Chetana, M., Sulochana, B., Devi, P.U. and Udupa, N. (2005). Effect of Dianex, a herbal formulation on experimentally induced diabetes mellitus. *Phytotherapy Research*, **19**: 409-415.

Ojha, S., Nandave, M., Kumari, S. and Arya, D.S. (2010). Dharamvir S Arya. Cardioprotection by *Inula racemosa* Hook in experimental model of myocardial ischemic reperfusion injury. *Indian Journal of Experimental Biology*, **48**: 918-924.

Ojha, S.K., Nandave, M., Arora, S., Mehra, R.D., Joshi, S., Narang, R. and Arya, D.S. (2008). Effect of *Commiphora mukul* extract on cardiac dysfunction and ventricular function in isoproterenol-induced myocardial infarction. *Indian Journal of Experimental Biology*, **46**: 646-652.

Pari, L. and Latha, M. (2002). Effect of *Cassia auriculata* flowers on blood sugar levels, serum and tissue lipids in streptozotocin diabetic rats, *Singapore Medical Journal*, **43**: 617-622.

Pari, L., Ramakrishan, R. and Venkateswaran, S. (2001). Antihyperglycemic effect of Diamed, a herbal formulation in experimental diabetes in rats. *Journal of Pharmacy and Pharmacology*, **53**: 1139-1143.

Patel, V., Banu, N., Ojha, J.K., Malhotra, O.P. and Udupa, K.N. (1982). Effect of indigenous drug (Pushkarmula) on experimentally induced myocardial infarction in rats. *Acta Nerva Superior* (*Praha*), Suppl **3**(Pt 2): 387-394.

Prince, P.S., Menon, V.P. and Pari, L. (1998). Hypoglycemic activity of *Syzium cumini* seeds: effect on lipid peroxidation in alloxan diabetic rats. *Journal of Ethanopharmacology*, **61**: 1-7.

Rai, V., Iyer, U. and Mani, U.V. (1997). Effect of Tulsi (*Ocimum sanctum*) leaf powder supplementation on blood sugar levels, serum lipids and tissue lipids in diabetic rats. *Plant Foods for Human Nutrition*, **50**: 9-16.

Rajak, S., Banerjee, S.K., Sood, S., Dinda, A.K., Gupta, Y.K., Gupta, S.K. and Maulik, S.K. (2004). *Emblica officinalis* causes myocardial adaptation and protects against oxidative stress in ischemic-reperfusion injury in rats. *Phytotherapy Research*, **18**: 54-60.

Ram, A., Lauria, P. and Gupta, R. (1997). Hypocholesterolaemic effects of *T. arjuna* tree bark. *Journal of Ethnopharmacology,* **55:** 165-169.

Rao, G.N., Kumar, P.M., Dhendapani, S. and Ramakrishna, T. (2000). Constituents of *C. auriculata, Fitoterapia,* **71**: 82-83.

Rege, N.N., Thatte, U.M. and Dahanukar, S.A. (1999). Adaptogenic properties of six rasayana herbs used in Ayurvedic medicine. *Phytotherapy Research,* **13:** 275-291.

Row, L.R., Murthy, P.S. and Subbha, G.S.R. (1970). Chemical examination of Terminalia species: Part XII – Isolation and structure determination of arjunic acid, a new trihydroxytriterpene carboxylic acid from *Terminalia arjuna* bark. *Indian Journal of Chemistry,* **8:** 716-721.

Sangeetha, T. and Darlin, Q.S. (2006). Preventive effects of s-allyl cysteine sulfoxide (allinin) on cardiac marker enzymes and lipids in isoproterenol induced myocardial injury. *Journal of Pharmacy and Pharmacology,* **58**: 617-623.

Saravanan, R. and Pari L. (2005). Antihyperlipidemic and antiperoxidative effect of Diasulin, a polyherbal formulation in alloxan-induced hyperglycemic rats. *BMC Complementary and Alternative Medicine,* **5**: 14-18.

Sarkar, A., Lavania, S.C., Pandey, D.N. and Pant, M.C. (1994). Changes in the blood lipid profile after administration of *Ocimum sanctum* leaves in the normal albino rabbits. *Indian Journal of Physiology and Pharmacology,* **38**: 311-312.

Satyavati, G.V., Gupta, A.K. and Ghatla, N. (1984). *Ocimum sanctum* Linn (Lamiaceae labiateae) in medicinal plants of India (*Indian Council of Medical Research,* New Delhi) Vol. **2**: 354-358.

Seth, S.D., Maulik, M., Katiyar, C.K. and Maulik, S.K. (1998). Role of Lipistat in protection against isoproterenol induced myocardial necrosis in rats: a biochemical and histopathological study. *Indian Journal of Physiology and Pharmacology,* **42**: 101-106.

Sharma, M., Kishore, K., Gupta, S.K., Joshi, S. and Arya, D.S. (2001). Cardioprotective potential of *Ocimum sanctum* in isoproterenol induced myocardial infarction in rats. *Molecular and Cellular Biochemistry,* **225**: 75-83.

Sharma, S.B., Nasir, A., Prabhu, K.M. and Murthy, P.S. (2003). Hypoglycemic and hypolipidemic effect of ethanolic extract of seeds of *Eugenia jambolana* in alloxan induced diabetic models of rabbit. *Journal of Ethnopharmacology,* **85**: 201-206.

Shukla, S.K., Dwivedi, S., Singh, U.R. and Sharma, S.B. (2009). Studies on antioxidant activity of water and hydroalcoholic extract of *Terminalia arjuna* in normal male albino rats. *Journal of Advance Researches in Biological Sciences,* **1**: 6-11.

Shukla, S.K., Gupta, S., Ojha, S.K. and Sharma, S.B. (2010). Cardiovascular friendly natural products: a promising approach in the management of CVD. *Natural Product Research,* **24**: 873-898.

Singh, B.B., Vinjamury, S.P., Der-Martirosian, C., Kubik, E., Mishra, L.C., Shepard, N.P., Singh, V.J., Meier, M. and Madhu, S.G. (2007). Ayurvedic and collateral herbal treatments for hyperlipidemia: a systematic review of randomized controlled trials and quasi-experimental designs. *Alternative Therapeutics and Health Medicine*, **13**: 22-28.

Singh, D.K. and Porter, T.D. (2006). Inhibition of sterol 4-methyl oxidase is the principal mechanism by which garlic decreases the cholesterol synthesis. *Journal of Nutrition*, **136**: 7595-7645.

Singh, G., Singh, A.T., Abraham, A., Bhat, B., Mukherjee, A., Verma, R., Agarwal, S.K. and Jha, S. (2008). Protective effects of *Terminalia arjuna* against doxorubicin-induced cardiotoxicity. *Journal of Ethnopharmacology*, **117**: 123-129.

Singh, R.B., Nian, M.A. and Ghosh, S. (1994). Hypolipidemic and antioxidant effects of *Commiphora mukul* as an adjunct to dietary therapy in patients with hypercholesterolemia, *Cardiovascular Drugs and Therapy*, **8**: 659-664.

Singh, S., Majumdar, D.K. and Rehan, H.M. (1996). Evaluation of anti-inflammatory potential of fixed oil of *Ocimum sanctum* (Holy Basil) and its possible mechanism of action. *Journal of Ethnopharmacology*, **54**: 19-26.

Sirtori, C.R., Galli, C., Anderson, J.W., Sirtori, E. and Arnoldi, A. (2009). Functional foods for dyslipidemia and cardiovascular risk prevention. *Nutrition Research Reviews*, **22**: 244-61.

Sood, S., Narang, D., Dinda, A.K. and Maulik, S.K. (2005). Chronic oral administration of *Ocimum sanctum* Linn augments cardiac endogenous antioxidants and prevents isoproterenol- induced myocardial necrosis in rats. *Journal of Pharmacy and Pharmacology*, **57**: 127-133.

Su, B.N., Takaishi, Y., Yabuuchi, T., Kusumi, T., Tori, M., Takaoka, S., Honda, G., Ito, M., Takeda, Y., Kodzhimatov, O.K. and Ashurmetov, O. (2001). Sesquiterpenes and monoterpenes from the bark of *Inula macrophylla*. *Journal of Natural Products*, **64**: 466-471.

Sumitra, M., Manikandan, P., Kumar, D.A., Arutselvan, N. and Balakrishna, K. (2001). Experimental myocardial necrosis in rats: role of arjunolic acid on platelet aggregation, coagulation and antioxidant status. *Molecular and Cellular Biochemistry*, **224**: 135-142.

Surana, S.J., Gokhale, S.B., Jadhav, R.B., Sawant, R.L. and Wadekar, J.B. (2008). Antihyperglycemic activity of various fractions of *Cassia auriculata* Linn. in alloxan diabetic rats. *Indian Journal of Pharmaceutical Sciences*, **70**: 227-229.

Thomson, M., Al-Qattan, K.A., Bordia, T. and Muslim, A. (2006). Significance of garlic and its constituents in cancer and cardiovascular disease. *Journal of Nutrition*, **136**: 8005-8025.

Tripathi, V.K., Pandey, V.B. and Udupa, K.N. (1992). Arjunolitin, a triterpene glycoside from *Terminalia arjuna*. *Phytochemistry*, **31**: 349-351.

Tripathi, Y.B. (2009). BHUx: a patented polyherbal formulation to prevent hyperlipidemia and atherosclerosis. *Recent Patents in Inflammation, Allergy and Drug Discovery*, **3**: 49-57.

Tripathi, Y.B., Tripathi, P. and Upadhyay, B.N. (1988). Assessment of the adrenergic beta-blocking activity of *Inula racemosa*. *Journal of Ethnopharmacology*, b: 3-9.

Visioli, F., Borsani, L. and Claudio, G. (2000). Diet and prevention of coronary heart disease: the potential role of phytochemicals. *Cardiovascular Research*, **47**: 419-425.

Wang, X., Greilberger, J., Ledinski, G., Kager, G. and Paigen, B. (2004). The hypolipidemic natural product *Commiphora mukul* and its component guggulsterone inhibit oxidative modification of LDL. *Atheroscelrosis*, **172**: 239-246.

Wang, Y.F., Wang, X.Y., Ren, Z., Qian, C.W., Li, Y.C., Kaio, K., Wang, Q.D., Zhang, Y., Zheng, L.Y., Jiang, J.H., Yang, C.R., Liu, Q., Zhang, Y.J. and Wang, Y.F. (2009). Phyllaemblicin B inhibits Coxsackie virus B3 induced apoptosis and myocarditis. *Antiviral Research*, **84**: 150-158.

Wattanapitayakul, S.K., Chularojmontri, L., Herunsalee, A., Charuchongkolwongse, S., Niumsakul, S. and Bauer, J.A. (2005). Screening of antioxidants from medicinal plants for cardioprotective effect against doxorubicin toxicity. *Basic and Clinical Pharmacology and Toxicology*, **96**: 80-87.

Yokozawa, T., Kim, H.Y., Kim, H.J., Okubo, T., Chu, D.C. and Juneja, L.R. (2007). Amla (*Emblica officinalis* Gaertn.) prevents dyslipidaemia and oxidative stress in the ageing process. *British Journal of Nutrition*, **97**: 1187-1195.

Traditional and Folk Herbal Medicine: Recent Researches Vol. 1 (2012) *Pages* **175–190**
Editor: **V.K. Gupta**
Published by: **DAYA PUBLISHING HOUSE, NEW DELHI**

8

Recent Investigations on Therapeutic Potential of *Clitoria ternatea* Linn.

Y.B. Solanki[1]* and S.M. Jain[1]

ABSTRACT

Clitoria ternatea is one of important medicinal plant of traditional Ayurvedic system of medicine, employed in variety of disease conditions like bronchitis, asthma, inflammation, swollen joints, enlargement of abdominal viscera, eye ailments, and animal stings. It has been reported to contain fixed oil, water-soluble mucilage- delphinidin-3,3,5-triglucoside, three unidentified trypsin inhibitors, anthocyanins - ternatins, p-hydroxycinnamic acid, flavonol-3-glycoside, ethyl-α-D-galactopyranoside, adenosine, 3,5,7,4′-tetrahydroxyflavone-3-rhamnoglucoside, a polypeptide, hexacosanol, β-sitosterol, γ-sitosterol, anthoxanthin glucoside, flutulene, finotin, and flavonol glycoside. It has well documented brain tonic, anxiolytic, anticonvulsant, analgesic, anti-pyretic, anti-inflammatory, anti-diabetic, anti-microbial, diuretic, and local anesthetic activities. Recently, it has been reported to have anti-hyperlipidemic, hepatoprotective, immunosuppressant, and wound healing properties. Further, it has been proposed that inhibition of inflammatory and immune components might be responsible for these activities along with anti-oxidant properties. These studies provided evidences for the traditional uses of the plant and suggested new therapeutic potential in chronic low inflammatory diseases associated with age.

Keywords: Anti-hyperlipidemic, Butterfly pea, Hepatoprotective, Immunosuppressive, Wound healing.

1 Department of Pharmacology, L.M. College of Pharmacy, Navrangpura, Ahmedabad – 380 009, Gujarat, India.

* *Corresponding author*: E-mail: dy_solanki@yahoo.co.in; Tel: +91-79-26302746

Introduction

Clitoria ternatea Linn. (Family: Fabaceae) is popularly known as butterfly pea in western countries and as Aparajita in the traditional Ayurvedic system of medicine. Traditionally, *Clitoria ternatea* (CT) is employed in variety of disease conditions as laxative, diuretic, purgative, and as antidote to animal stings. The roots of blue-flowered variety are aphrodisiac, cure dysentery, sever bronchitis, and asthma. It is used in the treatment of various conditions like indigestion, constipation, fever, arthritis, and eye ailments. Root and seeds are demulcent and given in chronic bronchitis, enlargement of abdominal viscera, sore throat, and skin diseases. The seeds are considered for colic, dropsy, and swollen joints. It is one of the constituents of Ayurvedic preparations "*Misraka sneha*" and "*Vata rakta antaka rasa*". In India, it is sold under the name of "Sankhapushpi" and used as a brain tonic.

Being important medicinal plant of Ayurvedic system of medicine, it has long been explored scientifically for its therapeutic potential for diuretic, local anesthetics, analgesic, anti-pyretic, anti-inflammatory, nootropic, anxiolytic, anti-depressant, anti-convulsant, sedative and anti-stress properties (Mukherjee *et al.*, 2008). Various parts of Plant have been investigated for anti-diabetic, platelet-aggregation-inhibition, and vascular smooth muscle relaxing properties. Recently, the plant has been explored for its anti-hyperlipidemic, hepatoprotective, immunosuppressive, and wound healing activities. The current review focused on the recent pharmacological investigations of CT.

Botanical Source and Geographical Distribution

Clitoria ternatea Linn. is perennial climber belonging to family: Fabaceae (alt. Leguminoceae) and subfamily: *Faboideae*. It is also placed in family: Papilionaceae. It occurs as two varieties- blue one and white one. It is also known as *Clitoria albiflora* Mattei, *Clitoria bracteata* Poir., *Clitoria mearnsii* De Wild., *Clitoria tanganicensis* Micheli, *and Clitoria zanzibarensis* Vatke. Botanically the plant can be classified as shown in Table 8.1.

Table 8.1: Botanical classification of *Clitoria ternatea* Linn.

Kingdom	*Plantae*
Sub-kingdom	*Tracheobionta*
Super-kingdom	*Spermatophyta*
Division	*Magnoliophyta*
Class	*Magnoliopsoda*
Sub-class	*Rosidae*
Order	*Fabales*
Family	*Fabaceae*
Genus	*Clitorea*
Species	*Ternatea*

The plant is popularly known as Butterfly pea or conch flower. In India, it is also known as Sankhapuspi, Girikan, and Visnukrana. In the traditional Indian system of medicine, it is known as Aparajita, Koyala (Hindi), Gokarni (Gujarati), Aparajita (Bengali), Dintena (Telugu), Sangu pushpam (Malayalam), and Kakkattan (Tamil). The plant is also known as Blue-pea (Australia); Honte (French); Blaue klitorie (German); Clitoria-azul (Portugese); Bejuco de conchitas (Spanish); Cunha (Brazil); Pukingan (Phillippine); Bunga telang (Malaysia). The plant is distributed throughout the tropical counties of word including America, Africa, southern Asia, and Australia.

Morphological and Microscopic Description

Root system consists of a fairly stout taproot with few branches and many slender lateral roots. The root is woody, cream white with a few lenticels united to form transverse cracks. The fresh root is slightly bitter and acrid in taste. The transverse section shows outermost phloem composed of 12–25 rows of thin walled longitudinally elongated cells some of which are compressed and a few exfoliating. Phellogen is single layered and phelloderm is two to three layered, some cells contain rhomboidal crystals of calcium oxalate. Cortex is composed of 10–12 layers of thin-walled almost polygonal or tangentially elongated cells, packed with mostly compound starch grains (Shah and Bole, 1961). Some of the cortical cells contain rhomboidal crystals of calcium oxalate. Central core consists of vascular elements. Phloem appears as conical strands separated by narrow medullary rays. Phloem fibers in groups of two to eight or a few solitary fibers are present. Some of the phloem parenchyma cells contain starch grains and a few others contain calcium oxalate crystals. Woody elements form the major and central part of the root consisting of vessels, wood parenchyma, wood fibers and mostly uniseriate to triseriate medullary rays. Three rays which start from the center is wider and four to five seriate. All the ray cells are fully packed with starch grains and few contain calcium oxalate crystals (Anonymous, 2001; Kalamani and Michael, 2001, 2003). Stems are fine twining, sparsely pubescent, suberect at base, 0.5-3 m long. The pods are 5–10 cm long, flat, nearly straight, sharply beaked and 6–11 seeded (Karandikar and Satakopan, 1959; Pillai, 1976). Seeds are subglobose, oblong or oval, somewhat flattened, 4.5-7 mm long, 3-4 mm wide, 6-11 per pod, brown to almost black, shiny, often mottled, minutely pitted; 23,000 seeds/kg (Figure 8.1). The plant possesses imparipinnate leaves consisting of five to seven leaflets, 6–13 cm long. The leaflets are ovate or oblong, 2–5cm long and subcoriaceous, rubiaceous stomata with wavy cell walls are present on both upper and lower epidermis of the leaflets. Multicellular trichomes, with two basal cells smaller than the terminal cells, are present. In transverse section leaf shows a dorsiventral structure. The lower epidermal cells of midrib region are papillose whereas those of lamina are not so noticeably papillose. All along the veins prismatic crystals of calcium oxalate are present (Mukherjee *et al.*, 2008).

Phytochemical Review

The roots form nodules, which contain higher amount of plant growth substance such as indole acetic acid, kinetin and gibberelic acid. Rajagopalan (1964) investigated the presence of free amino acids and amides in the root nodules; the root nodules contains glycine, alanine, valine, leucine, α-aminobutyric acid, γ-aminobutyric acid, aspartic acid, glutamic acid, γ-methyleneglutamic acid, arginine, ornithine, histidine and γ-aminobutyric acid. Banerjee and Chakravarti (1963, 1964) reported isolation and identification of pentacyclic triterpenoids, taraxerol and taraxerone from the roots (Banerjee and Chakravarti, 1963, 1964). Kumar *et al.* (2008) standardized root of *C. ternatea* for taraxerol content by high performance thin layer chromatography.

The seeds contain greenish-yellow fixed oil (Tiwari and Gupta, 1957). The fatty acid content of CT includes palmitic, stearic, oleic, linoleic, and linolenic acids (Vianni and Souto, 1971; Debnath *et al.*, 1975; Joshi *et al.*, 1981; Husain and Devi, 1998). The

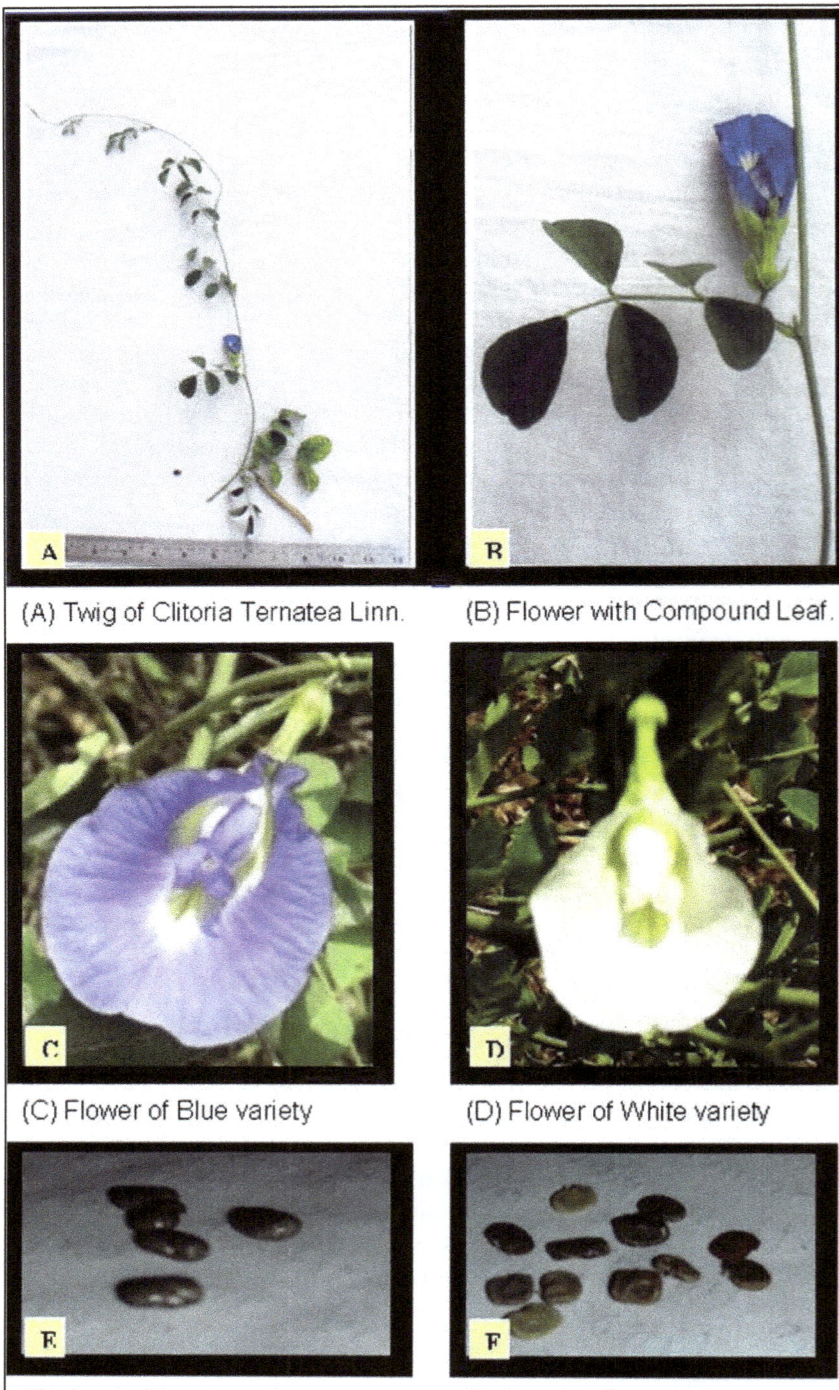

(A) Twig of Clitoria Ternatea Linn.

(B) Flower with Compound Leaf.

(C) Flower of Blue variety

(D) Flower of White variety

(E) Seeds (Good quality)

(F) Seeds (Poor quality)

Figure 8.1: Photograph of *C. ternatea* plant showing morphological features.

Figure 8.2: Secondary metabolites of *C. ternatea*.

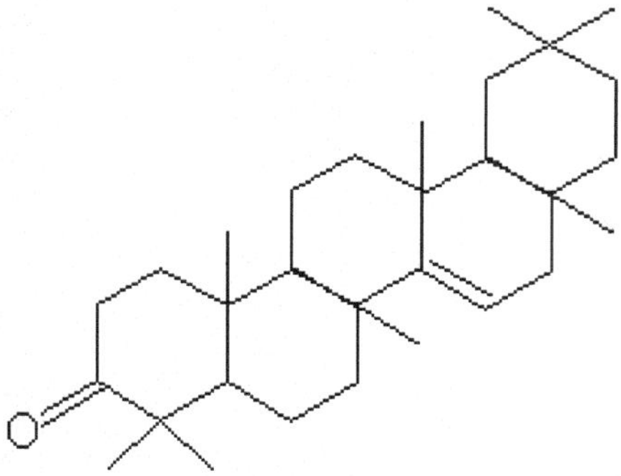

Taraxerone

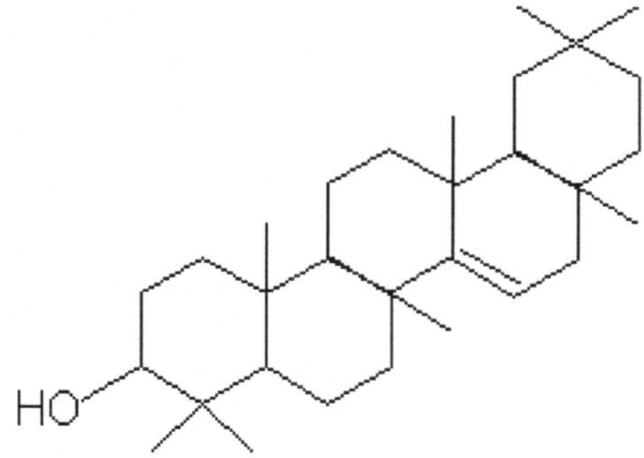

Taraxerol

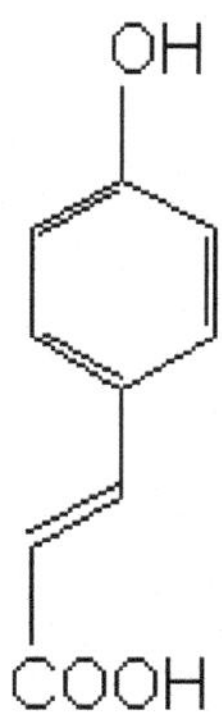

p-Hydroxycinnamic acid

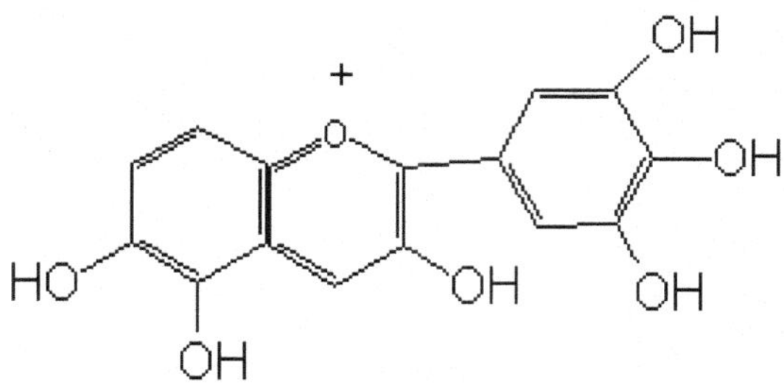

Delphinidin

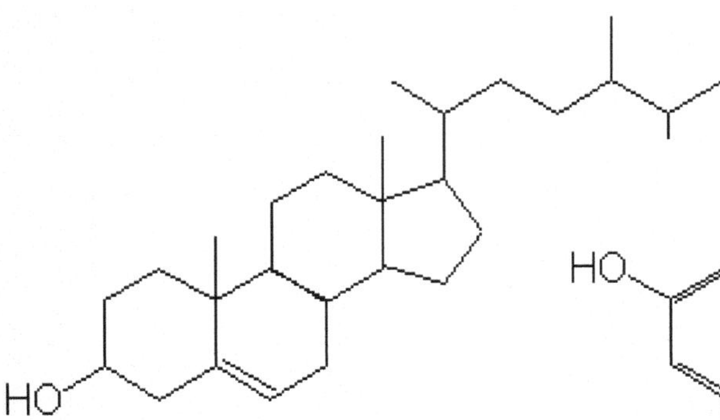

B-sitosterol

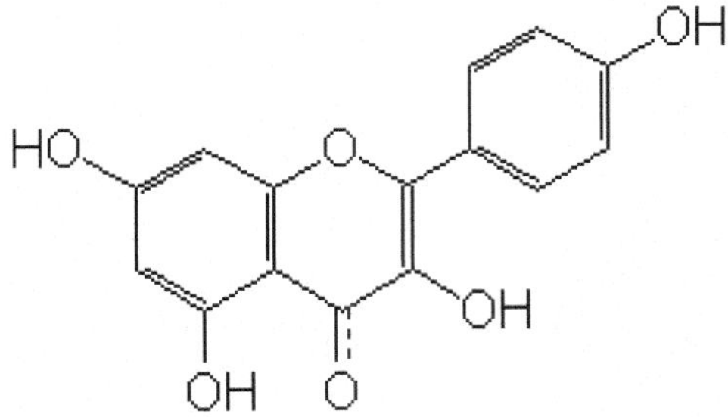

Kaempferol

Contd...

Figure 8.2–*Contd...*

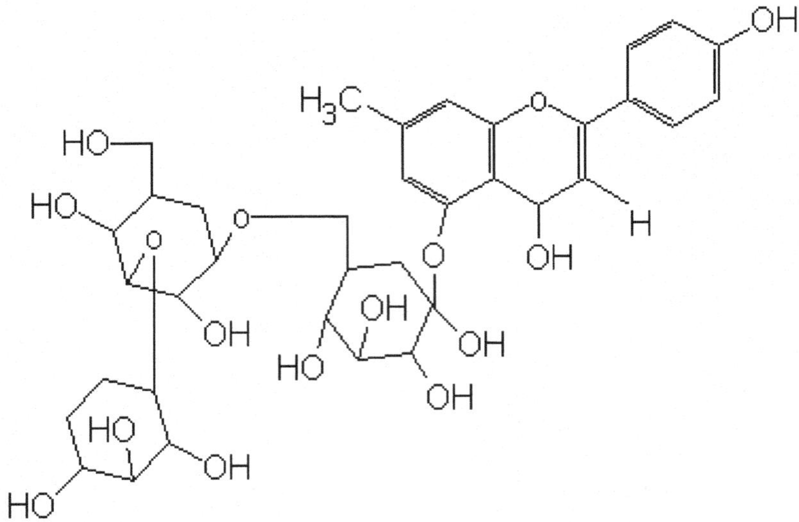

Flavonol glycoside 3,5,4'-trihydrocy-7-methoxyflavonol-3-O-β-D-xylopyranosyl-(1,3)-O-β-D-galactopyranosyl (1,6)-O-β-D-glucopyranoside.

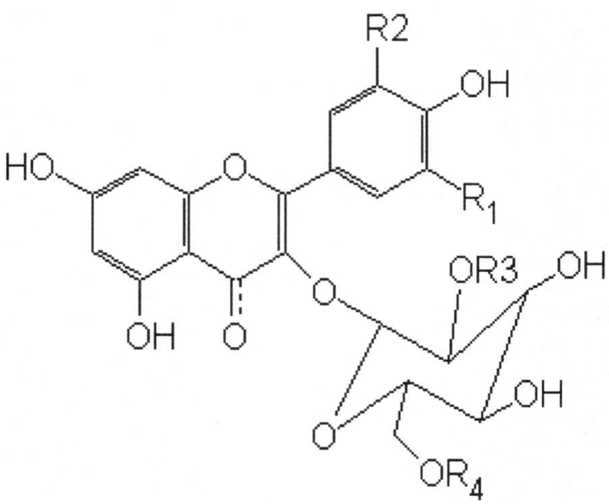

Flavonoid glycosides
(R1 and R2 may be H and/or OH)
R3 and R4 may be Sugar moiety)

seeds also contain a water-soluble mucilage, delphinidin 3,3,5-triglucoside useful as a food dye, besides, three unidentified trypsin inhibitors (Macedo and Xavier-Filho, 1992). Other substances present in the seeds are *p*-hydroxycinnamic acid, flavonol-3-glycoside, ethyl-α-D-galactopyranoside, adenosine, 3,5,7,4′-tetrahydroxyflavone, 3-rhamnoglucoside, α-polypeptide, hexacosanol, β-sitosterol, γ-sitosterol and an anthoxanthin glucoside (Sinha, 1960a,b; Kulshrestha and Khare, 1967, 1968; Gupta and Lal, 1968). The seeds contain oligosaccharides or flutulene (Revilleza *et al.*, 1990). Kelemu *et al.* (2004) isolated antimicrobial and insecticidal protein finotin from seeds of CT.

Yadava and Verma (2003) isolated anti-microbial flavonol glycoside-3,5,4′-trihydroxy-7-methoxyflavonol-3-O-β-D-xylopyranosyl(1,3)-O-β-D-galacto-pyranosyl(1,6)-O-β-dglucopyrano-side from the ethyl acetate soluble fraction of the defatted seeds. The flowers were reported to contain anthocyanins called ternatins (Terahara *et al.*, 1989a,b, 1990a,b,c,d; Kondo *et al.*, 1990). Ranaganayaki and Singh (1979) reported kaempferol isolation and identification from the flowers of CT and Saito *et al.* (1985) detected kaempferol, kaempferol 3-2G-rhamnosylrutinoside, kaempferol 3-neohesperidoside, kaempferol 3-rutinoside, kaempferol 3- glucoside, quercetin, quercetin 3-2G-rhamnosylrutinoside, quercetin 3-neohesperidoside, Quercetin 3-rutinoside, quercetin 3-glucoside, Myricetin 3-neohesperidoside, Myricetin 3-rutinoside and Myricetin 3-glucoside. Three flavanol glycosides, kaempferol 3-O-(2″-O-α-rhamnosyl-6″-O-malonyl)-β-glucoside, quercetin 3-O-(2″-O-α-rhamnosyl-6″-O-malonyl)-β-glucoside, and myricetin 3-O-(2″,6″-di-O-α-rhamnosyl)-β-glucoside were isolated from the petals of CT together with eleven known flavanol glycosides (Kazuma *et al.*, 2004). The leaves contain β-sitosterol, kaempferol-3-monoglucoside, kaempferol-3-rutinoside, kaempferol-3-neohesperiodoside, kaempferol- 3-O-rhamnosyl-(1,6)-glucoside, kaempferol-3-O-rhamnosyl-(1, 6)-galactoside and kaempferol-3-O-rhamnosyl-(1,2)-Ochalmnosyl-(1,2)-O-[rhamnosyl-(1,6)]-glucoside. Lactones aparajitin and clitorin from leaves were also reported (Tiwari and Gupta, 1959; Morita *et al.*, 1977; Ripperger, 1978).

Medicinal and Non-medicinal Uses

The leaves and roots of CT are used in the treatment of a number of ailments including body aches, infections, urinogenital disorders, and as an anthelmentic and antidote to animal stings. The roots have purgative, laxative and diuretic properties. The root of blue-flowered variety has all the properties of the white-flowered variety, in addition, it is aphrodisiac, cures dysentery, sever bronchitis, and asthma. The root is used in the treatment of various diseases, like indigestion, constipation, fever, arthritis and eye ailments. It is also employed in cases of ascetics, enlargement of the abdominal viscera, sore throat, and skin diseases (Anonymous, 1995). They are also demulcent and given in chronic bronchitis. Though they are purgative, they cause griping and tenderness, and hence are not recommended (Nadkarni, 1976). They are, however, administered with honey and ghee as a general tonic to children for improving mental activities, muscular strength, as complexion tonics, and in epilepsy as well as insanity (Anonymous, 1976). The root-juice of the white-flowered variety is blown up the nostrils as a remedy for hemicrania. The decoction or powder of root is given in rheumatism, and ear-diseases. Powdered seeds are mixed with ginger and given as laxative; however, this action is accompanied by griping in lower abdomen. The seeds are considered for colic, dropsy and enlargement of abdominal viscera, and are also used in swollen joints (Morris, 1999; Anonymous, 2001). The root, stem, and flower are recommended for the treatment of snakebite and scorpion sting in India (Kirtikar and Basu, 1935). It is one of important constituents of Ayurvedic preparations "*Misraka sneha*", and "*Vata rakta antaka rasa*". In India, the roots and seeds are sold under the name of "*Shankhpushpi*" and are used as a "*brain tonic*".

In Cuba, decoction of roots and flowers is considered emmenagogue. The roots are also used to promote menstruation and induce uterine contractions. The preparation of flowers and roots known as *medicamento magnífico* (magnificent medicine) is used to treat clorosis (probably anemia), liver disorders, and intestinal problems. Seeds are said to be laxative, vermifugal, and slightly emetic. Fantz (1991) reported economic uses for 23 species of *Clitoria*. The seeds are used as antihelminthic, diuretic, antidote in poison, and refrigerant (Duke, 1986) as well as in reproductive disorders.

CT is a highly palatable forage legume exhibiting excellent re-growth after cutting or grazing within short period and produce high yields. It yields a useful green fodder throughout the year (Anonymous, 2001). It is also used as a cover crop and green manure. Due to its attractive flower color, it is also grown as an ornamental plant (Michael and Kalamani, 2003). The young shoots, leaves, flowers and tender pods are eaten as vegetable in India and Philippines. In Malaysia, the leaves are employed to impart a green color to food, and the flowers to impart a bright blue color to rice cakes.

Pharmacological Review

CT has been widely screened for its various pharmacological activities. It has relatively well documented neuropharmacological actions such as enhancing acetylcholine content, nootropic, anti-stress, anxiolytic, antidepressant, anticonvulsant, tranquilizing and sedative activities which justify its use in CNS diseases in Ayurvedic system of medicine. It has antimicrobial, antipyretic, anti-inflammatory, analgesic, diuretic, local anesthetic, anti-diabetic, insecticidal, blood platelet aggregation inhibiting and vascular smooth muscle relaxant properties. Mukherjee *et al.* (2008) have discussed the pharmacological activities of CT in detail (Table 8.2).

Recent Investigations on Therapeutic Potentials

Anti-hyperlipidemic Activity

The seed and root extracts of blue-flowered variety were investigated for their anti-hyperlipidemic activity against both poloxamer-407-induced acute hyperlipidemia, and diet-induced hyperlipidemia in rats (Solanki, 2010). In poloxamer-407 induced acute hyperlipidemia model, CT seed and root extracts (500 mg/kg, p.o.) decreased serum triglyceride levels both at 15 and 24 h. These effects

Table 8.2: Pharmacological screening of *C. ternatea* Linn.

Si.No.	*Plant Part Extract/Compound(s)*	*Activity*	*Reference*
1	Roots Alcoholic extract	Diuretic activity in dogs	Piala *et al.* (1962)
2	Aerial parts Alcoholic extract	Local anesthetic effect	Kulkarni *et al.* (1988)
3	Roots, and Methanolic extract, aerial parts alcoholic extract	Antipyretic acitivty	Kulkarni *et al.* (1988), and Parimaladevi *et al.* (2003, 2004).
4	Roots Methanolic extract,	Anti-inflammatory and analgesic Activity	Kulkarni *et al.* (1988), and Parimaladevi *et al.* (2003).
5	Roots and aerial Alcoholic extract, parts methanolic extract	Nootropic, anxiolytic, antidepressant, sedative, tranquilizing, anticonvulsant, and anti-stress activity.	Kulkarni *et al.* (1988), Boominathan *et al.* (2003), and Jain *et al.* (2003).
6	Flowers Ethanol extract	Anti-diabetic activity	Sharma and Majumdar (1990).
7	Flowers Anthocyanins (Ternatin A1, A2, B1, B2, D1, and D2)	Blood-platelet aggregation-inhibiting and vascular smooth muscle relaxing activity.	Honda *et al.* (1991).
8	Seeds/Flavonoid glycoside	Anti-microbial activity	Osborn *et al.* (1995), Yadava and Verma (2003).
9	Roots and aerial aqueous extract	Learning and memory enhancing, enhancement of acetylcholine content.	Rai *et al.* (2000a, b, 2001, 2002, 2005), Taranalli and Cheeramkuczhi (2000), and Ng *et al.* (2006).
10	Seeds/Fitotin	Anti-microbial and insecticidal activity.	Kelemu *et al.* (2004).

were found to be comparable with that of atorvastatin and gemfibrozil. Since, poloxamer-407-induced acute hyperlipidemia is characterized by inhibition of vascular (heparin extractable) pool of lipoprotein lipase; it is likely that plant extracts affect lipoprotein lipase activity either by preventing its inhibition or increasing its activity. The same study also investigated effects of both the extracts on diet-induced hyperlipidemia in rats, and demonstrated significant anti-hyperlipidemic activity as evident from significant decrease in serum total cholesterol, triglyceride, low density lipoprotein, and atherogenic index. It is the oxidative modification of lipids especially low density lipoprotein, that is responsible for the further complications of hyperlipidemia *i.e.* atherosclerosis. In the same study, it has been observed that both the extracts significantly prevented lipid peroxidation in liver tissues and improved natural antioxidant mechanisms by decreasing superoxide dismutase (SOD), catalase, and increasing reduced glutathione (GSH). Vitamin-C is the only water soluble anti-oxidant vitamin of the body and involved in the regeneration of vitamin-E. The effects on serum vitamin-C levels showed that L-ascorbic acid levels were significantly increased in extracts treated rats. Attempts have also been made to demonstrate possible mechanisms of anti-hyperlipidemic activity by measuring effects on 3-hydroxy-3-methyl-glutaryl-CoA (HMG CoA) reductase activity, hepatic microsomal enzyme activities, and fecal bile acid as well as cholesterol excretion. It has been observed that the seed and root extracts did not affect HMG CoA reductase or hepatic microsomal enzymes activities, but significantly increased fecal bile acid and cholesterol excretion. The plant sterols were reported to produce anti-hyperlipidemic activity by increasing fecal bile acid excretion and decreasing gastrointestinal absorption of cholesterol. CT has also been reported to contain sterols like β-sitosterol and γ-sitosterol. Therefore, it has been proposed to produce anti-hyperlipidemic effect by increasing fecal bile acid excretion, and decreasing gastrointestinal absorption of cholesterol. Since low grade inflammation, oxidative modification of lipid, and hyperlipidemia are closely associated with the development and progress of atherosclerosis; the CT could be promising herbal remedy in its treatment.

Hepatoprotective Activity

Liver is the largest organ of the body responsible for the metabolism of wide variety of substance, both internal as well as external. It is constantly exposed to environmental toxicants and liver diseases are leading cause of death. Hydroalcoholic extracts of CT seed and root have been investigated for their hepatoprotective action against paracetamol and CCl_4-induced liver injuries in rats (Solanki and Jain, 2011). CT seed and root extracts at 500 mg/kg, p.o. dose, showed significant protection against CCl_4-induced liver injuries. It has been found to decrease serum marker enzymes-aspartate animotransferase (AST), alanine animotransferase (ALT), and alkaline phosphatase (ALP). The serum direct and indirect bilirubin levels were also significantly decreased. The same study attempted to demonstrate possible mechanisms of liver protection. The leukocytes infiltration in liver tissues is the hallmark of liver injuries, which was suppressed by the root and seed extracts. They also decreased hepatic collagen content measured in terms of L-hydroxyproline levels and thereby prevented fibrogenesis. In addition to these, they also inhibited hepatic lipid peroxidation and improved natural anti-oxidant defense mechanisms, *i.e.*, decreased SOD and catalase activities, and increased GSH levels. The histopathological observations of liver tissues also supported these findings.

In case of paracetamol-induced liver injury, only CT seed extract showed protective effects by decreasing serum marker enzymes, direct and indirect bilirubin, and lipid peroxidation. The root extract although, inhibited lipid peroxidation, did not reduced serum marker enzymes. The phytochemical analysis of CT seed and root carried out at our Laboratory showed presence of flavonoid compounds especially kaempferol in glycoside forms in seed only (Solanki, 2010). This could partly explain why only seed extract showed hepatoprotection against both the toxicants. Suppression of leukocyte infiltration and liver collagen synthesis as well as prevention of liver lipid peroxidation could be possible mechanisms of hepatoprotective action of plant. Besides these, inhibition of inflammatory changes and improvement of natural antioxidant defense mechanisms could also be possible factors contributing to liver protective. Thus, CT could be used in the treatment of liver problems. However, it requires further exploration in humans.

Immunosuppressive Activity

Study by Solanki and Jain (2011) reported an immunosuppressive activity of CT seed and root extracts in male albino rats. It has been found that seed and root extracts (500 mg/kg, p.o.) decreased primary and secondary antibody titers in sheep red blood cell (SRBC) sensitized rats. The delayed type of hypersensitivity (DTH) response was also decreased by seed and root extracts in SRBC-sensitized rats. Further, the same study also reported significant decrease in neutrophil index and in vivo phagocytosis measured by carbon clearance method. These finding suggested that CT seed and root extracts inhibited humoral, cell-mediated, and non-specific immune responses like neutrophil activity and phagocytosis. Since the plant did not showed significant decrease in the total leukocytes counts, the inhibition of innate component of inflammatory response is like the possible mechanism of action. It has been proposed that the plant constituents like flavonoid and phenolic compounds might be responsible for the inhibition of inflammatory response and there by produce immunosupression. Further investigations are required to explore immuno-suppressive potential of the plant.

Wound Healing Activity

Wound healing is a complex and highly dynamic process that rapidly close a wound, and prevents infection after injury. It can be broadly divided into inflammatory, proliferative, and remodeling phases that involve numerous different cell types, some from the local area, while others are recruited upon injury (Gurtner *et al.*, 2008). Study at our laboratory (unpublished data) also showed that CT seed and root extracts increase wound healing process against excision, incision and dead space wound model. When applied topically or given orally, they significantly increased wound closure, tensile strength, and granuloma formation. It has been observed that plant produced effects on wound contracture, fibrogenesis, and collagen synthesis. The so observed anti-inflammatory and immunosuppressive activities could be responsible for the wound healing activity of plant.

Since, the plant has been reported to contain flavonoid compounds in glycoside form and phenolic compounds. The study at our laboratory also demonstrated presence of glycosidic form in seed only. Since root also showed similar activities as seeds, there might be other phytochemical constituents also responsible for these activities.

Summary

Clitoria ternatea Linn. possesses significant anti-hyperlipidemic, hepatoprotective, immunosuppressive and wound healing activities against experimental models. Inhibition of inflammatory and immune components might be responsible for these activities. Studies have provided evidence for traditional uses of CT in variety of disease conditions. Recent investigations on therapeutic potential of CT demonstrated efficacy of the plant in chronic low inflammatory diseases associated with age, opening new field for further exploration.

Acknowledgements

Authors are thankful to Gujarat Council for Science and Technology (GUJCOST), Gandhinagar for providing financial assistance for investigations on *Clitoria ternatea*.

References

Anonymous (1976). Medicinal Plants of India, Vol. I, Indian Council of Medical Research, New Delhi.

Anonymous (1995). Indian Medicinal Plants, Vol. 2, Orient Longman, Madras.

Anonymous (2001). The Wealth of India: A Dictionary of Indian Raw Materials and Industrial Products: First Supplement Series (Raw Materials), Vol. I, National Institute of Science and Communication, New Delhi.

Banerjee, S.K. and Chakravarti, R.N. (1963). Taraxerol from *Clitoria ternatea*. *Bulletine of the Calcutta School of Tropical Medicine*, **11:** 106-107.

Banerjee, S.K. and Chakravarti, R.N. (1964). Taraxerone from *Clitoria ternatea*. *Bulletin of the Calcutta School of Tropical Medicine*, **12:** 23.

Boominathan, R., Parimaladevi, B. and Mandal, S.C. (2003). Studies on neuropharmacological effects of *Clitoria ternatea* Linn. root extract in rats and mice. *Natatural Product Science*, **9:** 260-263.

Debnath, N.B., Chakravarti, D., Ghosh, A. and Chakravarti, R.N. (1975). Fatty acids of *Clitoria ternatea* seed oils. *Journal of the Institution of Chemists (India)*, **47:** 253-255.

Duke, J.A. (1986). Isthmian Ethnobotanical Dictionary, 3rd edition, Scientific Publishers, Jodhpur.

Fantz, P.R. (1991). Ethnobotany of *Clitoria* (Leguminosae). *Economic Botany*, **45:** 511-520.

Gupta, R.K. and Lal, L.B. (1968). Chemical components of the seeds of *Clitoria ternatea*. *Indian Journal of Pharmacy*, **30:** 167-169.

Gurtner, G.C., Werner, S., Barrandon, Y. and Longaker, M.T. (2008). Wound repair and regeneration. *Nature.* **453(7193):** 314–321.

Honda, T., Saito, N., Kusano, T., Ishisone, H., Funayama, N., Kubota, T. and Araogi, S. (1991). Isolation of anthocyanins (Ternatin A1, A2, B1, B2, D1, and D2) from *Clitoria ternatea* cv. (double blue) having blood platelet aggregation-inhibiting and vascular smooth muscle relaxing activities. *Japan Kokai Tokyo Koho.* 7.

Husain, S. and Devi, K.S. (1998). Fatty acid composition of three plant species: *Clitorea ternatea, Mandulea suberosa* and *Ruta chalapensis. Journal of Oil Technologist Association of India,* **30:** 162-164.

Jain, N.N., Ohal, C.C., Shroff, S.K., Bhutada, R.H., Somani, R.S., Kasture, V.S. and Kasture, S.B. (2003). *Clitoria ternatea* and the CNS. *Pharmacology Biochemistry and Behaviour,* **75:** 529-536.

Joshi, S.S., Shrivastava, R.K., Shrivastava, D.K. (1981). Chemical examination of *Clitoria ternatea* seeds. *Journal of American Oil and Chemical Society,* **58:** 714-715.

Kalamani, A. and Michael, G.S. (2001). Genetic variability in *Clitoria* spp. *Annals of Agricultural Research,* **22:** 243-245.

Kalamani, A. and Michael, G.S. (2003). Exploitation of new ornamental types in *Clitoria* (*Clitoria* spp.). *International Journal Mendel,* **20:** 41-42.

Karandikar, G.K. and Satakopan, S. (1959). Shankhpushpi – a pharmacognostic study-III. *Clitoria ternatea* Linn. *Indian Journal Pharmacology,* **21:** 327-331.

Kazuma, K., Kogawa, K., Noda, N., Kato, N. and Suzuki, M. (2004). Identification of delphinidin 3-O-(6″-O-malonyl)-β-glucoside-3″-O-β-glucoside, a postulated intermediate in the biosynthesis of ternatin C5 in the blue petals of *Clitoria ternatea* (butterfly pea). *Chemistry and Biodiversity,* **1:** 1762-1770.

Kelemu, S., Cardona, C. and Segura, G. (2004). Antimicrobial and insecticidal protein isolated from seeds of *Clitoria ternatea,* a tropical forage legume. *Plant Physiology and Biochemistry,* **42:** 867-873.

Kirtikar, K.R. and Basu, B.D. (1935). Indian Medicinal Plants. (Basu, L.M., Eds.), Allahabad, India.

Kondo, T., Ueda, M. and Goto, T. (1990). Structure of ternatin B1, a penta acylated anthocyanin substituted on the B-ring asymmetrically with two long chains. *Tetrahedron,* **46:** 4749-4756.

Kulkarni, C., Pattanshetty, J.R. and Amruthraj, G. (1988). Effect of alcoholic extract of *Clitoria ternatea* Linn. on central nervous system in rodents. *Indian Journal of Experimental Biology,* **26:** 957-960.

Kulshrestha, D.K. and Khare, M.P. (1967). Chemical investigation of the seeds of *Clitoria ternatea. Current Science,* **36:** 124-125.

Kulshrestha, D.K. and Khare, M.P. (1968). Chemiache Untersuchung der Samen von Clitoria ternatea L. *Chemische Berichte,* **101:** 2096-2105.

Kumar, V., Mukherjee, K., Kumar, S., Mal, M. and Mukherjee, P.K. (2008). Validation of PTLC method for the analysis of taraxerol in *Clitoria ternatea*. *Phytochemical Analysis*, **19:** 244-250.

Macedo, M.L.R. and Xavier-Filho, J. (1992). Purification and partial characterization of trypsin inhibitors from seeds of *Clitoria ternatea*. *Journal of Science Food and Agriculture*, **58:** 55-58.

Michael, G.S. and Kalamani, A. (2003). Butterfly pea (*Clitoria ternatea*): a nutritive multipurpose forage legume for the tropics – an overview. *Pakistan Journal of Nutrition*, **2:** 374-379.

Morita, N., Arisawa, M., Nagase, M., Hsu, H. and Chen, Y. (1977). Studies on the constituents of Formosan Leguminosae I. The constituents in the leaves of *Clitoria ternatea* L. *Yakugaku Zasshi*, **97:** 649-653.

Morris, J.B. (1999). Legume genetic resources with novel value added industrial and pharmaceutical use. *In*: Perspectives on Newcrops and New Uses, Ed. By Janick, J., ASHS Press, Alexandria, pp.196-201.

Mukherjee, P.K., Kumar, V., Kumar, N.S. and Heinrich, M. (2008). The Ayurvedic medicine *Clitoria ternatea*–from traditional use to scientific Assessment. *Journal of Ethnopharmacology*, **120:** 291-301.

Nadkarni, K.M. (1976). Indian Materia Medica, Popular Publication, Bombay.

Ng, T.B., Wang, H.X., Liu, F. and Xia, L.X. (2006). Plants beneficial to the aging brain. *Neuroembryology and Ageing*, **3:** 136-141.

Osborn, R.W., De Samblanx, G.W., Thevissen, K., Goderis, I., Torrekens, S., Van Leuven, F., Attenborough, S., Rees, S.B. and Broekaert, W.F. (1995). Isolation and characterization of plant defensins from seeds of Asteraceae, Fabaceae, Hippocastanaceae and Saxifragaceae. *FEBS Letters*, **368:** 257-262.

Parimaladevi, B., Boominathan, R. and Mandal, S.C. (2003). Anti-inflammatory, analgesic and antipyretic properties of *Clitoria ternatea* root. *Fitoterapia*, **74:** 345-349.

Parimaladevi, B., Boominathan, R. and Mandal, S.C. (2004). Evaluation of antipyretic potential of *Clitoria ternatea* L. extract in rats. *Phytomedicine*, **11:** 323-326.

Piala, J.J., Madissoo, H. and Rubin, B. (1962). Diuretic activity of roots of *Clitoria ternatea* L. in dogs. *Experientia*, **18:** 89.

Pillai, N.G. (1976). On the botanical identity of Sankhapushpi. *Journal of Research in Indian Medicine Yoga and Homoeopath*, **11:** 67-76.

Rai, K.S., Murthy, K.D., Karanth, K.S. and Rao, M.S. (2001). *Clitoria ternatea* Linn. Root extract treatment during growth spurt period enhances learning and memory in rats. *Indian Journa of Physiology and Pharmacology*, **45:** 305-313.

Rai, K.S., Murthy, K.D., Karanth, K.S., Nalini, K., Rao, M.S. and Srinivasan, K.K. (2002). *Clitoria ternatea* root extract enhances acetylcholine content in rat hippocampus. *Fitoterapia*, **73:** 685-689.

Rai, K.S., Murthy, K.D., Rao, M.S. and Karanth, K.S. (2000a). *Clitoria ternatea* (Linn) root extract treatment in rats during growth spurt period affects dendritic morphology of hippocampal CA3 neurons. *In*: Third Congress. Federation of Indian Physiological Societies (FIPS), Calcutta (Abstract no. 4.5), Calcutta, India, p. 45.

Rai, K.S., Murthy, K.D., Rao, M.S. and Karanth, K.S. (2005). Altered dendritic arborization of amygdala neurons in young adult rats orally intubated with *Clitorea ternatea* aqueous root extract. *Phytotherapy Research,* **19:** 592-598.

Rai, K.S., Rao, M.S., Karanth, K.S. and Murthy, K.D. (2000b). *Clitoria ternatea* enhances learning and memory - an experimental study on rats. International Congress on Frontiers in Pharmacology and Therapeutics in 21st Century, vol. 32, New Delhi, India, Abstracts. *Indian Journal of Pharmacology,* **150**.

Rajagopalan, N. (1964). Free amino acids and amides in legume root nodules. *Current Science,* **33:** 454-456.

Rajan, V. and Murray, R.Z. (2008). The duplicitous nature of inflammation in wound repair. *Wound Practice and Research,* **16(3):** 122-129.

Ranaganayaki, S. and Singh, A.K. (1979). Isolation and identification of pigments of the flowers of *Clitoria ternatea. Journal of the Indian Chemical Society,* **56:** 1037-1038.

Revilleza, M.J., Mendoza, E.M., Raymundo, L.C. (1990). Oligosaccharides in several Philippine indigenous food legumes: determination, localization and removal. *Plant Foods for Human Nutrition,* **40:** 83-93.

Ripperger, H. (1978). Isolation of stigmast-4-ene-3, 6-dione from Hamelia patens and *Clitoria ternatea. Pharmazie,* **33:** 82-83.

Saito, N., Abe, K., Honda, T., Timberlake, C.F. and Bridle, P. (1985). Acylated delphinidin glucosides and flavonols from *Clitoria ternatea. Phytochemistry,* **24:** 1583-1586.

Shah, V. and Bole, P.V. (1961). Botanical identity of Shankhapushpi. *Indian Journal of Pharmacology,* **23:** 223-224.

Sharma, A.K. and Majumdar, M. (1990). Some observations on the effect of *Clitoria ternatea* Linn. on changes in serum sugar level and small intestinal mucosal carbohydrase activities in alloxan diabetes. *Calcutta Medical Journal,* **87:** 168-171.

Sinha, A. (1960a). γ-sitosterol from the seeds of *Clitoria ternatea. Current Science,* **29:** 180-181.

Sinha, A. (1960b). Studies on the unsaponifiable matter of the seeds of *Clitoria tematea* Linn. and isolation of γ-sitosterol. *Proceedings of the National Academy of Sciences of India,* **29:** 23-26.

Solanki, Y.B. (2010). Investigation on therapeutic potential of *Clitoria ternatea* Linn. in hepatic dysfunction, inflammation and associated disorders using experimental animals. Ph.D. Thesis, Gujarat University. Ahedabad, pp. 46-111.

Solanki, Y.B. and Jain, S.M. (2010). Immunomodulatory activity of Ayurvedic plant Aparajita (*Clitoria ternatea* L.) in male albino rats. *Global Journal of Science Frontier Research*, **10(3):** 2-8.

Solanki, Y.B. and Jain, S.M. (2011). Hepatoprotective effects of *Clitoria ternatea* and *Vigna mungo* against acetaminophen and carbon tetrachloride-induced hepatotoxicity in rats. *Journal of Pharmacology and Toxicology*, **6(1):** 30-48.

Taranalli, A.D. and Cheeramkuczhi, T.C. (2000). Influence of *Clitoria ternatea* on memory and central cholinergic activity in rats. *Pharmaceutical Biology*, **38:** 51-56.

Terahara, N., Saito, N., Honda, H., Toki, K. and Osajima, Y. (1989a). Structures of *Clitoria ternatea* pigments. *Tennen Yuki Kagobutsu Toronkai Koen Yoshishu*, **31:** 324-331.

Terahara, N., Saito, N., Honda, T., Toki, K. and Osajima, Y. (1989b). Structure of ternatin D1, an acylated anthocyanin from *Clitoria ternatea* flowers. *Tetrahedron Letters*, **30:** 5305-5308.

Terahara, N., Saito, N., Honda, T., Toki, K. and Osajima, Y. (1990a). Acylated anthocyanins of *Clitoria ternatea* flowers and their acyl moieties. *Phytochemistry*, **29:** 949-953.

Terahara, N., Saito, N., Honda, T., Toki, K. and Osajima, Y. (1990b). Further structural elucidation of the anthocyanin, deacylternatin, from *Clitoria ternatea*. *Phytochemistry*, **29:** 3686-3687.

Terahara, N., Saito, N., Honda, T., Toki, K. and Osajima, Y. (1990c). Structure of ternatin A1, the largest ternatin in the major blue anthocyanins from *Clitoria ternatea* flower. *Tetrahedron Letters*, **31:** 2921-2924.

Terahara, N., Saito, N., Honda, T., Toki, K. and Osajima, Y. (1990d). Structure of ternatin A2, one of *Clitoria ternatea* flower anthocyanins having unsymmetrical side chains. *Heterocycles*, **31:** 1773-1776.

Tiwari, R.D. and Gupta, R.K. (1957). Chemical examination of the oil from the seeds of *Clitoria tematea* Linn. *Journal of Oil Technologist Association of India*, **13:** 9-13.

Tiwari, R.D. and Gupta, R.K. (1959). Chemical examination of the leaves of *Clitoria ternatea*. *Journal of Indian Chemical Society*, **36:** 243-246.

Vianni, R. and Souto, S.M. (1971). Oil content and fatty acidcomposition of *Clitoria ternatea* seeds. *Arquivos da Universidade Federal Rural do Rio de Janeiro*, **1:** 47-50.

Yadava, R.N. and Verma, V. (2003). Antimicrobial activity of a novel flavonol glycoside isolated from the roots of *Clitoria ternatea* Linn. *Asian Journal of Chemistry*, **15:** 842-846.

Traditional and Folk Herbal Medicine: *Pages* **191–208**
Recent Researches Vol. 1 (2012)
Editor: **V.K. Gupta**
Published by: **DAYA PUBLISHING HOUSE, NEW DELHI**

9

The Research Advance of Tibetan Medicine Dida

Zhen Zhang[1], PengFei Gao[1], Huai Xiao[1], Xiumei Wu[1], ChengGui Zhang[1], ShuiXian Li[1], Bei Jiang[1], Xi Xie[1], CongLong Xia[1*], DongMei Li[1], Guangming Liu[1*] and Yu Zhao[1]

ABSTRACT

Dida is one of the most representative Tibetan medicines. Due to the inconvenience of transportation and culture-exchange of the special area of Tibet, the botanical origins of this drug are in varieties and species used in clinic becomes more complicated. In order to well control the quality of Dida, to highly code its utilizing standards, and to scale its industrial development, this chapter reviews the research status of Tibetan Dida with focus on phytochemisty and bioactivity.

Keywords: Tibetan medicine, Dida, Swertia, Phytochemisty, Bioactivity.

As one of the most popular Tibetan medicine, Dida is widely used to treat icteric hepatitis, viral hepatitis, cholecystitis, and symptoms or diseases related to fever, liver and gallbladder diseases, and blood diseases (Maer, 1988). However, the diversities of original plants of Dida scattered in different Tibetan areas may not only influence its clinical application and efficacy, but also constraint its quality standardization and market supervision. To systematically review the research advance of Dida, topics will be discussed as follows: (1) nomenclature and botanical origin of Dida; (2) the chemical constituents and bioactivities of the representative

1 Key Laboratory of Yunnan Insect Drug R&D, Pharmacy School of Dali College, Wanhua Road, Dali 671000, China

* *Corresponding author*: E-mail: long7484@126.com (C.L. Xia) or lgm888999@yahoo.com.cn (G.M. Liu)

components of Dida; (3) bioactivity study related to six genuses of Dida origin plants during the last 5 years.

The Nomenclature and Botanical Origin of Dida

It is very common for Tibetan herbs that one strain owns different names and origins because of various collecting regions and local denomination. This is also the case for the popular Tibetan medicine Dida (Zhong *et al.*, 2008). By checking of the Chinese Pharmacopoeia, Tibetan literatures (Chinese version) and periodical literatures, it could be summarized to three sources introduced as follows:

The Nomenclature and Origins of Dida from the Source of Pharmacopoeia

According to "Drug Standard of Ministry of Public Health of the People's Republic of China Tibetan Medicine (volume 1)", Dida refers to a series of medicine including Tidajieba, Tidaniar, Tidaniaribu, Tidagubaribu, etc. They possess the efficacies on cholagogic and clearing heat, and are mainly used to treat cholecystitis and initial stage of jaundice hepatitis (Chinese Pharmacopoeia Commission of Ministry of Public Health of the People's of Republic of China, 1995). According to Chinese Pharmacopoeia (2005 edition, part one), Dida is also named as "Dieda", describing the dried whole herb of *Saxifraga tangutica* (Chinese Pharmacopoeia Commission, 2005).

Nomenclature and Origins of Dida Derived from Tibetan Literatures

In "Crystal Beads Material Medica (Jing Zhu Ben Cao)," short titled as "Photo Identification (Tu Jian)", a plant of Saxifragaceae families reported three Swertia species (India Swertia, Nepal Swertia, and Tibet Swertia) as "3-Swertia", whose popular name is "Three Tibetan Swertia". From them Tibet Swertia is named Dida (Maer, 1988). According to the book of "Blue-Coloured Glaze/Continuing Explaining Theory (LanLiuLi/LunShuoXu)", Dida was classified as three types including Jadi, Wadi, and Wodi (Des, 1982).

Nomenclature and Origins of Dida Recorded in the Journal Literatures

Echangba Zhuomadongzhu (Ecangba and Liu, 1996) collated Dida as two families, which include five genuses, thirteen species, and one variant altogether. The two families are Gentuanaceae and Saxifragaceae, while the five genuses are *Swertia, Halenia, Gentianopsa, Lomatogonium,* and *Saxifraga*. Dida can also be divided into three types including rgyatig (Tibetan language, meaning Swertia growing in India), the whole herb of *Swertia chirayita;* baltig (meaning Swertia growing in Nebal), the whole herb of *Swertia ciliata;* as well as bodtig (Tibetan language, means Swertia growing in Tibet), which means Swertia species. Among them, bodtig has six sub classes containing: (1) sumcuting (Tibetan language), referring mainly to the whole herb of *Saxifraga umbellulata* (Saxifragaceae). Besides, the dried whole herb of *Saxifraga unbellulata* var *pectinnata* and *Saxifraga tangutica* can also be used as sumcuting; (2) gsertig (Tibetan language), referring to the dry whole herb of *Saxifraga montanella* (Saxifragaceae); (3) dngultig (Tibetan language), referring to dry whole herb of *Lomatogonium oreocharis* (Gentuanaceae); (4) sangstig, dry whole herb of *Swertia mussoti* (Gentuanaceae); (5) lcagstig (Tibetan language), which were further divided into three categories: (5.1) lcagstigkrpo, the dry whole herb of *Gentianopsa barbata* (Gentuanaceae); (5.2) lcagstignagpo (Tibetan language), dry whole herb of *Gentianopsa paludosa* (Gentuanaceae); (5.3) lcagstigrwamgo, the dry whole herb of *Halenia elliptica* (Gentuanaceae); (6) gurtig (Tibetan language), the dry whole herb of *Saxifraga nigroglandulifera* (Saxifragaceae).

Tang *et al.* (2008) reported that Zang Yin Chen, whose Tibetan name is Dida, has the Tibetan medical names of Sangdi (also named Tidajabu, Zireya), Jadi, Songdi (also named Songjüdou, Songjiudida, Bendi, etc.), Jihedou or Sangdou, according to plant origins.

Based on the discussion and confirmation with the Tibetan physicians and Tibetan pharmacists, Zhong *et al.* (2010) drew the following conclusions: (1) Dida is often called Zang Yin Chen as a Chinese name in Tibetan area (the medicinal material Dida is commonly named as Zang Yin Chen by local salesmen). Meanwhile, the classification of Dida based on the books of "Crystal Beads Material Medica (Jing Zhu Ben Cao)" and "Tibetan Medicine Pharmacopoeia (Zang Yao Zhi)" are basically approved by all of the interviewed people. (2) Dida from different origins has different efficiencies and various clinical applications. Three varieties are classified according to its plant origins as Indian Dida, Nepal Dida, and Tibetan Dida. Tibetan Dida can also be sub-classified. The imported Dida and Chinese domestic Dida are also separated in the market mainly according to their botanical origins. *Saxifraga chirayita* is the botanical origin of Indian Dida and Nebal Dida, while *Saxifraga chirayita* from Nepal is regarded as certified product (Top grade 3). Two types of Dida, Indian original *Saxifraga chirayita*and the local *Saxifraga umbellulata* are both well sold in local Tibetan medicine market, but with different clinical efficacies. The former (*S. chirayita*) is traditionally used to treat cholecystitis, whilst the latter (*S. umbellulata*) mainly used to treat liver disease. Apart from that, *Halenia elliptica, Sonchus oleraceus* (used in Tibetan Lhasa area) and *Saussurea japonica* (used in Changdou area) were also found to be used as alternatives of Dida. It should be noticed that there are large discrepancy over the Dida plant origins in China and consistence have not been met among different Tibetan areas and related pharmacologically professional staff. However, although it is commonly believed that there are a variety of Dida medicines, only two or three species are actually used in Tibet. For example, *Swertia* species were commonly used as Nepal Swertia or Indian Swertia. In fact, *Saxifraga chirayita* is rare in the market of Tibetan community of Qinghai, Sichuang, and Ganshu provinces, since the people there use the locally abundant *Swertia, Saxifraga mussoti, Swertia franchetiana,* and *Halenia elliptica* as a substitution. Moreover, some Tibetans in these provinces also use mixed local plants which belongs to genus *Swertia* or *Saxifraga* as Dida medicine. In addition, the common Dida origin plants used in Tibetan area in Yunnan province are *Saxifraga punicea, Saxifraga cincta, Saxifraga mileensis, Saxifraga yunnanensis,* and *Saxifraga marcosperma.*

The Chemical Constituents and Bioactivities of the Representative Components of Dida

Though the nomenclature and the plant origins of "Dida" seem rather variant and somewhat chaotic, all of the so-called Tibetan medicine "Dida" showed certain therapeutic effects. According to the literature concerning pharmacological investigation (Liang and Gao, 1979), various types of components such as xanthones, glucosides, triterpenes, and flavonoids isolated from *Swertia yunnanensis* Burk., to varying degrees, exhibited the functionalities of hepatic protection and may decrease the activities of hepatic enzyme, such as alanine aminotransferase (ALT) and aspartate aminotransferase (AST). Moreover, it exhibited improved efficacies when these four types of compounds used together. Therefore, the representative compounds and their bioactivities of these four major components attributable to Dida were reviewed as follows, respectively:

Xanthones

Modern phytochemical and pharmacological investigations of Dida showed that xanthone is one class of the major active components for anti-hepatitis and anti-cirrhotic existed in it. Xanthones and its glycosides were reported to have cholagogic, diuretic and liver-protecting effects (Zhao, 2008). The xanthone glycosides, however, demonstrated their activities through transformation via their aglycon forms (Gao *et al.*, 2004). Pharmacological research found that the C-glycosides of xanthone are effective for the treatment of cholecystitis while its O-glycosides are useful in curing hepatitis (Li *et al.*, 2008). The structure-activity relationships (SAR) investigations disclosed that xanthone with a methyl group at C-8 is more cytotoxic than that with a hydroxy group. However, the placement of a C-glycoside at C-1 of xanthones reduced cytotoxicity because of the spatial hindrance (Zhang, 2009; Ding *et al.*, 2010). Furthermore, the existence of glycosides and the lack of methoxyl groups in xanthones may increase the protective efficacy against liver damage.

One of the representative xanthone compounds in Dida, mangiferin, showed cholagogic and diuretic effects. It is the highest content of xanthones reported in Dida, which confirmed the reasonability to use it as a standard of the quality control in most of Tibetan Dida preparations (Deng *et al.*, 2006). The pharmacological activities of it were concluded as following three aspects:

Relieving Internal Heat and Detoxication Effect

Deng *et al.* (2006) reported that mangiferin has obvious capability of relieving fever induced by endotoxin, while the efficacy has positive correlation with the dose increase.

Anti-inflammatory and Antibiosis Effect

Oral administration of mangiferin (50-200 mg/kg) can attenuate the auricular swelling in mice induced by arachidonic acid and PMA (phorbol-12-myristate-13-acetate). In an *in vitro* study, mangiferin (at dosage of 17.2 μg/mL and 2.1 μg/mL) could inhibit the release of PEG (2) and LTB (4) when stimulating LPS-TNF gamma and calcium ion A23187 in macrophage (Ministry of Public Health of General Logistics Department Chinese People's Liberation Army, 2007).

Hepatoprotective Effect

Huang *et al.* (2009) reported that mangiferin can significantly decrease the serum content of ALT, AST, hyaluronic acid (HA), procollagen III (PCIII), hydroxyproline (Hyp), total protein (TP) and albumin (ALB) ($P<0.01$ or $P<0.05$), it can also increase the activities of superoxide dismutase (SOD) and glutathione peroxidase (GSH-PX) in hepatic tissue homogenate ($P<0.01$ or $P<0.05$), while reduce the content of malondialdehyde (MDA) ($P<0.01$ or $P<0.05$). Mangiferin can inhibit the expression of transforming growth factor-beta1 (TGF-beta1), attenuate the degree of damage of chronic liver injury. The protective mechanism of mangiferin may be related to the its capabilities of anti-lipid peroxidation and decreasing expression of TGF- beta1 in the hepatic tissue.

Iridoid and Iridoid Glycosides

The representative compounds in Dida are swertiamarin, gentiopicroside, etc.

Anti-inflammatory Effect

Chen *et al.* (2003) found that gentiopicroside showed certain anti-inflammatory effects. It could decrease auricular edema in mice induced by dimethyl benzene, raising permeability of celiac blood capillary induced by acetic acid and paw swelling induced by zymosan A and carrageenan in rats, but has little effect on the inflammatory model induced by nystatin. Its mechanism may relate to the inhibitory functionalities on many kinds of inflammatory medium.

Liver-protection Effect

Liu *et al.* (2002) reported that gentiopicroside shows remarkably liver-protection effect on mice model. Gentiopicroside (240 mg/kg, i.p.) was administered to the mice with hepatic injury induced by GALN and CCL_4–, the serum levels of ALT and AST of the treated group were obviously reduced, whilst the GSH-Px activity was promoted, as well as the bile secretion bettered, and the concentration of bilirubin in the bile was increased when compared with the model group. Additionally, in an *in vitro* study, the gentiopicroside was added to the cultured hepatic cells of the rat injured by CCl_4, while several indexes such as GPT, isolated cell DNA, hepatin, adipose staining were examined, as well as the cell morphology was observed under electronic microscope, it could be found that the injury in the administrated group is significantly less when compared with the model group. The liver-protective mechanism of gentiopicroside is closely related to anti-free radical lipid peroxidation. In addition, Inouye *et al.* (1967) described that another iridoid glycoside, swertiamarin, could promote the reparation of injured hepatic cell and remarkably inhibit the increasing of cAMP in the plasma induced by D-Ga1N.

Actions on Human Digestive System

Luo *et al.* (2010) reported that gentiopicroside could increase the bile secretion of rats and promote gall bladder concentration of the test animal.

Triterpenoids

The representative triterpenoids in Dida are ursolic acid and oleanolic acid, which exhibited liver-protective activity, though their action mechanism is still somewhat unclear. Some investigations supposed that its bioactivity may be related with both of its antioxidative/anti-inflammatory effect and of its impact on the drug metabolic enzymes (Li *et al.*, 2009).

Oleanolic acid (OA) has been used as a liver-protective medicine to treat with both acute jaundice hepatitis and virus hepatitis. Early in 1970s it was found that OA administration significantly decreased the serum level of ALT in rats poisoned by CCl_4. Later in 1980s, the liver-protective effect of OA was further confirmed through histopathological examination. It could remarkably reduce the acute liver damage of mice induced by many chemicals such as CCl_4, bromobenzene, paracetamol, nicorol, thioacetamide, phalloidin, colchicine, cadmium chloride, D-galactosamine and endotoxin. Meanwhile, OA administration can also reduce the increasing serum levels of aminotransferase and iditol dehydrogenase. In addition, the liver-protective effect of OA is strictly correlated to the administration dosage, for a lower dosage demonstrated liver-protective effect whilst a higher dosage may induce cholestasis and hepatotoxicity (Liu, 2005).

Nevertheless, by the investigation of Liu *et al.* (2008), OA has been proved to be effective in treatment of acute hepatic intoxication by reducing the amounts of the toxic metabolites and by increasing glucose enzyme to elevate the detoxication effect, therefore, OA was regarded as one of the chief active ingredient in Dida to treat with hepatitis. Moreover, the in vitro study showed that OA has also notable anti-virus effect. The active functional groups existed in OA could competitively combine with the necessary enzymes involved in the growing and multiplication of hepatitis virus, and therefore inhibits the virus's proliferation.

Martin-Aragón *et al.* (2001) found that ursolic acid displayed protective effect against the variations of the antioxidative enzymes induced by CCl_4. Pretreatment of ursolic acid may significantly increase the survival rate of rats' hepatic cells, and it was observed that it could inhibit cholestasis, while both of the rats' bile flow and its content were increased. Saraswat *et al.* (2000) found that ursolic acid demonstrated *in vitro* protective effect to the cytotoxicity of hepatic cells induced by ethanol. Xiong *et al.* (2004) verified that ursolic acid could remarkably reduce the serum levels of glutamic-pyruvic transaminase (SGPT) and hepatic glyceride, apparently increase the serum glyceride and hepatic hepatin contents, whilst attenuate denaturation and necrosis of hepatocytes in the test rats with acute hepatic injury, the authors pointed out that ursolic acid possesses significant protective and curing effects against the acute hepatic injury.

Flavonoids

The representative flavonoids existed in Dida are luteolin and swertisin.

Liver-protection Effect

Wang *et al.* (1994) reported that luteolin has various effects of antibacterial, anti-virus and reducing blood-fat, cholesterol *in vivo*. Zhao *et al.* (2002a) indicated that luteolin decreases the liver fibrosis induced by CCl_4 through both of scavenging of free radicals and inhibiting the expression of collagen protein. Zhao *et al.* (2002b) also disclosed that swertisin could reduce the rising level of aminotransferase induced by CCl_4.

Anti-inflammatory and Antibacterial Effects

By the investigation of He *et al.* (2000), luteolin could inhibit proliferation and collagen synthesis of hepatic stellate cells cultured *in vitro*. It also exhibited the anti-virus function against Cox B_3 virus. Luteolin may obviously inhibit ankle swelling induced by carrageen or ferment and granuloma induced by croton oil on rat model (Dai *et al.*, 1985). Zheng *et al.* (1990) described that luteolin could clearly restrain H_2O_2 release of rats' peritoneal macrophage activated by zymosan, and the authors also suggested that these inhibiting effects of luteolin may be the action mechanism of its anti-inflammatory efficacy.

Comparisons of the Chemical Constituents and Efficacy of the Plant Origins of Dida

The above following four types of compounds all exhibited functionalities of hepatic protection and decrease the activities of hepatic enzymes, such as ALT and AST. The chemical constituents, efficacy, medicinal parts and sources of the plant origins of Dida in Tibetan area are summarized in Table 9.1. It could be observed that there are nearly 20 kinds of plants in genus *Swertia* (Gentuanaceae), among which the phytochemical examination over this genus is relatively more concentrated. However, those Dida origin plants attributable to genus *Comastama* (Gentuanaceae) and genus *Saxifraga* (Saxifragaceae) were comparatively insufficient (Zheng *et al.*, 1990; Yang and Liu, 2005; Tian *et al.*, 2005; Hu *et al.*, 2003; Ji *et al.*, 2004; Zhang *et al.*, 2006; Xu and Zhang, 2008; Yu *et al.*, 2010; Liu, 2006; Zuo *et al.*, 2005).

Bioactivity Study Related to Six Genuses of Dida Origin Plants during the Last 5 Years

From the literatures appeared in the recent five years (Geng *et al.*, 2009a; Geng *et al.*, 2009b; Li, 2008; Lv *et al.*, 2010; Rana and Rawat, 2005; Zhang *et al.*, 2010; Nikolaev *et al.*, 2003; Nikolaev *et al.*, 2001; Gao *et al.*, 2004; Huang *et al.*, 2010; Yuan *et al.*, 2010; Myagmar and Aniya, 2000; Zuo *et al.*, 2005; EP 1213027, 2004), we reviewed either active compounds or active fractions of six genuses of Dida origin plants concerned with its clinical efficacy (Table 9.2). Among the above research, it could be noticed that the plants of genus *Swertia* were paid more attention comparatively. However, all of the six genuses of Dida origin plants were reported to contain effective fractions for clinical utilization (Table 9.2).

Discussion

Dida is actually a total name of category of the closed-medicine, which tastes bitter, and was used to treat the liver and gallbladder diseases. But with lapse of time, the Tibetan medicines Dida, as many other Chinese medicinal materials, were replaced and forged by some other plants. Thus, there are many counterfeit and inferior

Table 9.1: Chemical constituents, efficacies, therapeutic effects and resource distributions of Dida origin plants.

Family	*Genus*	*Species*	*xan*	*iri*	*tri*	*lia*	*Medicinal Parts*	*Efficacy*	*Therapeutic Effect*	*Resources*
Gentuanaceae	*Swertia*	*S. chirayita*	√	√	√	√	Whole herb	Clearing liver-heat, cholagogue and antipyretic	Treatment of icterohepatitis, hepatitis and cholecystitis	Plenty
		S. ciliata	√	√	√	√	Whole herb	Clearing heat, soothing the liver and cholagogic	Treatment of icterohepatitis, hepatitis and cholecystitis	Plenty
		S. racemosa	√	√	√	√	Whole herb	Soothing the liver and cholagogic	Treatment of icterohepatitis, hepatitis and cholecystitis	Plenty
		S. mussoti	√	√	√	√	Whole herb	Soothing the liver and cholagogic	Treatment of icterohepatitis and cholecystitis	Plenty
		S. franchetiana	√	√	√	√	Whole herb	Soothing the liver and cholagogic	Treatment of icterohepatitis and cholecystitis	Rich
Gentuanaceae	*Swertia*	*S. cincta*	√	√		√				Plenty
		S. erthrosticta	√	√	√	√		Clearing heat, detoxicating, strengthening the stomach and pesticide	Treatment of cough due to pathogenic wind-heat, jaundice, etc.	Rich
		S. bifolia	√					Calming the Liver, cardiotonic, nourishing blood, regulating menstruation	Treatment of blood deficiency, dizziness, neurasthenia, hypertension, abnormal menstruation, etc.	Rich
		S. marcosperma	√	√	√	√	Medicinal plants	Liver-protective, cholagogue, clearing heat, detoxication, and strengthening the stomach		Plenty

Contd...

Table 9.1–*Contd...*

Family	Genus	Species	xan	iri	tri	lia	Medicinal Parts	Efficacy	Therapeutic Effect	Resources
Gentuana-ceae	*Swertia*	*S. wolfangiana*								Scant
		S. punicea	√	√	√					Scant
		S. tetraptera	√	√	√	√				Plenty
		S. dilatata								Plenty
		S. mussotii var. *flavescens*								Scant
Gentuana-ceae	*Swertia*	*S. elata*								Scant
		S. pzrewalskii	√	√	√	√			Liver-protective, hpyerglycemic and calmness	Scant
		S. angustifolia var. *pulchella*								Rich
		S. yunnanensis	√	√	√		Whole herb	Clearing liver-heat, cholagogue, clearing heat and eliminating dampness	Acute icterohepatitis, and cholecystitis	
		S. mileensis	√	√	√	√	Whole herb	Clearing liver-heat, cholagogue, clearing heat and eliminating dampness	Jaundice, red urine, relin with acerbity and pain	
Gentuana-ceae	*Halenia*	*H. elliptica*	√	√	√	√	Aerial part	clearing heat, draining dampness, soothing the liver and cholagogic	Acute icterohepatitis of children, hepatitis B, cholecystitis, headache and toothache	Plenty

Contd...

Table 9.1–*Contd...*

Family	Genus	Species	xan	iri	tri	lia	Medicinal Parts	Efficacy	Therapeutic Effect	Resources
	Gentia-nopsa	*G. paludosa*	√	√	√	√	Whole herb	clearing heat and drying dampness	Fever and diarrhea of children resulted from influenza and cholic disease	Plenty
		G. barbata	√				Whole herb	Clearing heat and cholagogic	Icterohepatitis and plague	Plenty
		G. paludosa var. *alpina*								Rich
Gentuana-ceae	*Gentia-nopsa*	*G. paludosa* var. *ovata-deltoidea*								Rich
	Comastoma	*C. pulmonarium*								Plenty
		C. pedunlulatum	√		√	√	Whole herb			Rich
		C. falcatum								Rich
	Lomato-gonium	*L. carinthiacum*	√	√	√	√	Whole herb	Clearing heat, strengthening the stomach and callus		Plenty
Gentuana-ceae	*Lomato-gonium*	*L. rotatum*	√	√	√	√	Whole herb		Disease of liver, gall bladder and spleen	Rich
		L. corditolium								Scant
		L. gamosepalum								Rich
		L. oreocharis					Whole herb	Clearing heat and cholagogic	cholecystitis and tumor	Rich
Gentuana-ceae	*Lomato-gonium*	*L. perenne*								Scant

Contd...

Table 9.1–*Contd...*

Family	Genus	Species	xan	iri	tri	fla	Medicinal Parts	Efficacy	Therapeutic Effect	Resources
Saxifraga-ceae	*Saxifraga*	*S. tangutica*					Whole herb	Clearing heat, draining soothing the liver, cholagogic and discharging pus	Hepatitis, cholecystitis, high fever, etc.	Rich
		S. unbellulata var. *pectinn-ata*				√	Whole herb	Clearing heat, draining soothing the liver, cholagogic and discharging pus	Hepatitis, cholecystitis, high fever, etc.	Rich
		S. litangensis								Scant
Saxifraga-ceae	*Saxifraga*	*S. umbellulata*					Whole herb	Clearing heat, draining dampness, soothing the liver, cholagogic and discharging pus	Hepatitis, cholecystitis, high fever, etc.	Scant
		S. nigroglan-dulifera					whole herb	Clearing heat and cholagogic	Cholecystitis, sores fever, pulse fever, etc.	Scant
		S. montanella			√	√	whole herb	Clearing heat and cholagogic	Cholecystitis and plague	Scant
Saxifraga-ceae	*Saxifraga*	*S. melanolentra*				√				Plenty
		S. pseudohirculus								Rich
		S. tibetica								Scant
		S. implicans								Scant
		S. sibirica								Scant
		S. punctulatoides								Scant

xan: Xanthones; iri: Iridoids; tri: Triperpenoids; fla: Flavonoids.

Table 9.2: Active compounds and active parts study related to six genus of Dida origin plants of clinical efficacy during the last 5 years.

Genus	*Active Compounds*	*Active Parts*	*Literature Number*	*Remarks*
Swertia	Swerilactone A, Swerilactone C, Swerilactone D, 1,8-hydroxy-2,3,4,5-methoxyl xanthone, 1- hydroxy-3,5,8-methoxyl xanthone *, 1,7- hydroxy-3,8-methoxyl xanthone *, ursolic acid *, 1,5,8-trihydroxy-3-methoxy-9H-xanthen-9-one*, 3-O-demethylswertipunicoside*	SME-a, SME-c, SME-d, extract of acetone, methanol	6	
Gentianopsa	Decussatin#, gentiacaulein #	extract of *Gentianopsa, Swertia* #	2	
Haienia	1,7-hydroxy-2,3,5-methoxyl xanthones, 1,7-hydroxy-2,3,4,5-methoxyl xanthones, 2,3,5-methoxyl-1-O-oxlip glycosyloxy xanthone#*	extract of methanol	2	
Comastoma		Ethyl acetate extract of 70 per cent ethanol extraction	1	Nine compounds of xanthones
Lomatogonium		Extract of *Lomatogonium*	1	
Saxifraga	1,2,3,4,6-penta-O-galloyl-beta-D-glucosidE* 11-O-(4-O-methylgalloyl) bergenin	Extract of *Saxifraga*	2	

*: Known compound; #: 2001-2004 literatures.

categories or names, including several origin-complicated herbs, whose botanical origins get involved in multiple plants attributable to two families (Gentianaceae and Saxifragaceae). We supposed the reasons were caused as followings:

Regional Aspect

The western part in China contains ten provinces, municipality, and autonomous regions including Shaanxi, Ganshu, Qinhai, Ningxia, Sichuang, Chongqing, Yunnan, Xizang, etc., the area of which is 5410,000 square kilometers, occupying 56 per cent of the whole national land area. The Tibetan concentrated areas are in Xizang autonomous region and provinces of Qinhai, Gansu, Sichuang, and Yunnan (half of above 10 western China regions). The characteristics of Tibetan areas are in large domain, sparsely population, less developed, high altitude with inconvenience of transportation. Meanwhile, the climate types display great difference between south and north of these regions; therefore, the diversity of plants is obvious. These reasons made the use of Tibetan medicines adjust to local conditions and the Tibetan physicians has to obtain herbal medicine materials from local sources. In fact, *Saxifraga chirayita* is distributed in Dingjie county of Tibet and *Saxifraga ciliata* is distributed and in great deposit in Dingjie county, Longri county and Pulan county of Tibet, but because of poor means of transportation, the local herbs are not mainly used, whilst the *Saxifraga chirayita* imported from India as well as *Saxifraga ciliata* imported from Nepal become more popular in Tibet Dida markets.

Furthermore, the Tibetan area in Qinghai and Sichuang mainly use the local resourceful herbs such as *Saxifraga mussoti* (Gentuanaceae). Diqing Tibetan antonymous region of Yunnan Tibetan area use Saxifraga punicea and many other Swertia plants as Dida medicines.

Inconvenience of Cultural Exchange and Limitation to Inheritance Way of Education in Tibetan Medicines

The mainly inheritance way of education in Tibetan medicines is the apprenticeship between teacher and his students. Traditional Tibetan Medicine and pharmaceutical education in Tibetan medicinal higher education is relative backward when comparing to traditional Chinese medicine education and western pharmaceutical education in non-Tibetan regions of China. The disciplines of Tibetan medicine have been set up very recently at Qinghai University, Tibet Tibetan Medicine College, Chengdu University of Traditional Chinese Medicine Ethnic Medical School. The earliest one started to recruit students with major in Tibetan medicine is Tibetan Medical College of Qinghai University in the year of 2000. Shortness of professional Tibetan medicine staff lead to the confusion of nomenclature of Dida origin plants, such as the record of Swertia Nepal, one of "Three Swertias" in the books of "Spectrogram Identification (Tu Jian)" and "Crastal Pearl Material (Zhen Zhu Bai Cao)" confirmed later as *Saxifraga ciliata* (Gentuanaceae).

As academician Wu Cheng-yi (Kunming Institute of Botany, Chinese Academic of Sciences) said, "There are about 10,000 vegetable herbs in China. Since Chinese medicine thesaurus is in bulky content and has sealed for many years, it is rather difficult to master the usual Chinese medicines, not to mention the whole 10,000 herbal medicine species, their name rectification is still necessary, even the usually used sesame and flax have been confused for more than one thousand years, to say nothing of other herbs". The identification of Tibetan Dida is as complex as that of traditional Chinese medicines, including its morphology identification, anatomic identification, chemical identification and molecular marker techniques, etc.

Shortage of Plant Resources of Dida

Large-scale cultivation of *Saxifraga mussoti* was once carried out in Qinghai province, but at present, all series of Dida original herbal medicine come from natural sources, which make the apparent lack of natural resources. Like *Saxifraga mussoti,* although it has the production of 50 tons per year, this number lowly meets the needs of demand of 500 tons per year (Yang and Liu, 2005). It has been found several times that three plants (*Swertia nervosa, Saxifraga punicea* Hemsl and *Halenia elliptica* D.Don) have been faked as *Saxifraga mileensis.*

More investigation have been accomplished about the chemical constituents and correlative bioactivity of *Swertia* species but less work has been reported on the other genus of "Dida", the reason was supposed to be as followings:

Swertia has lots of genus and species and it permeates the whole world. It shows that there are about 100 species of *Swertia* scattered in the world and 70 species distributed in the north and south of China. Since 1970s, besides China, phytochemical and pharmacological investigations of *Swertia* have been carried out also in India and Japan, such as the investigation of *Swertia pseudochinesis* of Japan.

Swertia such as *Swertia delavayi, Swertia decora,* etc. was used to treat hepatitis and cholecystitis in some civil areas of China. Their folk therapeutic effects were paid more attention by Chinese herbal physicians. For example, *Swertia mileensis,* a new medicine was explored from civil uses of Mile and Kaiyuan counties, Honghe Hani and Yi Autonomous Prefecture, Yunnan province. It has been written into the Chinese Pharmacopoeia, published in 2005 and republished in 2010, respectively. At present, simple recipe and compound recipe pharmaceutics refined from the *Swertia mileensis* are Tabellae Swertlae Mileensis, Yugan Tablet, Swertia Mileensis Injection, Ganfukang Tablet, Zhuzi Gantai Capsule, Danweikang Capsule, Zhenxiongdan Pill, Ligankang Tablet, and Icterus Hepatitis Tablet, etc.

References

Chen, C.X., Liu, Z.W., Sun, Z.W., Song, C.Q. and Hu, Z.B. (2003). Studies on anti-inflammatory effect of gentiopicroside. *Chinese Traditional and Herbal Drugs,* **34(9)**: 814-816.

Chinese Pharmacopoeia Commission of Ministry of Public Health of the People's of Republic of China (1995). Drug Standard of Ministry of Public Health of the People's Republic of China• Tibetan Medicine (Volume One). pp 11, 27, 59, 60, 88, 195.

Chinese Pharmacopoeia Commission. Pharmacopoeia of the People's Republic of China (2005). Beijing: Chemical Industry Press, 2005: Appendix p 23.

Dai, L.M., Cheng, H., Wei, P., Liu, Q., Chen, M.C. and Xu, S.Y. (1985). The influence of Luteolin on experimental inflammatory model in rat. *Acta Universitatis Medicinalis Anhui.*, **20(1)**: 1-3.

Deng, J.G., Zhang, Z.W. and Yang, K. (2006). Mangiferin endotoxin induced fever in rabbits body temperature. *Chinese Journal of Experimental Traditional Medical Formulae*, **12(2)**: 72-73.

Des. Lan Liu Li/Lun Shuo Xu. (1982). Tibet: Tibetan People's Publishing House. p 281.

Ding, L., Liu, B., Zhang, S.D., Hou, Q., Qi, L.L. and Zhou, Q.Y. (2010). Cytotoxicity, apoptosis-inducing effects and structure-activity relationships of four natural xanthones from Gentianopsis paludosa Ma in HepG2 and HL-60 cells. *Journal of Asian Natural Products Research*, **23**: 1-15.

EcangBa• Zhuomadongzhu and Liu, H.Q. (1996). Variety finishing of Tibetan Medicine "Dida". *Journal of Chinese Medicinal Materials*, **19(10)**: 494-496.

EP 1213027. The liver-protective effect of Saxifragaceae or its extract. (2004). *World Phytomedicines*, **19(2)**: 83.

Gao, J., Wang, S.J., Fang, F., Si, Y.K., Yang, Y.C., Liu, G.T. and Shi, J.G. (2004). Xanthones from Tibetan medicine *Halenia elliptica* and their antioxidant activity. *Acta Academiae Medicinae Sinicae*, **26(4)**: 364-367.

Gao, J., Wang, S.J., Xu, R.M., Yang, Y.C., Zhang, S.R., Si, Y.K. and Shi, J.G. (2004). Studies on chemical constituents of *Halenia elliptica* guided by *in vivo* absorption and distribution of the constituents. *Acta Pharmaceutica Sinica*, **39(3)**: 198-203.

Geng, C.A., Jiang, Z.Y., Ma, Y.B., Luo, J., Zhang, X.M., Wang, H.L., Zhang, X.M., Zuo, A.X., Zhou, J. and Chen, J.J. (2009 a). Swerilactones A and B, anti-HBV new lactones from a traditional Chinese herb: *Swertia mileensis* as a treatment for viral hepatitis. *Organic Letters*, **11(18)**: 4120-4123.

Geng, C.A., Zhang, X.M., Zhang, X.M., Zuo, A.X., Liu, J.F., Ma, Y.B., Luo, J., Zhou, J., Jiang, Z.Y. and Chen, J.J. (2009 b). Swerilactones C and D, anti-HBV New Lactones from a Traditional Chinese Herb: *Swertia mileensis. Organic Letters*, **11(21)**: 4838-4841.

He, L.N., He, S.B. and Yang. J. (2000). *In vitro* anti-Coxsackie B3 virus effect of Luteolin. *Chinese Journal of Modern Applied* Pharmacy, **7(5)**: 362-265.

Hu, F.Z., Song, Y.L., Liu, M. and Shi, Z.X. (2003). Analysis of medicinal bioactive composition of Gentianaceae in Qinghai-Tibet plateau by high performance liquid chromatography. *Chinese Journal of Chromatography*, **21(1)**: 63-65.

Huang, B., Ban, X., He, J., Zeng, H., Zhang, P. and Wang Y. (2010). Hepatoprotective and antioxidant effects of the methanolic extract from *Halenia elliptica. Journal of Ethnopharmacology*, **131(2)**: 276-281.

Huang, X.O., Dang, J.G. and Chen, Z. (2009). Protective effect of mangiferin dropping pills on chronic liver injury in rats. *China Pharmacist*, **12(6)**: 701-704.

Inouye H., Ueda S. and Nakamura Y. Z. (1967). Biosynthese derbitteren glucoside dergentianazeen, desgentiopicrosids, desswer tiamarins and dessweroside. *Tetrahedron Letter*, p. 3221-3226.

Ji, L.J., Bao, Y., Chen, G.C., Zhao, M.C. and Sun, H.F. (2004). Determination of active constituents in fifteen species Swertia of genus by high performance liquid chromatography. *Acta Botanica Boreali-occidentalia Sinica*, **24(7)**: 1298-1302.

Li, H.Y., Liu, G.M., Liu, F., Zhang, F.Q. and Li, X.L. (2009). Research of ursolic acid and similar pentacyclic triterpenoid. *Journal of Hunan University of Technology*, **23(5)**: 18-21.

Li, X. S. (2008). Study on chemical constituents and anti-HBV activities of *Swertia mileensis*. *Dali Univercity*, Master thesis.

Li, Y.L., Ding, C.X., Wang, H.L., Y.R., Suo. and You, J.M. (2008). Xanthone glycosides in Gentianceae of Qinghai-Tibet plateau. *Journal of Analytical Chemistry*, **63(6)**: 574-579.

Liang, Q.X. and Gao, X.Y. (1979). Study on flavonoids of anti-hepatitis constituents in *Swertia mileensis*. *Letters in Chinese Traditional and Herbal Drugs*, **9**: 1-4.

Liu, J. (2005). Oleanolie acid and ursolic acid: Research perspectives. *Journal of Ethnopharmacology*, **100**: 92-94.

Liu, J.X. (2006). Study on chemical constituents from Tibetan herbal medicine *Ligularia myriocephala* Ling, *Saxifraga montana* H. and *Ixeridium gramineum*. *Lanzhou Institute of Chemical Physics of the Chinese Academy of Sciences*, PhD thesis.

Liu, N.N. (2008). Synthesis and characterization of ursolic acid and oleanolic acid derivatives. *Journal of Liaoning Normal University*, Master thesis.

Liu, Z.W., Chen, C.X., Jin, R.M., Shi, G.Q., Song, C.Q. and Hu, Z.B. (2002). Studies on liver-protection and promoting bile secretion of gentiopicroside. *Chinese Traditional and Herbal Drugs*, **33(1)**: 48.

Luo, Y.Y., Lu, C.Y., Chen, G.X. and Gui, K.Y. (2010). Research progress in secoiridoid compounds. *Food Science*, **31(21)**: 431-436.

Lv P., Wei L.X., Du Y.z., Yang H.X. and Peng M. (2010). Hepatoprotective and toxic characteristics of the whole herb of traditional Tibetan folk medicine *Swertia mussotii* Franch. *Journal of Medicinal Plants Research*, **4(8)** : 706-709.

Maer • DanZengPengCuo. Compiling by Mao J.Z., Luo S.D. and Wang Z.H. (1988). Jing Zhu Ben Cao. *Shanghai Science and Technology Press*, **79-81**: 181.

Martin-Aragon, S., De las Heras, B. and Sanchez-Reus, M.I. (2001). Pharmacological modification of endogenous antioxidant enzymes by ursolic acid on tetrachloride-induced liver damagein rats and primary cultures of rat hepatocytes. *Experimental and Toxicologic Pathology*, **53(2/3)** : 199-206.

Ministry of Public Health of General Logistics Department Chinese People's Liberation Army. Good Pharmacy Practice of Chinese People's Liberation Army. (2007). Beijing: People's Medical Publishing House, p. 115.

Myagmar, B.E. and Aniya, Y. (2000). Free radical scavenging action of medicinal herbs from Mongolia. *Phytomedicine*, **7(3)**: 221-229.

Nikolaev, S.M., Sambueva, Z.G., Tsyrenzhapov, A.V., Nikolaeva, G.G., Tankhaeva, L.M., Glyzin, V.I. and Dargaeva, T.D. (2003). Comparative choleretic properties of natural xanthone compounds from *Gentianopsis barbata*. *Eksperimental'naia i klinicheskaia farmakologiia*, **66(4)**: 29-31.

Nikolaev, S.M., Tsyrenzhapov, A.V., Sambueva, Z.G., Nikolaeva, G.G., Ratnikova, G.V. and Tankhaeva, L. (2001). Hepatoprotective effect of granules of the dry extract obtained from *Gentianopsis barbata* (Froel) Ma. *Eksperimental'naia i klinicheskaia farmakologiia*, **64(1)**: 49-52.

Rana, V.S. and Rawat, M.S. (2005). A new xanthone glycoside and antioxidant constituents from the rhizomes of *Swertia speciosa*. *Chemistry and Biodiversity*, **2(10)**: 1310-1315.

Saraswat, B., Visen, P.K. and Agarwal, D.P. (2000). Ursolic acid isolated from *Eucalyptus tereticornis* protect against ethanol toxicity in isolated rat hepatocytes. *phytotherapy research*, **14(3)**: 163-166.

Tang, L., Zhu, Z.H., Wang, J. and Cui, J. (2008). Summary of original plants of Tibetan medicine Zangyinchen. *Chinese Journal of Ethnomedicine and Ethnopharmacy*, **6**: 8-10.

Tian, W., Chen, Z.H., Zhai, J., Chen, L.R. and Li, Y.M. (2005). HPLC fingerprinting of total glycosides of *Swertia franchetiana*. *Acta Pharmaceutica Sinica*, **40(5)**: 447-452.

Wang, X.K. (1994). Natural Pharmaceutical Chemistry (first edition). Beijing: People's Medical Publishing House, pp. 272-289.

Xiong, X.J., Chen, W., Xiao, X.H., Xu, L.Y. and Cui, J.L. (2004). The Comparative protective effect of ursolic acid and oleanolic acid on experimental liver injury in mice. *Journal of Jiangxi Normal University (Natural Sciences Edition)*, **28(6)**: 540–543.

Xu, M. and Zhang, Z.Q. (2008). Study on the specific chromatogram of *Swertia mileensis* and determination of target components by HPLC. *Chinese Journal of Hospital Pharmacy*, **28(20)**: 1751-1755.

Yang, H. and Liu, J.Q. (2005). Seven constituents in nine species used as Tibetan medicine "Zangyinchen". *Chinese Traditional and Herbal Drugs*, **36(8)**: 1233-1237.

Yang, H.L. and Liu, J.Q. (2005). Seed Germination of *Swertia mussotii*, an Important Application in Tibetan Folk Medicine. *Acta Botanica Yunnanica*, **27(3)**: 295-300.

Yuan, Y., Cui, B.S., Zhang, Y. and Li, S. (2010). Xanthones of *Comastoma pedunlulatum*. *China Journal of Chinese Materia Medica*, **35(12)**: 1577-1599.

Yu, Y., Wang, S.S., Ding, F.J., Zhu, J.B. and Zhao, W.J. (2010). Isolation and identification of the chemical constituents from *Swertia yunnanensis* Burk. *Chinese Journal of Medicinal Chemistry*, **20(2)**: 125-127.

Zhang J. (2009). Studies on chemical constituents and bioactivities in *Halenia corniculata* (L.) Cornaz. *Shenyang Pharmaceutical University*, PhD thesis.

Zhang, S.P., Du, X.G. and Pu, X.P. (2010). 3-O-demethylswertipunicoside protects against oxidative toxicity in PC12 cell. *Biological and Pharmaceutical Bulletin*, **33(9)**: 1529-1533.

Zhang, X.Q., Tian, L.Y., Hou, J.J., Chen, J.C. and Liu, Y.W. (2006). Study on HPLC fingerprint of *Swertia punicea*. *Chinese Journal of Hospital Pharmacy*, **26(8)**: 922-924.

Zhao J.F. (2008). The resource investigation and pharmacognostical study of Tibetan Medicine Zangyinchen. *Beijing University of Chinese Traditional Medicine*, Master thesis.

Zhao, W.X., Chen, Z.M., Hou, H., Liang, C.L., Pang, R.Q., Zhao, B. and Chen,Z.L. (2002a). Luteolin decreases CCl_4 induced liver fibrosis is in rats. *World Chinese Journal of Digestology*, **10(7)**: 780-782.

Zhao, W.X., Liang, C.L., Chen, Z.M., Pang, R.Q., Zhao, B. and Chen, Z.L. (2002b). Luteolin inhibits proliferation and collagen synthesis of hepatic stellate cells. *Chinese Journal of Hepatology*, **10(3)**: 204-206.

Zheng, Y.W., Ma, D.L. and Chen M.Z. (1990). Effects of luteolin on H_2O_2 release of peritoneal macrophages in rat. *Chinese Pharmacological Bulletin*, **6(1)**: 56-58.

Zhong, G.Y., Wang, C.H., Liu, X., Qin, S.Y., Zhou, H.R. and Luo, W.Z. (2010). The resources and usage status of the commonly used tibetan medicinal crop "Dida". *World Science and Technology (Modernization of Traditional Chinese Medicine and Materia Medica)*, **12(1)**: 122-128.

Zhong, G.Y., Wang, C.H., Zhou, H.R. and Qing SY. (2008). Tibetan Medicine: Pharmacognosy and Species Consolidation. *World Science and Technology (Modernization of Traditional Chinese Medicine and Materia Medica)*, **10(2)**: 29-32.

Zuo, G.Y., Li, Z.Q., Chen, L.R. and Xu, X.J. (2005). *In vitro* anti-HCV activities of Saxifraga melanocentra and its related polyphenolic compounds. *Antiviral Chemistry and Chemotherapy*, **16(6)**: 393-398.

Zuo, G.Y., Zhang, Z.J., Chen, L.R. and Xu, X.J. (2005). Chemical constituents of Tibetan herbal medicine *Saxifraga melanocentra*. *Acta Botanica Yunnanica*, **27(6)**: 691-694.

Traditional and Folk Herbal Medicine: Recent Researches Vol. 1 (2012) *Pages* **209–250**
Editor: **V.K. Gupta**
Published by: **DAYA PUBLISHING HOUSE, NEW DELHI**

10

Euphorbia Latex: A Magic Potion or Poison

Papiya Bigoniya[1]

ABSTRACT

Euphorbia is one of the most diverse genera in the plant kingdom belonging to the family Euphorbiaceae. The plants are annual or perennial herbs, woody shrubs or trees with a caustic, poisonous milky sap (latex). Spurges are large and variable worldwide genus of more than 2,000 plants that include small trees, shrubs, vines, herbaceous plants and succulents. The presence of polycyclic diterpene esters in Euphorbia plants sap makes it highly irritant.

The milky sap of many Euphorbias are notably corrosive, contact with skin may cause burning pain temporary or permanent blindness. Swallowing the sap is potentially fatal causing severe inflammation of the walls of the stomach and intestine with perforation in some cases. The active component of Euphorbia latex is Euphorbon, a cytotoxic protein. Traditionally Euphorbias are used as purgative, analgesic, anti-inflammatory, antipyretic, antimicrobial, antiparasitic, in the treatment of cough, asthma, rheumatism, cancer and other maladies as folk remedy. Euphorbia latex and its isolated compounds have reported molluscicidal, cytotoxic, mitogenic, pesticidal, anti-inflammatory and anti-arthritic activity. Latex contains diterpene esters of the phorbol, ingenol and 12-deoxyphorbol esters known to be highly active cocarcinogenic and tumor promoting agents.

Recently some of the diterpene esters described in Euphorbias have been reported to exhibit antileukemogenic activity. Quite a number of Euphorbia species are used in folk medicine, in agriculture, as resources for certain technologies and as ornamental plants in horticulture. The toxic nature of the Euphorbia phytochemicals is discouraging their uses despite the possible manifold therapeutic potentials. This review work compiles details of

---

1 Department of Pharmacology, Radharaman College of Pharmacy, Fatehpur Dopra, Ratibad, Bhopal – 462 002, M.P., India.
E mail: p_bigoniya2@hotmail.com; Ph: +91-0755-2477941, 91-0755-2896237, 09827011258.

traditional uses, ethnic reports of toxicities and lethalities, phytochemical and pharmacological studies, and future possibilities as potential sources of new compounds with promising therapeutic values of Euphorbia latex.

Keywords: *Euphorbia*, Poisonous latex, Irritation, Toxins, Euphorbon, Terpenes, Cocarcinogenic, Cytotoxic.

Introduction

Euphorbia is a plant genus belonging to the family Euphorbiaceae. *Euphorbia* is one of the most diverse genera in the plant kingdom. Members of this family and genus are sometimes referred to as *Spurges*. The genus is primarily found in the tropical and subtropical regions of Africa and America, but also in temperate zones worldwide. Succulent species originate mostly from Africa, America and Madagascar.

The botanical name *Euphorbia* derives from Euphorbus, the Greek physician of king Juba II of Numidia (52-50 BC–23 AD). He was reported to have used a certain plant, possibly Resin Spurge (*E. resinifera*), as an herbal remedy when the king suffered from a swollen belly. *Carolus* Linnaeus assigned the name *Euphorbia* to the entire genus in the physician's honor.

Euphorbia regisjubae (King Juba's *Euphorbia*) was named to honor the king's contributions to natural history and his role in bringing the genus to notice. In view of extensive ethinopharmacological importance of this plant species all over the world, this review is an effort to compile all the information reported on phytochemical and pharmacological activities of latex. Exploration of the therapeutic benefits of the plant will help the under privileged and poor population of the acrid tribal region to use the plant for alleviation of commonly occurring ailments.

Family Euphorbiaceae - Spurges

The Euphorbiaceae is a very large, widely distributed family including around 300 genera and 7700 species, mainly non-succulent herbs, shrubs and trees found in temperate, sub-tropical and tropical climates. The Euphorbiaceae have very distinctive small cup-shaped compound flowers. The flower is technically known as a cyathium, consisting of a cup-like involucre containing a single pistil (female flower) surrounded by male flowers, each contributing a single stamen. The cyathium may be surrounded by brightly coloured bracts that give the appearance of a large flower. However, some species of Euphorbia have separate male and female plants and some species *e.g. Euphorbia obesa* may produce a mixture of male, female and bisexual flowers.

The milky saps of many species are called latex. *Euphorbia* latex is commercially exploited, with *Hevea brasiliensis* (rubber tree), the main species grown in plantations throughout the tropics, especially Malaysia and South East Asia. Other important Euphorbiaceaes include *Ricinus communis* (Castor Oil Plant) and *Manihot esculenta* (Cassava) cultivated for its starchy root and *Euphorbia pulcherrima* (Poinsettia) widely cultivated for the horticultural trade.

Euphorbiaceae Juss. includes *Acalypheae* (*Acalyphaceae*) J.G. Agardh, *Bertyaceae* J.G. Agardh, *Columellaceae* Dulac (p.p.), *Crotonaceae* J.G. Agardh, *Hippomaneae*

(*Hippomanaceae*) J.G. Agardh, *Micrantheae* (*Micrantheaceae*) J.G. Agardh, *Oldfieldiaceae* auctt., *Phyllantheae* (*Phyllanthaceae*) J.G. Agardh, *Putranjiveae* (*Putranjivaceae*) Endl., *Porantheraceae* (Pax) Hurusawa, *Pseudanthaceae* Endl., *Ricinaceae* Barkley, *Ricinocarpaceae* (Pax) Hurusawa, *Scepaceae* Lindl., *Tithymaloideae* (*Tithymalaceae*) Vent., *Treviaceae* Bullock, *Trewiaceae* Lindl.

The plants of family Euphorbiaceae are mostly monoecious herbs, shrubs or trees, sometimes succulent and cactus-like, laticiferous or non-laticiferous and without colored juice (*e.g.* Phyllanthoideae) or with colored juice (rarely). 'Normal' plants or 'switch'-plants, often with the principal photosynthesizing function transferred to stems or phyllodineous, or 'cactoid' (often). Plants are succulent or non-succulent, self supporting or climbing, mesophytic or xerophytic. Leaves minute to large, mostly alternate but may be opposite or whorled and they are simple or compound or distichous, herbaceous or leathery, fleshy or membranous, modified into spines, petiolate to sessile, non-sheathing, gland dotted or not dotted, not peltate, when compound palmate. Lamina entire, pinnately veined or palmately veined. Stipules are generally present but may be reduced to hairs, glands or spines. Leaves are without a persistent basal meristem. Mucilaginous epidermis are present or absent. Stomata mainly confined to one surface or on both surfaces; anomocytic or anisocytic or paracytic. Urticating hairs present (in a few lianes) or absent. Lamina are dorsiventral or isobilateral or centric, without secretory cavities. The mesophyll with sclerenchymatous idioblasts or are without sclerenchymatous idioblasts. Minor leaf veins are without phloem transfer cells (*Euphorbia*).

Cork cambium are present, initially deep-seated (rarely) or superficial. Nodes are tri-lacunar or unilacunar. Primary vascular tissue are bicollateral or centrifugal. Cortical bundles present (occasionally) or absent. Medullary bundles present (occasionally) or absent. Internal phloem is present (occasionally) or absent. Secondary thickening develop from a conventional cambial ring or anomalous, from a single cambial ring. The secondary phloem stratified into hard (fibrous) and soft (parenchymatous) zones (very rarely. *e.g.* in Bridelia) or not stratified. 'Included' phloem present (occasionally, *e.g.* Dalechampia) or absent. Xylem is with libriform fibres. Vessel end-walls are scalariform or simple (usually). Vessels are with vestured pits (rarely) or without vestured pits. Wood parenchyma apotracheal or paratracheal (or very sparse, or absent).

The flowers are unisexual and usually actinomorphic. They may be highly reduced by suppression of parts, in the extreme form consisting of a naked stamen as a male flower and a naked pistil as a female flower. Flowers aggregated in 'inflorescences'. The terminal inflorescence unit is nearly as always cymose (commonly the first branching racemose, with all the subsequent branching cymose). Inflorescences are terminal or axillary, with involucral bracts or without involucral bracts, pseudanthial or not pseudanthial. Flowers are bracteate or ebracteate, minute or small or medium-sized and regular. Free hypanthium is absent. Hypogynous disk present (commonly) or absent in separate members or annular.

A specialized type of miniature inflorescence called a 'cyathium' occurs in about 1,500 species comprising the genera *Euphorbia* and *Chamaesyce*. The cyathium consists of a single naked pistillate flower surrounded by cymes of naked staminate flowers, each consisting of a single stamen. These flowers are all enclosed in a cup-like involucre that is provided typically with peripheral nectaries and petaloid appendages such that the whole aggregation closely resembles a single flower. In other members of this family, the flowers and inflorescences are more ordinary in appearance, with male and female flowers typically bearing a 5-merous calyx and corolla of distinct segments, although the corolla is sometimes absent. In these forms, the androecium most commonly consists of 5, 10 or sometimes numerous distinct or monadelphous stamens. The gynoecium of female flowers consists of a single compound pistil of typically 3 carpels, an equal number of styles or primary style branches and a superior ovary with typically 3 locules, each bearing 1 or 2 collateral, axile-apical pendulous ovules. The fruit is usually a capsular schizocarp.

Fruit is non-fleshy (usually) or fleshy, dehiscent or indehiscent or a schizocarp (usually). Mericarps when schizocarpic are 2 to 3 (usually dehiscent) or 4. Fruit when non-schizocarpic are a capsule or a drupe, elastically dehiscent (schizocarpic capsules often splitting elastically) or passively dehiscent. Seeds are endospermic (nearly always). Endosperm is oily. Cotyledons are 2 (usually wider than the radical), flat or folded. Embryo is chlorophyllous (4/6) or achlorophyllous (8/10 *Euphorbia* being variable), straight or curved.

Euphorbia Genus

The genus *Euphorbia* is one of the largest and most complex genera of flowering plants and several botanists have made unsuccessful attempts to subdivide the genus into numerous smaller genera. According to the phylogenetic studies, *Euphorbia* can be divided into 4 subgenera, each containing several but not yet sufficiently studied sections and groups.

Table 10.1: Plants of *Euphorbia* spurge genus containing irritant and poisonous latex.

Category		*Euphorbia Plants*
Mouse ear irritancy		*E. cooperi, E. genistoides, E. helioscopia, E. hermentiana, E. ingens, E. lateriflora, E. lathyris, E. neriifolia, E. nivulia, E. paralias, E. striata* and *E. tirucalli*
Intensity	**Strong**	
	Moderate	*E. coerulescens, E. hypericifolia, E. ledienii, E. mauritanica, E. striata* and *E. triangularis*
	Weak	*E. franckiana, E. hermentiana, E. milli, E. nubica,* and *E. orientalis*
Eye irritant		*E. antiquorum, E. cotinifolia, E. drupifera, E. gorgonis, E. grandidens, E. ingens, E. lacteal, E. lathyris, E. obtusifolia, E. officinarum, E. systyloides, E. tirucalli* and *E. trigona*
Skin irritant		*E. antiquorum, E. gomerifera, E. guadichaudii, E. helioscopia, E. ingens, E. lateriflora, E. laterifolia, E. lathyris, E. myrsinites, E. poissonii, E. trigona, E. unispina* and *E. venenifica*
Lethally poisonous		*E. canariensis, E. candelabrum, E. cereiformis, E. ferox, E. genistoides, E. grantii, E. heptagone, E. heterophylla, E. ledienii, E. mauritanica, E. neglecta, E. pugniformis, E. reinhardtii, E. striata, E. venenata* and *E. virosa*

Spurges (genus *Euphorbia*) are a large and variable worldwide genus of more than 2,000 plants that include small trees, shrubs, vines, herbaceous plants and succulents (Merani *et al.*, 2007). They range from annual weeds to trees. Spurges contain acrid, poisonous milky latex, a unique flower structure and some of them are armed with thorns. A significant percentage is succulent, but they are mostly originating from Africa and Madagascar. The roots are fine or thick and fleshy or tuberous. The main stem and also mostly the side arms of the succulent species are thick and fleshy, 15–91 cm (6–36 inches) tall. The deciduous leaves are opposite, alternate or in whorls. In succulent species, the leaves are mostly small and short-lived. The stipules are mostly small, partly transformed into spines or glands, or missing. Similar to the members of the family Euphorbiaceae, all spurges have unisexual flowers. In *Euphorbia* these are greatly reduced and grouped into pseudanthia called cyathia. The majority of species are monoecious (bearing male and female flowers on the same plant), although some are dioecious with male and female flowers occurring on different plants. It is not unusual for the central cyathia of a cyme to be purely male, and for lateral cyathia to carry both sexes. Sometimes young plants or those growing under unfavourable conditions are male only, and only produce female flowers in the cyathia with maturity or as growing conditions improve. The bracts are often leaf-like, sometimes brightly coloured and attractive, reduced to tiny scales. The fruits have three (rarely two) compartment capsules, sometimes fleshy but almost always ripen to a woody container that then splits to open (explosive dehiscence). The seeds are 4-angled, oval or spherical, and in some species have a caruncle.

Succulent species are found in the genera *Euphorbia*, and the obsolete genera *Endadenium*, *Jatropha*, *Monadenium*, *Pedilanthus*, *Phyllanthus*, and *Synadenium* which have been merged into *Euphorbia*s. In the genus *Euphorbia*, succulence in the species has often evolved divergently to different degrees. About 850 species are succulent in the strictest sense. If one includes slightly succulent and xerophytic species, this figure rises to about 1,000, representing about 45 per cent of all *Euphorbia* species. Some of the succulent species are very spiny, resembling *Cactaceae* at a first glance despite being unrelated. Most of the spurges yield powerful emetic and cathartic products. Some of the succulent species are very spiny, resembling *Cactaceae* at a first glance. The *Euphorbias* are named after a Greek surgeon called Euphorbus. He was physician of Juba II who was the Romanised king of a North African kingdom in 18 AD and was supposed to use their milky latex as an ingredient for his potions (Cooper and Johnson, 1984). Many studies have suggested that these plants have not only therapeutic relevance but that they also display toxicity (Hohmann and Molnár, 2004).

Cultivation

Most succulent *Euphorbias* require warmth and full to partial sun. Watering should be very sparing during the winter, especially for caudiciform species unless they can be kept warm and in growth. Very gritty compost suits most species, especially caudiciforms and those with fleshy roots. However, succulent *Euphorbias*, especially those with leaves, appreciate a dose of high nitrogen fertiliser at the start of their growing season. At other times a balanced fertiliser is suitable. Cuttings of the Mexican tree, *Euphorbia pulcherrima* (Poinsettia) are widely cultivated in the horticultural trade as leafy non-succulent plants with brightly coloured bracts. Bracts appear in response to strict short day lengths of less than ten hours for ten weeks, are usually induced artificially under controlled lighting and are quite difficult to reproduce under domestic conditions.

Species and Varieties of *Euphorbias* Common in India with Widely Used Latex

- *Euphorbia acaulis* Roxb. Parts used: juice.
- *Euphorbia antiquorum* Linn. (Tredhara Sehund), Parts used: root, latex and juice.
- *Euphorbia atoto* Forst. f. Parts used: latex.
- *Euphorbia canariensis* Parts used: leaf and latex.
- *Euphorbia characias* Parts used: leaf and latex.
- *Euphorbia coerulescens* Parts used: leaf and latex.
- *Euphorbia corollata* Parts used: leaf, root and latex.
- *Euphorbia cotinifolia* Parts used: leaf and latex.
- *Euphorbia cyparissias* Parts used: leaf, root and latex.
- *Euphorbia fortissimo* Parts used: whole herb, root and latex.
- *Euphorbia gorgonis* Parts used: leaf, root and latex.
- *Euphorbia hermentiana* Parts used: leaf, root and latex.
- *Euphorbia hirta* Linn. (Dudhi) Parts used: leaf, latex, root and stem.
- *Euphorbia ingens* Parts used: leaf, root and latex.
- *Euphorbia lactea* Parts used: whole plant, root and latex.
- *Euphorbia lateriflora* Parts used: latex.
- *Euphorbia lathyris* Linn. Parts used: leaf, latex, seeds and capsules.
- *Euphorbia ledienii* Parts used: leaf, root and latex.
- *Euphorbia longifolia* D. Don. Parts used: juice.
- *Euphorbia milii* Parts used: leaf, root and latex.
- *Euphorbia neriifolia* Parts used: leaf, root and latex.
- *Euphorbia nivulia* Buch.- Ham. (Katathohar), Parts used: latex, root, stem and juice.
- *Euphorbia paralias* Parts used: whole herb, root and latex.
- *Euphorbia pilosa* Linn. Parts used: root and juice.
- *Euphorbia poisonii* Parts used: leaf, root and latex.
- *Euphorbia resinifera* Parts used: root and latex.
- *Euphorbia royleana* Boiss. (Thor, Shakarpitan), Parts used: latex.
- *Euphorbia serrata* Parts used: leaf, root and latex.

Figure 10.1: Photograph of some commonly occurring *Euphorbia* species.

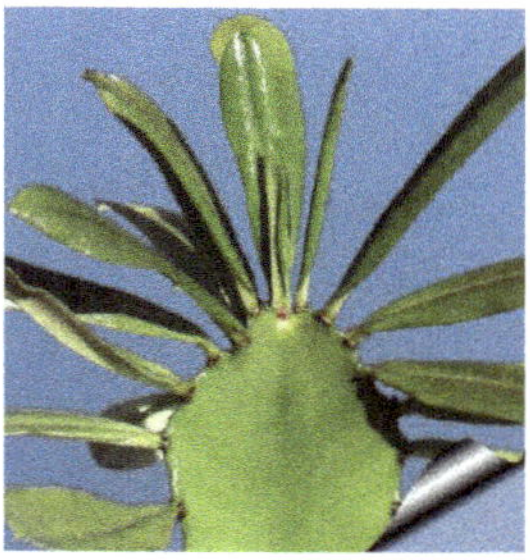

Euphorbia antiquorum

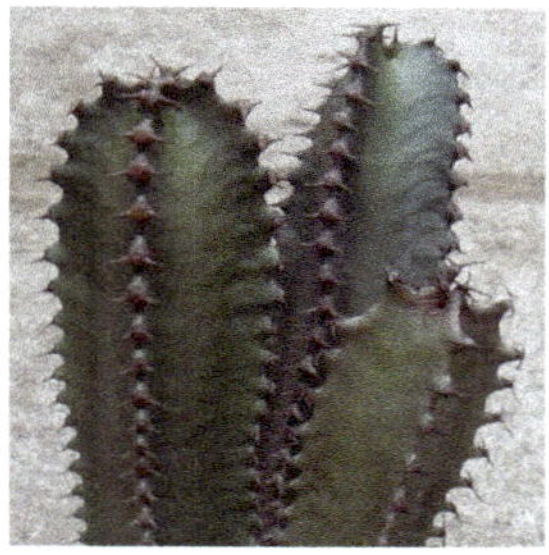

Euphorbia canariensis

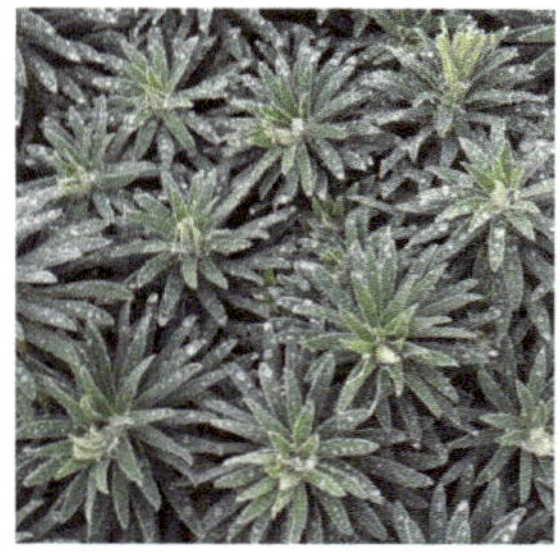

Euphorbia characias

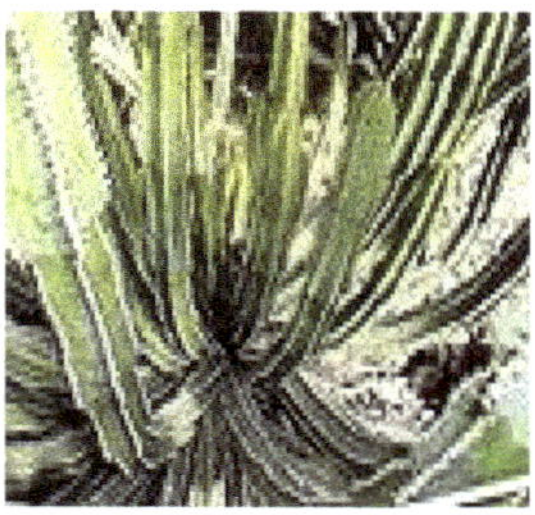

Euphorbia coerulescens

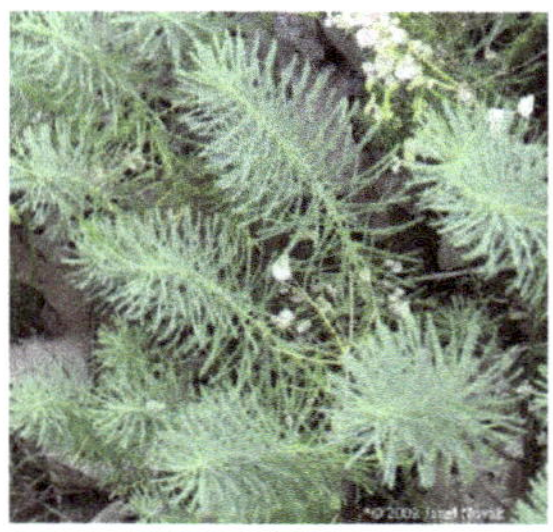

Euphorbia cyparissias

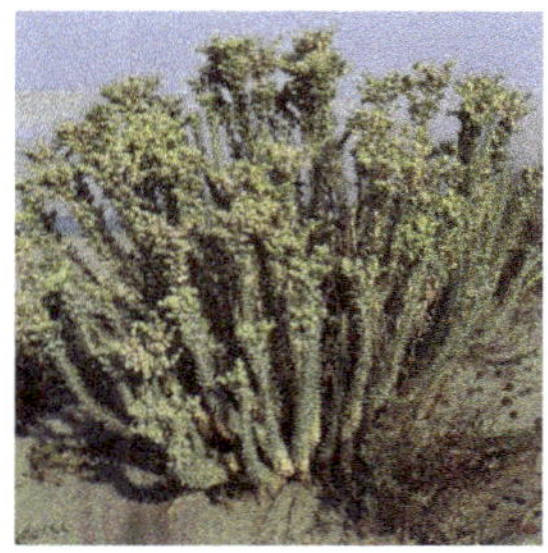

Euphorbia fortissima

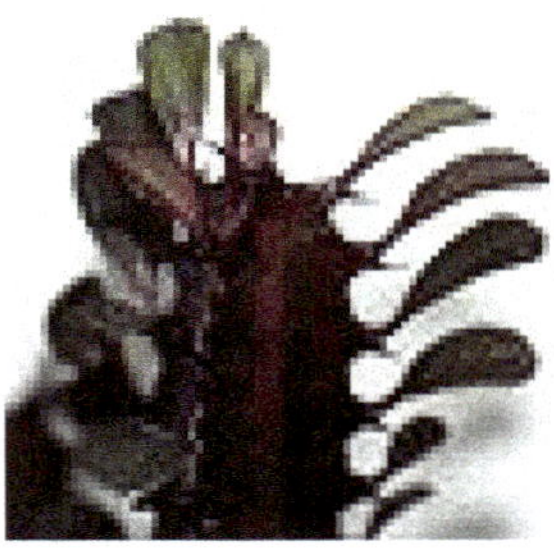

Euphorbia hermentiana

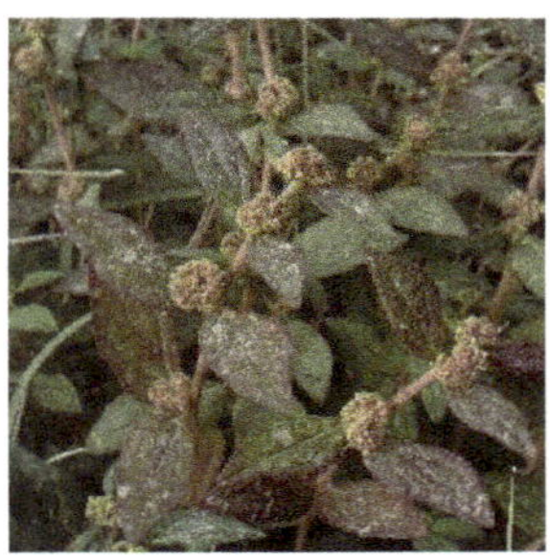

Euphorbia hirta

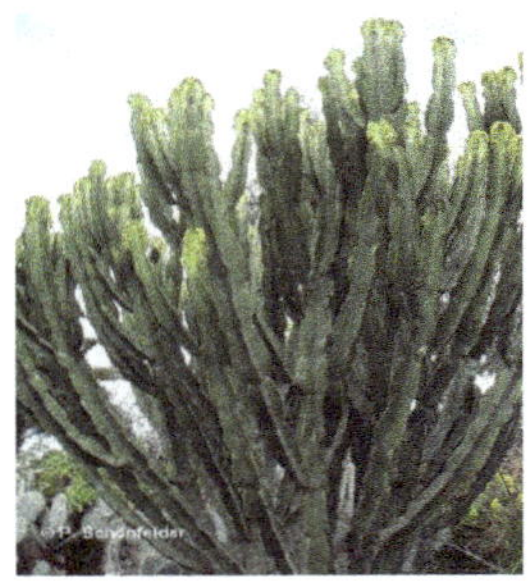

Euphorbia ingens

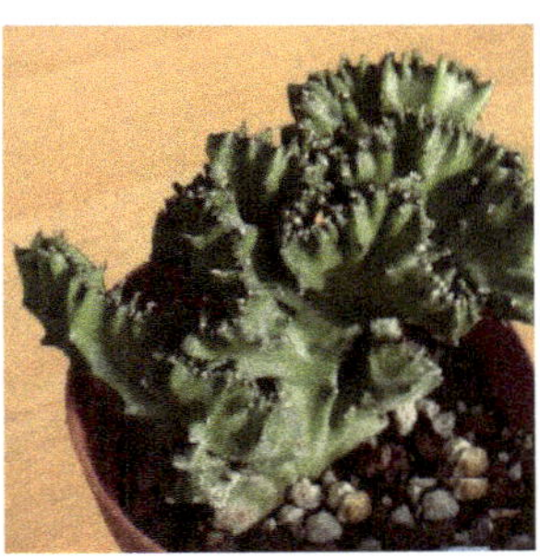

Euphorbia lactea

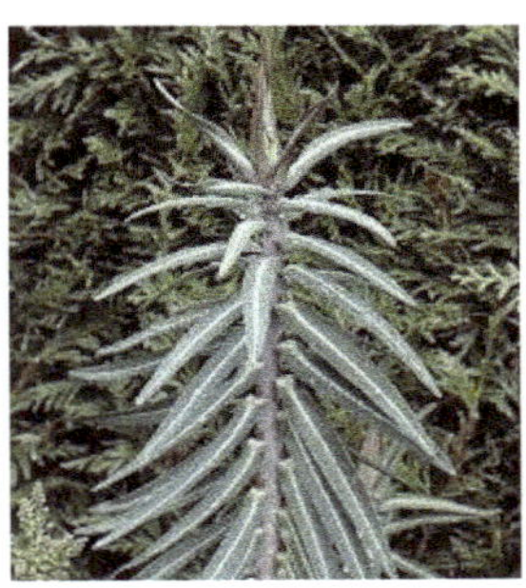

Euphorbia lathyris

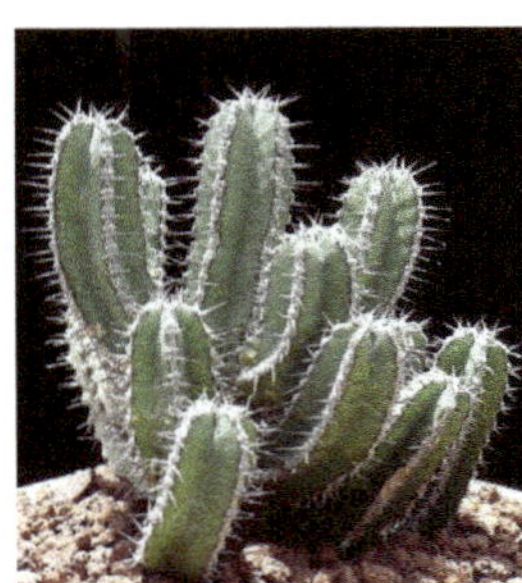

Euphorbia ledienii

Contd...

Figure 10.1–*Contd...*

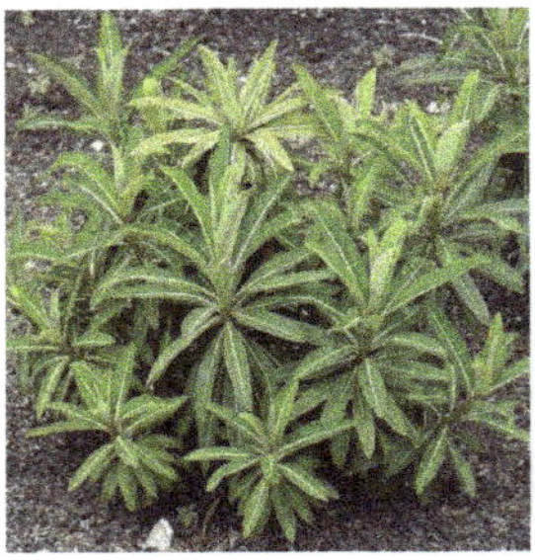

Euphorbia longifolia

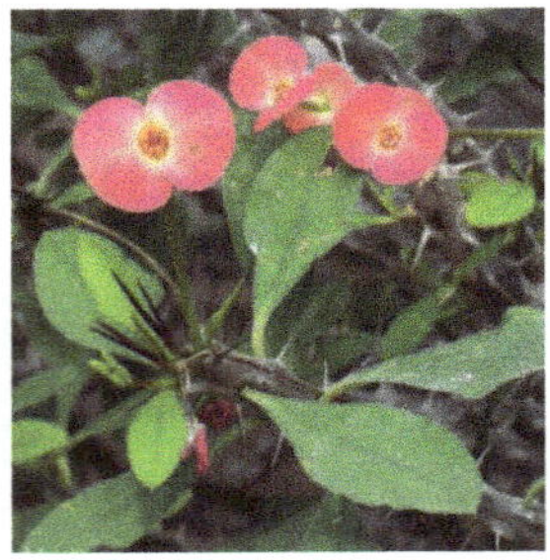

Euphorbia milii

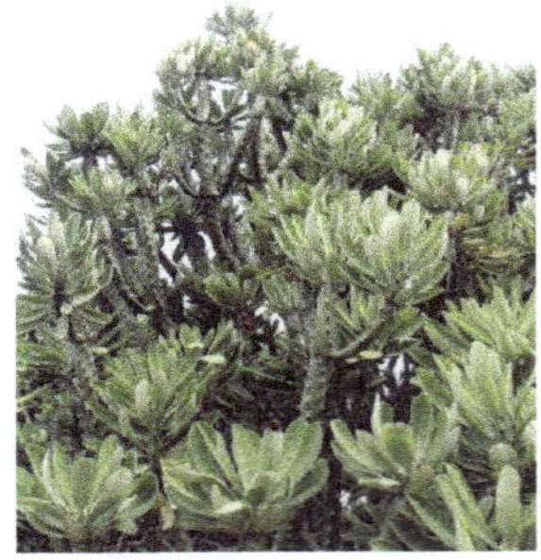

Euphorbia neriifolia

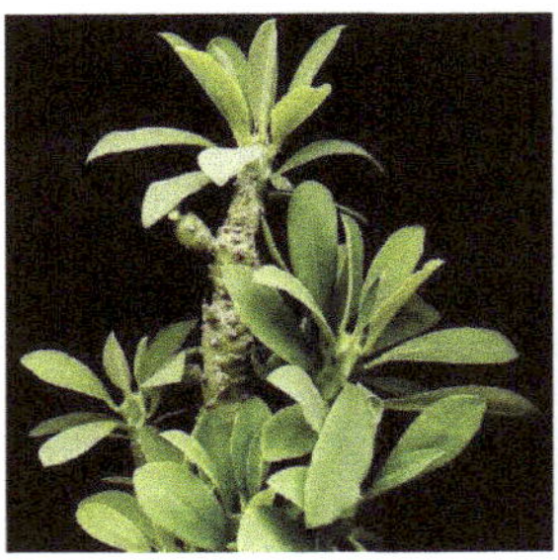

Euphorbia nivulia

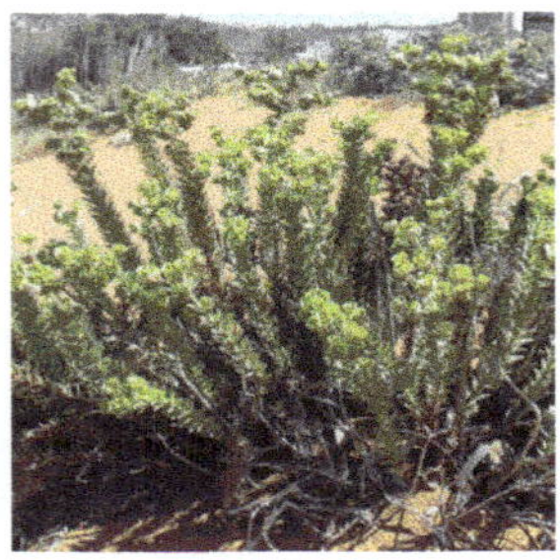

Euphorbia paralias

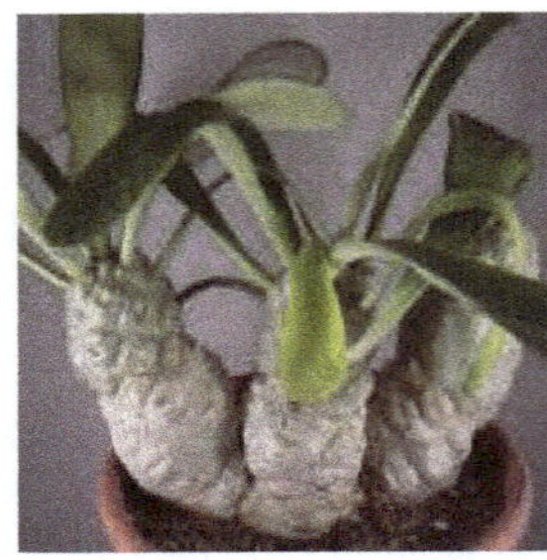

Euphorbia poisonii

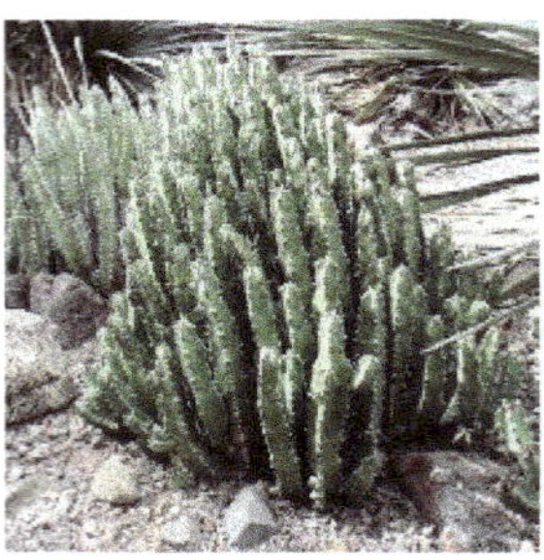

Euphorbia resinifera

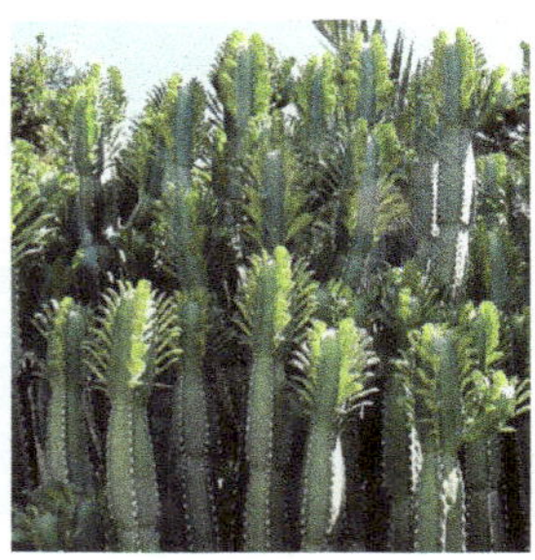

Euphorbia royleana

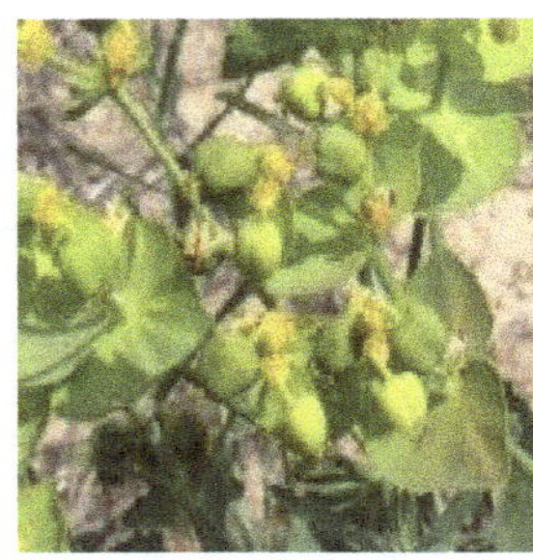

Euphorbia serrata

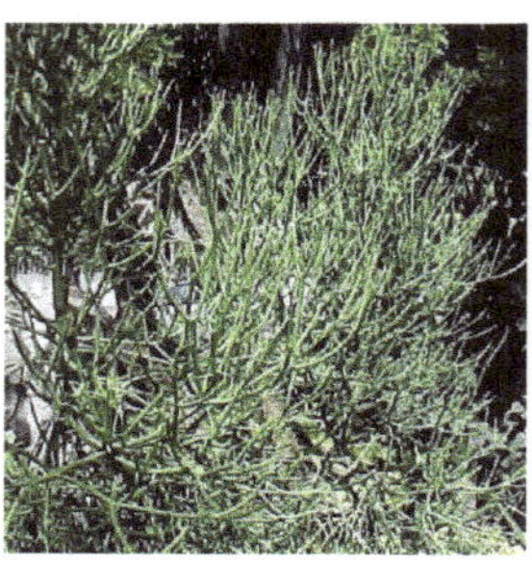

Euphorbia tirucalli

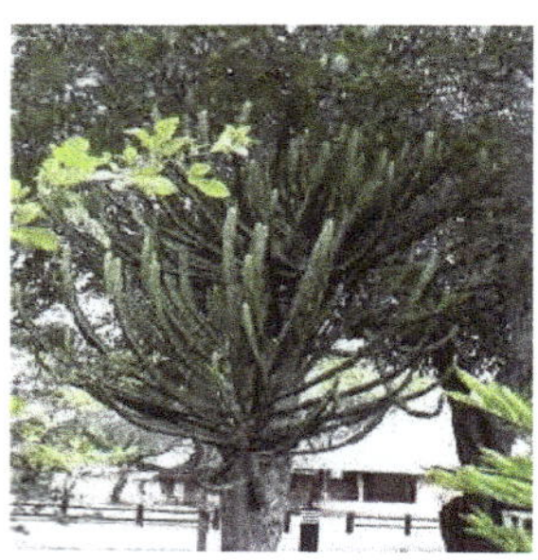

Euphorbia triangularis

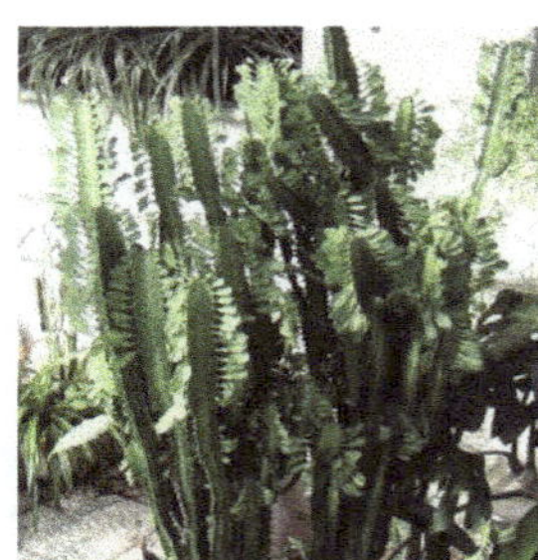

Euphorbia trigona

- *Euphorbia striata* Parts used: latex.
- *Euphorbia tirucalli* Linn. (Konpal Sehund, Milk Bush, Indian Spurge), Parts used: latex.
- *Euphorbia triangularis* Parts used: whole herb, root and latex.
- *Euphorbia trigona* Haw. Parts used: leaf and latex.

Latex

Latex is produced by 20,000 species from over 40 families occurring in multiple lineages in both dicotyledonous and monocotyledonous plants. It is also found in conifers and pteridophytes. Several members of the fungal kingdom also produce latex upon injury. Latex is the stable dispersion of polymer microparticles in an aqueous medium. Latex as found in nature is a milky sap-like fluid occurring in 10 per cent of all flowering plants and angiosperms. It is a complex emulsion consisting of proteins, alkaloids, starches, sugars, oils, tannins, resins, and gums that coagulates on exposure to air. It is usually exuded after tissue injury. In most plants, latex is white, but some have yellow, orange, or scarlet latex. Since the 1600s, latex has been used as a term for the fluid substance in plants. It serves mainly as defense against small insects. The latex are found in cells or vessels, which make up the laticiferous system, These systems are formed in two very different ways. In many plants, the laticiferous system is formed from rows of cells laid down in the meristem of the stem or root. The cell walls between these cells are dissolved so that continuous tubes, called latex vessels, are formed. In the milkweed and spurge families, the laticiferous system is formed quite differently. Early in the development of the seedling, latex cells differentiate and as the plant grows, these latex cells grow into a branching system extending throughout the plant. In the mature plant, the entire laticiferous system is descended from a single cell or group of cells present in the embryo.

This sap provides a defence against grazing and helps to reduce water loss by lowering the potential loss of water through evaporation. The toxic sap averse native herbivores to consume these plants and aids in the control of invasion in bushland.

Latex contains 50–1000 per cent higher concentrations of defense substances than other plant tissues. These toxins also include ones that are toxic to the plant itself and consist of a diverse range of chemicals that are either poisonous or "antinutritive". The clotting latex is functional in these defenses since it prevents wastage and its stickiness traps insects and their mouthparts.

Euphorbia milky sap contains irritant toxins including diterpene esters, alkaloids, glycosides, and ricin-type protein toxins. Terpenes, including diterpenes and triterpenes, have been frequently found in *Euphorbia* species. Steroids, cerebrosides, glycerols, phenolics and flavonoids were also isolated from plants of the genus, but the compounds most relevant to the toxicity and considerable biological activities in *Euphorbia* are diterpenes, especially those with abietane, tigliane, and ingenane skeletons (Shi *et al.*, 2008a). Usually latex is white, but in rare cases (*e.g. E. abdelkuri*) yellow also. As it is under pressure, it runs out from the slightest wound and congeals within a few minutes of contact with the air. *Euphorbia* plants are used as raw material for rubber, castor oil and tapioca. The saps of *Euphorbia* plants are highly irritant due to the presence of polycyclic diterpene esters (Webster, 1986), which is a deterrent to insects and herbivores (Singh and Agarwal, 1988). Besides, they may possess extreme proinflammatory and tumor promoting toxicities (Vogg *et al.*, 1999; Baloch *et al.*, 2005).

Traditional Uses of Some *Euphorbia* Latex

The latex (milky sap) of spurges acts as a deterrent for herbivores as well as a wound healer. It is a severe irritant especially for mucous membranes and that should primarily be the reason for its molluscicidity.

E. gorgonis

Many good uses are attributed to this *Euphorbia* including the use of the latex to aid the removal of warts, relief of toothache, sores, skin eruptions and cancerous skin conditions.

E. hypericifolia

While used medicinally in Angola and Mozambique, it is reported as being poisonous. In East Africa the latex is used as a purgative and as a caustic on skin lesions.

E. lateriflora

The latex of *E. lateriflora* is taken in Northern Nigeria along with milk or cereals to cause purging and sometimes vomiting. Because the latex is a drastic purge, it is used in the treatment of syphilis and also as a remedy for head lice and ringworm on the scalp (Burkill, 1994).

Euphorbia neriifolia L. (*Elaeophorbia neriifolia* A. Chev., *Euphorbia ligularia* Roxb)

It is called as Cactus Hedge, Indian Spurge Tree, Sesudu. This spiny plant is used as a hedge plant in India. The milky juice (latex) is pungent, laxative, and good for abdominal troubles, tumors and leucoderma. It is liable to cause dermatitis. The milky juice is locally rubefacient and irritant to the skin and throat (Nadkarni, 1976). It has been used to destroy warts and cutaneous eruptions. It is used as a purgative, expectorant and rubefacient. The milky juice is also employed in earache, when mixed with soot it is applied in ophthalmia. It is used in the composition of most of the drastic purgatives. The tribal population of Chattishgarh region uses the milky latex as an ingredient of aphrodisiac mixture (Oudhia, 2005).

The juice of the plant is used in Gujarat for smearing cuts made by tapers on *Borassus flabellifer* (Linn) in order to prevent the palm from the attack of red weevil. Stem juice is also used to deroot skin warts and in earache. Stem or leaf juice is used in case of cough and cold mixed with honey. A succus consisting of equal parts of the juice of this plant and simple syrup was prepared and administered in doses of 10 to 20 minims three times a day in case of asthama. Thura milk in castor oil after adding salt is used externally on soles. The milk of Thura is also used commonly like aloe gel in case of burns. Thura milk can be used successfully for healing of wounds. Thura milk boiled in neem oil is applied externally on affected parts in treatment of rheumatism (Pandey, 1992).

Euphorbia resinifera (Euphorbium officinarum)

It is the spurge of antiquity, known to the Greek physician Dioscorides (40-90 AD) and Pliny the Elder (23-79 AD) for its purgative sap (euphorbium). This is probably a thorny plant growing in Mount Atlas of Morocco recorded by King Juba II of Mauretania, said to have named the plant "*Euphorbea*", after the famous Greek physician, Euphorbus. The name was later formalised by Linnaeus (*Species Plantarum,* 1753) as the official botanical name for the genus. All parts are toxic, but the resin (dried sap) has been used as an extreme purgative and wood preservative.

E. striata

Many good uses are attributed to this plant. The powdered root is used as an emetic in stomach disorders. However, at the same time, in Southern Rhodesia a species of *Euphorbia* (which is probably this one) is one of the materials used in making poison arrows. Handling of a fresh plant has been known to produce irritation of the skin.

Euphorbia tirucalli (Pencil Tree, Milk Bush, Kraalmelkbos)

The plant *E. tirucalli* Linn is popularly known, as Aveloz, is native from Africa and Asia, where its latex is used as folk medicine, in rheumatism treatment, neuralgia, asthma and gastric disorders.

The milky sap of many *Euphorbias* is generally an irritant, but that of *E. tirucalli* is notably corrosive. Contact with skin may cause burning pain and irritation. Contact with the eyes may cause temporary or permanent blindness. In case of eye contamination flushing with water for several times and immediate medical assistance are recommended. Swallowing the sap is potentially fatal. *E. tirucalli* is used as an impenetrable hedge or livestock encosure. Fishermen put mascerated branches into streams, where the sap stuns the fish making them easy to catch. The empirical utilization of *E. tirucalli* Linn. for cancer treatment has diffused around the Porto Alegre city and also in Montenegro, a city from Rio Grande do Sul State, in South Brazil, mainly associated directly or indirectly with religious leaders or entities (Cataluna and Rates, 1999).

How Dangerous are *Euphorbia* Latex

The name *Euphorbia* is considered by many in South Africa as synonymous with poison. Post-mortem examination of people killed by *Euphorbia* latex has revealed severe inflammation of the walls of the stomach and intestine and in some cases, the wall of the stomach has been perforated. The active component of *Euphorbia* latex that gives its toxicity is **Euphorbon**, a protein called cytotoxin (glycoside and resin). *Euphorbias* also contain haemolytic saponins.

The terpen ester composition determines how caustic and irritating to the skin it is. In contact with mucous membranes (eyes, nose, mouth), the latex can produce extremely painful inflammation. In experiments with animals it was found that the terpen ester resiniferatoxin had a 10,000 to 1,00,000 times stronger irritating effect than capaicin, the "hot" substance found in chili peppers. Severe pain and inflammation can result from contact with the eyes, nose, mouth and even skin, which may be due to the activation of protein kinase C enzyme (Kedei *et al.*, 2004). The toxic constituents of *Euphorbia* species were considered to be a kind of specific diterpenes, globally called phorboids, which comprise tigliane, ingenane and daphnane diterpene derivatives (Cateni *et al.*, 2003; Zayed *et al.*, 2001).

The latex of *E. drupifera* species is very dangerous. It has been used to remove warts and can cause blindness. *E. genistoides* is suspected of producing urethritis and strangury in the horse and the bovine. The plant is capable of producing cystitis and inflammation of the skin. Death cases are reported from drinking a decoction of the plant *E. gorgonis* and *E. grandidens*. *E. tetragona* latex is an irritant possibly causing blindness. *E. grantii* is suspected to produce serious irritant poisoning with fatalities. It has been used in the treatment of snake bite and as a powerful purgative. *E. ferox* and *E. ledienii* is said to be highly poisonous. *E. helioscopia* latex is highly irritant and toxic. Severe ulceration of the skin has been reported. The latex of *E. striata, E. mauritanica, E. heptagonea, E. virosa, E. cereiformis, E. candelabrum,* and *E. mauritanica* is used to make poison arrows. Both leaf and latex of *E. heterophylla* are poisonous and reportedly have been the cause of a child's death (personal communication). *E. ingens* latex is highly toxic causing blisters of the skin (acute bullous dermatitis). A single drop can cause intense pain and temporary blindness. Cattle driven through dense bush containing this tree may suffer severe burns on the eyes, lips and facial skin leading to fatality. *E. neglecta* has been the cause of an irritant poisoning among the staff of a Mission Station in Tanganyika, East Africa. This species is also thought to have caused human death, the symptoms being those of a gastro-intestinal irritation.

E. neriifolia, E. scheffleri and *E. pulcherrima* (Poinsettia) latex is used as a fish poison in the Netherlands, Indies and Indonesia. *E. poissonii, E. unispina,* and *E. venenifica* latex is a powerful irritant, used in de-hairing hides. These three species are often confused as they are having similar appearance. *E. pugniformis* latex is a violent emetic and purgative and is used by natives of Africa for the purpose of curing indigestion and constipation. However, the sap is highly irritant. A case of death from drinking a concoction of the plant is on record. A yellow resinous substance in this plant called **Euphorbin** can cause terrible blistering of the skin. The Nyamwezi (Tanzania) use the latex of *E. reinhardtii* as an emetic in snake bite. The latex is described as being acrid and toxic and in sufficient dose can fatally poison someone. *E. systyloides* latex can cause irritation and even blisters and produces severe irritation and even blindness when accidentally applied to the eyes.

E. tirucalli is probably the most dangerous *Euphorbia* and one of the most readily available plants, even in supermarkets of China and Singapore. Latex can cause skin rash, blisters and intense burning. Latex from this plant has been the cause of death, blindness and severe illness. The latex produces excruciating pain and severe inflammation if it gets into a cut on the skin or on to the eyeball. It produces intense irritation of the eyeball than on the skin, which may result in ulcerations on the eyeball (Anonymous, 2010). *E. cotinifolia* [red spurge] latex is highly irritant causing blindness. *E. lactea* [candelabra cactus] latex can cause rash, blisters, intense burning and temporary blindness. *E. laterifolia* sap is corrosive, produces white patches on skin. *E. guadichaudii* and *E. gomerifera* latex causes acute dermatitis. *E. trigona* (African milk tree) sap is irritating to skin and eyes, causes swelling to tongue, mouth and throat. *E. venenata* is said to be an extremely poisonous producing severe inflammation of the mucosae. *E. canariensis* and *E. virosa* latex is an irritant and highly poisonous, also used as an arrow poison. *E. drupifera* (*Elaeophorbia drupifera*) tree from tropical Africa should be handled with extreme care. The latex is very dangerous, can cause blindness. *E. antiquorum* from Peninsula India is a succulent in the family Euphorbiaceae. It can irritate the skin to varying degrees and if squirted accidentally into the eyes can cause blindness. In particular, if the plant is injured or cut, the sap gives off a very strong and lingering odour, not great for the lungs. Exposure of *E. myrsinites* causes pediatric erythematous papulovescicular skin eruptions. *E. lateriflora* juice is corrosive to flesh, acting as an escharotic, and has been used to produce white patches on the skin in order to simulate leprosy. *E. lathyris* plant does not lose its toxicity on drying or storage. The juice is irritant and when applied to the skin causes redness, itching, pimples, dermatitis and sometimes gangrenes. The milky sap is irritant to the human eye producing kerato-conjunctivitis (Lampe and Fagerstrom, 1968). The two most commonly encountered *Euphorbia* species are *E. milii* Des Moul. (syns *E. bojeri* Hooker) and *E. milii var splendens* Ursch and Leandri. The plant is grown as a garden subject in Western Australia. The sap is corrosive to the skin and eyes, causing temporary blindness. The latex of *Euphorbia poisonii* known as 'Tinya' (Hausa), has been used in West Africa as an arrow poison for fishing. It produces lesions of the skin in both man and livestock which are susceptible to secondary infections. It also causes inflammation of the mucus membrane of the nose, mouth and eyes, and in the chronic exposure leads to blindness (Evans and Kinghorn, 1975). A biological screen of the Nigerian produced succulent, although produces acute inflammation initially, it did not persist for more than 24 hour as is normally the case for diterpene stimulated hyperplasia.

E. myrsinites L. is a creeping spurge causes dermatitis by the irritant latex. Contact with the latex produced swelling and blistering of the skin, usually on the face, which appeared from two to eight hours after exposure and mostly cleared over the next four days (Spoerke and Temple, 1979). This species has been reported to contain esters of ingenol and 5-deoxyingenol (Evans and Kinghorn, 1974; Evans and Kinghorn, 1977) to which the irritant properties of the latex may be ascribed (Evans and Kinghorn, 1975; Upadhyay *et al.*, 1980b). *E. obtusifolia* and *E. officinarum* L. Latex is irritant to wounds and eyes.

Ocular inflammation caused by *Euphorbia* plant saps is a well known occupational hazard in lumber workers in tropical regions, and people who work in area clearing and grass cutting operations. Ocular injuries by *Euphorbia* plant saps were frequently reported in Western literatures. Joshi and Shingal (2008) reported the largest series of ocular injuries from the sap of *Euphorbia* plant involving 12 patients. They got ocular exposure from splashes of the plant sap while performing grass cutting work. Apart from direct splashing into the eyes, there are other mechanisms of ocular contact with the sap, *i.e.* rubbing the eyes with contaminated hands or intentional application of the sap for the treatment of eyelid wart.

Ocular toxicities caused by *Euphorbia* plants comprise of a spectrum of severity ranging from burning pain, lacrimation, photophobia, conjunctivitis, keratitis, uveitis, cornea ulceration, corneal oedema, sloughing off of the central corneal epithelium to even blindness (Scott and Karp, 1996; Eke *et al.*, 2000; Hsueh *et al.*, 2004). It has been noticed, when cutting large succulent spurges in a greenhouse, vapours from the latex spread and can cause severe irritation to the eyes and air passages several meters away (Lam *et al.*, 2009). Precautions, including sufficient ventilation are required. Small children and domestic pets should be kept away from contact with spurges.

There is a difference in some people's reactions to the toxicity of various *Euphorbias*. The different clinical presentations may be related to the different chemical substances present in the saps of different *Euphorbia* species. Besides the species of plant, the severity of ocular inflammation certainly is also related to the concentration of the sap and the duration of contact (Shi *et al.*, 2008a). Exposed to eyes, *Euphorbia* latex causes acute burning and painful irritation, lately eye becomes swelled and inflamed. The hallmark of ocular injuries by the saps of *Euphorbia* plants is intense pain immediately upon exposure. Apparently, *Euphorbia* latex can burn right through the outer layers of the eye, causing lasting eye damage and blindness. Copious irrigation with tap water at scene should be started as soon as possible in order to reduce the concentration of the sap in the eyes and the duration of exposure. Rinsed repeatedly with water, saline water and milk are beneficial to remove the toxin. The milk is the best, but still there will be bad pain and distraction. Supposing the toxic effects of the latex are lost when mixed with acetic acid. Though the pH values of *E. milii*, *E. tirucalli*, *E. antiquorum*, *E. neriifolia* and *E. lactea* saps were found to be acidic with a pH of 5 (Lam *et al.*, 2009), application of vinegar with a cotton swab around eye, lips, hands and other exposed parts may stop pain very quickly. Application of olive oil, or other oily things increases the level of discomfort may be due to trapping latex in. Early recognition of the symptoms and appropriate treatment is the cornerstone in the management of ocular injuries by *Euphorbia* plant saps. Corticosteroid eye drop can decrease the swelling very fast and eyesight comes to normal within a day. Topical antibiotic and steroid can reduce the risk of bacterial infection and corneal oedema. Topical cycloplegics for pain relieve should also be considered. Early follow-up by the ophthalmologist should be arranged for detailed evaluation. Apart from direct ocular toxicity from the *Euphorbia* plant sap, secondary bacterial infection also plays a very important role on visual loss and complications. Corneal ulcer with scarring or even perforation can result from delayed management. The symptoms may worsen several hours to days after the exposure (Scott and Karp, 1996; Eke *et al.*, 2000).

Among the potential poisonous plants, a succulent enthusiast might encounter, latex from various *Euphorbia* species is likely to be the most frequently encountered "nasty" poison. The *Euphorbia* sap on contact with skin causes painful rashes, red patches, burning, irritation and mild to severe blistering which may last for few days. These rashes may be confused with allergies, infections, or other dermatoses. The fumes from the latex evolve incredibly strong smell making the eyes tear, and usually give headache lasting an hour or two. Symptoms followed two to eight hours after exposure to the irritant sap. They increased in severity for an additional 12 hours and resolved within three to four days. There is no treatment other than to give pain relief medication, prevent infection and try to reduce swelling.

Therefore, spurges should be handled with caution. Latex coming in contact with the skin should be washed off immediately and thoroughly and eye contamination should be avoided at any costs. Some species bleed latex easily and it is advisable to spray plants with water before moving them to stop spillage. Partially or completely congealed latex is often no longer soluble in water, but can be removed with an emulsion (milk, hand-cream). A physician should be consulted regarding any inflammation of a mucous membrane, especially the eyes, as severe eye damage including possible permanent blindness may result from acute exposure to the sap (Pitts *et al.*, 1993).

The people who are very sensitive to *Euphorbia* latex should take precautions while handling the plant. If the skin has no reaction to the sap, more tender areas such as eyes might have a totally different reaction. The serious reaction experienced by some people to the sap of this plant is well documented over a great many years. These plants should be kept away from small children and pats. Animals should never be allowed anywhere near *Euphorbias*. It is also never a good idea to let animals drink dirty water or water running out of plant pots. It is certainly possible the pets may somehow ingest poison from the plant through the water. It doesn't take more than a pinprick of the skin to produce the poisonous latex.

Phytochemistry and Pharmacology of *Euphorbia* Latex

Various plants belonging to family Euphorbiaceae have been studied worldwide for their cytotoxic constituents. *E. elbractolata, Croton laciferus, E. pekinensis, Jatropa curcas* and *E. lathyris* are a few of the Euphorbiaceae plants that have been studied for toxic constituents (Akihisa *et al.*, 2002; Bicchi *et al.*, 2001). Carcinogenic properties of latex of Euphorbiaceae plants have been reported by various scientists although there are no records on co-carcinogenic activities of constituents of the latex of *E. antiquorum* (Evans and Kinghorn, 1973; Salah *et al.*, 1998). Latex of 75 *Euphorbia* species has been examined for their triterpenes. Fifteen tetra and penta cyclic triterpenes have been identified by TLC and GLC (Ponsinet and Ourisson, 1968). The groups constituted by chemical criteria can be correlated with sections and sub-sections and particularly with morphological, morphogenetical and geographical characteristics.

Euphorbia acrurensis

The latex of *E. acrurensis* yielded in addition to widespread triterpenes euphol and euphorbol, two ingenane and nine lathyrene diterpene derivatives (Marco *et al.*, 1998).

Euphorbia antiquorum

The acetone extract of the latex of *E. antiquorum* showed considerable irritant activity of the mouse ear in comparison to 12-*O*-tetradecanoyl-phorbol-13-acetate (TPA). The acetone extract exhibited only weak tumor promoting activity on the mice

Table 10.2: Pharmacological effect and traditional use profile of *Euphorbia* spurge genus latex.

Species	*Latex Characters*	*Activity*	*Reference*
E. characias	Isolated Ca^{++} dependent peroxidase	Ca^{++} Calmodulin Binding	Mura *et al.*, 2005
E. gorgonis	Folk medicine	Tooth ache, wart deroot	Traditional use
E. lactea	Tirucallol (triterpene)	Topical anti-inflammatory	Fenandez-Arche *et al.*, 2010
E. lathyris	Euphorbain (Proteolytic enzyme)	Proteolytic activity	Lennox and Ellis, 1945; Lynn and Clevette-Radford, 1983
E. maddeni	Euphornin (polycyclic diterpene)	Anticancer	Sahai *et al.*, 1981
E. milii	Crude	Molluscicide, schistosomal	Souza *et al.*, 1997
	Lyophilized latex (500 mg/kg)	Embryolethal and maternally toxic	Souza *et al.*, 1997
E. neriifolia	Traditional use	Laxative, abdominal troubles, tumors and leucoderma	Nadkarni, 1976
	Purified	Mitogenic	Seshagirirao and Prasad, 1995
	Aqueous extract	Wound healing	Rashik *et al.*, 1996
	Fresh latex (Kshaarasootra)	Anal-fistula	ICMR report, 1991
	Fresh	Oral antiarthritic	Amin *et al.*, 1995
	Nerifolione (triterpene)	Topical antiarthritic	Ilyas *et al.*, 2003
	Fresh	Pesticidal, cytotoxic (Brine shrimp)	Guerreroy and Guzman, 2004
E. nivulia	Ingenane diterpene esters	Activity (Colo205, MT2, and CEM cell lines), PGE_2 inhibition	Ravikanth *et al.*, 2002; Ravikanth *et al.*, 2003
E. obtusifolia	Tigliane diterpenes (4-deoxy phorbol)	Inhibition of mammalian mitochondrial respiratory chain	Betancur-Galvis *et al.*, 2003
E. paralias	Ingenane diterpene esters	Inhibition of ^{3}H-thymidine uptake by TLX/5 mouse lymphoma cells	Sayed *et al.*, 1980

Contd...

Table 10.2–*Contd...*

Species	*Latex Characters*	*Activity*	*Reference*
E. royleana	Cycloart-24-en-3β-ol (triterpene)	Piscicide	Tiwari *et al.*, 2008
	Ethyl acetate fraction of 85 per cent ethanol extract	Oral analgesic and antipyretic	Bani *et al.*, 1998
	Hydrosoluble fraction	Oral anti-inflammatory and anti-arthritic	Bani *et al.*, 2000
E. resinifera	Dried	Extreme purgative	Traditional use
	Resiniferatoxin	Desensitizer and analgesic in diabetic neuropathic pain	Appendino and Szallasi, 1997
E. tirucalli	Aqueous solution	Leucopenia and the higher placental weight	Silva *et al.*, 2007
	Folk medicine	Rheumatism treatment, *neuralgia*, asthma and gastric disorders	Traditional use

back skin but it causes severe ulceration and necrosis above 2 mg dose per application. After sequential countercurrent distributions and separations of acetone fraction by preparative TLC, an irritant fraction was characterize as 3-*O*-angeloylingenol (Adolf *et al.*, 1983).

Three new triterpenes euphol-3-*O*-cinnamate, antiquol A and antiquol B, together with known triterpene, euphol, 24-methylenecycloartenol and cycloeucalenol were isolated (Gewali *et al.*, 1990). Insecticidal properties of *E. antiquorum* latex was studied using the samples collected from Eppawala (Anuradhapura District), Ibbagala (Kurunegala District) and Halloluwa (Kandy District) of Sri Lanka. Seven solvents *i.e.* dichloromethane, petroleum ether, acetone, methanol, n-hexane, distilled water and xylene were used and it was found that the insecticidal components were best extracted by xylene. Potters sprayer method was found to be better for insect bioassays than the leaf-dip, hand sprayer and microapplicator methods. Xylene-latex extract was tested against six insect pest species, two predatory coccinellid species and a predatory spider. The latex extract was highly toxic to soft bodied insects/spiders. Latex collected from dry, intermediate and wet zones showed significant differences in insecticidal activities. Insecticidal activity was independent of seasonal variations. Crude extracts and cheap detergents were used to make formulations which were effective in field usage (De Silva *et al.*, 2008).

Euphorbia biglandulosa

Four new cerebrosides were isolated from the latex of *E. biglandulosa* Desf. by normal and reverse-phase flash chromatography was effective for the isolation of the cerebrosides. FAB-MS spectrometry, ^{1}H-NMR, ^{13}C-NMR, DQF-COSY and HMQC experiments and chemical reactions were used to elucidate their structure (Falsone *et al.*, 1994).

Euphorbia broteri

Six polycyclic diterpenes have been isolated from the latex of *E. broteri*. Two had a tigliane skeleton: 12-*O*-(2Z,4E-octadienoyl)-4-deoxyphorbol-13,20-diacetate and 12-*O*-(2Z,4E-octadienoyl)-phorbol-13,20-diacetate and four with ingenane skeleton: 20-acetyl-ingenol-3-decadienoate, 3-*O*-tetradecanoyl-ingenol, 20-*O*-tetradecanoyl-ingenol and 5-*O*-tetradecanoyl-ingenol (Urones *et al.*, 1988).

Euphorbia canariensis

The latex of *E. canariensis* contains known ingenol esters, the ingenane derivatives are ingenol-3-angelate-5,20-diacetate and 5-deoxyingenol-3-angelate-20-acetate, and the lathyrane derivatives 2,3-di*epi*ingol-7,12-diacetate-8-benzoate, 2,3-di*epi*ingol-7,12-diacetate-8-isobutyrate and 2-*epi*ingol-3,7,12-triacetate-8-benzoate (Marco *et al.*, 1997).

Euphorbia cauducifolia

Mice back skin tumor-promoting seven diterpene esters of 17-hydroxyingenol group was isolated from Latex of *E. cauducifolia* (Baloch *et al.*, 2005)

Euphorbia characias

Calmodulin (CaM) is a ubiquitous Ca^{++} sensor found in all eukaryotes, where it participates in the regulation of diverse calcium-dependent physiological processes.

In response to fluctuations of the intracellular concentration of Ca^{++}, CaM binds to a set of unrelated target proteins and modulates their activity. In plants, a growing number of CaM-binding proteins have been identified that apparently do not have a counterpart in animals. Some of these plant-specific Ca^{++}/CaM-activated proteins are known to tune the interaction between calcium and H_2O_2 in orchestrating plant defenses against biotic and abiotic stresses. A calcium-dependent peroxidase isolated from the latex of the Mediterranean shrub *Euphorbia characias* peroxidase is a CaM-binding protein activated by the Ca^{++}/CaM system. (Mura *et al.*, 2005).

A lipase from the latex of *E. characias* was purified using a method involving extraction with apolar solvent and adsorption chromatography on silica gel. The lipase (specific activity, 1500 international units/mg of protein) was eluted from silica gel complexed with a lipid. The main protein fraction, which had a molecular mass of 38 kDa, was inactive when dissociated from the lipid fraction. When the lipid and protein fractions were reassociated, 72 per cent of the lipolytic activity was recovered. This lipolytic activity was inhibited by diethyl-p-nitrophenyl phosphate, which was shown to bind the lipase with a molar ratio of 0.75. High specific activities (1000 international units/mg) were measured for the lipase of *E. characias* on lipid extracts rich in galactldiacylglycerols. The apolipase was sequenced up to residue 23. The B chain of ricin has a strong homology (43.5 per cent) with that sequence and cross-reacted with antibodies raised against the purified lipase from *E. characias*. The activity of the B chain of ricin was comparable (54 international units/mg) to that of the apolipase of *E. characias* (100 international units/mg) mixed with the same lipid cofactor complex. The primary structure (residues 68-72) of the B chain of ricin contains the lipase consensus sequence Gly-Xaa-Ser-Xaa-Gly. Its reactivity with diethyl p-nitrophenyl phosphate indicates the presence of an activated serine that, in addition to its well-documented lectin activity for glactosides, suggests that the B chain of ricin may be a galactosyl diacylglycerol lipase, closely analogous to the lipase from *E. characias* (Moulin *et al.*, 1994).

Euphorbia coerulescens

E. coerulescens (Fam. Euphorbiaceae) or Sweetnoors is indigenous to South Africa, where it is used as an animal feedstuff. Several more toxic species including *E. franckiana, E. ledienii* and *E. cooperi* are indigenous to the same area. The relative toxicities of these latices were assessed by a mouse ear irritancy test. *E. coerulescens, E. franckiana* and *E. ledienii* were of low toxicity, while *E. cooperi* was more potent. Its latex is extremely irritant, as breathing near a wounded plant produces burning in throat. Six esters isolated from the latex of *E. coerulescens* produced inflammation of mice ears. Two esters were identified as diesters of 12-deoxyphorbol and three were monoesters of the same diterpene. The sixth compound was a triester of phorbol. Comparisons of toxicity of the pure esters after both 4 and 24 hour suggested that compounds with a free C-20 hydroxyl group and a trans A-B ring fusion were more potent and their effects are long lasting. The low toxicity of Sweetnoors latex is due to low yields of esters and to the fact that C-20 acylated diterpenes predominated in the mixture (Evans, 1978).

Euphorbia cyparissias

Ocular inflammations caused by the latex of Euphorbiaceae manifest themselves as lid swelling, severe conjunctivitis, keratitis and iritis with hypopyon. As therapy, the local application of corticosteroids is recommended. In severely affected cases, intravenous injections of Sandosten-Calcium and Solu-Dacortin (Merck) are indicated. The caustic effect of *Euphorbia cyparissias* is caused by phorbol acid and phorbol acid ester. On the Canary Islands, the latex of Euphorbiaceae is quickly removed by drop application of the sap of *Aeonium lindleyi*, family Crassulaceae. In the past, the sap of *Sempervivum tectorum* which also belongs to the Crassulaceae, was used as a household remedy for ocular inflammations (Lisch, 1980).

Euphorbia hermentiana

Nine compounds, based on four biogenetically-related polycyclic diterpene skeletons were subjected to open and closed patch testing on human volunteer subjects. The tigliane esters phorbol-12,13,20-triacetate, 12-*O*-2*Z*-4*E*-octadienoyl-4-deoxyphorbol-13-acetate and 12-*O*-tigloyl-4-deoxyphorbol-13-isobutyrate, in increasing order of potency, produced symptoms of toxicity in closed patch tests, the dose of the most potent compound in this series being 0.5 µg in 5 µl acetone. Phorbol, a tigliane alcohol, was inactive in closed tests at a dose level of 50 µg/5 µl. The daphnane derivative, daphnetoxin produced bullae and vesiculation in closed patch tests, but daphnetoxin-5,20-diacetate was devoid of these effects when applied at 10 times the dose of daphnetoxin. The ingenane compounds, ingenol-3,5,20-triacetate, 20-deoxy-16-hydroxyingenol-3,5,16-triacetate, and the lathyrane compound, ingol-3,7,8,12-tetraacetate were obtained from the hydrolyzed, acetylated irritant latex of *E. hermentiana*. At the doses tested, ingenol-3,5,20-triacetate was the only compound derived from this plant exhibit irritant activity in closed patch tests. In contrast, this compound is inactive as an irritant to the mouse ear at doses up to 250 µg/ear. Only three compounds, 12-*O*-2*Z*-4*E*-octadienoyl-4-deoxyphorbol-13-acetate, 12-O-tigloyl-4-deoxyphorbol-13-isobutyrate and daphnetoxin produced dermatological toxicity in open patch tests at the doses used. Inflammatory signs and symptoms for several of the compounds under test persisted for over four days in open patch tests and for a week or more after application in closed patch testing (Hickey *et al.*, 1981).

Euphorbia ingens

Zechmeister *et al.* (1970) reported chemical characterization of a new irritant and cocarcinogenic hexadecanoic acid monoester of ingenol from the latex of *E. ingens*. Presence of ingenol-3-palmitate in the latex is responsible for its high skin irritant activity and weak cocarcinogenic activity inducing fibrosarcoma like conditions (Upadhyay *et al.*, 1976).

Euphorbia lactea

Latex from *E. lactea*, a native Dominican medicinal plant is claimed to be useful in the treatment of inflammation. Topical application of tirucallol, a tetracyclic triterpene isolated from *E. lactea* latex, suppressed ear edema in the mouse model in a dose-dependent manner, as well as affecting the influx of polymorphonuclear cells in response to topical application of TPA in the mouse ear. In addition, the effect of

Figure 10.2: Isolated phytochemical constituents of *Euphorbia* spurge genus latex.

Tigliane Diterpenes

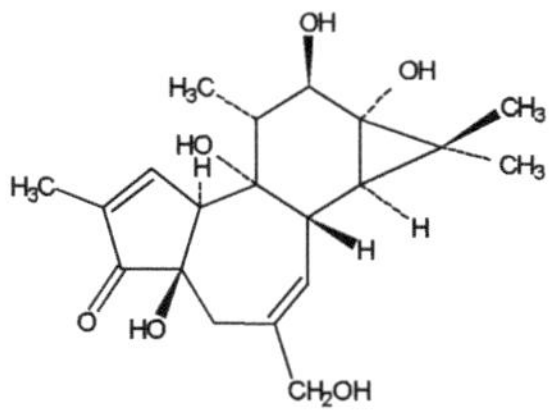

Phorbol

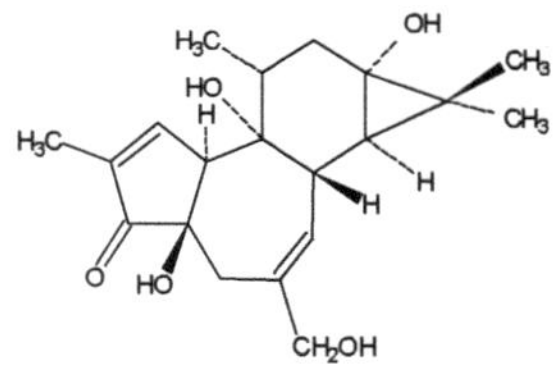

12-deoxy phorbol

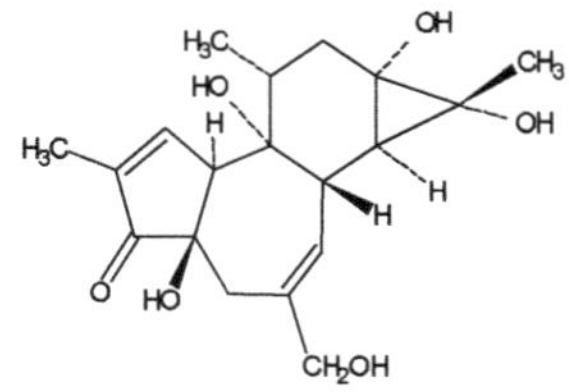

12- deoxy-16-hydroxy phorbal

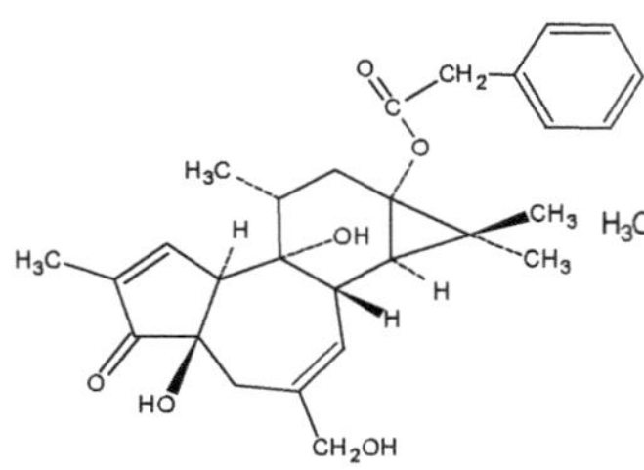

12-deoxy-4 β OH phorbol-13 phenyl acetate

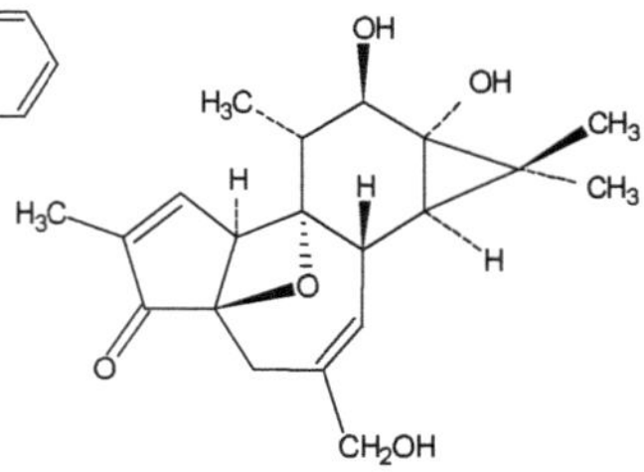

4-epi-4-deoxyphorbol

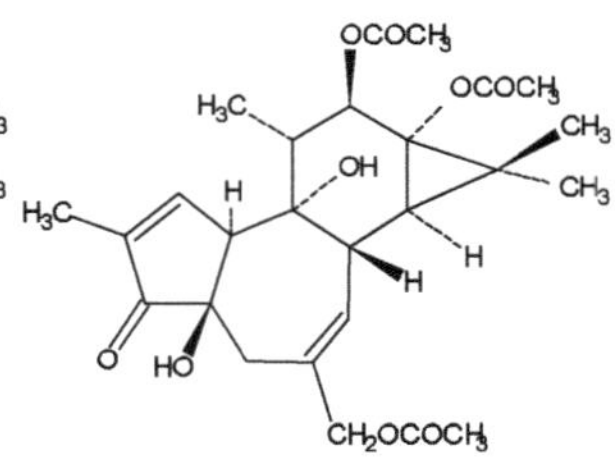

Phorbol-12,13,20-triacetate

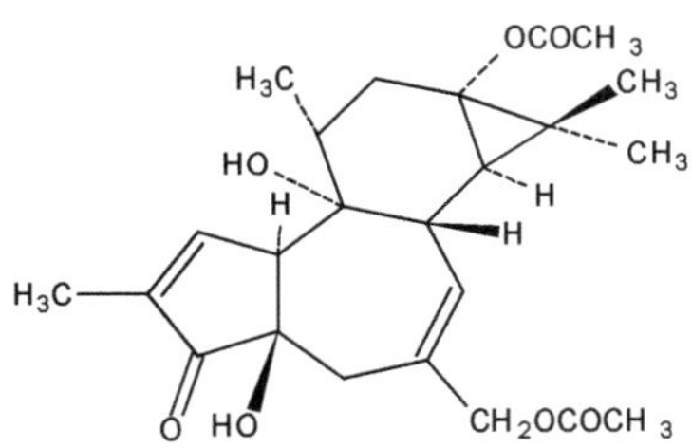

12-deoxy phorbol-13,20-diacetate

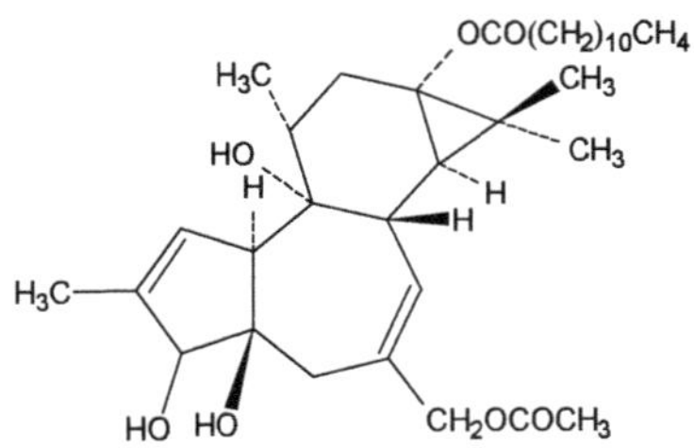

12-deoxy-3β-OH-phorbol-13-dodecenoate-20-acetate

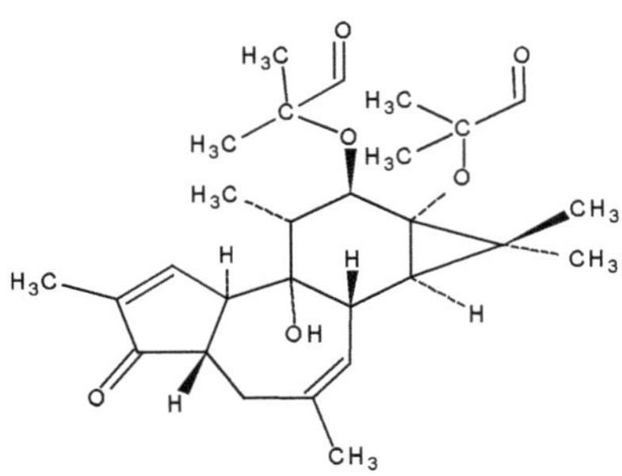

4,20-dideoxyphorbol-12,13-bis-isobutyrate

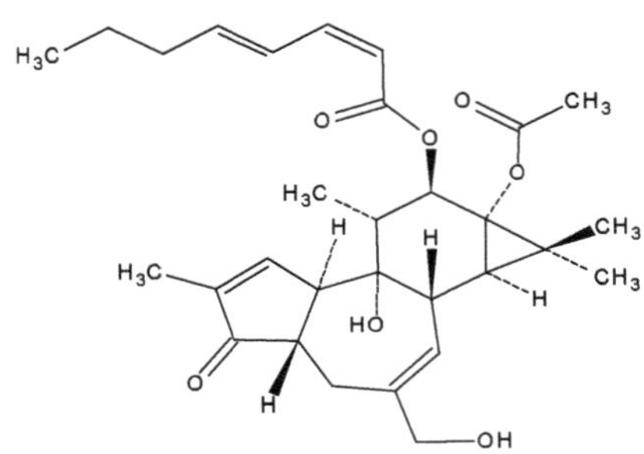

12-O-2Z-4E-Octadienoyl-4-deoxy phorbol-13-acetate

Contd...

Figure 10.2–*Contd...*

Ingenane Diterpenes

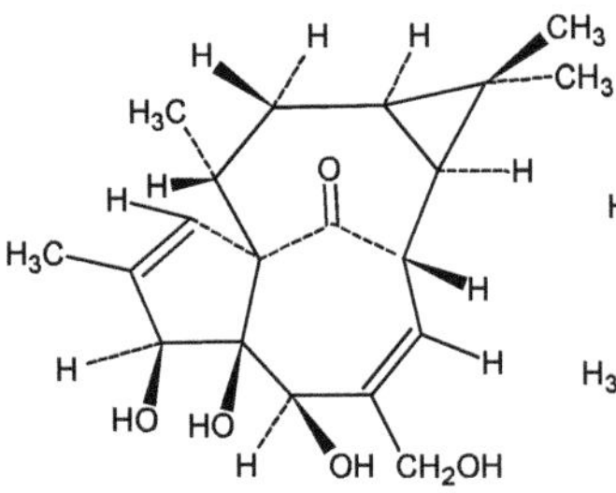

Ingenol

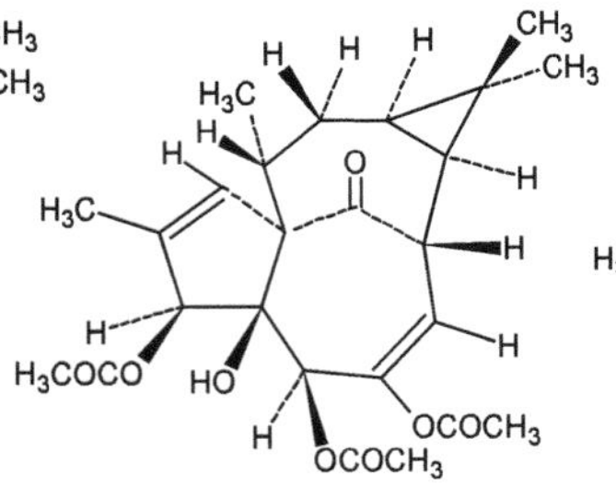

Ingenol triacetate

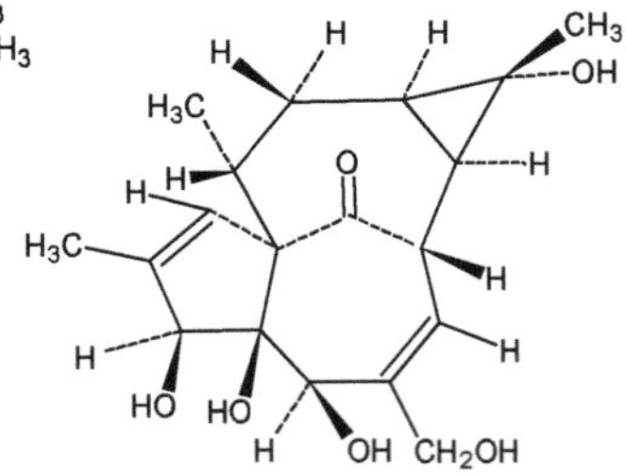

16-hydroxy ingenol

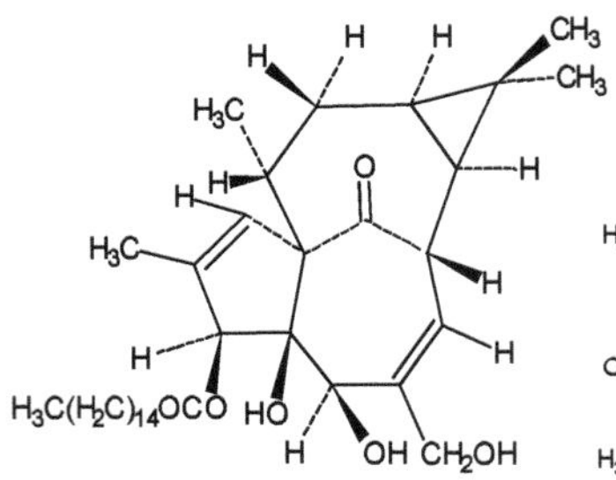

Ingenol -3-hexadecanoate

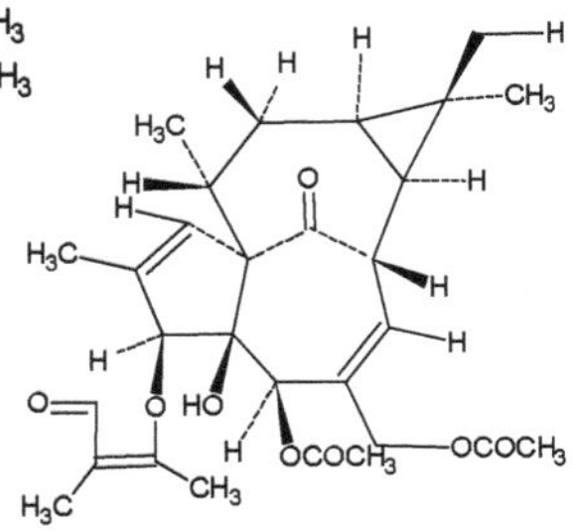

Ingenol-3-angelate-5,20-acetate

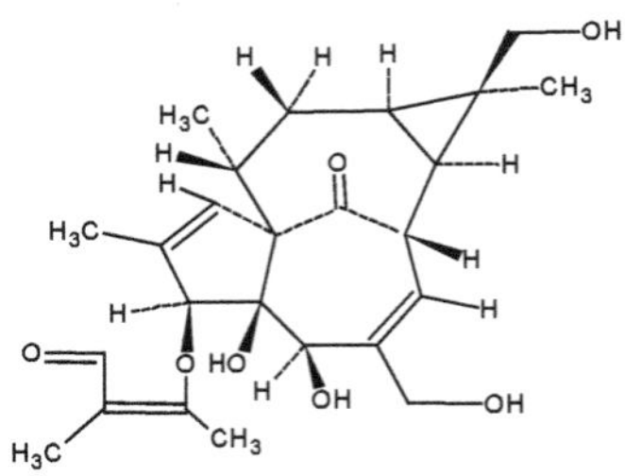

3-O-angeloyl-ingenol

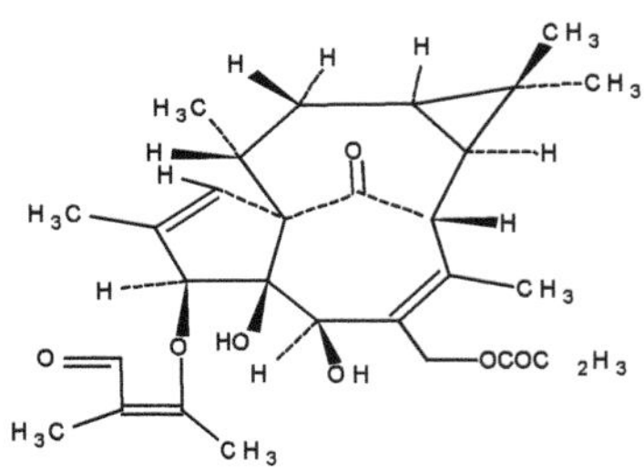

Milliamine

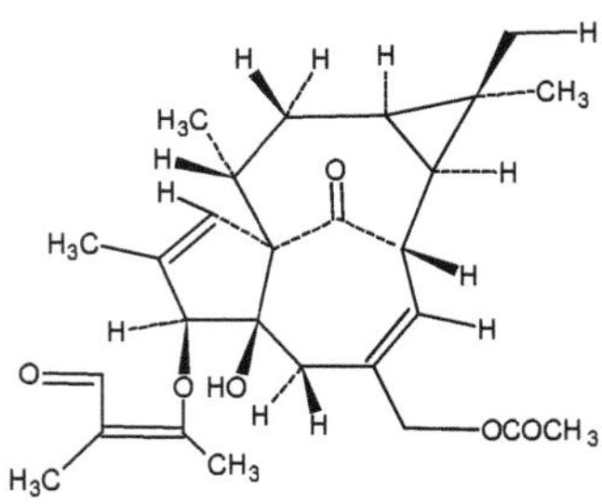

5 deoxyingenol-3-angelate-20-acetate

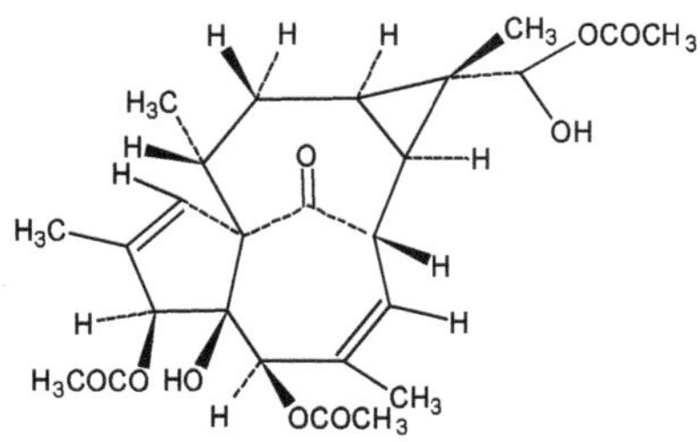

20-deoxy-16-hydroxyingenol-3,5,16-triacetate

Contd...

Figure 10.2–*Contd...*

Lathyrane Diterpenes

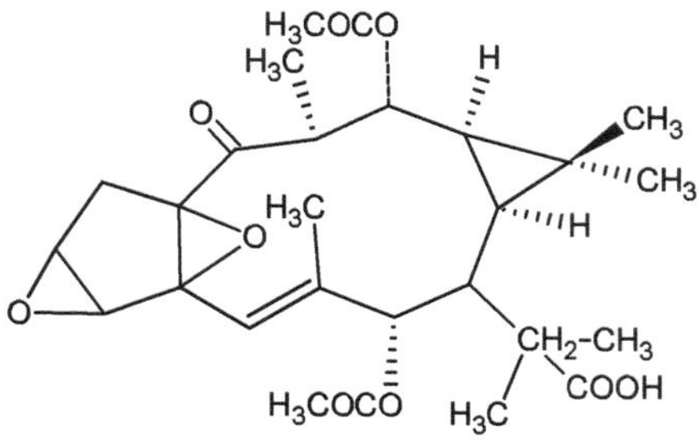

2,3-di*epi*ingol-7,12-diacetate-8-isobutyrate

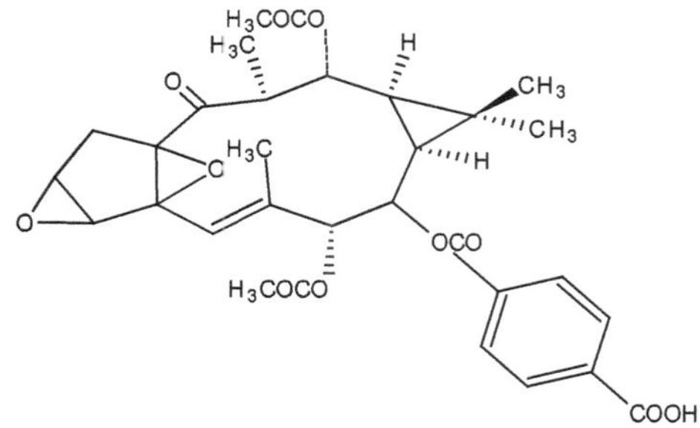

2,3-di*epi*ingol-7,12-diacetate-8-benzoate

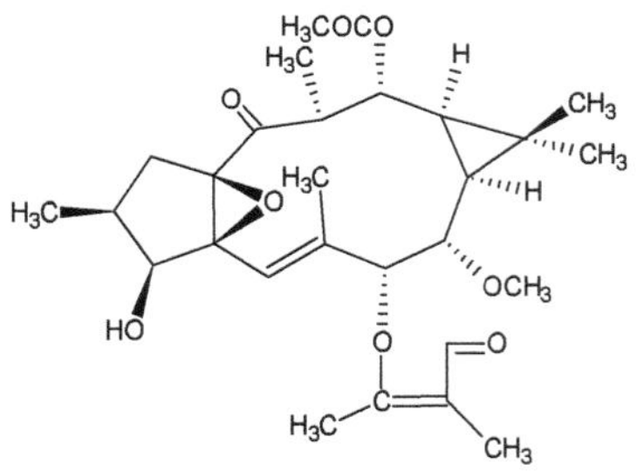

7-angeloyl-12-acetyl-8-methoxyingol

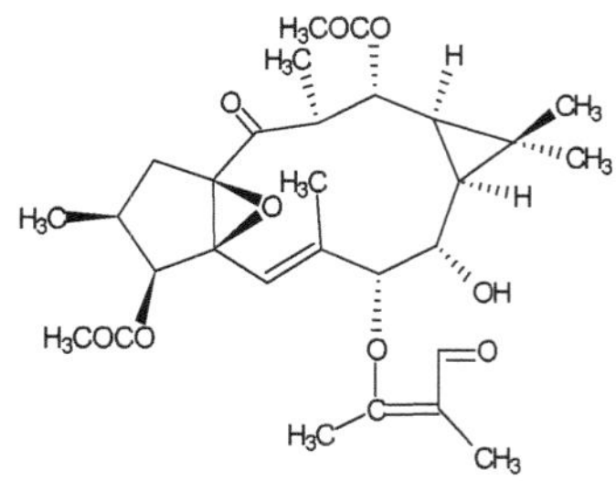

3,12-diacetyl-7-angolyl-8-hydroxyingol

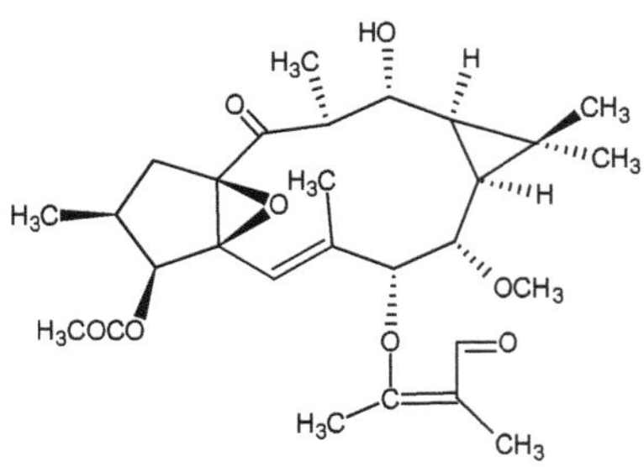

3-acetyl-8-methoxy-7-angolyl-12-hydroxyingol

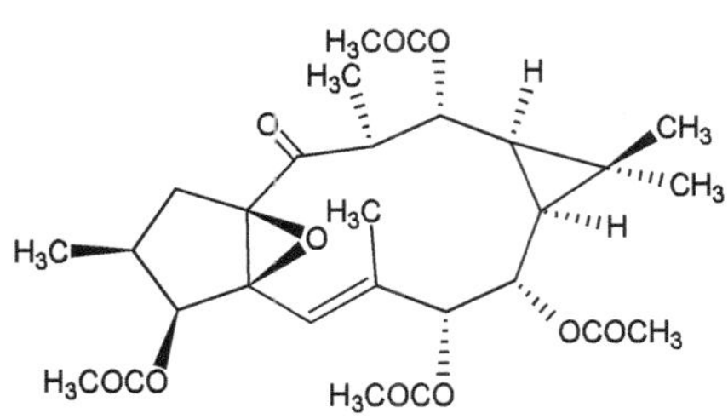

Ingol-3,7, 8,12-tetraacetate

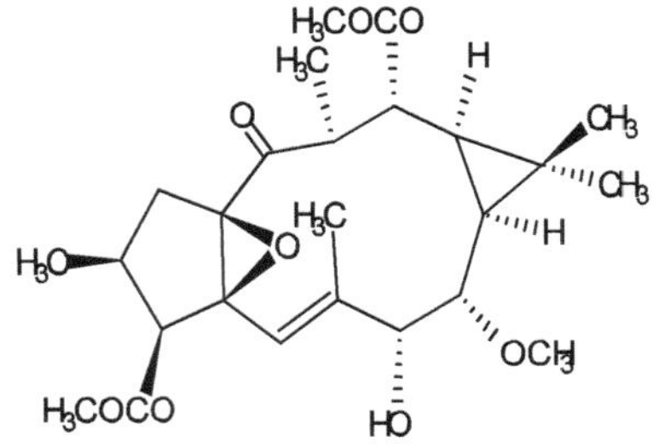

3,12-diacetyl-7-hydroxy-8-methoxyingol

Contd...

Figure 10.2–*Contd...*

Daphnane Diterpenes

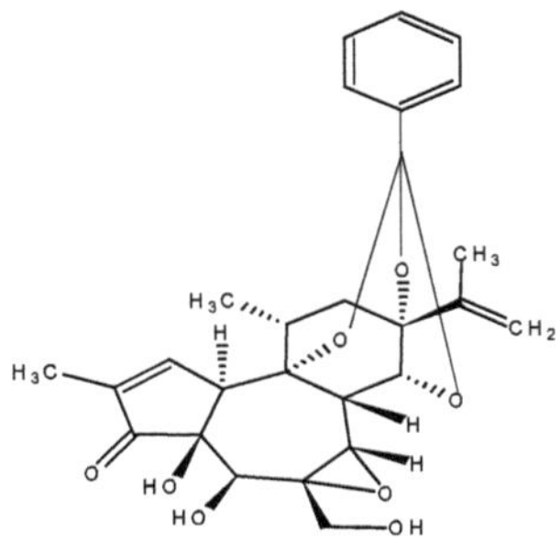

Daphnetoxin

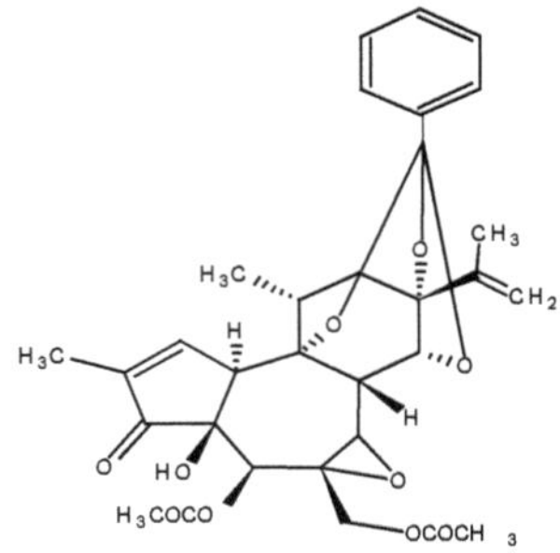

Daphnetoxin-5,20-diacetate

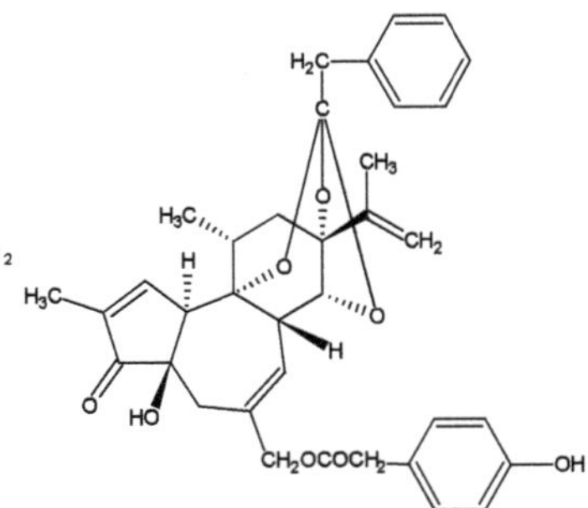

Tinyatoxin

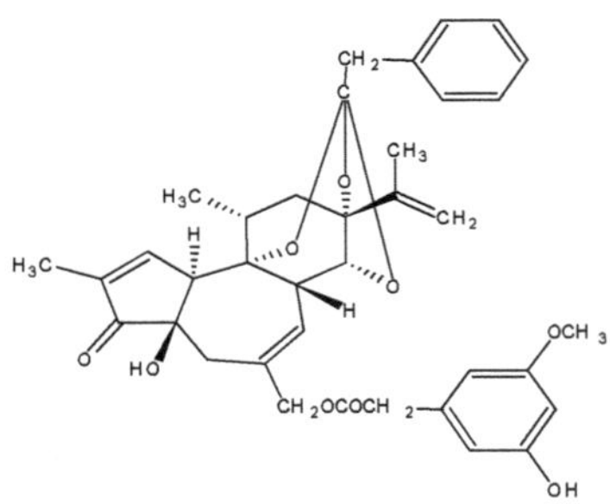

Resiniferatoxin

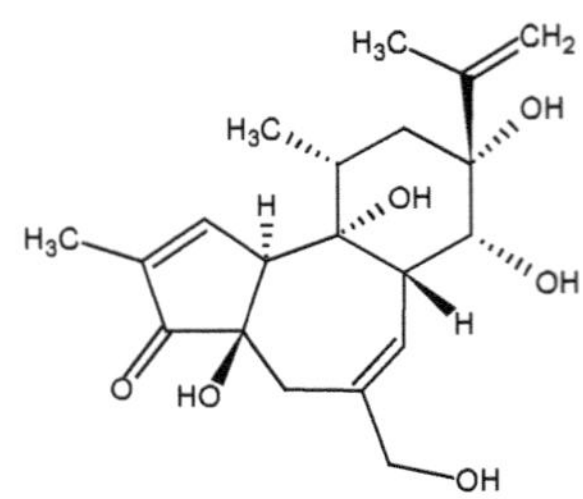

Resiniferonol

Jatrophone Diterpenes

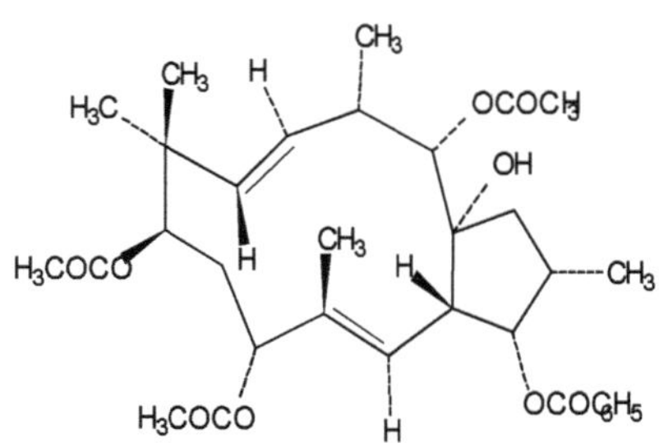

Euphornin

Contd...

Figure 10.2–*Contd...*

Triterpene alcohols

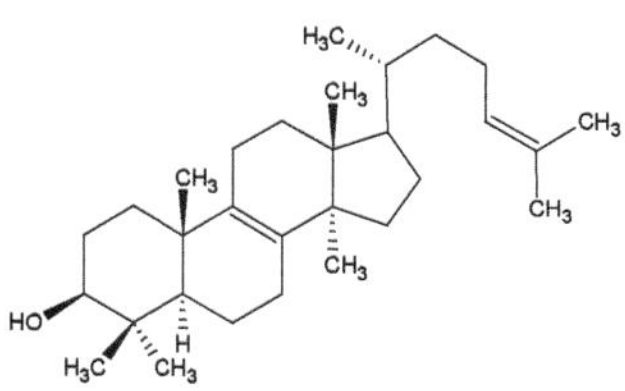

Euphol

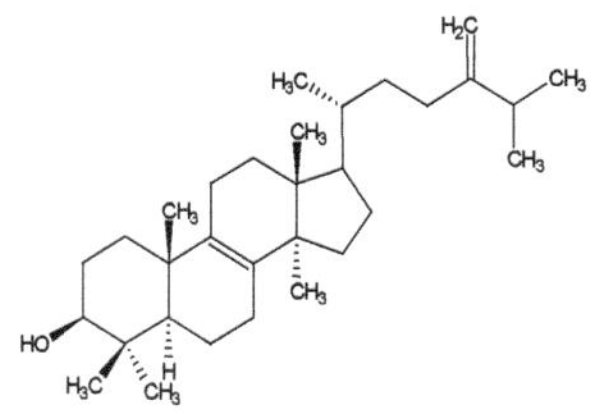

Euphorbol

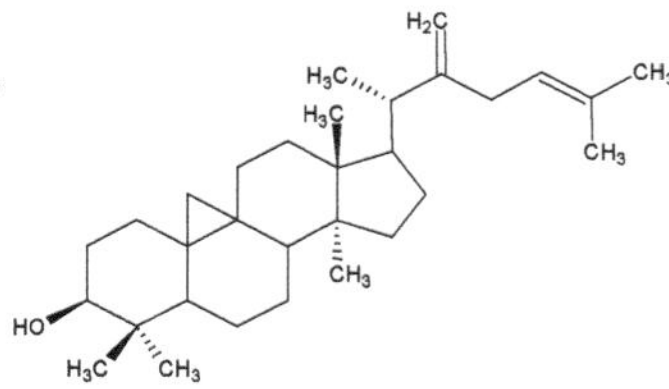

Nerifoliene

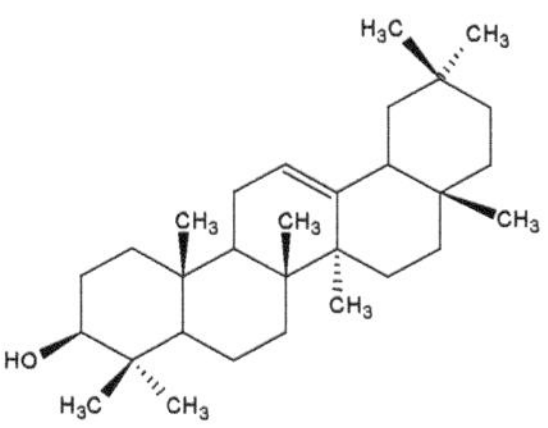

β-amyrin

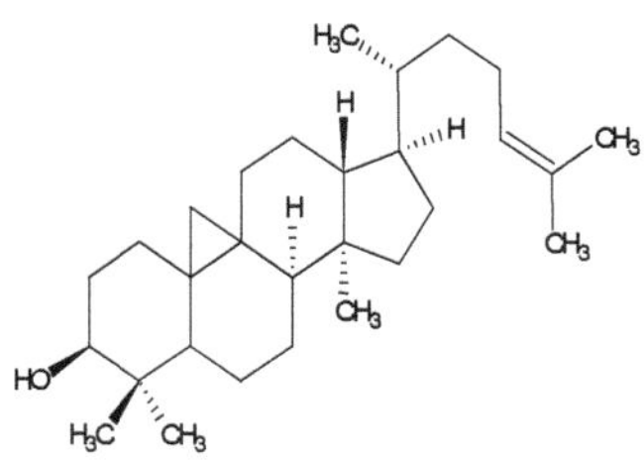

Cycloartenol

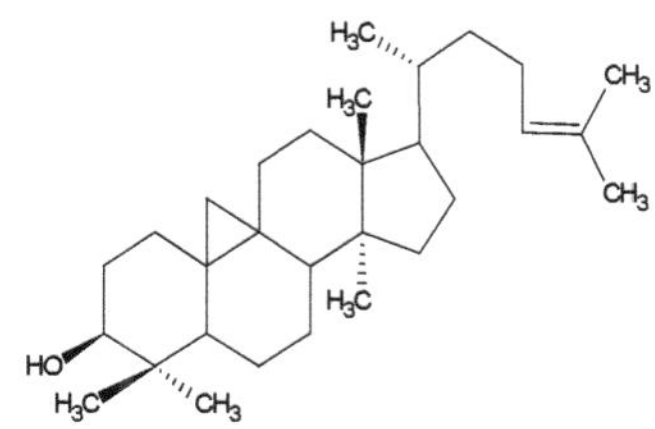

Cycloart-24-en-3β-ol

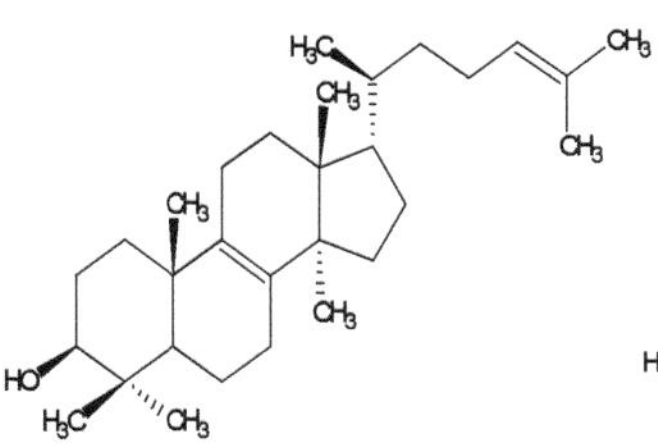

Tirucallol

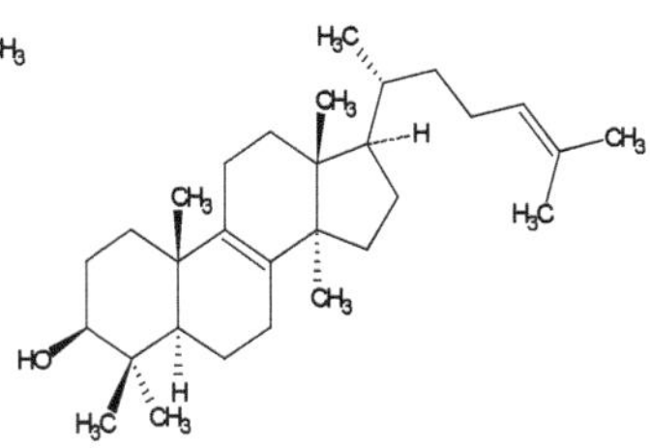

Lanosterol

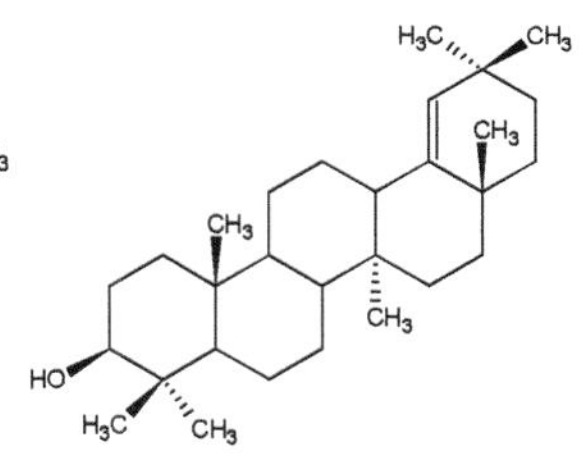

Germanicol

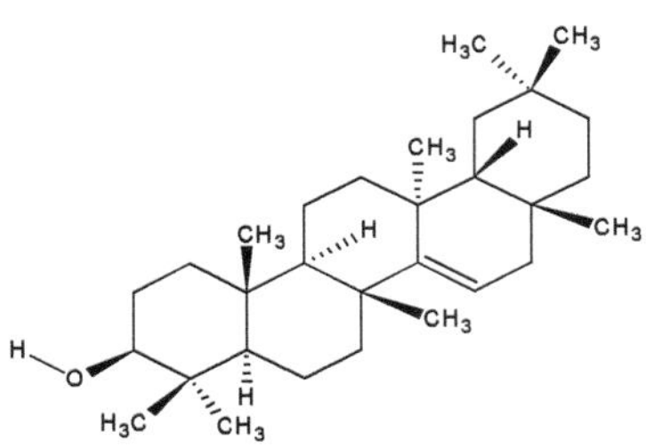

Taraxerol

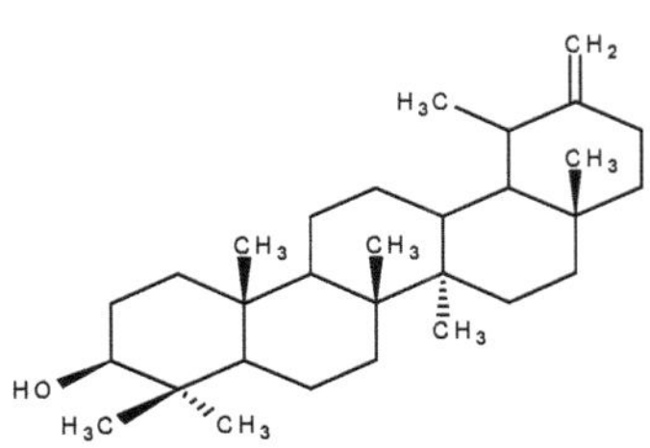

Taraxasterol

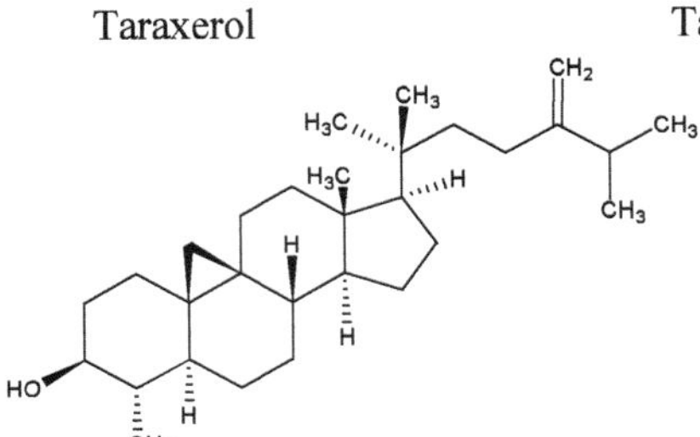

Cycloeucalenol

Contd...

Figure 10.2–*Contd...*

Triterpene

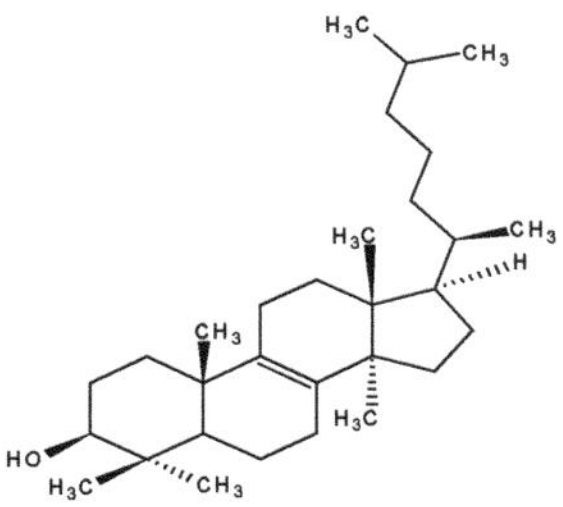

Lanostenol

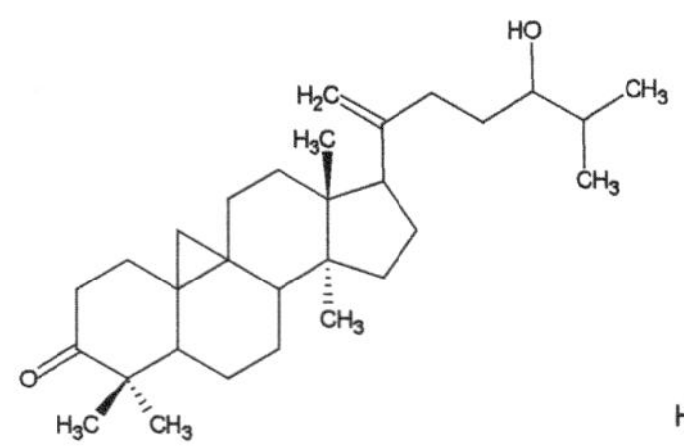

Neriifolione

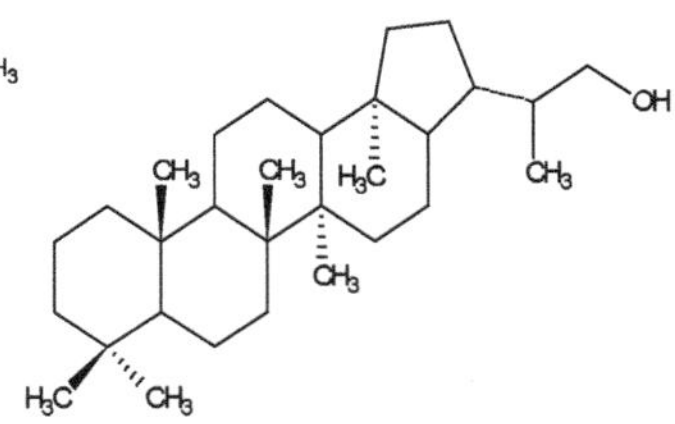

Nerifoliol

Miscellaneous

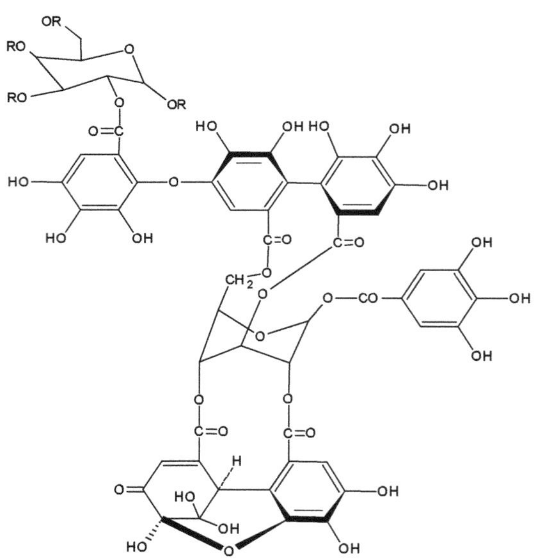

Euphorbin

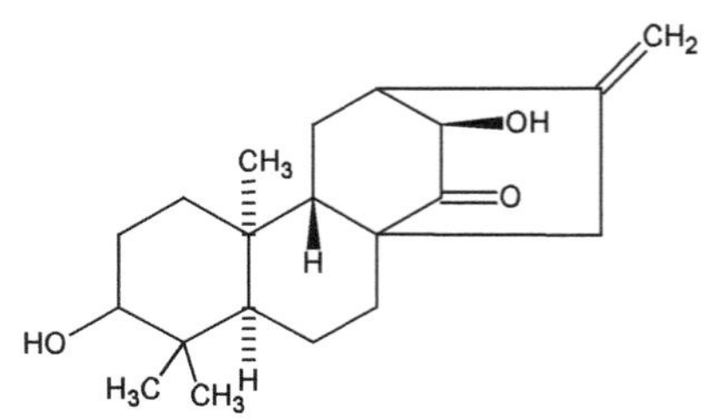

Neriifolene

tirucallol on some macrophage functions was analyzed in vitro. Non-toxic concentrations of tirucallol potentially inhibited nitrite production in lipopolysaccharide-stimulated macrophages. Western blot analysis showed that nitric oxide reduction was a consequence of the inhibition of inducible nitric oxide synthetase expression although tirucallol slightly affected to prostaglandin E_2 generation. Tirucallol present in *Euphorbia lactea* latex, exerts a topical anti-inflammatory effect *in vivo*, via a mechanism of action related to the neutrophil migration and reduction in production of Nitric Oxide (Fenandez-Arche *et al.*, 2010).

Euphorbia lathyris

A proteolytic enzyme named euphorbain has been reported to occur in the latex of this species (Lennox and Ellis, 1945; Lynn and Clevette-Radford, 1983). The plant contains irritant ingenol esters (Furstenberger and Hecker, 1972) and 16-hydroxyingenol (Adolf and Hecker, 1975).

Euphorbia leuconeura

Latex of *E. leuconeura* exhibited Epstein-Barr-virus (EBV) inducing activity comparable to TPA, a well known tumor promoter. Three ingenol esters with EBV inducing activity were isolated and identified belonging to milliamine type of diterpene esters containing aromatic peptidyl group (Vogg *et al.*, 1999).

Euphorbia maddeni

Euphornin, a polyacylated diperpene of jatrophone was isolated from the anticancer active chloroform extract (Sahai *et al.*, 1981).

Euphorbia milii

The crude latex of Crown-of-Thorns (*Euphorbia milii* var. hislopii) is a potent plant molluscicide and a promising alternative to the synthetic molluscicides used in schistosomiasis control. Lyophilized latex of *E. milii* (125, 250 and 500 mg/kg) in corn oil was given by orally to Wistar rats from days 6 to 15 of pregnancy and cesarean sections were performed on day 21 of pregnancy. *E. milii* latex was maternally toxic over the dose range tested, latex-induced embryolethality was not noted at the lowest dose but the resorption rate was markedly increased at 250 mg/kg (62.5 per cent) and 500 mg/kg (93.4 per cent). Signs of delayed ossification indicated fetal growth retardation and minor skeletal malformations at higher dose. Molluscicide *E. milii* latex poses no teratogenic hazard at low dose or that this possibility is of a considerably low order of magnitude, but with higher dose it can be toxic to fetal growth (Souza *et al.*, 1997).

Euphorbia neriifolia

Seshagirirao and Prasad (1995) attempted purification and partial characterization of *E. neriifolia* latex which have mitogenic property and agglutinate trypsinized human and rabbit erythrocytes, but does not inhibit protein synthesis in rabbit reticulocyte lysate. Aqueous extract of *E. neriifolia* latex shows wound healing activity in guinea pig by increasing epithelization, angiogenesis, tensile strength and DNA content in wounds (Rashik *et al.*, 1996).

E. neriifolia is one of the constituents of "Kshaarasootra", which is used in Indian medicine to heal anal-fistula. "Kshaarasootra" is prepared by smearing a surgical linen thread with fresh latex of snuhi or *E. neriifolia* and a specially prepared alkaline powder (kshaara) from plant *Achyranthes aspera* and turmeric powder from the dried rhizomes of *Curcuma longa*. "Kshaarasootra" is quiet effective in treating various fistulous tracks. Indian Council of Medical Research has carried out a multicentric randomized controlled trial to evaluate the efficacy of "Kshaarasootra" in the management of fistula-in-ano. (265 patients) The results revealed that the long term out come with "Kshaarasootra" (recurrence in 4 patients) was better than with the surgery (recurrence in 11 patients) although the initial healing time was longer (8 week with out and 4 week with surgery). "Kshaarasootra" offered an effective, ambulatory and safe treatment for patients with fistula-in-ano. In chronic non-healing milk fistula "Kshaarasootra" gives encouraging result with 'stannadi-vrana' and additional application of 'Jatyaditaila' and oral administration of 'Shigru Guggulu' (2 tablets t.i.d) during the course of treatment (ICMR report, 1991).

The latex contains water and water solubles (69.4 – 93.3 per cent) and caoutchouc (0.2 – 2.6 per cent). Neriifolione, a triterpene have been isolated from dried latex of *E. neriifolia* (Ilyas *et al.*, 1998). Euphol and nerifoliol along with cycloartenol and nerifolione was reported in latex. A new tetracyclic triterpene 9,19-cyclolanost-22(22'), 24-diene-3beta-ol, named as nerifoliene along with euphol were isolated from the fresh latex of *E. neriifolia* (Mallavadhani, 2004).

The latex of *E. neriifolia* is used for arthritis in Tibbe-e-Unani (Unani Medicine) as Sheer-e-Zaqqoom. Amin *et al.* (1995) reported its efficacy and safety on adjuvant arthritis.

Ilyas *et al.* (2003) reported anti-inflammatory and antiarthritic activity of a novel triterpene (Nerifolione) isolated from the latex of *E. neriifolia* along with total extract of latex in acetone.

Guerreroy and Guzman (2004) evaluate the Brine shrimp lethality test in microplate (BSLT) bioassay. *E. neriifolia* latex showed significant concentration related activity with LC_{50} values below 100 mg/mL (LC_{50} 77 μg/mL tested in a range of 154.0-43.0 μg/mL). According to this result, there is a good probability that metabolites of this plant may have cytotoxic and/or pesticidal activities.

Euphorbia nivulia

The latex of *Euphorbia nivulia* afforded three new ingol diterpenes, 3-acetyl-8-methoxyl-7-angolyl-12-hydroxylingol, 3,12-diacetyl-7-hydroxy-8-methoxylingol, and 3,12-diacetyl-7-angolyl-8-hydroxylingol along with five known ingol diterpenes and a known triterpene cyclonivulinol. Their structures were established by means of spectroscopic analysis. Diterpenes were evaluated for their cytotoxic activity and 3,12-diacetyl-7-angeloyl-8-methoxyingol, 7-angeloyl-12-acetyl-8-methoxyingol and 3,12-diacetyl-7-hydroxy-8-methoxyingol showed significant cytotoxic activity against Colo205, MT2, and CEM cell lines and others showed moderate or no activity (Ravikanth *et al.*, 2003). Prostaglandin E_2 inhibitory activity (IC_{50} 0.003μM) was also reported in 7-angeloyl-12-acetyl-8-methoxyingol (Ravikanth *et al.*, 2002).

Euphorbia obtusifolia

Six diterpenes isolated from the latex of *E. obtusifolia* were evaluated for their inhibition of the NADH oxidase activity in submitochondrial particles from beef heart. 4,20-dideoxyphorbol-12,13-bis-isobutyrate was the most potent inhibitor with IC_{50} 2.6 mM. This study indicates natural product derivatives of 4-deoxyphorbol esters have inhibitory activity of the mammalian mitochondrial respiratory chain (Betancur-Galvis *et al.*, 2003).

The latex yielded twelve new dipertene polyesters. Seven of them displayed the jatrophane framework and five were 4-deoxy phorbol esters. A tigliane diterpene derivative, 4-*epi*-4-deoxyphorbol was also reported (Marco *et al.*, 1999).

Euphorbia paralias

The irritant and cytotoxic constituents of the latex of *E. paralias* L. were separated from the hydrocarbon fraction by means of solvent partition. Three new ingenane esters were identified from the toxic ether fraction. The major compound was 3-angelyl-20-deoxyingenol and the two minor compounds were 3-hexanoyl-20-deoxyingenol and 3-*O*-angelylingenol. These compounds were of a similar potency to podophyllin in the inhibition of ^{3}H-thymidine uptake by TLX/5 mouse lymphoma cells. In addition the compounds produced a persistent erythema of the mouse ear in sub-microgram doses (Sayed *et al.*, 1980).

Euphorbia poisonii

A series of toxins were isolated from the latex out of which two were newly found natural product, diterpene orthoesters resiniferatoxin and tinyatoxin along with phorbol derivatives (Evans and Schmidt, 1976).

Steriostructure and cytotoxicity of a new diterpene 3,12-diacetyl-8-nicotinyl-7-phenylacetyl-19-acetoxyingol was reported by Fatope *et al.* (1996).

Euphorbia resinifera

Topical resiniferatoxin isolated from dried latex of *E. resinifera* (Euphorbium) was clinically trialled in human volunteers having diabetic neuropahthic pain, which showed promising analgesic effect. Resiniferatoxin binds with specific membrane recognition sites (referred as venilloid receptors) expressed by sensory neurone mediating pain perception and neurogenic inflammation. Binding with this receptors resiniferatoxin causes desensitization of neurones thus inhibition of pain perception. It is a more potent counter irritant than capsaisin, working in the same pathway but interacting with different modulators (Appendino and Szallasi, 1997).

Euphorbia royleana

Cycloart-24-en-3β-ol isolated from *E. royleana* latex possesses potent piscicidal activity against freshwater predatory fish *Channa punctatus*. Their short as well as long-term exposure causes significant ($P<0.05$) time and dose-dependent reversible alteration in the oxidative metabolism of the fish *Channa punctatus*. Cycloart-24-en-3β-ol also shifts fish respiratory pathway, inhibits energy production and AchE activity reversibly may be advantageous for their use as environmentally safe piscicide for controlling the predatory fish *C. punctatus* population from carp culture ponds (Tiwari *et al.*, 2008).

The ethyl acetate fraction from the residue of an 85 per cent ethanol extract of the latex of *E. royleana* showed a dose related peripheral analgesic effect. The ethyl acetate fraction exhibited a significant antipyretic effect in hyperthermic rats and rabbits. The oral LD_{50} was more than 2 g/kg in rats and mice (Bani *et al.*, 1998).

The hydrosoluble fraction of *E. royleana* latex, administered by gavage at doses of 50 - 200 mg/kg showed dose-dependent anti-inflammatory and anti-arthritic effects in different acute and chronic test models in rats and mice. It reduced the exudate volume and the migration of leukocytes and showed a poor inhibitory effect on the granuloma formation induced by cotton pellets, while it had a low ulcerogenic score. The oral LD_{50} was more than 1500 mg/kg in both rats and mice (Bani *et al.*, 2000).

Euphorbia serrata

E. serrata latex contains ingenol-3-palmitate, which is responsible for the irritant and cocarcinogenic activity of the latex on mouse ear and mouse back skin. Cocarcinogenic activity was determined using standard 7,12-dimethyl benz (a) anthracene (DMBA) initiated method which give out multiple squamous cell papilloma like conditions (Upadhyay *et al.*, 1976).

Euphorbia tirucalli

The sap is irritating to the human eye (Lampe and Fagerstrom 1968) and causes kerato-conjunctivitis and uveitis from the sap of this plant. The eyes of dogs exposed to the sap of *E. tirucalli* causes corneal opacities which appeared after 24-36 hours and cleared in one to three weeks.

Developmental toxicity of latex aqueous solution on Wistar rats was assessed for maternal variables like; clinical signs of toxicity, body weight, ovaries, liver and kidneys weight, and number of corpora lutea and identification of embryos. Apart from the leucopenia and the higher placental weight observed in treated rats, no other significant differences were observed. The latex aqueous solution of *E. tirucalli* did not interfere with tubaric embryo development or with implantation, but it seems to alter the placenta morphology (Silva *et al.*, 2007).

A new ingol ester named 3,7,12-tri-*O*-acetyl-8-isovaleryl-ingol, was isolated from and undried latex of *E. tirucalli* (Khan and Malik, 1990).

Euphorbia ledienii, *Euphorbia coerulescens*, and *Euphorbia triangularis*

The lattices of the three South African species *E. ledienii*, *E. coerulescens*, and *E. triangularis*, belonging to a group of cactiform *Euphorbias* locally called "Noors". They contain four 12 mono and three 13,20-diesters of the tetracyclic tigliane type parent alcohol 12-deoxyphorbol, in addition, two 13,16-di- and two 13,16,20-triesters of the related 12-deoxy-16-hydroxyphorbol. Ester groups are made up of acetic, isobutyric, tiglic, angelic, and 2-methylbutyric acid. On the mouse ear, all esters isolated showed moderate irritant activity as compared to the tigliane-type standard TPA. The latex of *E. ledienii* also contains five esters of the macrocylic, lathyrane type parent alcohol ingol. In the two triesters and three tetraesters the ester groups were made up of acetic, tiglic, and 2-methylbutyric acid. None of the ingol esters showed irritant activity (Sosath, 1988).

Schmidt and Evans (1980) studied the skin irritant potencies of several naturally occurring and semisynthetic polyfunctional esters based on the tigliane hydrocarbon skeleton using a mouse ear irritancy assay. The compounds used were esters of 12-

deoxyphorbol, 12-deoxy-16-hydroxyphorbol and phorbol. These compounds were applied in acetone solution to the mouse ear, and the attainment of maximum response was observed on different time intervals and found to be moderately irritant.

By means of a combination of partition and chromatographic methods six irritant constituents were isolated from the fresh latex of *E. fortissima*. Compounds were di-esters of the common parent diterpene 12-deoxyphorbol and mono-esters of the same diterpene. Di-esters are short-acting irritants reaching a maximum activity within 4 hour of application to the skin, whilst the monoesters maintained potent irritant effects for up to 24 hour. Selective hydrolysis of the di-esters at the C-20 primary ester group also produced mono-esters of greater potency after 24 hour. An increase in the length of the fatty acid located at C-13 produced greater biological activity in both the mono and diester groups (Kinghorn and Evans., 1975).

Thirteen esters were isolated from the fresh latex of four species of the genus *Euphorbia*. All the isolated esters produced inflammation of the skin and their relative potencies were assessed by the mice ear test for irritant effects. These compounds were of two types. The first group were diesters with an acetate moiety at carbon 20 and a long chain fatty acid at C-13; the second group were mono-esters with a free C-20 hydroxyl group. The mono-esters were more potent than the diesters and their irritant effects were long lasting. Within both groups of irritants an increase in the chain length of the C-13 acid produced an increase in the irritation of the esters. It was also observed that unsaturation in the side chain increased the potency (Evans *et al.*, 2009).

Evans and Kinghorn (1977) demonstrated the irritancy of *E. lateriflora*, *E. tirucalli*, *E. neriifolia*, *E. lathyris*, *E. nivulia* Buch.-Ham. (syn. *E. varians* Haw.) latex in a mouse ear irritancy assay, and subsequently also reported the presence of ingenol esters, 12-deoxyphorbol and 12-deoxy-16-hydroxyphorbol esters in the plant. Latex of *E. milii* collected in Nigeria to be only weakly irritant in a mouse ear irritancy assay. *E. nubica* N.E. Br. latex was found to be non-irritant in a mouse ear irritancy assay (Evans and Kinghorn, 1975). Upadhyay *et al.* (1980a) found *E. orientalis* L. (syn. *Tithymalus orientalis* Hill) to be weakly irritant in mouse ear irritancy assay.

Discussion

Plants are considered to be promising source of medicine in the traditional health care system. The efficacy and safety of herbal medicine have turned the major pharmaceutical population towards medicinal plant's research. Medicinal plants form the backbone of traditional medicine, which have been subjected to intense pharmacological studies in the last few decades. Chemical investigation of some anticancer plants of *Euphorbia* species has been done by Baslas and Agarwal (1980). Extracts of *E. hirta* Linn. (aerial parts in pet. ether, roots in pet. ether, pet. ether defatted roots in ethanol), *E. neriifolia* Linn. (bark in pet. ether, roots in pet. ether) and *E. thymifolia* (whole plant in pet. ether) shows cocarcinogenic activity due to presence of several terpenes, anthocyanins, alcohols and steroids like 12-deoxy-4β-phorbol-13-dodecanoate-20-acetate, 12-deoxyphorbol-13,20-diacetate, ingenol triacetate, euphorbol, euphol and reiniferonol derivatives. Diterpene esters of the phorbol, ingenol and 12-deoxyphorbol esters types are known to be highly active tumor promoting agents that typically occur in members of the Euphorbiaceae (Vogg *et al.*, 1999; Hergenhahn *et al.*, 1984).

E. tirucalli is an environmental risk factor for Burkitt's lymphoma, also presents pharmacological activities. In the northeast region of Brazil, its latex is used as an antimicrobial, antiparasitic in the treatment of coughs, rheumatism, cancer and other maladies as folk remedy (Silva *et al.*, 2007). Despite the irritant properties on skin, the *E. tirucalli* sap is traditionally used as a remedy for warts, calluses and verrucas (Scott and Karp, 1996). The prevalent constituents of this plant latex are diterpenes from the tigliane (phorbol esters) and from the ingenane types (ingenol esters). The cancer management with *E. tirucalli* has no scientifical basis, at least up to the moment, since the phorbol esters have already presented tumor-promoting ability. Good examples of the last statement are the prototypes TPA and 4-deoxyphorbol that induce the Burkitt's lymphoma and the nasopharyngeo carcinoma development, which are endemical in places where the plant *E. tirucalli* is widely distributed in the environment. (Cataluna and Rates, 1999).

The naturally occurring polyfunctional diterpene parent alcohols like phorbol, ingenol and resiniferonol respectively are structurally related to diterpene parent hydrocarbons: the tetracyclics tigliane and ingenane and the tricyclic daphnane. The toxic and irritant principles are esters in case of the daphnane type orthoesters (of such parent alcohols with a variety of aliphatic, aromatic and aromatic/aliphatic acids). Ingenane diterpenoids widely reported in many *Euphorbia* species have antinematodal and termiticidal activity (Shi *et al.*, 2008c; Shi *et al.*, 2008b). There were also reports about toxicity such as tumor promoting and proinflammatory activity (Vogg *et al.*, 1999; Matsumoto *et al.*, 1992; Sayed *et al.*, 1980). Studies on the relationships between structure and irritant activity indicate that presence of a hydroxyl on C-20 is crucial for stimulatory properties. Introduction of an acetyl group in the C-20 position results in lowing of irritancy (Marston and Hecker, 1983). Some 20-deoxyingenol diterpenes induced cell cleavage arrest, but this activity became weak when C-16 had an acyl residue (Wang *et al.*, 2003). Acetylation in the 5-position resulted into a considerable depression of irritancy (Marston and Hecker, 1983). The skin tumor promoting and irritant activities of the ingenol-3-esters depend on the length of the aliphatic chain in their ester moiety (Sorg *et al.*, 1987). In addition, the presence of one free hydroxy group at C-3 or C-5 may play an important role in the antinematodal activity (Shi *et al.*, 2008c).

Most of the irritant esters also exhibit cocarcinogenic or tumor promoting activity in the skin of mice. Therefore, in the etiology of human tumors such cocarcinogenic diterpene esters may be considered second order carcinogenic risk factors. Most tigliane diterpenoids have polyhydroxy groups located on C-4, C-9, C-13 and C-20. C-3 forms a carbonyl group. C-1,2 and C-6,7 form double bonds, respectively. Like the abietane and the ingenane diterpenoids, the hydroxyl groups of tigliane diterpenoids are easily esterified, as in compounds. This type of macrocyclic deterpene, which is widespread in the latex of *Euphorbia* genus, is the main toxic constituent causing irritant, proinflammatory and tumor promoting activity (Betancur-Galvis *et al.*, 2003; Wu and Sorg, 1994). When the C-12 OH and C-13 OH were esterified as a bis-ester, the tumor promoting activity was reinforced at the same time. For example TPA is well-known as a tumor promotor. The diterpene ester with a saturated aliphatic long chain acyl group exhibited high irritant activity and high tumor promoting activity, and the highly unsaturated analogue exhibits high irritant activity, but very weak tumor promoting activity, suggesting that the irritant activity but not the tumour promoting activity of these diterpenoids is related to the degree of unsaturation of the aliphatic long chain (Marston and Hecker, 1984). The absence of a C-20 OH is known to be important for the irritant and tumor promoting activities of phorbol esters (Wu *et al.*, 2009). Introduction of an acetyl group in the 20-position gives rise to a lower irritancy (Furstenberger and Hecker, 1986).

Certain *Euphorbia* species have been reported to possess antitumor activity and have been recommended for use as anticancer remedies (Ahmad *et al.*, 1988; Duarte *et al.*, 2006). Recently some of the diterpene esters described have been reported to exhibit antileukemogenic activity. Their antitumor activity was mainly attributed to the presence of abietane diterpene derivatives, most of which contain lactone structures reported to possess potent antineoplastic activity towards various cancer cell lines (Yan *et al.*, 2008; Duarte *et al.*, 2008). Euphorbiaceae plants are also having potent anticancer activity. Euphol, a triterpene alcohol from the roots of *Euphorbia kansui* have inhibitory activity on mice skin tumour (Yasukawa *et al.*, 2000). Ingenol-3-hexadecanoate from *Euphorbia lathyris* showed antitumor activity against sarcoma (Itokawa *et al.*, 1989) Diterpenoid ingenol 3,20-dibenzoate and phorbol-12-tiglate-13-decanoate isolated from Euphorbiaceae plants showed antileukemic activity against the P-388 lymphocytic leukemia in mice (Kupchan *et al.*, 1976).

Conclusions

From centuries, plants and plant materials of species of the family Euphorbiaceae (spurge family) have been known to be poisonous to human beings. Often they are held responsible for the poisoning of live stock and are used as constituents of arrow poisons. Quite a number of species are used in folk medicine as drugs and raw materials for pharmaceutical preparations, in agriculture, as resources for certain technologies and as ornamental plants in horticulture. Some constituents of *Euphorbia* species may be promising lead compounds for drug development. Moreover, some *Euphorbia* species have been used as medicinal plants for the treatment of skin diseases, gonorrhea, migraines, intestinal parasites, warts and for mediating pain perception (Shi *et al.*, 2008a; Singla and Pathak, 1990; Appendino and Szallasi, 1997). Many researchers have shown that *Euphorbia* species possess antiproliferative activity (Xu *et al.*, 1998), cytotoxicity (Fatope *et al.*, 1996), antimicrobial activity (Murugan *et al.*, 2007), antipyretic-analgesic activity (Ma *et al.*, 1997), inhibition of HIV-1 viral infection (Hezareh, 2005), inhibitory activity on the mammalian mitochondrial respiratory chain (Betancur-Galvis *et al.*, 2003) etc.

Detailed phytochemical exploration of *Euphorbia* latex revealed presence of phorboids which, comprise of tigliane, ingenane, lathyrane and daphnane diterpene derivatives and triterpenes. The diterpenes of phorbol and ingenol esters are considered to be extensively toxic without any possible medicinal use. The 20-deoxy or 20-hydroxylated tigliane and ingenane diterpines are less irritant on the other hand di or triacetylation decreases the irritancy profile but an increase in the length of the fatty acid located at C-13 produced greater biological activity in both the mono and diester groups. Mono esters with C-16, C-17 or C-20 hydroxyl, or C-4 or C-12 deoxy substitution make them highly potent irritant. Most having polyhydroxy groups on C-4, C-9, C-13 and C-20 or with C-12 OH and C-13 OH esterified as a bis-ester have reinforcing irritant, proinflammatory and tumor promoting activity. The skin tumor promoting and irritant activities of the diterpene esters depend on the length of the aliphatic chain (fatty acid) in their ester moiety and degree of saturation. 20-deoxy, 19-acetoxy or 8-methoxy ingenane or lathyrane diterpines act as cytotoxic by inducing cell cleavage arrest. Euphornin (antitumor activity) and resiniferatoxin (capsaicine like counter irritant and analgesic in neuropathic pain) display toxicity but in adequate dose they have very promising therapeutic perspectives (Seiji *et al.*, 2003; Sahai *et al.*, 1981; Appendino and Szallasi, 1997). Non-phorboid type diterpenes with macrocyclic or polycyclic structure like, lathyrane and jatrophane have very less or no toxic effect with interesting biological activities like antiproliferative, antileukaemic and tubulin-interacting effects (Engi *et al.*, 2007; Lage *et al.*, 2010). *Euphorbia* latex triterpenes euphol, euphorbol, tirucallol, nerifolione etc have anti-inflammatory, analgesic and antiarthritic effect (Fenandez-Arche *et al.*, 2010; Ilyas *et al.*, 2003). In view of manifold uses of *Euphorbias* it appears surprising that the toxic nature of the phytochemical principles of the plants is discouraging their uses. The biological nature of their toxic chemicals and their possible therapeutic value remained relatively undefined until recently (Hecker, 1977). Systematic study on phytoconstituents of *Euphorbia* plants latex, as potential sources of new compounds will lead to promising natural product based new drug development. There arises a need, therefore to screen *Euphorbia* bioactive compounds for detailed and intensive pharmacological properties.

Acknowledgements

The author is grateful to the CDRI, Lucknow and NISCAIR, New Delhi, India for providing library facility. I owe my sincere thanks to Mr. Chandra Shekar Singh, Lecturer, Radharaman College of Pharmacy, for active help in drawing the structures of the compounds.

References

Adolf, W. and Hecker, E. (1975). On the active principles of the spurge family-III. *Krebsforschung und Klinische Onkologie Cancer Research and Clinical Oncology,* **84(3):** 325-344.

Adolf, W., Chanai, S. and Hecker, E. (1983). 3-*O*-angeloylingenol, the toxic and skin irritant factor from latex of *Euphorbia antiquorum* L. (Euphorbiaceae) and from a derived Thai purgative and anthelimintic (vermifuge) drug. *Journal of Scientific Society of Thailand*, **9:** 81-88.

Ahmad, W., Nazir, M. and Khan, S.A. (1988). The chemical composition of various *Euphorbia* species for industrial applications. Part II. Neutral lipids of *Euphorbia cauducifolia*. *Pakistan Journal of Science and Research*, **31:** 85-89.

Akihisa, T., Wijeratne, K., Tokuda, H., Enjo, F., Toriumi, M., Kimura, Y., Koike, K., Nikaido, T., Tezuka, Y. and Nishino H. (2002). Eupha-7, 9 (11), 24-trien-3β-ol ("antiquol C") and other triterpenes from *Euphorbia* latex and their inhibitory effects on Epstein-Barr virus activation. *Journal of Natural Products*, **65(2):** 158-162.

Amin, A.L., Faridi, M.A., Asif, M. and Khan, N.A. (1995). The efficacy and safety of *Euphorbia neriifolia* – Unani antiarthritic drug. *Indian Journal of Pharmacology*, **27:** 60.

Anonymous. (2010). *Euphorbia tirucalli*. Wikipedia, the free encyclopedia. Available at: http: //en.wikipedia.org/wiki/Euphorbia_tirucalli.

Appendino, G. and Szallasi, A. (1997). Euphorbium: Modern research on its active principle, resiniferatoxin, revives an ancient medicine. *Life Science*, **60:** 681-696.

Baloch, I.B., Baloch, M.K. and Saqib, Q.N. (2005). Tumor-promoting diterpene esters from Latex of *Euphorbia cauducifolia* L. *Helvetica Chimica Acta*, **88:** 3145-3150.

Bani, S., Kaul, A., Jaggi, B.S., Suri, K.A., Suri, O.P. and Sharma, O.P. (2000). Anti-inflammatory activity of the hydrosoluble fraction of *Euphorbia royleana* latex. *Fitoterapia*, **71(6):** 655-662.

Bani, S., Suri, K.A., Suri, O.P. and Sharma, O.P. (1998). Analgesic and antipyretic properties of *Euphorbia royleana* latex. *Phytotherapy Research*, **11(8):** 597–599.

Baslas, R.K. and Agarwal, R. (1980). Chemical investigation of some anticancer plants of *Euphorbia* genus. *Indian Journal of Chemistry*, **19B:** 717-718.

Betancur-Galvis, L., Palomares, E., Marco, J.A. and Estornell, E. (2003). Tigliane diterpenes from the latex of *Euphorbia obtusifolia* with inhibitory activity on the mammalian mitochondrial respiratory chain. *Journal of Ethnopharmacology* **85:** 279-282.

Bicchi, C., Appendino, G., Cordero, C., Rubiolo, P., Ortelli, D. and Veuthey, J.L. (2001). HPLC-UV and HPLC-positive-ESIMS analysis of the diterpenoid fraction from caper spurge (*Euphorbia lathyris*) seed oil. *Journal of Phytochemistry*, **12(4):** 255-262.

Burkill, H.M. (1994). The useful plants of West Tropical Africa. *Royal Botanical Garden of Kenya*, **2:** 12-150.

Cataluna, P. and Rates, S.M.K. (1999). The traditional use of the latex from *Euphorbia tirucalli* Linnaeus (Euphorbiaceae) in the treatment of cancer in South Brazil. Phytomedicine, Toxicology. ISHS Acta Horticulturae 501: II WOCMAP Congress Medicinal and Aromatic Plants, Part 2: Pharmacognosy, Pharmacology, Phytomedicine, Toxicology.

Cateni, F., Falsone, G. and Zilic, J. (2003). Terpenoids and glycolipids from Euphorbiaceae. *Mini-Reviews in Medicinal Chemistry*, **3:** 425-437.

Cooper, M.R. and Johnson, A.W. (1984). Poisonous plants in Britain and their effects on animals and man. *London*, **HMSO:** 113-116,161.

De Silva, W.A.P.P., Manuweera, G.K. and Karunaratne, S.H.P.P. (2008). Insecticidal activity of *Euphorbia antiquorum* L. latex and its preliminary chemical analysis. *Journal of the National Science Foundation of Sri Lanka*, **36(1):** 15-23.

Duarte, N., Gyémánt, N., Abreu, P.M., Molnár, J. and Ferreira, M.J.U. (2006). New macrocyclic lathyrane diterpenes, from *Euphorbia lagascae,* as inhibitors of multidrug resistance of tumour cells. *Planta Medica,* **72:** 162-168.

Duarte, N., Loge, H. And Ferreira, M.J.U. (2008). Antiproliferative activity of *ent*-abietane lactones against resistant human cancer cell lines. *Planta Medica,* **74:** 984-985.

Eke, T., Al-Husainy, S. and Raynor, M.K. (2000). The spectrum of ocular inflammation caused by *Euphorbia* plant sap. *Archives of Ophthalmology,* **118(1):** 13-16.

Engi, H., Vasas, A., Redei, D., Molnar, J. and Hohmann, J. (2007). New MDR modulators and apoptosis inducers from *Euphorbia* species. *Anticancer Research,* **7(5A):** 3451-3458.

Evans, F.J. (1978). The irritant toxins of Blue Euphorbia (*Euphorbia coerulescens* Haw.). **16(1):** 51-57.

Evans, F.J. and Kinghorn, A.D. (1973). Proceedings: A screening procedure for *Euphorbia* co-carcinogenic irritants. *Journal of Pharmacy and Pharmacology,* **25:** 145-146.

Evans, F.J. and Kinghorn, A.D. (1974). A new ingenol type diterpene from the irritant fractions of *Euphorbia myrsinites* and *Euphorbia biglandulosa. Phytochemistry,* **13:** 2324-2325.

Evans, F.J. and Kinghorn, A.D. (1975). The Succulent Euphorbias of Nigeria. *Lioydia,* **38:** 363-365.

Evans, F.J. and Kinghorn, A.D. (1977). A comparative phytochemical study of the diterpene of some species of the genera *Euphorbia* and *Elaeophorbia. Botanical Journal of the Linnean Society,* **74:** 23-35.

Evans, F.J., Kinghorn, A.D. and Schmidt R.J. (2009). Some naturally occurring skin irritants. *Acta Pharmacologica et Toxicologica,* **37(3):** 250–256

Evans, J.E. and Schmidt R.J. (1976). Two new toxins from the latex of *Euphorbia poisonii. Phytochemistry,* **15:** 333-335.

Falsone, G., Cateni, F., Visintin, G., Lucchini, V., Wagner, H. and Seligmann, O. (1994). Constituents of Euphorbiaceae 12. Comm. (1). Isolation and structure elucidation of four new cerebrosides from *Euphorbia biglandulosa* Desf. *Farmaco,* **49(3):** 167-174.

Fatope, M.O., Zeng, L., Ohayaga, J.E., Shi, G. and McLaughlin, J.L. (1996). Selectively cytotoxic diterpenes from *Euphorbia poisonii. Journal of Medicinal Chemistry* **39:** 1005-1008.

Fatope, O.M., Zeng, L., Ohayagha, E.J. and Mclaughlin L.J. (1996). New 19-acetoxyingol diterpenes from the latex of *Euphorbia poisonii* (Euphorbiaceae). *Bioorganic and Medicinal Chemistry,* **4:** 1679-1683.

Fenandez-Arche, A., Saenz, M.T., Arroyo, M., de la Puerta, R. and Garcia, M.D. (2010). Topical anti-inflammatory effect of tirucallol, a triterpene isolated from *Euphorbia lactea* latex. *Phytomedicine,* **17(2):** 146-148.

Furstenberger, G. and Hecker, E. (1972). Zum Wirkungsmechanismus cocarcinogener Pflanzeninhaltsstoffe. *Planta Medica*, **22(3):** 241-266.

Furstenberger, G. and Hecker, E. (1986). On the active principles of the Euphorbiaceae, XII. Highly unsaturated irritant diterpene esters from *Euphorbia tirucalli* originating from Madagascar. *Journal of Natural Product*, **49:** 386-397.

Gewali, M.B., Hattori, M., Tezuka, Y., Kikuchi, T. and Namba, T. (1990). Contituents of the latex of *Euphorbia antiquorum*. *Phytochemistry*, **29(5):** 1625-1628.

Guerreroy, O.A. and Guzman, A.L. (2004). Bioactivities of latexes from selected tropical plants. *Review of Cubana Planta Medica* **9(1):** Available at: http: //scielo.sld.cu/ scielo.php?pid=S1028-47962004000100015&script=sci_arttext.

Hecker, E. (1977). New toxic, irritant and cocarcinogenic diterpene esters from Euphorbiaceae and Thymelaeaceae. Pure and Applied Chemistry. Vol. 49, Pergamon Press, Great Britain, pp.1423-1431.

Hergenhahn, M., Kusumoto, S. and Hecker, E. (1984). On the active principles of the spurge family (Euphorbiaceae). V. Extremely skin-irritant and moderately tumor-promoting diterpene esters from *Euphorbia resinifera* Berg. *Cancer Research and Clinical Oncology*, **108(1):** 98-109.

Hezareh, M. (2005). Prostratin as a new therapeutic agent targeting HIV viral reservoirs. *Drug News Perspective*, **18:** 496-500.

Hickey, T.A., Worobec, S.M., West, D.P. and KinghornIr, A.D. (1981). Irritant contact dermatitis in humans from phorbol and related esters. *Toxicon*, **19(6):** 841-850.

Hohmann, J. and Molnár, J. (2004). Euphorbiaceae diterpenes: Plant toxins or promising molecules for the therapy? *Acta Pharmaceutica Hungarica*, **74:** 149-157.

Hsueh, K.F., Lin, P.Y., Lee, S.M. and Hsieh, C.F. (2004). Ocular injuries from plant sap of genera *Euphorbia* and *Dieffenbachia*. *Journal of the Chinese Medical Association*, **67(2):** 93-98.

ICMR report. (1991). Multicentric randomized controlled clinical trial of Kshaarasootra (Ayurvedic medicated thread) in the management of fistula-in-ano. *Indian Journal of Medical Research(B)*, **94:** 177-185.

Ilyas, M., Perveen, M., Muhaisen, H.M.H. and Basudan, O.A. (2003). A novel triterpene (Nerifolione) a potent anti-inflammatory and antiarthritic agent from *Euphorbia neriifolia*. *Hamdard-Medicine (Pakistan)*, **46(2):** 97-102.

Ilyas, M., Praveen, M. and Amin, K.M.Y. (1998). Neriifolione, a triterpene from *Euphorbia neriifolia*. *Phytochemistry*, **48:** 561-563.

Itokawa, H., Ichihara, Y., Watanabe, K. and Takeya, K. (1989). An antitumor principle from *Euphorbia lathyris*. *Planta Medica*, **55(3):** 271-272.

Joshi, D. and Shingal, P. (2008). Ocular injuries from plant sap in army soldiers. *Medical Journal Armed Forces India*, **64:** 293-294.

Kedei, N., Lundberg, D.J., Toth, A., Welburn, P., Garfield, S.H. and Blumberg, P.M. (2004). Characterization of the interaction of ingenol 3-angelate with protein kinase C. *Cancer Research*, **64:** 3243-3255.

Khan, Q.A. and Malik, A. (1990). A new macrocyclic diterpene ester from the latex of *Euphorbia tirucalli. Journal of Natural Products*, **53:** 728-731.

Kinghorn, A.D. and Evans, F.J. (1975). Skin irritants of *Euphorbia fortissima. Journal of Pharmacy and Pharmacology*, **27(5):** 329-333.

Kupchan, S.M., Uchida, I., Branfman, A.R., Dailey, R.G. Jr. and Fei, B.Y. (1976). Antileukemic principles isolated from euphorbiaceae plants. *Science*, **191(4227):** 571-572.

Lage, H., Duarte, N., Coburger, C., Hilgeroth, H. and Ferreira, M.J. (2010). Antitumor activity of terpenoids aganist classical and atypical multidrug resistant cancer cell. *Phytomedicine*, **17(6):** 441-448.

Lam, T.S.K., Wong, O.F., Leung, C.H. and Fung, H.T. (2009). A case report of ocular injury by *Euphorbia* plant sap. *Hong Kong Journal of Emergency Medicine*, **16(4):** 267-270.

Lampe, K. and Fagerstrom, R. (1968). Plant toxicity and dermatitis. Williams and Wilkins, Baltimore.

Lennox, F.G. and Ellis, W. J. (1945). Euphorbain, a protease occurring in the latex of the weed *Euphorbia lathyris. Biochemical Journal*, **39:** 465-470.

Lisch, K. (1980). The effect of the sap of Euphorbiaceae on the eye. *Klin Monatsbl Augenheilkd*, **176(3):** 469-471.

Lynn, K.R. and Clevette-Radford, N.A. (1983). Isolation and charecteraisation of Euphorbain proteinase from the latex of *Euphorbia lathyris. Biochemica et Biophysica Acta*, **746:** 154.

Ma, Q.G., Liu, W.Z., Wu, X.Y., Zhou, T.X. and Qin, G.W. (1997). Chemical studies of Lang-Du, a traditional Chinese medicine. 1. Diterpenoids from *Euphorbia fischeriana. Phytochemistry* **44:** 663-666.

Mallavadhani, U.V., Satyanarayana, K.V., Mahapatra, A. and Sudhakar, A.V. (2004). A new tetracyclic triterpene from the latex of *Euphorbia neriifolia. Natural Product Research*, **18(1):** 33-37.

Marco, A.J., Sanz-Cervera, J.F. and Yuste, A. (1997). Ingenane and lathyrane diterpenes from the latex of *Euphorbia canariensis. Phytochemistry*, **45(3):** 563-570.

Marco, A.J., Sanz-Cervera, J.F., Ropero, F.J., Checa, J. and Fraga B.M. (1998). Ingenane and lathyrane diterpenes from the latex of *Euphorbia acrurensis. Phytochemistry*, **49(4):** 1095-1099.

Marco, J.A., Sanz-Cervera, J.F., Checa, J., Palomares, E. and Fraga, B.M. (1999). Jatrophane and tigliane diterpenes from the latex of *Euphorbia obtusifolia. Phytochemistry*, **52:** 479-485.

Marston, A. and Hecker, E. (1983). On the active principles of the Euphorbiaceae. 6. Isolation and biologicalactivities of seven milliamines from *Euphorbia milii. Planta Medica*, **47**: 141-147.

Marston, A. and Hecker, E. (1984). Active principles of the Euphorbiaceae. 7. Milliamine H and milliamine I, peptide esters of 20-deoxy-5ξ-hydroxyphorbol from *Euphorbia milii. Planta Medica*, **50:** 319-322.

Matsumoto, T., Cyong, J.C. and Yamada, H. (1992). Stimulatory effects of ingenols from *Euphorbia kansui* on the expression of macrophage Fc receptor. *Planta Medica*, **58:** 255-258.

Merani, R., Sa-Ngiampornpanit, T., Kerdraon, Y., Billson, F. and McClellan, K.A. (2007). *Euphorbia lactea* sap keratouveitis: case report and review of the literature. *Cornea*, **26(6):** 749-752.

Moulin, A., Teissere, M., Bernard, C. and Pieroni, G. (1994). Lipases of the Euphorbiaceae family: Purification of a lipase from *Euphorbia characias* latex and structure-function relationships with the B chain of ricin. *Proceedings of Natural Academy of Science USA*, **91:** 11328-11332.

Mura, A., Medda, R., Longu, S., Floris, G., Rinaldi, A.C. and Padiglia, A. (2005). A Ca2+/calmodulin-binding peroxidase from *Euphorbia* latex: novel aspects of calcium-hydrogen peroxide cross-talk in the regulation of plant defenses. *Biochemistry*, 1; **44(43):** 14120-14130.

Murugan, S., Anand, R., Uma-Devi, P., Vidhya, N. and Rajesh, K.A. (2007). Efficacy of *Euphorbia milli* and *Euphorbia pulcherrima* on aflatoxin producing fungi (*Aspergillus flavus* and *Aspergillus parasiticus*). *African Journal of Biotechnology*, **6:** 718-719.

Nadkarni, K.M. (1976). Indian Materia Medica, Vol. 1, Popular Prakashan, Bombay, India, pp.1296.

Oudhia, P. (2005). Medicinal herbs of Chhattisgarh. India having less known traditional uses, VII. Thura (*Euphorbia neriilofia*, family: Euphorbiaceae). Available at: http: //www.botanical.com/site/column_poudhia/ 160_thura.html.

Pandey, G.S. (1992). Bhavaprakasa Nighantu. In: B. Mishra (Eds.), Indian Materia Medica, Chaukhambha Bharti Academy, Varanashi, pp.308 and 76-77.

Pitts, J.F., Barker, N.H., Gibbons, D.C. and Jay, J.L. (1993). Manchineel keratoconjunctivitis. *British Journal of Ophthalmology*, **77(5):** 284-288.

Ponsinet, G. and Ourisson, G. (1968). Chemotexonomical studies of the family Euphorbiaceae. *Phytochemistry*, **7:** 89-98.

Rashik, A.M., Shukla, A., Patnaik, G.K., Dhawan, B.N. and Kulshrestha, D.K. (1996). Wound healing activity of latex of *Euphorbia neriifolia* Linn. *Indian Journal of Pharmacology*, **28(2):** 107-109.

Ravikant, V., Reddy, V.L.N., Reddy, A.V., Ravinder, K., Rao, T.P., Shiva Ram, T., Kumar, K.A., Vamanarao, D.P. and Venkateswarlu, Y. (2003). Three New Ingol

Diterpenes from *Euphorbia nivulia*: Evaluation of cytotoxic activity. *Chemical and Pharmaceutical Bulletin*, **51(4):** 431-434.

Ravikanth, V., Reddy, V.L.N., Rao, T.P., Diwan, P.V., Ramakrishna, S. and Venkateswarlu, Y. (2002). Macrocyclic diterpenes from *Euphorbia nivulia*. *Phytochemistry*, **59:** 331-335.

Sahai, R., Rastogi, R.P., Jakupovic, J. and Bohlmann, F. (1981). A diterpene from *Euphorbia maddeni*. *Phytochemistry*, **20:** 1665-1667.

Salah, M.A.D., Zayed M.F., Hubert G. and Erich H. (1998). Dietary cancer risk conditional cancerogens in produce of livestock fed on species of spurge (Euphorbiaceae). *Journal of Cancer Research and Clinical Oncology*, **124(4):** 131-140.

Sayed, M.D. Riszk, A., Hammouda, F.M., El-Missiry, M.M., Williamson E.M. and Evans F.J. (1980). Constituents of Egyptian Euphorbiaceae. IX. Irritant and cytotoxic ingenane esters from *Euphorbia paralias* L. *Cellular and Molecular Life Sciences*, **36(10):** 1206-1207.

Schmidt, R.J. and Evans, F.J. (1980). Skin irritant effects of esters of phorbol and related polyols. *Archives of Toxicology*, **44(4):** 279-289.

Scott, I.U. and Karp, C.L. (1996). Euphorbia sap keratopathy: Four cases and a possible pathogenic mechanism. *British Journal of Ophthalmology*, **80(9):** 823-826.

Seiji, K., Shinwa, K. and Shosuke, Y. (2003). Euphornin have antitumor activity (Synthetic study on euphornin, a toxic diterpene from Euphorbia L (II). *Hiyoshi Review of Natural Science*, **33:** 1-17.

Seshagirirao, K. and Prasad, M.N. (1995). Purification and partial characterization of a lectin from *Euphorbia neriifolia* latex. *Biochemistry and Molecular Biology International*, **35(6):** 1199-1204.

Shi, J.S., Li, Z.X., Nitoda, T., Minoru, I., Hiroshi, K., Naomichi, B., Kazuyoshi, K. and Shuhei, N. (2008c). Antinematodal activities of ingenane diterpenes from *Euphorbia kansui* and their derivatives against the pine wood nematode. *Zeitschrift fur Naturforschung-Section C: Biosciences*, **63:** 59-65.

Shi, J.X., Li, Z.X., Izumi, M., Baba, N. and Nakajima, S. (2008b). Termiticidal activity of diterpenes from the roots of *Euphorbia kansui*. *Zeitschrift fur Naturforschung-Section C: Biosciences*, **63:** 51-58.

Shi, Q.W., Su, X.H. and Kiyota, H. (2008a). Chemical and pharmacological research of the plants in genus *Euphorbia*. *Chemical Review*, **108(10):** 4295-327.

Silva, A.C.P., De Faria, D.E.P., Do Espírito Santo Borges, N.B., De Souza, I.A., Peters, V.M. and De Oliveira Guerra, M. (2007). Toxicological screening of *Euphorbia tirucalli* L.: Developmental toxicity studies in rats. *Journal of Ethopharmacology*, **1109(1):** 154-159.

Singh, A. and Agarwal, R.A. (1988). Possibility of using latex of euphorbiales for snail control. *Science of the Total Environment*, **77:** 231-236.

Singla, A.K. and Pathak, K. (1990). Phytoconstituents of *Euphorbia* species. *Fitoterapia*, **61:** 483-516.

Sorg, B., Schmidt, R. and Hecker, E. (1987). Structure activity relationships of polyfunctional diterpenes of the ingenane type. I. Tumor-promoting activity of homologous, aliphatic 3-esters of ingenol and of Δ7,8-isoingenol-3-tetradecanoate. *Carcinogenesis*, **8:** 1-4.

Sosath, S., Ott, H.H. and Hecker, E. (1988). Irritant principles of the spurge family (Euphorbiaceae). XIII. Oligocyclic and macrocyclic diterpene esters from latices of some *Euphorbia* species utilized as source plants of honey. *Journal of Natural Product*, **51(6):** 1062-1074.

Souza, C.A., de-Carvalho, R.R., Kuriyama, S.N., Araujo, I.B., Rodrigues, R.P., Vollmer, R.S., Alves, E.N. and Paumgartten, F.J. (1997). Study of the embryofeto-toxicity of Crown-of-Thorns (*Euphorbia milii*) latex, a natural molluscicide. *Brazilian Journal of Medicine and Biological Research*, **30(11):** 1325-1332.

Spoerke, D.G. and Temple, A.R. (1979). Dermatitis after exposure to a garden plant (*Euphorbia myrsinites*). *American Journal of Diseases of Children*, **133(1):** 28-29.

Tiwari, S., Pandey, R.P. and Singh, A. (2008). Effect of cycloart-24-en-3β-ol from *Euphorbia royleana* latex on meuroenzyme AChE and oxidative metabolism of freshwater fish, *Channa punctatus*. *African Journal of Traditional, Complement and Alternative Medicine*, **5(4):** 332–339.

Upadhyay, R.R., Ansarin, M., Zarintan, M.H. and Shakui, P. (1976). Tumor promoting constituent of *Euphorbia serrata* L. Latex. *Experientia*, **32(9):** 1196-97.

Upadhyay, R.R., Bakhtavar, F., Mohseni, H., Sater, A.M., Saleh, N., Tafazuli, A., Dizaji, F.N. and Mohaddes, G. (1980a). Screening of *Euphorbia* form Azarbijan for skin irritant activity and for diterpenes. *Planta Medica*, **38:** 151-154.

Upadhyay, R.R., Zarintan, M.H. and Ansarin, M. (1980b). Isolation of ingenol from the irritant and cocarcinogenic latex of *Euphorbia seguieriana*. *Planta Medica*, **30:** 32-34.

Urones, J.G., Barcala, P.B., Cuadrado, M.J.S. and Marcos, I.S. (1988). Diterpenes from the latex of *Euphorbia broteri*. *Phytochemistry*, **27:** 207-212.

Vogg, G., Mattes, E., Rothenburger, J., Hertkorn, N., Achatz, S. and Sandermann, H. Jr. (1999). Tumor promoting diterpenes from *Euphorbia leuconeura* L. *Phytochemistry*, **51(2):** 289-295.

Vogg, G., Mattes, E., Rothenburger, J., Hertkorn, N., Stefan, A. and Sandermann, J.H. (1999). Tumor promoting diterpenes from *Euphorbia leuconeura* L. *Phytochemistry*, **51:** 289-295.

Wang, L.Y., Wang, N.L., Yao, X.S., Miyata, S. and Kitanaka, S. (2003). Studies on the bioactive constituents in Euphorbiaceae, part 3. Diterpenes from the roots of *Euphorbia kansui* and their in vitro effects on the cell division of Xenopus. *Chemical and Pharmaceutical Bulletin*, **51:** 935-941.

Webster, G.L. (1986). Plant dermatitis. Irritant plants in the spurge family (Euphorbiaceae). *Clinical Dermatology*, **4(20):** 36-45.

Wu, D.G., Sorg, B. and Hecker, E. (1994). Oligocyclic and macrocyclic diterpenes in thymelaeaceae and Euphorbiaceae occurring and utilized in Yunnan (Southwest China). 6. Tigliane type diterpene esters from latex of *Euphorbia prolifera. Phytotherapy Research*, **8:** 95-99.

Wu, D.G., Sorg, B., Adolf W., Seip, E.H. and Hecker, E. (2009). Oligo- and macrocyclic diterpenes in thymelaeaceae and Euphorbiaceae occurring and utilized in Yunnan (Southwest China). 2. Ingenane type diterpene esters from *Euphorbia nematocypha* Hand.-Mazz. *Phytotherapy Research*, **6:** 237-240.

Xu, Z.H., Sun, J., Xu, R.S. and Qin, G.W. (1998). Casbane diterpenoids from *Euphorbia ebracteolata. Phytochemistry*, **49:** 149-151.

Yan, S.S., Li, Y., Wang, Y., Shen, S.S., Gu, Y., Wang, H.B., Qin, G.W. and Yu, Q. (2008). 17–Acetoxyjolkinolide B irreversibly inhibits I-kappaB kinase and induces apoptosis of tumor cells. *Molecular Cancer Therapeutics*, **7**: 1523-1532.

Yasukawa, K., Akihisa, T., Yoshida, Z.Y. and Takido, M. (2000). Inhibitory effect of euphol, a triterpene alcohol from the roots of *Euphorbia kansui*, on tumour promotion by 12-O-tetradecanoylphorbol-13-acetate in two-stage carcinogenesis in mouse skin. *Journal Pharmacy and Pharmacology*, **52(1):** 119-124.

Zayed, S., Farghaly, M., Soliman, S.M., Gotta, H., Sorg, B. and Hecker, E. (2001). Dietary cancer risk from conditional cancerogens (tumor promoters) in produce of livestock fed on species of spurge (Euphorbiaceae) V. Skin irritant and tumor-promoting diterpene ester toxins of the tigliane and ingenane type in the herbs *Euphorbia nubica* and *Euphorbia helioscopia* contaminating fodder of livestock. *Journal of Cancer Research and Clinical Oncology*, **127:** 40-47.

Zechmeister, K., Brandl, F. And Hoppe, W. (1970). Structure determination of the new tetracyclic diterpene ingenol-triacetate with triple product methods. *Tetrahedron Letters*, **47:** 4075-4078.

Traditional and Folk Herbal Medicine: Recent Researches Vol. 1 (2012) *Pages* **251–277**
Editor: **V.K. Gupta**
Published by: **DAYA PUBLISHING HOUSE, NEW DELHI**

11

Traditional Medicine: Its Role in the Search for New Antiparasitic Candidates

F. Benoit-Vical[1,2]* and P. Njomnang Soh[1,2]

ABSTRACT

Malaria, African trypanosomiasis and leishmaniasis kill more than 2 million people each year. Developing countries that are particularly affected by these parasitic diseases face a number of problems relating to treatment of the diseases, such as cost, availability of the treatment, side-effects, safety issues and practicability of the drug regimens. Natural products have played a key role throughout the history of drug discovery as a source of both novel compounds and as molecular scaffolding for the development of blockbuster drugs such as taxol, morphine, antibiotics, and antimalarial drugs like quinine and artemisinin. The authors here present an overview of plants from endemic areas studied against these 3 pathologies for the last ten years and showing the most in vitro and/or in vivo and/or clinical promising results. According to our criteria, 22 plants species traditionally used were here selected for their anti-Leishmania properties reported but only two (Maesa balansae and Vernonia polyanthes) demonstrated both high activities and very weak toxicities. The most active molecule against Leishmania was 7-deacetyl-7-oxogedunin extracted from Pseudocedrela kotschyi. On the other hand, Alchornea cordifolia is the most promising plant among the only 12 plants concerned by anti-Trypanosoma activities but the only one interesting anti-

1 CNRS; LCC (Laboratoire de Chimie de Coordination); 205, route de Narbonne, F-31077 Toulouse, France et Université de Toulouse III; UPS, INP; LCC; F-31077 Toulouse, France.

2 Service de Parasitologie-Mycologie, Centre Hospitalier Universitaire de Rangueil, TSA 50032, 31059 Toulouse 9, France and Faculté de Médecine de Rangueil, Université de Toulouse III, 31059 Toulouse Cedex 9, France.

* *Corresponding author*: Parasitology-Mycology Dept, Toulouse-Rangueil Hospital University, 1 av Poulhès, TSA50032, 31059 Toulouse 9, France.
E-mail: francoise.vical@inserm.fr; Ph.: ++33 561323446; Fax: ++33 561322096.

Trypanosoma molecule identified seems to be saringosterol isolated from Strychnos spinosa. In fact, the plants were more studied these last ten years for their antimalarial properties and several active molecules were so explored. That's the case of ellagic acid, cryptolepine, gedunin, harman, febrifugine and simalikalactone E which showed real antiplasmodial drug-candidates status. It is interesting to note that Meliaceae appears to be a family of plants particularly interesting for its antileishmanial, antitrypanosomial and also antimalarial properties.

The development of partnerships whereby the tools of modern science are used to evaluate traditional knowledge provides some hope that the intractable problems of the treatment regimens for these "neglected diseases" could be perhaps solved in the not-too distant future.

Keywords: Active principles, Drug, Leishmania, Malaria, Plasmodium, Ethnopharmacology, Traditional medicine, Trypanosoma.

Introduction

The Burden of Malaria, African Trypanosomiasis and Leishmaniasis

Malaria, African trypanosomiasis and leishmaniasis are a group of tropical diseases that are especially endemic in low-income populations in developing regions of Africa, Asia and South America. Together, these three parasitic diseases cause up to an estimated 2 million deaths annually. In terms of morbidity, currently half a million people in Sub-Saharan Africa are affected by African trypanosomiasis or sleeping sickness (WHO, 2005a), 1.5-2 million by leishmaniasis (cutaneous, visceral, mucocutaneous) and 300 million people around the world by malaria. In the countries where these diseases are endemic, they are a major public health problem and represent in addition a dramatic loss of income for the affected populations. No vaccines exist for any of these 3 diseases.

For a long time, the global strategy for further reducing the burden of disease concerned only HIV-AIDS, tuberculosis and malaria. The DNDI's (Diseases Neglected Drugs Initiatives) analysis clearly showed that this strategy has ignored the « most neglected diseases » such as, in particular, sleeping sickness and leishmaniasis, which are highly debilitating and invariably fatal if untreated. Furthermore, the global drug market for these diseases is virtually non-existent. Indeed, drugs are available for these diseases: meglumine antimoniate, sodium stibogluconate, amphotericin B, miltefosine, paromomycin and sitamaquine for leishmaniasis; pentamidine, suramin, melarsoprol and eflornithine for African trypanosomiasis (Croft *et al.*, 2005) and chloroquine, amodiaquine, quinine, mefloquine, atovaquone-proguanil, sulfadoxine-pyrimethamine and artemisinin and its derivatives for malaria (Lalloo *et al.*, 2007). However, a frequent characteristic of all these current drugs is their numerous severe side-effects and, for some of these drugs, resistances developed by parasite are spread. Resistance has appeared principally for pentavalent antimony (Sundar *et al.*, 2000), miltefosine (Seifert *et al.*, 2007), chloroquine (Wellems and Plowe, 2001), amodiaquine (Mutabingwa *et al.*, 2005), quinine (Zalis *et al.*, 1998), mefloquine (Olliaro, 2001), atovaquone-proguanil (Kuhn *et al.*, 2005), and sulfadoxine-pyrimethamine (Gatton *et al.*, 2004). Furthermore, the risk of resistance is ever-present for all drugs and some sporadic resistance has been recently reported even with ACT (artemisinin-based combination therapy) (Lim *et al.*, 2009; Rogers *et al.*, 2009), the last antimalarial treatment still effective against all *P. falciparum* strains.

Paromomycin, an aminoglycoside antibiotic with good antileishmanial activity (den Boer and Davidson, 2006) and artemisinin, a sesquiterpene lactone with high antiplasmodial activity (Woodrow *et al.*, 2005) show less toxicity than other current drugs but because of their cost, these molecules and derivatives are often unavailable to most of the populations concerned (Mutabingwa, 2005). Combination therapies have been suggested as the way forward to increase treatment efficacy, prevent the development of drug resistance, and reduce treatment duration, but with combinations, treatments costs are actually increased and thus remain a major problem (White and Olliaro, 1996).

The Double Interest of Ethnopharmacology in the Antiparasitic Drugs Research

Briefly, these three fatal diseases have thus less than ten interesting drugs really available and at the same time it is developing countries with poor populations that are particularly affected by these diseases. Urgent research is therefore necessary to find new effective and affordable alternative treatments with few side-effects. However, investing in drug development for these diseases is not thought to be marketable or profitable by the pharmaceutical industry. Since conventional drug development typically requires at least ten years, incurring research and development costs of US $300 to 500 million, many pharmaceutical companies are unwilling to invest in drug programs against diseases in developing countries (Benoit-Vical, 2005). In this context, all new inexpensive therapies that are simple to use and that will limit the cost of drug research are strong justifications for the ethnopharmacological approach.

Given the pressure of the price of drugs, and given established customs, accessibility and availability, in some endemic areas traditional medicines are still the only treatments really used. In a review discussing the controversy concerning malaria treatment by traditional medicine and the poor and slow efficacy in comparison with standard drugs, Willcox said "An imperfect treatment given rapidly, to prevent deterioration to severe malaria and death, may be preferable to the « gold standard » modern treatment if there is a significant delay in obtaining it" (Willcox, 2009).

What is more, WHO currently cooperates with its Member States to promote the use of traditional medicine for health care to (*i*) support and integrate traditional medicine into national health systems in combination with national policy and regulation for products, practices and providers to ensure safety and quality; (*ii*) ensure the use of safe, effective and quality products and practices, based on available evidence; (*iii*) acknowledge traditional medicine as part of primary health care, to increase access to care and preserve knowledge and resources; and (*iv*) ensure patient safety by upgrading the skills and knowledge of traditional medicine providers (WHO, 2008). For these populations it is thus crucial to rapidly define the interest, the impact and the consequences of scientifically-proven ancestral knowledge on local, national and international policies. Indeed, scientific research demonstrates more and more that many plants used either as food or in traditional medicine are potentially toxic (Schimmer *et al.*, 1988; Fennell *et al.*, 2004; Waako *et al.*, 2005; Elufioye *et al.*, 2009). Traditional medicine is not without risk and unscrupulous traders and healers may jeopardize the safety of patients. Additionally, the lack of trade regulations and quality control to monitor plant medicines is becoming a significant problem and must be addressed. The possibility of standardization of harvesting, extraction, plants used, and plant combinations proposed by traditional healers needs to be considered (Benoit-Vical, 2005).

The goals of ethnopharmacology in the case of these neglected diseases are both the evaluation of indigenous pharmacopoeias but also the contribution to pharmaceutical innovation. In this last case, research for new molecules from plants with their high chemical diversity, biochemical specificity and specifically of ones, based on ancestral traditional medicine, appears as a major pipeline for new molecules. Several well-known antiparasitic drugs, such as quinine, an alkaloid from *Cinchona pubescens* (Rubiaceae) and artemisinin, a sesquiterpene lactone from *Artemisia annua* (Asteraceae) used as antimalarial agents or emetine, an alkaloid from *Cephaelis ipecacuanha* (Rubiaceae) used to treat amoebiasis have their origins in natural products. From the 14 current chemical entities or medicinal indications against parasite diseases, recently reported by Newman and Cragg from 01/1981 to 06/2006, 2 (artemisinin and ivermectin) were isolated from natural products, 5 (arteether, artemether, artesunate, eflornithine, mefloquine) derived from a natural product, 2 (atovaquone, bulaquine) made by total synthesis but the pharmacophore is from a natural product and only 4 were totally synthetic (Newman and Cragg, 2007).

Finally, the enormous potential of natural products as modulators of biomolecular function has become increasingly evident with the advent of chemical genomics, and increasingly powerful high-throughput screening methodologies.

Therefore as ethnopharmacology is transdisciplinary, we present here the combination of two points of view, cultural and biological, in a complementary approach.

Methodology: Search Strategy and Selection Criteria

For these 3 selected diseases, the most representative ethnopharmacological studies *i.e.* with data from both ethnobotany, *in vitro*, *in vivo* and clinical assays showing the best promising pharmacology results are here reported. As the reference database for pharmacology and medicinal fields is PubMed (http://www.ncbi.nlm.nih.gov), we conducted hand-search investigations and analyzed articles extracted firstly from this database and then from all the relevant associated citations. Search terms were "Malaria OR Plasmodium AND Africa AND plant", "Leishmania AND Africa AND plant", "Trypanosoma AND Africa AND plant". English language papers were reviewed. Articles up to July, 2010, were included.

Results and discussion

In vitro and *In vivo* Methods for Antiparasitic Assays in Laboratory

Different ethnopharmacological methodologies can be used to gain information about plant species used as medicines within a society (Bourdy *et al.*, 2008) however this field requires methodology standards to be comparable (Heinrich *et al.*, 2009). To validate the use in folk medicine of selected plants, *in vitro* and *in vivo* rigorous assays are performed in laboratory. These tests used for the assessment of antiparasitic activity can be grouped into two categories: those testing the parasite response *in vitro* on cell models and/or *in vivo* on animal models (all results in this manuscript used this strategy), and those based on the inhibition of a specific target that plays a key function in parasite life.

Requirements and methods for research and evaluation of the safety and efficacy of herbal medicines are more complex than those for conventional pharmaceuticals. A single medicinal plant may contain hundreds of natural constituents, and a mixed herbal medicinal product may contain several times that number. If every active ingredient was to be isolated from every herb, the time and resources required would be tremendous (WHO, 2005b). Methodologies to evaluate *in vitro* and *in vivo* the antiparasitic activities of selected extracts and compounds used by different laboratories of which the results are here reported are thus briefly summarized below.

Many general characteristics are shared by the parasites of Trypanosomiasis and Leishmaniasis, including subcellular structures such as the kinetoplast and glycosomes, but each disease is transmitted by a different insect and has its own life-cycle characteristics, different target tissues, and distinct disease pathogenesis in their mammalian host (El-Sayed *et al.*, 2005).

In vitro anti-parasite tests are now possible since the protozoan pathogens, *Leishmania* sp., *Trypanosoma brucei* sp. and *Plasmodium falciparum*, can be maintained in continuous culture. To evaluate the plant extracts or the purified molecules in a disease context, *in vivo* murine assays existed for many years but are less used than *in vitro* tests because of the constraints of assays with animals such as ethic, price, administrative regulations, infrastructures, facilities, technical skills of the team, etc.

Promastigote forms of *Leishmania* are cultured at 25°C in Warren's medium with 10 per cent heat-inactivated fetal bovine serum. Axenic amastigote cultures, obtained by *in vitro* transformations of infective promastigotes are incubated at 32°C in Schneider's insect medium pH 4.6, with 20 per cent fetal bovine serum (Gachet *et al.*, 2010). *Leishmania* sp. promastigotes are checked under the microscope for flagellar motility as an indication of viability before each experiment (Ueda-Nakamura *et al.*, 2001). Activity of plant extracts against promastigotes is determined by both microscopic counting or the colorimetric cell viability XTT (2,3-bis[2-methyloxy-4-nitro-5-sulfophenyl]-2H-tetrazolium-5-carboxanilide) assay (El-On *et al.*, 2009).

The antileishmanial activity against amastigotes in mouse macrophages at 37°C can also be determined. To estimate parasite load in *Leishmania braziliensis* infected mice, a murine test system with a closer resemblance to the American cutaneous leishmaniasis was recently developed (de Moura *et al.*, 2005): Stationary-phase promastigotes are inoculated into mice. One-month post infection, parasitemia is confirmed in arbitrarily selected animals. The molecule to be tested is administered daily for five consecutive days. One week post treatment, mice are killed. Smears of liver, spleen, and bone marrow are prepared, fixed in methanol, and Giemsa-stained to examine the parasite burden (Dutta *et al.*, 2008).

Bloodstream forms of *Trypanosoma brucei brucei* are cultured at 37°C under a 5 per cent CO_2 atmosphere in Minimal Essential Medium with 25 mM Hepes, Earlìs salts supplemented with nonessential amino acids, l-glutamine, glucose, l-cysteine, pyruvic acid-Na-salt, thymidine; hypoxanthine, and 15 per cent heat-inactivated horse serum. Trypomastigotes are cultured for the determination of the *in vitro* trypanocidal activity of crude extracts by the Long Incubation Low Inoculation Test (LILIT) (Brun and Lun, 1994) as a screening test whereas the Long-Term Viability Assay (LTVA) (Kaminsky *et al.*, 1989) is subsequently used for confirmation. *In vivo* experiments against *Trypanosma brucei* (Casero *et al.*, 1982) are largely used.

Dose-response and increase-in-life-span studies are performed with *Trypanosoma* sp. in various mice.

Plasmodium falciparum is maintained in continuous culture in human red blood cells (Trager and Jensen, 1976) in a CO_2 atmosphere (Van Huyssen and Rieckmann, 1993). The WHO *in vitro* antiplasmodial activity tests uses radio-isotopes (Desjardins *et al.*, 1979): Parasite growth is estimated by ^{3}[H]hypoxanthine incorporation. Concentrations of molecules that inhibited 50 per cent of the parasite growth (IC_{50} values) are graphically determined by plotting the molecule concentration *versus* percentage of parasite growth inhibition at 48 h of incubation (Benoit-Vical *et al.*, 2007).

For *Plasmodium* the reference *in vivo* test is the Peter's 4-day suppressive test (Peters, 1980): after intravenous route for parasite inoculation, selected molecules are then administered (in general by intraperitoneal or oral administration) to the animals, daily for 4 consecutive days. Parasitemia in control mice generally reached as high as 20-50 per cent on day 5. In order to detect recrudescence, parasitemia was followed over 60 days.

For whatever the antiprotozoan activities investigated, assays of cytotoxicity are carried out *in vitro* using established cell lines and *in vivo* toxicity is tested with healthy and infected murine models (Benoit-Vical *et al.*, 2007).

Antiprotozoal Plant Extracts from Ethnomedicine

The resurgence of plant-based drugs, as is evidenced by the number of such drugs in clinical trials, is certainly promising (Saklani and Kutty, 2008). Nevertheless, the combination of anti-*Leishmania*, anti-*Trypanosoma* and natural substances is still a marginal research theme. As an objective component representing the impact of the ethnopharmacology approach for each disease, a search of the PubMed database, without any date limitations, shows that the keywords "Malaria OR Plasmodium AND Africa AND plant" produces 421 references (322 references for the last 10 years); while "Leishmania AND Africa AND plant" and "Trypanosoma brucei AND Africa AND plant", produce only 34 (32 for the last 10 years) and 47 references (38 for the last 10 years), respectively. We have deliberately chosen to report here only the plant extracts with activities particularly relevant pharmacologically against *Leishmania* (Table 11.1), *Trypanosoma* (Table 11.2), and *Plasmodium* (Soh and Benoit-Vical, 2007). Indeed, several natural products can exhibit promising biological activities *in vitro* and *in vivo* but limitations such as toxicity, low bioavailability and/

or poor solubility can restrict their scope of use in humans (Kaur *et al.*, 2009). Indeed, a lot of publications report the activity of plant extracts against parasites but with active concentrations that are too high to have a pharmacological interest. In this context, we preferred to present selectivity indices (SI) (or therapeutic indices *in vivo*) as the ratio between toxicity and activity in order to select extracts with real pharmacological properties against parasite diseases. In the chapter entitled « Guidelines for nonclinical evaluation of the efficacy of traditional antimalarials » of the book "Traditional Medicinal Plants and Malaria" (Willcox *et al.*, 2004), it is suggested that to select plants with interesting biological properties the SI value superior to 300 is recommended. But this threshold appears very high. Indeed antimalarial drugs currently used clinically have of course high SI values, for example chloroquine has SI values ranged from 270 to 2000 according to the sensitivity of the strains tested (Benoit-Vical *et al.*, 2007). However, it is necessary to distinguish SI for crude extracts and SI for purified molecules since activity is (or must be) better (since not diluted with other components of the mixture) for this last one category. Objectively, plant extracts with SI lower than 20 (Soh and Benoit-Vical, 2007) seemed few interesting for their pharmacological properties as such (Tables 11.1 and 11.2). In contrast molecules extracted from crude extracts which the SI was weak can nevertheless show their own and promising SI. When ethnopharmacology is used to isolate new molecules as active principle, the SI value of the initial crude extracts has poor predictive value. That is why according to the goal of ethnopharmacology approach chosen the cut-off for the SI values will be really different for crude extracts used as such and for purified molecules in a drug candidate context.

Leishmaniasis

Human leishmaniasis is caused by obligate intracellular parasite eukaryote of the genus *Leishmania* (order Kinetoplastida); about 21 of the 30 species that infect mammals infect human beings. Human leishmaniasis is transmitted by the bite of infected female phlebotomine sand flies that become infected by taking a blood meal from an infected mammalian host. As for all tropical diseases, traditional medicine treatments of *Leishmania* are widely used in endemic zones, and studied (Iwu *et al.*, 1994; Anthony *et al.*, 2005). The examination of the crude extracts of plant species reported to be used in traditional medicine in Sub-Saharan Africa, and showing notable *in vitro* anti-*Leishmania* activities are itemized in Table 11.1. From the 32 publications found in this topic, a study reported the *in vitro* antileishmanial activity of plants based on the ethnomedical knowledge of local population and showed that *Annona muricata* proved to have good activity with a minimum effective concentration (MEC) value at 0.1 mg/ml whereas the reference substance meglumine antimoniate had MEC at 4.1 mg/ml (Jaramillo *et al.*, 2000). However, according to our criteria, this plant did not appear as the most promising. Indeed, its SI defined as the ratio of the cytotoxicity on cell-lines to the *in vitro* anti-*Leishmania* activity was only of 1 whereas *Maesa balansae*, *Vernonia polyanthes*, *Symphonia globuliferia*, and *Harungana madagascarensis* showed SI values of upper 80, 62, 24, and 17 respectively. The 3 families of plant Annonaceae, Meliaceae and Clusiaceae widely spread in endemic areas, contained the highest numbers of plants with promising *in vitro* antileishmanial activity (Table 11.1). A review done in 2005 inventoried one hundred plants which

Table 11.1: Plant extracts with *in vitro* antileishmanial activity: IC_{50} below 100 µg/ml and/or SI superior to 1.

Plants	*Extract*	*Species (Forms)*	*Cytotoxicity: IC_{50} in µg/ml (Cell-lines used)*	*In vitro Antileishmanial Activities : IC_{50} (µg/ml)*	*Security Index (Cytotoxicity/ activity ratio)*	*Countries of the Plant Extract*	*References*
Maesa balansae Mez. (Myrsinaceae)	Methanol	*L. donovani* (*amastigotes*)	>32µg/ml (MRC5)	0.4 µg/ml	>80	Vietnam	(Maes *et al.*, 2004)
Vernonia polyanthes Less. (Asteraceae)	Methanol	*L. amazonensis* (*promastigotes*)	250 µg/ml (J774)*	4 µg/ml*	62.5	Brazil	(Braga *et al.*, 2007)
Symphonia globulifera L.f. (Clusiaceae)	Ethyl acetate	*L. donovani* (*amastigotes*)	52.3 µg/ml (L6)	2.1 µg/ml	24.9	Cameroon	(Ndjakou Lenta *et al.*, 2007)
Harungana madagascariensis Lam. ex Poir. (Clusiaceae)	Ethyl acetate	*L. donovani* (*amastigotes*)	28.1 µg/ml (L6)	1.6 µg/ml	17.6	Cameroon	(Ndjakou Lenta *et al.*, 2007)
Pseudocedrela kotschyi (Schweinf.) Harms (Meliaceae)	Dichloromethane	*L. donovani* (*axenic*)	65.44 µg/ml (L6)	7.36 µg/ml	8.89	Mali	(Hay *et al.*, 2007)
Polyalthia suaveolens Engl. and Diels. (Annonaceae)	Dichloromethane	*L. infantum* (*amastigotes*)	4.4 µg/ml (TPH1)	1.8 µg/ml	2.44	Gabon	(Lamidi *et al.*, 2005)
Garcinia lucida Vesque (Clusiaceae)	Dichloromethane/ methanol	*L. donovani* (*axenic amastigotes*)	>50 µg/ml (Vero)	50 µg/ml	>1	Cameroon	(Fotie *et al.*, 2007)
Albizia zygia (DC.) Macbr. (Mimosaceae)	Ethyl acetate	*L. donovani* (*amastigotes*)	4.5 µg/ml (L6)	3.5 µg/ml	1.3	Cameroon	(Ndjakou Lenta *et al.*, 2007)
Annona muricata L. (Annonaceae)	Ethyl acetate	*L. braziliensis* (*promastigotes*)	100 µg/ml** (U937)	100 µg/ml**	1	Colombia	(Jaramillo *et al.*, 2000)

Contd...

Table 11.1–*Contd...*

Plants	*Extract*	*Species (Forms)*	*Cytotoxicity: IC_{50} in µg/ml (Cell-lines used)*	*In vitro Antileishmanial Activities : IC_{50} (µg/ml)*	*Security Index (Cyto-toxicity/activity ratio)*	*Countries of the Plant Extract*	*References*
Hymenocardia acida Tul. (Hymenocardiaceae)	Dichloromethane	*L. mexicana* (*promastigotes*)	19-56 µg/ml* (J774)	19-56 µg/ml*	1	Benin	(Hoet *et al.*, 2004a)
Combretum cuspidatum Planch. ex Benth. (Combretaceae)	Dichloromethane	*L. infantum* (*amastigotes*)	25.3 µg/ml (TPH1)	28.6 µg/ml	0.88	Gabon	(Lamidi *et al.*, 2005)
Ganophyllum giganteum (Chev.) Hauman (Sapindaceae)	Dichloromethane	*L. infantum* (*amastigotes*)	11.2 µg/ml (TPH1)	15.6 µg/ml	0.72	Gabon	(Lamidi *et al.*, 2005)
Augouardia letestui Pellegrin (Fabaceae)	Dichloromethane	*L. infantum* (*amastigotes*)	8 µg/ml (TPH1)	62.1 µg/ml	0.13	Gabon	(Lamidi *et al.*, 2005)
Cola lizae N.Hallé (Sterculaceae)	Ethanol	*L. infantum* (*amastigotes*)	2.5 µg/ml (TPH1)	35.4 µg/ml	0.07	Gabon	(Lamidi *et al.*, 2005)
Azadiracta indica A. Juss (Meliaceae)	Methanol	*L. major* (*promastigotes*)	ND	10.2 µg/ml	ND	Sudan	(Fatima *et al.*, 2005)
Khaya senegalensis (Deser.) A. Juss (Meliaceae)	Methanol	*L. donovani* (*amastigotes*)	ND	9.8 µg/ml	ND	Guinea-Bissau	(Abreu *et al.*, 1999)
Anthostema senegalense A. Juss. (Euphorbiaceae)	Methanol	*L. donovani* (*amastigotes*)	ND	9.1 µg/ml	ND	Guinea-Bissau	(Abreu *et al.*, 1999)
Guarea kunthiana A. Juss (Meliaceae)	Hexane	*L. donovani* (*promastigotes*)	ND	7.9 µg/ml	ND	Brazil	(Mesquita *et al.*, 2005)

Contd...

Table 11.1–*Contd...*

Plants	*Extract*	*Species (Forms)*	*Cytotoxicity: IC_{50} in µg/ml (Cell-lines used)*	*In vitro Antileishmanial Activities : IC_{50} (µg/ml)*	*Security Index (Cytotoxicity/ activity ratio)*	*Countries of the Plant Extract*	*References*
Himatanthus succuba (Spruce ex. Müll.Arg) (Apocynaceae)	Ethanol	*L. infantum* (*amastigotes*)	ND	5 µg/ml	ND	Peru	(Castillo *et al.*, 2007)
Annona crassiflora Mart. (Annonaceae)	Ethanol	*L. donovani* (*promastigotes*)	ND	3.7 µg/ml	ND	Brazil	(Mesquita *et al.*, 2005)
Psychotria klugii Standl. (Rubiaceae)	Ethanol	*L. donovani* (*promastigotes*)	ND	3.3 µg/ml	ND	Peru	(Muhammad *et al.*, 2003)
Annona foetida Martius. (Annonaceae)	Dichloromethane	*L. guyanensis* (*promastigotes*)	ND	2 µg/ml	ND	Brazil	(Costa *et al.*, 2006)

*: MIC: Minimal Inhibition Concentration; **MEC: Minimal Effective Concentration.

Table 11.2: Plant extracts from Sub-Saharan Africa with *in vitro* anti-*Trypanosoma brucei* activity: IC_{50} below 10 μg/ml and/or SI superior to 1.

Plants	*Extract*	*Species*	*Cytotoxicity (Cell-lines)*	*In vitro Antitrypanosomal Activities: IC_{50} (μg/ml)*	*Security Index (Cytotoxicity/activity ratio)*	*Countries of the Plant Extract*	*References*
Alchornea cordifolia (Schumach. and Thonn.) (Euphorbiaceae)	Methanol	*T. brucei*	>64 μg/ml (MRC5)	0.7 μg/ml	> 92	DRC	(Mesia *et al.*, 2008)
Albizia zygia (DC.) Macbr. (Fabaceae)	Ethyl acetate	*T. brucei*	4.5 μg/ml (L6)	0.2 μg/ml	22.5	Cameroon	(Ndjakou Lenta *et al.*, 2007)
Strychnos spinosa Lam. (Loganiaceae)	Dichloromethane	*T. brucei*	25.9 μg/ml (L6)	1.5 μg/ml	17.3	Benin	(Hoet *et al.*, 2004a)
Garcinia lucida Vesque. (Clusiaceae)	Dichloromethane/ Methanol	*T. brucei*	>50 μg/ml (Vero)	4.9 μg/ml	>10	Cameroon	(Fotie *et al.*, 2007)
Alstonia macrophylla Wall. (Apocynaceae)	Methanol	*T. brucei*	49.8 μg/ml (KB)	7.8 μg/ml	6.38	Nigeria	(Camacho *et al.*, 2003)
Pseudocedrela kotschyi Schweinf. (Meliaceae)	Dichloromethane	*T. brucei*	65.44 μg/ml (L6)	8.89 μg/ml	7.36	Mali	(Hay *et al.*, 2007)
Hymenocardia acida Tul. (Euphorbiaceae)	Dichloromethane	*T. brucei*	12.2 μg/ml (L6)	5 μg/ml	2.4	Benin	(Hoet *et al.*, 2004b)
Trichilia emetica Vahl. (Meliaceae)	Dichloromethane	*T. brucei*	8.9 μg/ml (L6)	8.6 μg/ml	1	Benin	(Hoet *et al.*, 2004b)
Terminalia avicenoides Gull and Parr. (Combretaceae)	Water	*T. brucei*	ND	10 μg/ml	ND	Mali	(Bizimana *et al.*, 2006)
Acacia nilotica (L.) Willd. ex Del. (Mimosaceae)	Water	*T. brucei*	ND	5 μg/ml*	ND	Mali	(Bizimana *et al.*, 2006)
Cussonia zimmermannii Harms. (Araliaceae)	Ether of petroleum	*T. brucei*	ND	4.8 μg/ml	ND	Tanzania	(Senn *et al.*, 2007)
Afzelia africana Smith ex Persoon (Leguminaceae)	Methanol	*T. brucei*	ND	1 μg/ml*	ND	Mali	(Bizimana *et al.*, 2006)

*: MIC: Minimal Inhibition Concentration.

extracts showed antileishmanial activity (Rocha *et al.*, 2005). Surprisingly only 3 of these plants (*Annona muricata, Azadirachta indica* and *Khaya senegalensis*) were selected by our criteria as plants of interests because of their pharmacological properties.

Much *in vivo* anti-*Leishmania* activity concerned Asia and South America rather than Africa. An interesting example is the case of the *Aloe vera* leaf from India in a *L. donovani* BALB-C mouse model, where administration by the oral or subcutaneous route (15 mg/kg/d body weight x 5 days) reduced parasitemia by > 90 per cent (Dutta *et al.*, 2008).

African Trypanosomiasis

We have deliberately limited our report to ethnopharmacological studies focusing on African trypanosomiasis which is a vector-borne parasitic disease that is causing major health and economic problems in rural sub-Saharan Africa (and to exclude *T. cruzi*). The etiologic agents of the disease are flagellated eukaryotes *Trypanosoma brucei rhodesiense* and *T. brucei gambiense* that cause human sleeping sickness. The parasites are transmitted between vertebrate hosts by the tsetse fly (*Glossina* spp). Publications reporting the activity of plant extracts on African trypanosomes are not abundant compared with other parasite eukaryote diseases such as malaria. As *T. brucei* is only endemic in Africa, we were interested only in plants used in traditional medicine from Sub-Saharan Africa (Table 11.2). From the 38 studies on plants used against *Trypanosoma* published these last 10 years, only 12 plants can be selected according to our criteria. The most interesting SI values of crude extracts concerned *Alchornea cordifolia, Albizia zygia* and *Strychnos spinosa* with values of more than 92, 22 and 17, respectively. Meliaceae and Euphorbiaceae seem to be the families of plants containing the most of species with antitrypanosomial activity (Table 11.2).

Antitrypanosomial studies were also evaluated *in vivo* with infected mice treated by plant extracts. *Terminalia avicennoides* significantly reduced the parasitemia in animals orally and intraperitoneally treated with 150 mg/kg/d (Bizimana *et al.*, 2006). Additionally *in vivo* activity against *T. brucei* of *Annona senegalensis* extracts showed a 50 per cent suppression of parasitemia at 200 mg/kg/d (Ogbadoyi *et al.*, 2007).

Malaria

Malaria in humans is caused by parasite eukaryote of the genus *Plasmodium*, which is transmitted through bites by female mosquitoes of the genus *Anopheles*. Five species, namely *P. falciparum, P. vivax, P. ovale P. malariae*, and very recently *P. knowlesi*, are known to cause malaria in humans. However, the most severe malaria, with about 90 per cent of malaria deaths, is caused by *P. falciparum*, which is the predominant parasite species in Africa.

The publication of O'Neill (O'Neill *et al.*, 1985) was among the first to study plant extracts in an *in vitro* assay against *P. falciparum* in the search for new antimalarial drugs. A large survey of about 600 species of higher plants representing more than 120 families were tested in 1947 but against avian malaria (Spencer *et al.*, 1947). Since then, many other studies researching antiplasmodial molecules in plant extracts have been carried out in all malaria endemic areas. A recent review analyzing ethnopharmacological publications describing research into antimalarial treatments in West Africa, showed that from 610 crude extracts from plants checked and tested *in vitro* for their antiplasmodial activities, only 15 per cent of these extracts really showed very good antiplasmodial activity ($IC_{50} < 5$ µg/ml) (Soh and Benoit-Vical, 2007). Among them, nine plant species had interesting SI, greater than 25, namely: *Anogeissus leiocarpus* (Vonthron-Senecheau *et al.*, 2003), *Azadiracta indica* (Benoit *et al.*, 1996; Isah *et al.*, 2003), *Cochlospermum planchonii* (Benoit *et al.*, 1995; Benoit-Vical *et al.*, 1999; Benoit-Vical *et al.*, 2003b), *Cochlospermum tinctorium* (Ballin *et al.*, 2002), *Guiera senegalensis* (Fiot *et al.*, 2006), *Nauclea latifolia* (Benoit-Vical *et al.*, 1998; Zirihi *et al.*, 2005), *Rauvolfia vomitoria* (Zirihi *et al.*, 2005), *Terminalia schimperiana* (Kamanzi Atindehou *et al.*, 2004) and *Vernonia colorata* (Benoit *et al.*, 1996). These plants belong to four families (Combretaceae, Euphorbiaceae, Meliaceae and Rubiaceae) and are used against fever and/or malaria by the populations of the three continents: Africa, South America and Asia (Willcox and Bodeker, 2004). For example the Meliaceae *Azadiracta indica* shows a wide traditional use confirmed by interesting *in vitro* and *in vivo* antiplasmodial activity of extracts coming from various endemic areas very far such as Sudan (El-Tahir *et al.*, 1999), Ivory Coast (Benoit *et al.*, 1996), India (Dhar, 1998) and Thailand (Rochanakij *et al.*, 1985). Its large distribution is confirmed by recent ethnopharmacological studies reporting, for this plant, numerous citations from indigenous households in Ghana (Asase *et al.*, 2010) and in Kenya (Nguta *et al.*, 2010).

In summary, many crude extracts from plants are daily used all over the word to treat parasitic diseases and a lot of works reported their biological activities but often without enough data to be comparable. From all these studies, few crude extracts can really show both good biological activities and a weak toxicity.

Active Principles Extracted from Bio-guided Fractionation

In a drug-candidate approach, when biological activities of some plant extracts, widely used as traditional antiparasitic treatment by population, have been confirmed by laboratory investigations, it is interesting to go on the exploration to find new active compounds. This research way denotes that the higher structural diversity of natural products should be used in addition to libraries of synthetic compounds formed by combinatorial chemistry. Indeed, about 40 per cent of the chemical skeletons found in published databases of natural products could not be found in libraries of synthetic chemicals (Benoit-Vical, 2005). Unfortunately, one of the major problems is the complexity of the crude extracts as it is necessary to identify the compound(s) responsible for the biological activity with a bio-guided process. Natural product drug discovery is therefore slow and expensive, requiring numerous investigations to conduct chemical fractionations and evaluate biological activities of isolated compounds (Benoit-Vical, 2005). However, a renewed interest in natural product research is currently in progress (Butler, 2004) because of inadequate number of lead compounds for antiparasitic diseases and consequently of an important chemical space occupied by natural products.

In this way to find new antiparasitic drugs candidates, collaborative works grouping chemists and biologists can carry out the pharmaco-modulation of hits isolated from plants to reduce the cytotoxicity and improve the biological activity.

Leishmaniasis

The present review shows that despite the wide use of traditional medicine in endemic areas, relatively few plants (only 22 according to our criteria) presenting real *in vitro* antileishmanial potentialities of their crude extracts were reported (Table 11.1). On the contrary, many molecules obtained from bio-guided fractionation of these extracts showed interesting antileishmanial activity (for review, see Rocha *et al.*, 2005) with weak toxicity (SI > 15) as alkaloids (Bringmann *et al.*, 2008; Di Giorgio *et al.*, 2006; Fotie *et al.*, 2007) and limonoids (Krief *et al.*, 2006), compared with the pentamidine drug control (SI ≥ 10) (Fotie *et al.*, 2007). In this way, the dichloromethane extract of *Pseudocedrela kotschyi* (Meliaceae) root from Malian traditional medicine had been selected for its anti-*Leishmania* activity. The phytochemical investigations of this active extract yielded three limonoid orthoacetates. From these, one limonoid, the 7-deacetyl-7-oxogedunin (Figure 11.1a) exhibited good *in vitro* activities against both *Leishmania donovani* and *Trypanosoma rhodesiense* with SI superior to 90 and to 26 respectively (Hay *et al.*, 2007).

A plant program for potential leishmanicides was initiated for the first time in 1984 in French Guinea and South America. Fourteen plants used topically in folk medicine were collected in Bolivia and in the rain forest occupied by Chimane Indians. *In vitro* tests on the leishmanicidal activity of plants showed that *Galipea longiflora* was the most active against *Leishmania amazonensis* with an IC_{50} of 100 µg/ml. But the most interest of this work was the activity of this guided-fractionation extract by chromatography affording thirteen active compounds identified as 2-substituted quinoline alkaloids. An international patent was filed in 1991 by the IRD (Institut National de Recherche pour le Développement) in collaboration with traditional healers, on molecules isolated from these studied plants (Fournet *et al.*, 1994; Sauvain *et al.*, 1994).

Some chemically defined molecules such as isotetrandrine (Fournet *et al.*, 1993) or argentilactone (Waechter *et al.*, 1997) were also tested against *Leishmania*-infected mice and showed the same efficacy as the reference drug.

African Trypanosomiasis

Bioactive-guided fractionation of interesting trypanocidal plant extracts yielded a series of active compounds with weak cytotoxicity (SI > 20) which included alkaloids (Bringmann *et al.*, 2008), quinolines (Ferreira *et al.*, 2006), terpenes (Hoet *et al.*, 2007), coumarines (del Rayo Camacho *et al.*, 2004) and flavonoids (Sulsen *et al.*, 2007) as well. Fractionations of the antitrypanosomal leaf extract from *Strychnos spinosa* (Loganiaceae), led to the isolation of eight triterpenoids and sterols from which saringosterol (Figure 11.1b, IC_{50} of 3.3 µg/ml) was found to possess the best ratio *in vitro* antitrypanosomal and weak toxicity with SI superior to 30 (Hoet *et al.*, 2007).

Malaria

A lot of studies report antimalarial compounds isolated from plants (for review, see Kaur *et al.*, 2009). Studies on medicinal plants from Ivory Coast showed that the ethanol extract of *Sida acuta* Burm (Malvaceae) permitted the purification of cryptolepine (Figure 11.1c, IC_{50} of 0.2-0.6 µM) as the active antiplasmodial constituent of this plant (Banzouzi *et al.*, 2004). From these results, a series of crytolepine derivatives were synthesized and permitted an increase of the selectivity ratio (Lavrado *et al.*, 2008). Other molecules, like ellagic acid (Figure 11.1e, IC_{50} of 0.26-0.56 µg/mL) from *Alchornea cordifolia* (Euphorbiaceae) (Banzouzi *et al.*, 2002), gedunin (Figure 11.1d, IC_{50} of 0.02-0.04 µg/mL) from *Azadiracta indica* (Meliaceae) (MacKinnon *et al.*, 1997), harman (Figure 11.1f, IC_{50} of 1.3-2.2 µg/mL) and tetrahydroharman (Figure 11.1g, IC_{50} of 1.4-3.9 µg/mL) from *Guiera senegalensis* (Combretaceae) (Ancolio *et al.*, 2002) were isolated from some of selected antiplasmodial plants and are currently under complementary pharmacological investigations according to their promising malaria properties.

Among them, ellagic acid appears as one of the most promising. This active principle is a natural compound extracted from several plants as *Alchornea cordifolia* or *Melia composita* (Meliaceae) but also from fruits and vegetables and showed a very high activity against malaria parasites *in vitro* (IC_{50} values ranging from 105 to 330 nM according to the *P. falciparum* strain used corresponding to selectivity indices superior to 495) and *in vivo* (ED_{50} by the intraperitoneal route inferior to 1 mg/kg/d) without any toxicity (Soh *et al.*, 2009). However, with a weak oral bioavailability, an improvement in the oral antimalarial efficacy seems necessary before that this molecule becomes a new antimalarial drug candidate (Soh *et al.*, 2009). Very recently, simalikalactone E (Figure 11.1h, IC_{50} of 0.024-0.068 µM), a new quassinoid from *Quassia amara* L. (Simaroubaceae) showed both *in vitro* antiplasmodial activity and promising efficacy against murine *Plasmodium*-infected mice but also activity against gametocyte forms (Cachet *et al.*, 2009).

Antiplasmodial properties of febrifugine (Figure 11.1i, IC_{50} of 0.76-1.9 ng/mL and *in vivo* ED_{50} of 2.3 mg/kg/d), an alkaloid obtained from *Dichroa febrifuga* lour. is studied since the 1950s. However, febrifugine shown adverse side effects and even strong liver toxicity needing, to decrease this toxicity, many SAR (structure activity relationships) studies reported in more than 40 publications on this topic. From febrifugine chemical structure, the pharmaco-modulation assays permitted to increase for at least 40-fold the security and therapeutic indexes (TI: 128) and thus to obtain potent antimalarial analogs (Jiang *et al.* 2005; Zhu *et al.*, 2009).

Finally, it is interesting to note that the family of plant Meliaceae appears particularly interesting since gathers plants which crude extracts possess antileishmanial properties but also antitrypanosomial and antimalarial activities as demonstrated in chapter 3.2. This data is confirmed by the high number of active principles coming from the Meliaceae family like limonoids (Krief *et al.*, 2004; Maneerat *et al.*, 2008), limonoid peroxide (Yin *et al.*, 2007), triterpenoids (Murata *et al.*, 2008) (Benoit-Vical *et al.*, 2003a), flavonoids (Ambrozin *et al.*, 2004; Kamanzi Atindehou *et al.*, 2004) and polyphenols (Soh *et al.*, 2009), and that seem have real antiparasitic properties. A particular attention should be carried in the plants of this family that are used in traditional medicine and not yet studied since they appear as an interesting pipeline for new antiparasitic compounds.

Ethnopharmacology and Clinical Trials on Malaria, African Trypanosomiasis and Leishmaniasis

In 2002, WHO launched a strategy on traditional medicine to help countries to explore its potential for people's health and well-being, while minimizing the risks of

Antileishmanial compound

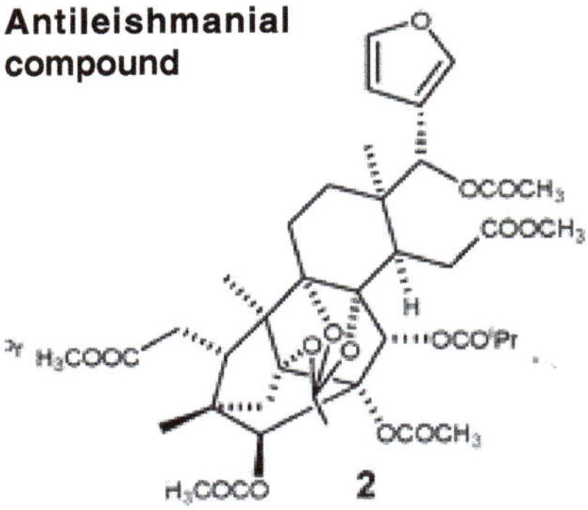

a. 7-deacetyl-7-oxogedunin

Antitrypanosomal compound

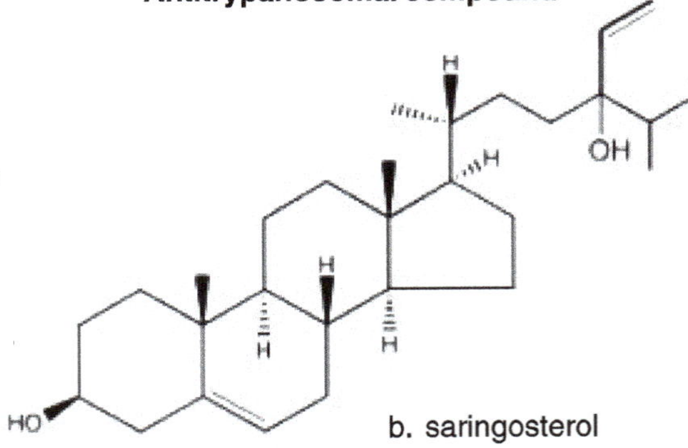

b. saringosterol

Antiplasmodial compound

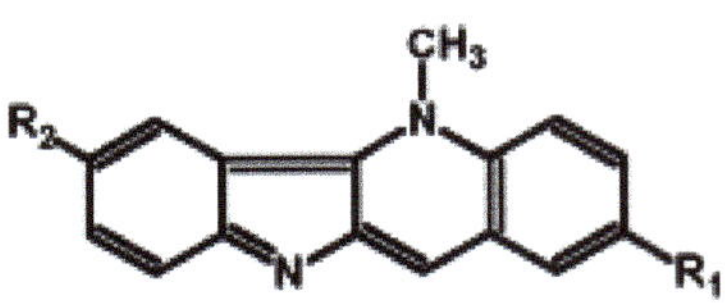

R1 = R2 = H,

c. cryptolepine

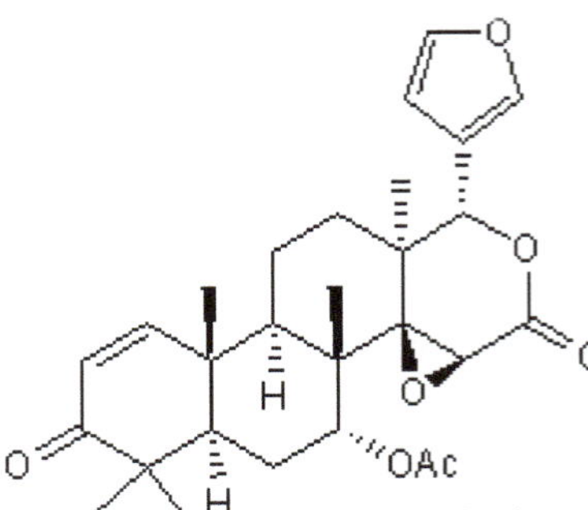

d. gedunin

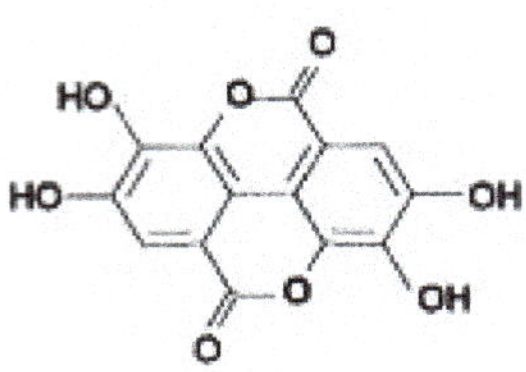

e. ellagic acid

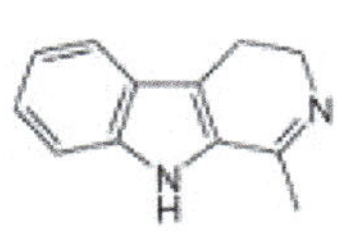

f. harman

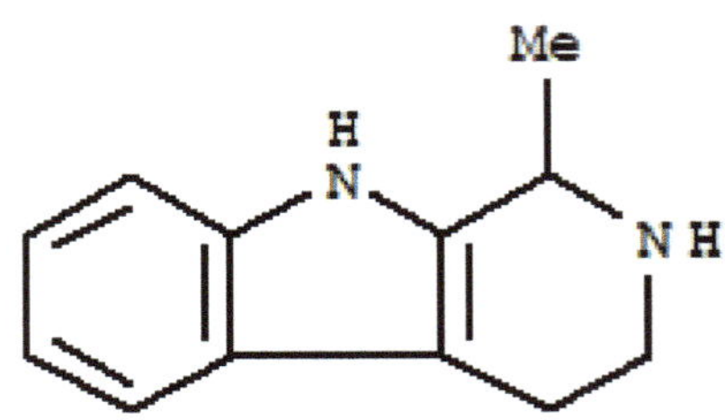

g. tetrahydroharman

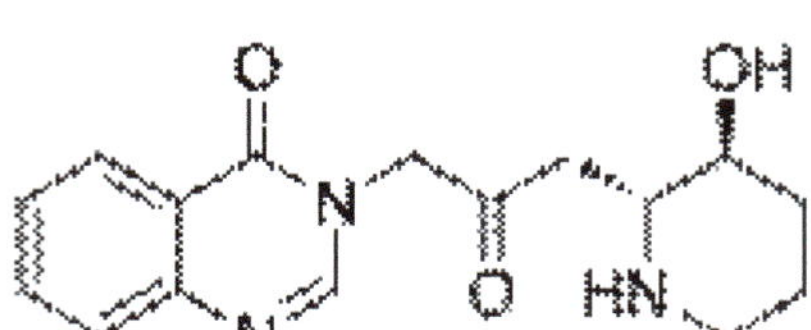

h. simalikalactone E

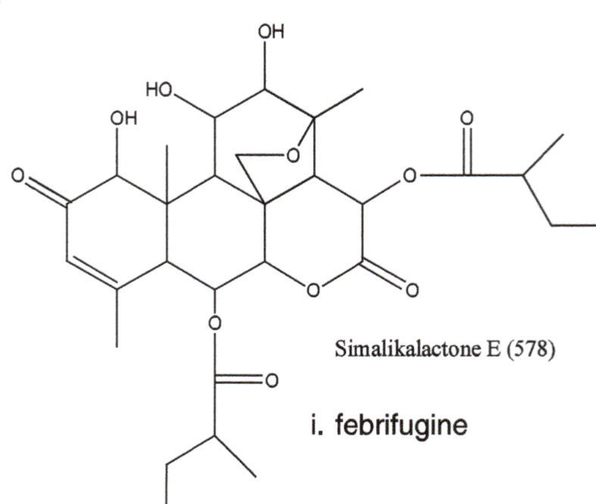

i. febrifugine

Figure 11.1: Molecules with antiparasitic activities isolated from plants.

unproven or misused remedies. Pharmacological investigations of plant extracts are thus needed to validate the efficacy and prevent the toxicity risks. All clinical trials with plant extracts pose the problem of ethic with studies in humans. However, as traditional medicines are already used daily by populations, the pharmacological investigations before clinical trials of the crude extracts traditionally used are lesser than for whatever other drug tested. The scope and design of such studies should be based on information on traditional use obtained from official national compendia and relevant literature, and/or by consultation with traditional medical practitioners (WHO, 2000). The absence of any reported or documented side-effects is not an absolute assurance of safety for herbal medicines. However, a full range of toxicological tests may not be necessary. Tests which examine effects that are difficult or even impossible to detect clinically should be encouraged (WHO, 2000). Only when there is no documentation of long historical use of an herbal medicine, or when doubts exist about its safety, additional toxicity studies should be performed (WHO, 2000). In this context, all *in vitro* and *in vivo* pharmacological assays for activity and toxicity investigations carried out before all clinical trials are thus an advantage.

To the best of our knowledge, no clinical trials on humans have been initiated to evaluate the effects on *Leishmania* or *Trypanosoma* of crude extracts or molecules isolated from plants even from extracts with promising *in vitro* and *in vivo* properties. In the case of malaria, very few authors have reported clinical trials to evaluate the efficacy of plants. No purified molecules, except of course quinine or artemisinin, extracted from plant fractions were studied in clinical trials. Only crude extracts of *Cochlospermum planchonii* (Benoit-Vical *et al.*, 2003b), *Artemisia annua* (Mueller *et al.*, 2004), "AM", an infusion of leaves from a protected Ugandan tree (Willcox, 1999), *Argemone mexicana* (Willcox *et al.*, 2007; Graz *et al.*, 2010;) and *Vernonia amygdalina* (Challand and Willcox, 2009) known to be routinely used in traditional medicine and that, for some of them, had previously shown high activities *in vitro* and/or *in vivo* on murine models have been tested on humans. Even if some recrudescence occurred (Mueller *et al.*, 2004), all these plant extracts, often used by populations as a first treatment when access to other antimalarials is delayed, appeared effective in human malaria treatment in these conditions. These first studies confirm the real interest as a future basis for research of these plants to discover new candidates in the arsenal of drugs, and particularly to fight malaria. However, most of these studies were carried out in adult patients that probably had some degree of immunity to malaria and these results cannot be extrapolated to young children in whom immunity is less developed. More clinical investigations would be thus needed to investigate, in health centers with regimens highly supervised, the efficacy of these treatments in patients little or not immune like children or pregnant women.

Conclusions

Due to the development of drug resistance and the diminishing resources within health systems, the control of parasitic diseases of humans is an urgent priority. In the case of the treatment of malaria, African trypanosomiasis and leishmaniasis, few drugs are available. As stated in the Roll Back Malaria project, a new drug needs to be discovered every 5 years. In Asia and Africa, 80 per cent of the population always uses plant medicines in primary healthcare (WHO, 2008). Scientific and clinical researches to investigate the quality, safety and efficacy of these herbal therapies need thus to be reinforced. The field of anti-*Leishmania*, anti-*Trypanosoma* and natural substances is still a marginal research theme. The investigation of the antiparasitic properties of plants, in particular from Sub-Saharan Africa showed that some families of plants and in particular Meliaceae are widely used to treat these diseases, and whatever the localization of the endemic regions, they have provided interesting antiparasitic activity in their crude extracts. Also a wide variety of molecules, such as alkaloids, limonoids and terpenoids found in these plants, exhibited interesting pharmacological properties and could become promising drug candidates to increase the low number of current antiparasitic molecules. The use of the tools of modern science to evaluate traditional medicine could provide some solutions to the many problems of treating neglected parasitic diseases.

References

Abreu, P. M., Martins, E. S., Kayser, O., Bindseil, K. U., Siems, K., Seemann, A., Frevert, J. (1999). Antimicrobial, antitumor and antileishmania screening of medicinal plants from Guinea-Bissau. *Phytomedicine*, **6:** 187-195.

Ambrozin, A. R., Vieira, P. C., Fernandes, J. B., da Silva, M. F., de Albuquerque, S. (2004). Trypanocidal activity of Meliaceae and Rutaceae plant extracts. *Memorias do Instituto Oswaldo Cruz.*, **99:** 227-231.

Ancolio, C., Azas, N., Mahiou, V., Ollivier, E., Di Giorgio, C., Keita, A., Timon-David, P., Balansard, G. (2002). Antimalarial activity of extracts and alkaloids isolated from six plants used in traditional medicine in Mali and Sao Tome. *Phytotherapy Research*, **16**: 646-649.

Anthony, J. P., Fyfe, L., Smith, H. (2005). Plant active components - a resource for antiparasitic agents? *Trends Parasitology*, **21:** 462-468.

Asase, A., Akwetey, G. A., Achel, D. G. (2010). Ethnopharmacological use of herbal remedies for the treatment of malaria in the Dangme West District of Ghana. *Journal of ethnopharmacology*, **129**: 367-376.

Ballin, N. Z., Traore, M., Tinto, H., Sittie, A., Molgaard, P., Olsen, C. E., Kharazmi, A., Christensen, S. B. (2002). Antiplasmodial compounds from *Cochlospermum tinctorium*. *Journal of natural products*, **65:** 1325-1327.

Banzouzi, J. T., Prado, R., Menan, H., Valentin, A., Roumestan, C., Mallie, M., Pelissier, Y., Blache, Y. (2002). *In vitro* antiplasmodial activity of extracts of *Alchornea cordifolia* and identification of an active constituent: ellagic acid. *Journal of Ethnopharmacology*, **81:** 399-401.

Banzouzi, J. T., Prado, R., Menan, H., Valentin, A., Roumestan, C., Mallie, M., Pelissier, Y., Blache, Y. (2004). Studies on medicinal plants of Ivory Coast: investigation of *Sida acuta* for *in vitro* antiplasmodial activities and identification of an active constituent. *Phytomedicine*, **11**: 338-341.

Benoit-Vical, F. (2005). Ethnomedicine in malaria treatment. *IDrugs*, **8:** 45-52.

Benoit-Vical, F., Imbert, C., Bonfils, J. P., Sauvaire, Y. (2003a). Antiplasmodial and antifungal activities of iridal, a plant triterpenoid. *Phytochemistry*, **62**: 747-751.

Benoit-Vical, F., Lelievre, J., Berry, A., Deymier, C., Dechy-Cabaret, O., Cazelles, J., Loup, C., Robert, A., Magnaval, J., Meunier, B. (2007). Trioxaquines: new antimalarial agents active on all erythrocytic forms including gametocytes. *Antimicrobial Agents and Chemotherapy*, **51**: 1463-1472.

Benoit-Vical, F., Valentin, A., Cournac, V., Pelissier, Y., Mallie, M., Bastide, J. M. (1998). *In vitro* antiplasmodial activity of stem and root extracts of *Nauclea latifolia* S.M. (Rubiaceae). *Journal of Ethnopharmacology*, **61**: 173-178.

Benoit-Vical, F., Valentin, A., Da, B., Dakuyo, Z., Descamps, L., Mallie, M. (2003b). N'Dribala (*Cochlospermum planchonii*) versus chloroquine for treatment of uncomplicated *Plasmodium falciparum* malaria. *Journal of Ethnopharmacology*, **89**: 111-114.

Benoit-Vical, F., Valentin, A., Mallie, M., Bastide, J. M., Bessiere, J. M. (1999). *In vitro* antimalarial activity and cytotoxicity of *Cochlospermum tinctorium* and *C. planchonii* leaf extracts and essential oils. *Planta Medica*, **65**: 378-381.

Benoit, F., Valentin, A., Pelissier, Y., Diafouka, F., Marion, C., Kone-Bamba, D., Kone, M., Mallie, M., Yapo, A., Bastide, J. M. (1996). *In vitro* antimalarial activity of vegetal extracts used in West African traditional medicine. *American Journal of Tropical Medicine and Hygyene*, **54**: 67-71.

Benoit, F., Valentin, A., Pelissier, Y., Marion, C., Dakuyo, Z., Mallie, M., Bastide, J. M. (1995). Antimalarial activity *in vitro* of *Cochlospermum tinctorium* tubercle extracts. *Transactions of the Royal Society of Tropical Medicine and Hygiene*, **89**: 217-218.

Bizimana, N., Tietjen, U., Zessin, K. H., Diallo, D., Djibril, C., Melzig, M. F., Clausen, P. H. (2006). Evaluation of medicinal plants from Mali for their *in vitro* and *in vivo* trypanocidal activity. *Journal of Ethnopharmacology*,**103**: 350-356.

Bourdy, G., Willcox, M. L., Ginsburg, H., Rasoanaivo, P., Graz, B., Deharo, E. (2008). Ethnopharmacology and malaria: New hypothetical leads or old efficient antimalarials. *International Journal Parasitology*, **38**: 33-41.

Braga, F. G., Bouzada, M. L., Fabri, R. L., de, O. M. M., Moreira, F. O., Scio, E., Coimbra, E. S. (2007). Antileishmanial and antifungal activity of plants used in traditional medicine in Brazil. *Journal of Ethnopharmacology*, **111**: 396-402.

Bringmann, G., Brun, R., Kaiser, M., Neumann, S. (2008). Synthesis and antiprotozoal activities of simplified analogs of naphthylisoquinoline alkaloids. *European Journal of Medicinal Chemistry*, **43**: 32-42.

Brun, R., Lun, Z. R. (1994). Drug sensitivity of Chinese *Trypanosoma evansi* and *Trypanosoma equiperdum* isolates. *Veterinary Parasitology*, **52**: 37-46.

Butler, M. S. (2004). The role of natural product chemistry in drug discovery. *Journal of natural products*, **67**: 2141-2153.

Cachet, N., Hoakwie, F., Bertani, S., Bourdy, G., Deharo, E., Stien, D., Houel, E., Gornitzka, H., Fillaux, J., Chevalley, S., Valentin, A., Jullian, V. (2009).

Antimalarial activity of simalikalactone E, a new quassinoid from *Quassia amara* L. (Simaroubaceae). *Antimicrobial Agents and Chemotherapy*, **53**: 4393-4398.

Camacho, M. R., Phillipson, J. D., Croft, S. L., Solis, P. N., Marshall, S. J., Ghazanfar, S. A. (2003). Screening of plant extracts for antiprotozoal and cytotoxic activities. *Journal of Ethnopharmacology*, **89**: 185-191.

Casero, R. A., Jr., Porter, C. W., Bernacki, R. J. (1982). Activity of tunicamycin against *Trypanosoma brucei in vitro* and *in vivo*. *Antimicrobial Agents and Chemotherapy*, **22**: 1008-1011.

Castillo, D., Arevalo, J., Herrera, F., Ruiz, C., Rojas, R., Rengifo, E., Vaisberg, A., Lock, O., Lemesre, J. L., Gornitzka, H., Sauvain, M. (2007). Spirolactone iridoids might be responsible for the antileishmanial activity of a Peruvian traditional remedy made with *Himatanthus sucuuba* (Apocynaceae). *Journal of Ethnopharmacology*, **112**: 410-414.

Challand, S., Willcox, M. (2009). A clinical trial of the traditional medicine *Vernonia amygdalina* in the treatment of uncomplicated malaria. *Journal of alternative and complementary medicine*, 1231-1237.

Costa, E. V., Pinheiro, M. L., Xavier, C. M., Silva, J. R., Amaral, A. C., Souza, A. D., Barison, A., Campos, F. R., Ferreira, A. G., Machado, G. M., Leon, L. L. (2006). A pyrimidine-beta-carboline and other alkaloids from *Annona foetida* with antileishmanial activity. *Journal of natural products*, **69**: 292-294.

Croft, S. L., Barrett, M. P., Urbina, J. A. (2005). Chemotherapy of trypanosomiases and leishmaniasis. *Trends Parasitology*, **21**: 508-512.

de Moura, T. R., Novais, F. O., Oliveira, F., Clarencio, J., Noronha, A., Barral, A., Brodskyn, C., de Oliveira, C. I. (2005). Toward a novel experimental model of infection to study American cutaneous leishmaniasis caused by *Leishmania braziliensis*. *Infection and Immunity*, **73**: 5827-5834.

del Rayo Camacho, M., Phillipson, J. D., Croft, S. L., Yardley, V., Solis, P. N. (2004). *In vitro* antiprotozoal and cytotoxic activities of some alkaloids, quinones, flavonoids, and coumarins. *Planta Medica*, **70**: 70-72.

den Boer, M., Davidson, R. N. (2006). Treatment options for visceral leishmaniasis. *Expert Review Anti-Infective Therapy*, **4**: 187-197.

Desjardins, R. E., Canfield, C. J., Haynes, J. D., Chulay, J. D. (1979). Quantitative assessment of antimalarial activity *in vitro* by a semiautomated microdilution technique. *Antimicrobial Agents and Chemotherapy*, **16**: 710-718.

Dhar, R. (1998). Inhibition of the growth and development of asexual and sexual stages of drug-sensitive and resistant strains of the human malaria parasite *Plasmodium falciparum* by Neem (*Azadiracta indica*) fractions. *Journal of Ethnopharmacology*, **61**: 31-39.

Di Giorgio, C., Lamidi, M., Delmas, F., Balansard, G., Ollivier, E. (2006). Antileishmanial activity of quinovic acid glycosides and cadambine acid isolated from *Nauclea diderrichii*. *Planta Medica*, **72**: 1396-1402.

Dutta, A., Sarkar, D., Gurib-Fakim, A., Mandal, C., Chatterjee, M. (2008). *In vitro* and *in vivo* activity of *Aloe vera* leaf exudate in experimental visceral leishmaniasis. *Parasitology Research*, **102**: 1235-1242.

El-On, J., Ozer, L., Gopas, J., Sneir, R., Golan-Goldhirsh, A. (2009). *Nuphar lutea: In vitro* anti-leishmanial activity against *Leishmania major* promastigotes and amastigotes. *Phytomedicine*, **16**: 788-92.

El-Sayed, N. M., Myler, P. J., Blandin, G., Berriman, M., Crabtree, J., Aggarwal, G., Caler, E., Renauld, H., Worthey, E. A., Hertz-Fowler, C., Ghedin, E., Peacock, C., Bartholomeu, D. C., Haas, B. J., Tran, A. N., Wortman, J. R., Alsmark, U. C., Angiuoli, S., Anupama, A., Badger, J., Bringaud, F., Cadag, E., Carlton, J. M., Cerqueira, G. C., Creasy, T., Delcher, A. L., Djikeng, A., Embley, T. M., Hauser, C., Ivens, A. C., Kummerfeld, S. K., Pereira-Leal, J. B., Nilsson, D., Peterson, J., Salzberg, S. L., Shallom, J., Silva, J. C., Sundaram, J., Westenberger, S., White, O., Melville, S. E., Donelson, J. E., Andersson, B., Stuart, K. D., Hall, N. (2005). Comparative genomics of trypanosomatid parasitic protozoa. *Science*, **309**: 404-409.

El-Tahir, A., Satti, G. M., Khalid, S. A. (1999). Antiplasmodial activity of selected sudanese medicinal plants with emphasis on *Acacia nilotica*. *Phytotherapy Research*, **13**: 474-478.

Elufioye, T. O., Alatise, O. I., Fakoya, F. A., Agbedahunsi, J. M., Houghton, P. J. (2009). Toxicity studies of *Tithonia diversifolia* A. Gray (Asteraceae) in rats. *Journal of Ethnopharmacology*, **122**: 410-415.

Fatima, F., Khalid, A., Nazar, N., Abdalla, M., Mohomed, H., Toum, A. M., Magzoub, M., Ali, M. S. (2005). *In vitro* assessment of anti - cutaneous leishmaniasis activity of some Sudanese plants. *Turkiye Parazitol. Derg*. **29**: 3-6.

Fennell, C. W., Lindsey, K. L., McGaw, L. J., Sparg, S. G., Stafford, G. I., Elgorashi, E. E., Grace, O. M., van Staden, J. (2004). Assessing African medicinal plants for efficacy and safety: pharmacological screening and toxicology. *Journal of Ethnopharmacology*, **94**: 205-217.

Ferreira, V. F., Jorqueira, A., Souza, A. M., da Silva, M. N., de Souza, M. C., Gouvea, R. M., Rodrigues, C. R., Pinto, A. V., Castro, H. C., Santos, D. O., Araujo, H. P., Bourguignon, S. C. (2006). Trypanocidal agents with low cytotoxicity to mammalian cell line: a comparison of the theoretical and biological features of lapachone derivatives. *Bioorganic and Medicinal Chemistry*, **14**: 5459-5466.

Fiot, J., Sanon, S., Azas, N., Mahiou, V., Jansen, O., Angenot, L., Balansard, G., Ollivier, E. (2006). Phytochemical and pharmacological study of roots and leaves of *Guiera senegalensis* J.F. Gmel (Combretaceae). *Journal of Ethnopharmacology*, **106**: 173-178.

Fotie, J., Bohle, D. S., Olivier, M., Adelaida Gomez, M., Nzimiro, S. (2007). Trypanocidal and antileishmanial dihydrochelerythrine derivatives from *Garcinia lucida*. *Journal of natural products*, **70**: 1650-1653.

Fournet, A., Barrios, A.A., Munoz, V., Hocquemiller, R., Cave, A. (1993). Effect of some bisbenzylisoquinoline alkaloid on American *Leishmania* sp. in BALB/c mice. *Phytotherapy Research*, **7**: 281–284.

Fournet A., Barrios A. A., Munoz V. (1994). Leishmanicidal and trypanocidal activities of Bolivian medicinal plants. *Journal of Ethnopharmacology*, **41**: 19-37.

Gachet M. S., Lecaro J. S., Kaiser M., Brun R., Navarrete H., Munoz R. A., Bauer R., Schühly W. (2010). Assessment of anti-protozoal activity of plants traditionally used in Ecuador in the treatment of leishmaniasis. *Journal of Ethnopharmacology*, **128**: 184-197.

Gatton, M. L., Martin, L. B., Cheng, Q. (2004). Evolution of resistance to sulfadoxine-pyrimethamine in *Plasmodium falciparum*. *Antimicrobial Agents and Chemotherapy*, **48**: 2116-2123.

Graz, B., Willcox, M. L., Diakite, C., Falquet, J., Dackuo, F., Sidibe, O., Giani, S., Diallo, D. (2010). *Argemone mexicana* decoction versus artesunate-amodiaquine for the management of malaria in Mali: policy and public-health implications. *Transactions of the Royal Society of Tropical Medicine and Hygiene*, **104**: 33-41.

Hay, A. E., Ioset, J. R., Ahua, K. M., Diallo, D., Brun, R., Hostettmann, K. (2007). Limonoid orthoacetates and antiprotozoal compounds from the roots of *Pseudocedrela kotschyi*. *Journal of natural products*, **70**: 9-13.

Heinrich, M., Edwards, S., Moerman, D. E., Leonti, M. (2009). Ethnopharmacological field studies: a critical assessment of their conceptual basis and methods. *Journal of Ethnopharmacology*, **124**: 1-17.

Hoet, S., Opperdoes, F., Brun, R., Adjakidjé, V., Quetin-Leclercq, J. (2004)a. *In vitro* antitrypanosomal activity of ethnopharmacologically selected Beninese plants. *Journal of Ethnopharmacology*, **91**: 37-42.

Hoet, S., Opperdoes, F., Brun, R., Quetin-Leclercq, J. (2004b). Natural products active against African trypanosomes: a step towards new drugs. *Natural Products Reproduction*, **21**: 353-364.

Hoet, S., Pieters, L., Muccioli, G. G., Habib-Jiwan, J. L., Opperdoes, F. R., Quetin-Leclercq, J. (2007). Antitrypanosomal activity of triterpenoids and sterols from the leaves of *Strychnos spinosa* and related compounds. *Journal of natural products*, **70**: 1360-1363.

Isah, A. B., Ibrahim, Y. K., Iwalewa, E. O. (2003). Evaluation of the antimalarial properties and standardization of tablets of *Azadirachta indica* (Meliaceae) in mice. *Phytotherapy Research*, **17**: 807-810.

Iwu, M., Jackson, J., Shuster, B. (1994). Medicinal plants in the fight against Leishmaniasis. *Parasitology Today*, **10**: 65-68.

Jiang, S., Zeng, Q., Gettayacamin, M., Tungtaeng, A., Wannaying, S., Lim, A., Hansukjariya, P., Okunji, C.O., Zhu, S., Fang, D. (2005). Antimalarial activities and therapeutic properties of febrifugine analogs. *Antimicrobial Agents and Chemotherapy*, **49**: 1169-76.

Jaramillo, M. C., Arango, G. J., Gonzalez, M. C., Robledo, S. M., Velez, I. D. (2000). Cytotoxicity and antileishmanial activity of *Annona muricata* pericarp. *Fitoterapia*, **71**: 183-186.

Kamanzi Atindehou, K., Schmid, C., Brun, R., Kone, M. W., Traore, D. (2004). Antitrypanosomal and antiplasmodial activity of medicinal plants from Cote d'Ivoire. *Journal of Ethnopharmacology*, **90**: 221-227.

Kaminsky, R., Chuma, F., Zweygarth, E. (1989). *Trypanosoma brucei brucei*: expression of drug resistance *in vitro*. *Experimental Parasitology*, **69**: 281-289.

Kaur, K., Jain, M., Kaur, T., Jain, R. (2009). Antimalarials from nature. *Bioorganic and Medicinal Chemistry*, **17**: 3229-3256.

Krief, S., Huffman, M. A., Sevenet, T., Hladik, C. M., Grellier, P., Loiseau, P. M., Wrangham, R. W. (2006). Bioactive properties of plant species ingested by chimpanzees (*Pan troglodytes schweinfurthii*) in the Kibale National Park, Uganda. *American Journal of Primatology*, **68**: 51-71.

Krief, S., Martin, M. T., Grellier, P., Kasenene, J., Sevenet, T. (2004). Novel antimalarial compounds isolated in a survey of self-medicative behavior of wild chimpanzees in Uganda. *Antimicrobial Agents and Chemotherapy*, **48**: 3196-3199.

Kuhn, S., Gill, M. J., Kain, K. C. (2005). Emergence of atovaquone-proguanil resistance during treatment of *Plasmodium falciparum* malaria acquired by a non-immune North American traveller to West Africa. *American Journal of Tropical Medicine and Hygiene*, **72**: 407-409.

Lalloo, D. G., Shingadia, D., Pasvol, G., Chiodini, P. L., Whitty, C. J., Beeching, N. J., Hill, D. R., Warrell, D. A., Bannister, B. A. (2007). UK malaria treatment guidelines. *Journal of Infections*, **54**: 111-121.

Lamidi, M., DiGiorgio, C., Delmas, F., Favel, A., Eyele Mve-Mba, C., Rondi, M. L., Ollivier, E., Nze-Ekekang, L., Balansard, G. (2005). *In vitro* cytotoxic, antileishmanial and antifungal activities of ethnopharmacologically selected Gabonese plants. *Journal of Ethnopharmacology*, **102**: 185-190.

Lavrado, J., Paulo, A., Gut, J., Rosenthal, P. J., Moreira, R. (2008). Cryptolepine analogues containing basic aminoalkyl side-chains at C-11: synthesis, antiplasmodial activity, and cytotoxicity. *Bioorganic and medicinal chemistry letters*, **18**: 1378-1381.

Lim, P., Alker, A. P., Khim, N., Shah, N. K., Incardona, S., Doung, S., Yi, P., Bouth, D. M., Bouchier, C., Puijalon, O. M., Meshnick, S. R., Wongsrichanalai, C., Fandeur, T., Le Bras, J., Ringwald, P., Ariey, F. (2009). Pfmdr1 copy number and arteminisin derivatives combination therapy failure in falciparum malaria in Cambodia. *Malaria Journal*, **8**: 11.

MacKinnon, S., Durst, T., Arnason, J. T., Angerhofer, C., Pezzuto, J., Sanchez-Vindas, P. E., Poveda, L. J., Gbeassor, M. (1997). Antimalarial activity of tropical Meliaceae extracts and gedunin derivatives. *Journal of natural products*, **60**: 336-341.

Maes, L., Germonprez, N., Quirijnen, L., Van Puyvelde, L., Cos, P., Vanden Berghe, D. (2004). Comparative activities of the triterpene saponin maesabalide III and liposomal amphotericin B (AmBisome) against *Leishmania donovani* in hamsters. *Antimicrobial Agents and Chemotherapy*, **48**: 2056-2060.

Maneerat, W., Laphookhieo, S., Koysomboon, S., Chantrapromma, K. (2008). Antimalarial, antimycobacterial and cytotoxic limonoids from *Chisocheton siamensis*. *Phytomedicine*, **15**: 1130-1134.

Mesia, G. K., Tona, G. L., Nanga, T. H., Cimanga, R. K., Apers, S., Cos, P., Maes, L., Pieters, L., Vlietinck, A. J. (2008). Antiprotozoal and cytotoxic screening of 45 plant extracts from Democratic Republic of Congo. *Journal of Ethnopharmacology*, **115**: 409-415.

Mesquita, M. L., Desrivot, J., Bories, C., Fournet, A., Paula, J. E., Grellier, P., Espindola, L. S. (2005). Antileishmanial and trypanocidal activity of Brazilian Cerrado plants. *Memorias do Instituto Oswaldo Cruz*, **100**: 783-787.

Mueller, M. S., Runyambo, N., Wagner, I., Borrmann, S., Dietz, K., Heide, L. (2004). Randomized controlled trial of a traditional preparation of *Artemisia annua* L. (Annual Wormwood) in the treatment of malaria. *Transactions of the Royal Society of Tropical Medicine and Hygiene*, **98**: 318-321.

Muhammad, I., Dunbar, D. C., Khan, S. I., Tekwani, B. L., Bedir, E., Takamatsu, S., Ferreira, D., Walker, L. A. (2003). Antiparasitic alkaloids from *Psychotria klugii*. *Journal of natural products*, **66**: 962-967.

Murata, T., Miyase, T., Muregi, F. W., Naoshima-Ishibashi, Y., Umehara, K., Warashina, T., Kanou, S., Mkoji, G. M., Terada, M., Ishih, A. (2008). Antiplasmodial triterpenoids from *Ekebergia capensis*. *Journal of natural products*, **71**: 167-174.

Mutabingwa, T. K. (2005). Artemisinin-based combination therapies (ACTs): Best hope for malaria treatment but inaccessible to the needy! Acta Tropica, **95**: 305-315.

Mutabingwa, T. K., Anthony, D., Heller, A., Hallett, R., Ahmed, J., Drakeley, C., Greenwood, B. M., Whitty, C. J. (2005). Amodiaquine alone, amodiaquine+sulfadoxine-pyrimethamine, amodiaquine+artesunate, and artemether-lumefantrine for outpatient treatment of malaria in Tanzanian children: a four-arm randomised effectiveness trial. *The Lancet*, **365**: 1474-1480.

Ndjakou Lenta, B., Vonthron-Senecheau, C., Fongang Soh, R., Tantangmo, F., Ngouela, S., Kaiser, M., Tsamo, E., Anton, R., Weniger, B. (2007). *In vitro* antiprotozoal activities and cytotoxicity of some selected Cameroonian medicinal plants. *Journal of Ethnopharmacology*, **111**: 8-12.

Newman, D. J., Cragg, G. M. (2007). Natural products as sources of new drugs over the last 25 years. *Journal of natural products*, **70**: 461-477.

Nguta, J. M., Mbaria, J. M., Gakuya, D. W., Gathumbi, P. K., Kiama, S. G. (2010). Antimalarial herbal remedies of Msambweni, Kenya. *Journal of ethnopharmacology*, **128**: 424-432.

O'Neill, M. J., Bray, D. H., Boardman, P., Phillipson, J. D., Warhust, D. C. (1985). Plants as sources of antimalarial drugs Part 1: *In vitro* test method for the evaluation of crude extracts from plants. *Planta Medica,* **61**: 394-397.

Ogbadoyi, E. O., Abdulganiy, A. O., Adama, T. Z., Okogun, J. I. (2007). *In vivo* trypanocidal activity of *Annona senegalensis* Pers. leaf extract against *Trypanosoma brucei brucei. Journal of Ethnopharmacology,***112**: 85-89.

Olliaro, P. (2001). Mode of action and mechanisms of resistance for antimalarial drugs. *Pharmacology and Therapeutics,* **89**: 207-219.

Peters, W. (1980). Chemotherapy of Malaria. *In*: Malaria, Inc., vol. 1. Ed. by A. P., ISBN 0-12-426101-9, pp. 145-176.

Rocha, L. G., Almeida, J. R., Macedo, R. O., Barbosa-Filho, J. M. (2005). A review of natural products with antileishmanial activity. *Phytomedicine,* **12**: 514-535.

Rochanakij, S., Thebtaranonth, Y., Yenjai, C., Yuthavong, Y. (1985). Nimbolide, a constituent of *Azadirachta indica,* inhibits *Plasmodium falciparum* in culture. *Southeast Asian Journal of Tropical Medicine and Public Health,* **16**: 66-72.

Rogers, W. O., Sem, R., Tero, T., Chim, P., Lim, P., Muth, S., Socheat, D., Ariey, F., Wongsrichanalai, C. (2009). Failure of artesunate-mefloquine combination therapy for uncomplicated *Plasmodium falciparum* malaria in southern Cambodia. *Malaria Journal,* **8**: 10.

Saklani, A., Kutty, S. K. (2008). Plant-derived compounds in clinical trials. *Drug Discovery Today,* **13**: 161-171.

Sauvain, M., Dedet, J. P., Kunesch, N., Poisson, J. (1994). Isolation of flavins from the Amazonian shrub *Faramea guianesis. Journal of natural products,* **57**: 403-406.

Schimmer, O., Hafele, F., Kruger, A. (1988). The mutagenic potencies of plant extracts containing quercetin in *Salmonella typhimurium* TA98 and TA100. *Mutation Research,* **206**: 201-208.

Seifert, K., Perez-Victoria, F. J., Stettler, M., Sanchez-Canete, M. P., Castanys, S., Gamarro, F., Croft, S. L. (2007). Inactivation of the miltefosine transporter, LdMT, causes miltefosine resistance that is conferred to the amastigote stage of *Leishmania donovani* and persists *in vivo. International Journal of Antimicrobial Agents,* **30**: 229-235.

Senn, M., Gunzenhauser, S., Brun, R., Sequin, U. (2007). Antiprotozoal polyacetylenes from the Tanzanian medicinal plant *Cussonia zimmermannii. Journal of natural products,* **70**: 1565-1569.

Soh, P. N., Benoit-Vical, F. (2007). Are West African plants a source of future antimalarial drugs? *Journal of Ethnopharmacology,* **114**: 130-140.

Soh, P. N., Witkowski, B., Olagnier, D., Nicolau, M. L., Garcia-Alvarez, M. C., Berry, A., Benoit-Vical, F. (2009). *In vitro* and *in vivo* properties of ellagic acid in malaria. *Antimicrobial Agents and Chemotherapy,* **53**: 1100-1106.

Spencer, C. F., Koniuszy, F. R., Rogers, E. F., Shavel, J., Easton, N. R., Kaczka, E. A., Kuehl, F. A., Phillips, R. F., Walti, A., Folker, K. (1947). Survey of plants for antimalarial activity. *LLoydia,* **10-3**: 145-147.

Sulsen, V. P., Cazorla, S. I., Frank, F. M., Redko, F. C., Anesini, C. A., Coussio, J. D., Malchiodi, E. L., Martino, V. S., Muschietti, L. V. (2007). Trypanocidal and leishmanicidal activities of flavonoids from Argentine medicinal plants. *American Journal of Tropical Medicine and Hygiene*, **77**: 654-659.

Sundar, S., More, D. K., Singh, M. K., Singh, V. P., Sharma, S., Makharia, A., Kumar, P. C., Murray, H. W. (2000). Failure of pentavalent antimony in visceral leishmaniasis in India: *report from the center of the Indian epidemic*, **31**: 1104-1107.

Trager, W., Jensen, J. B. (1976). Human malaria parasites in continuous culture. *Science*, **193**: 673-675.

Ueda-Nakamura, T., Attias, M., de Souza, W. (2001). Megasome biogenesis in *Leishmania amazonensis*: a morphometric and cytochemical study. *Parasitology Research*, **87**: 89-97.

Van Huyssen, W., Rieckmann, K. H. (1993). Disposable environmental chamber for assessing the drug susceptibility of malaria parasites. *Tropical Medicine and Parasitology*, **44**: 329-330.

Vonthron-Senecheau, C., Weniger, B., Ouattara, M., Bi, F. T., Kamenan, A., Lobstein, A., Brun, R., Anton, R. (2003). *In vitro* antiplasmodial activity and cytotoxicity of ethnobotanically selected Ivorian plants. *Journal of Ethnopharmacology*, **87**: 221-225.

Waako, P. J., Gumede, B., Smith, P., Folb, P. I. (2005). The *in vitro* and *in vivo* antimalarial activity of *Cardiospermum halicacabum* L. and *Momordica foetida* Schumch. Et Thonn. *Journal of Ethnopharmacology*, **99**: 137-143.

Waechter, A.I., Ferreira, M.E., Fournet, A., Arias, A.R., Nakayama, H., Torres, S., Hocquemiller, R., Cave, A. (1997). Experimental treatment of cutaneous leishmaniasis with argentilactone isolated from *Annona haematantha*. *Planta Medica*, **63**: 433–435.

Wellems, T. E., Plowe, C. V. (2001). Chloroquine-resistant malaria. *Journal of Infectious. Diseases*, **184**: 770-776.

White, N. J., Olliaro, P. L. (1996). Strategies for the prevention of antimalarial drug resistance: Rationale for combination chemotherapy for malaria. *Parasitology Today*, **12**: 399-401.

WHO. (2000). General Guidelines for Methodologies on Research and Evaluation of Traditional Medicine. *In*: WHO/EDM/TRM/2000.1 (Ed.). WHO, Geneva.

WHO. (2005)a. Control of Human African Trypanosomiasis: a Strategy for the African Region. Geneva.

WHO (2005)b. National Policy on Traditional Medicine and Regulation of Herbal Medicines - Report of a WHO Global Survey. *In*: TDR-WHO (Ed.), http: // apps.who.int/medicinedocs/pdf/s7916e/s7916e.pdf, Geneva.

WHO (2008). Traditional medicine. *In*: centre, W. M. (Ed.). http: //www.who.int/ mediacentre/factsheets/fs134/en/, Geneva.

Willcox, M. (2009). *Artemisia* species: From traditional medicines to modern antimalarials - and back again. *Journal of Alternative and complementary medicine,* **15**: 101-109.

Willcox, M., Bodeker, G., Rasoanaivo, P. (2004). Traditional Medicinal Plants and Malaria Traditional Herbal Medicines for modern times, ed. R. Hardman. CRC Press: Boca Raton.

Willcox, M. L. (1999). A clinical trial of 'AM', a Ugandan herbal remedy for malaria. *Journal of Public Health Medicine,* **21**: 318-324.

Willcox, M. L., Bodeker, G. (2004). Traditional herbal medicines for malaria. *Bmj,* **329**, 1156-1159.

Willcox, M. L., Graz, B., Falquet, J., Sidibe, O., Forster, M., Diallo, D. (2007). *Argemone mexicana* decoction for the treatment of uncomplicated falciparum malaria. *Transactions of the Royal Society of Tropical Medicine and Hygiene,* **101**: 1190-1198.

Woodrow, C. J., Haynes, R. K., Krishna, S. (2005). Artemisinins. *Postgrad Medicinal Journal,* **81**: 71-78.

Yin, S., Wang, X. N., Fan, C. Q., Liao, S. G., Yue, J. M. (2007). The first limonoid peroxide in the Meliaceae family: walsuronoid A from *Walsura robusta. Organic letters,* **9**: 2353-2356.

Zalis, M., Pang, L., Silveira, M., Milhous, W., Wirth, D. F. (1998). Characterization of *Plasmodium falciparum* isolated from the amazon region of Brazil: evidence for quinine resistance. *American Journal of Tropical Medicine and Hygiene,* **58-5:** 630-637.

Zhu, S., Zhang, Q., Gudise, C., Wei, L., Smith, E., Zeng, Y. (2009). Synthesis and biological evaluation of febrifugine analogues as potential antimalarial agents. *Bioorganic and Medicinal Chemistry,* **17**: 4496-502.

Zirihi, G. N., Mambu, L., Guede-Guina, F., Bodo, B., Grellier, P. (2005). *In vitro* antiplasmodial activity and cytotoxicity of 33 West African plants used for treatment of malaria. *Journal of Ethnopharmacology,* **98**: 281-285.

Traditional and Folk Herbal Medicine: *Pages* **279–288**
Recent Researches Vol. 1 (2012)
Editor: **V.K. Gupta**
Published by: **DAYA PUBLISHING HOUSE, NEW DELHI**

12

A Simplified Procedure for Monitoring the Stability of Herbal Remedies for the Purpose of Manufacturing and Regulation

Sunday J. Ameh*[1], Obiageri O. Obodozie[1], Mujitaba S. Abubakar[2] and Magaji Garba[3]

ABSTRACT

The World Health Organization (WHO) had estimated that some 80 per cent of people in the developing countries relied on traditional herbal products for most of their health care needs in 2008. Unfortunately these products are poorly regulated in these countries because of the absence of proper policies on traditional remedies; and the dearth of reliable procedures for regulating their production and distribution. In this chapter we describe a simplified approach to the problem based, essentially on the methods of the WHO manual of 1998 on the quality control of medicinal plant materials. We here present some results based on the application of the said methods to the herb and extract of Mitracarpus scaber, believing that a wide application of the procedures to other herbal products will improve the regulatory status of herbal medicines

1 Department of Medicinal Chemistry and Quality Control, National Institute for Pharmaceutical Research and Development (NIPRD), PMB 21, Garki, Idu Industrial Area, Abuja, Nigeria.

2 Department of Pharmacognosy and Drug Development, Ahmadu Bello University, Zaria.

3 Department of Pharmaceutical and Medicinal Chemistry, Ahmadu Bello University, Zaria.

* *Corresponding author*: E-mail: sjitodo@yahoo.com; Telephone. 2348053691277.

generally, especially in the developing countries, where they are most used and needed. The present study was in particular carried out to facilitate the registration of NIPRIFAN® - an antifungal preparation derived from Mitracarpus scaber.

Keywords: Quality control, Procedures, Monitoring, Stability, *Mitracarpus scaber*, Regulatory, Traditional medicine, NIPRIFAN®.

Introduction

It is generally held that in many countries in Africa, Latin America and Asia, where the populations depend a great deal on herbal remedies, the products are hardly regulated, or at best, only poorly (Ameh *et al.*, 2010a). Even in the highly industrialized countries where such products, called dietary supplements in the United States (US), are more closely regulated, it is often thought by the general public that they are at best only loosely regulated (DSHEA, 1994; Ameh *et al.*, 2010b). This rather negative perception arises mainly from the fact that proof of efficacy is often not a requirement for registering an herbal remedy in most countries, including those belonging to the Organization for Economic Cooperation and Development (OECD) or to the International Conference on Harmonization (ICH). On the whole, most countries that regulate herbal remedies as closely as possible; either follow the WHO paradigm, or the ICH model, as in the European Union (EU). The main difference between the two models is that the ICH model, operative in the EU, US and Japan (De Smet, 2005; Ann Godsell Regulatory, 2008) is far more explicit, though not necessarily more exacting than the WHO model – applied in Nigeria and several other developing countries (WHO, 1998; Ameh *et al.*, 2010b). The approach described in this chapter for assessing and monitoring the physicochemical stability of medicinal plant materials is inexpensive, and is applicable to both the ICH and the WHO models. The most outstanding similarity between the two models, which is the most important issue, lies in their common emphasis on long term safe use in traditional medicine. Another key similarity is the emphasis beamed on quality control and good manufacturing practice (GMP) for herbal products (Ameh *et al.*, 2010b).

Experimental

An Introduction to the Experimental Approach

Agrawal and Paridhavi (2007) in their book titled *Herbal Drug Technology* aptly classified the tests prescribed by WHO (1998) for medicinal plant materials into the following categories: botanical/histochemical or phytochemical – such as sensory evaluations and tests for key phytochemicals; physicochemical tests – such as loss on drying and thin layer chromatography; pharmacological tests – such as the determinations of bitterness value, foaming index and hemolytic activity; and toxicological tests – such as test for heavy metals and specific pathogens. By monitoring changes that occur in these parameters during the storage of these materials, useful information can be gained regarding their stability or shelf life. Thus, where applicable, necessary or feasible, a battery of tests were carried out on the herb and ethylacetate extract of *Mitracarpus scaber*, as a means of gauging their keeping qualities in capped glass bottles, stored at room temperature and humidity (RTH). The tests were: Loss on drying; Extractability/Solubility; Light absorption/TLC; and pH/Foaming index. The assumptions were as follows: samples which stay dry within narrow limits of hygroscopicity tend to be stable to moisture induced spoilage; but significant changes in the extractability, solubility, light absorption, TLC characteristics, and pH/foaming indices may indicate spoilage. The absence of significant changes in these parameters may indicate some degree of stability.

Sampling and Treatment of Material

The aerial parts of *Mitracarpus scaber* obtained during the months of October and November from the botanical garden of the National Institute for Pharmaceutical Research and Development (NIPRD) were air-dried in a well-ventilated shade, designed for drying medicinal plant materials. The materials were subsequently comminuted to coarse powder with a grinding machine. The procedure for sampling was as follows (WHO, 1998):

Three (3) original samples from each batch or container were combined into a pooled sample and subsequently used to prepare the average sample. The average sample was prepared by "quartering" the pooled sample as follows: each pooled sample was mixed thoroughly, and constituted into a square-shaped heap. The heap was then divided diagonally into 4 equal parts. Any 2 diagonally opposite parts were taken and mixed carefully. This step was repeated 2 to 4 times to obtain the required quantity of sample. Any material remaining was returned to the batch. The final samples were obtained from an average sample by quartering, as described above. This means that an average sample gave rise to 4 final samples. Each final sample was divided into 2 portions. One portion was retained as reference material, while the other was tested in duplicate or triplicate.

Loss on Drying

This was carried out using a minimum of 0.5–1.0 g of material. Drying was effected in a Lindberg/Blue M gravity-convention oven maintained at 105–110 °C, for 3 h, after which the sample was allowed to cool to room temperature in a desiccator, and subsequently weighed. The time interval from the oven to point of weighing was usually about 30 minutes. The results are expressed as a range or as mean ± standard deviation.

Evaluation of Extractive Matter

About 4 g of accurately weighed coarsely powdered, air-dried sample was transferred into a glass-stoppered, 250-ml reflux conical flask, followed by the addition of 100 ml of solvent. The flask was weighed along with its contents, and recorded as W1. The flask was well shaken, and allowed to stand for 1 h. Subsequently a reflux condenser was attached to the flask, and gently brought to boiling and maintained thereat boiled 20 – 60 minutes depending upon the solvent. The mixture was subsequently cooled and weighed again. The weight was recorded as W2, and then readjusted to W1 with the solvent. The flask was shaken well once again and its contents rapidly filtered through a dry filter paper. By means a pipette, 25 ml of the filtrate was transferred to a previously dried and tarred glass dish and then gently evaporated to dryness on a hot plate. Subsequently, the dish was dried at 105 °C for 1-6 hours, cooled in a desiccator for 30 min, and weighed. The extractable matter was calculated as per cent w/w of the air-dried sample.

Determination of Solubility of Material in a Given Solvent–Methods I and II

The solubility of a material was determined at room temperature ~ 25°C and expressed in terms of "parts", representing the number of milliliter of solvent, in which 1 g of the material is soluble. Vials of appropriate sizes: ~4-ml, ~ 12-ml and ~20-ml capacities were used. The mixtures were thoroughly shaken for at least 30 min before inspection for un-dissolved solute. In methods I, each vial received 100 mg of sample and the volume of water indicated. In method II, a vial received 100 mg and increasing volumes of water. The methods give the same results. An example of results obtained in the determination of the solubility of ethylacetate extract of *Mitracarpus scaber* in acetone is presented in Table 12.1.

Table 12.1: An example of the presentation/interpretation of results from the determination of the solubility of the ethylacetate extract of *Mitracarpus scaber* in acetone.

Acetone (ml)	0.5	1.0	1.5	2.0	2.5	3.0
Extract (mg)	100	100	100	100	100	100
Inspect for presence (+) absence or (–) of un-dissolved solute	+	+	–	–	–	–
Inference: the volume of solvent (ml) in which 1g of the solute is soluble is termed the "relative solubility", expressed in "parts"	> 5	>10	≥ 15	< 20	< 25	< 30

The test was carried out at room temperature ~ 25°C and expressed in terms of "parts", representing the minimum number of milliliters (ml) of solvent, in which 1 g of the material is soluble. Vials of appropriate sizes ~2 to 10-ml capacity were used. The mixtures were thoroughly shaken for at least 30 min before inspection for any un-dissolved solute. In this method a vial received 100 mg and increasing volumes of solvent.

Light Absorption and Thin Layer Chromatography (TLC)

UV-VIS Spectrophotometer (Jenway or Shimadzu) and quartz 1-cm cells were used for the study. Solutions of herb and extract were made by thoroughly mixing 1 part of the solute and with 100 parts of solvent methanol: water 80:20, v/v filtering, and diluting the filtrates by 150x with the same solvent. Absorbencies were measured at λ227 nm, using the solvent as the blank. Florescent, precoated plates were used for normal phase TLC, utilizing silica K6, and hexane: ethylacetate as mobile phase. Solutions of analytes were prepared and applied as follows: To 1 mg of the analyte, 2 drops of ethanol were added and mixed well (~1 per cent w/v solution). The plates used were 5 cm wide x 20 cm long. With a ruler and a pencil, a distance of 5 mm was measured from the bottom of the plate, and a line of origin was lightly drawn across the plate, without disturbing the adsorbent. The analyte was applied to the origin as a 1 µl droplet. The spot was allowed to dry. Subsequently, the plate was developed in a developing tank saturated with the vapour of the solvent system to be used as mobile phase. The level of the solvent in the tank was adjusted to a level 2 to 3 mm below the line of origin on the plate. The plate was considered developed when the solvent front reached a predetermined line, not less than 5 mm below the top of the plate. The air-dried plate was visualized using a viewing cabinet (Cammag) and a UV-lamp (Cammag – equipped to emit light at 254 or 366 nm). The resulting chromatogram was photographed and subsequently drawn to scale.

Determinations of pH of Preparations–Herb and the Dry Extract

Determination of pH values was with a Jenway pH Meter. Standard pH solutions: 4, 7 and 10; and freshly distilled water were used for the study.

Determination of Foaming Indices of Preparations–Herb and the Dry Extract

Decoctions of plant materials foam due to the presence of saponins. This ability is measured as foaming index, and is an important quality control parameter. The requirements for the test include: conical flasks (500-ml); volumetric flasks (100-ml); test tubes (16cm × 16mm); ruler; and stop-clock. The procedure is as follows: Exactly 1.0 g of powdered material was accurately transferred into a 500-ml conical flask containing 100 ml of boiling water, and maintained at moderate boiling for 30 minutes. The mixture was then cooled and filtered into a 100-ml volumetric flask. The volume was made up to 100 ml with water. Successive portions of 1 ml, 2 ml, 3 ml etc up to 10 ml of the filtrate was poured into ten stoppered tubes having the following dimensions: height, 16 cm; diameter, 16 mm. Subsequently, the volume of each tube was adjusted to 10 ml with water, stoppered and shaken in lengthwise motion for 15 seconds, at 2 shakes per second. The tubes were allowed to stand for 15 minutes, and the height of the foam in each tube was measured. The results were assessed as follows:

1. Foaming index is less than 100, if the height of foam in all the tubes is less than 1 cm.
2. If a height of 1 cm is obtained in any tube, the volume [V] of the decoction in that tube, is used to determine the foaming index, as = 1000/V.
3. But if the tube above is the first or the second in the series, prepare an intermediate dilution to obtain a more precise result.
4. If the height of the foam is more than 1 cm in every tube, the foaming index is over 1000.
5. To obtain a more precise result, repeat the determination using a new series of dilutions of the decoction. Note the tube in which the height of foam is 1 cm, and the volume [V] of the decoction therein, and calculate the foaming index, as = 1000/V.

Foaming indices were expressed as a quantity [Q] per ml or as [Q]ml^{-1}.

Results

Effect of Storage on Moisture Content

The results in Table 12.2 show that storage of the herb and the extract in capped glass bottles at room temperature and humidity (RTH) for up to 39 months produced no consistent or statistically significant changes in moisture content.

Table 12.2: Real time stability of herb and ethylacetate extract of *Mitracarpus scaber* as monitored by loss on drying.

Months of Storage in Capped Glass Bottles at RTH	Loss on Drying (mean ± SD) per cent w/w	
	Herb	Extract
0	10.51 ± 0.74[a] (7)	14.07 ± 1.73[b] (7)
3	9.11 ± 0.67[a] (6)	12.84 ± 1.18[b] (6)
9	10.44 ± 0.51[a] (6)	13.14 ± 1.11[b] (7)
21	10.87 ± 0.73[a] (7)	14.84 ± 0.99[b] (5)
39	8.96 ± 0.59[a] (5)	11.44 ± 0.86[b] (6)

The results show that storage of both the herb and the extract in capped glass bottles at room temperature and humidity for up to 39 months, produced no consistent or statistically significant change in moisture content. Both ([a]) and ([b]) indicate that the figures bearing them do not differ statistically from one another ($P > 0.05$).

Effect of Storage on Appearance, Extractability and Solubility of the Herb and Extract in Different Solvents

Table 12.3 shows that the extractability of the herb in methanol, ethylacetate, ethanol, acetone and hexane were unchanged with storage, as was its appearance as wrinkled, brownish green leaves and twigs. However, the extractability of the herb in water fell slightly but significantly as from after the 3rd month of storage. Similarly, Table 12.4 shows that neither the solubility profile nor the appearance of the extract and its solutions in different solvents changed with storage.

Table 12.3: Real time stability of *Mitracarpus scaber* herb as monitored by appearance and extractability in different solvents.

Months of Storage of Herb in Capped Glass bottles at RTH	Extractability of Herb in Different Solvents					
	Methanol	Ethylacetate	Ethanol	Acetone	Hexane	Water
0	15.4 ± 1.5	14.0 ± 1.8	11.7 ± 0.8	7.2 ± 0.9	4.2 ± 0.5	36.4 ± 2.2[a]
3	16.3 ± 1.5	15.3± 2.2	11.3 ± 0.8	6.9 ± 1.1	4.3 ± 0.5	33.9 ± 2.4
9	14.8 ± 1.2	14.4 ± 1.9	12.1 ± 0.8	7.6 ± 0.9	4.1 ± 0.5	30.2 ± 2.4
21	16.0 ± 1.8	14.7 ± 1.9	12.8 ± 0.8	8.0 ± 0.9	4.1 ± 0.5	30.7± 2.2[a]
39	15.5 ± 1.5	15.0 ± 2.0	12.4 ± 0.8	7.8 ± 0.9	4.3 ± 0.5	27.8 ±2.0[a]

The appearance of the herb did not change with storage, in that, at every stage of storage (0-39 months) the herb appeared as wrinkled brownish green leaves and twigs. The extractabilities of the herb in methanol, ethylacetate, ethanol, acetone and hexane, in descending order of extractability, were unchanged with storage, but the extractability in water fell slightly but significantly as from after the 3rd month of storage. The results represent means ± SDs (n=2–7). The difference between the 0th and the 21st/39th months, denoted by ([a]), was significant at $P < 0.05$ (n=7).

Table 12.4: Real time stability of ethylacetate extract of *Mitracarpus scaber* as monitored by appearance and solubility in different solvents.

Months of Storage of Extract in Capped Glass bottles at RTH	*Solubility of Extract in Different Solvents*					
	Acetone	*Methanol*	*Ethylacetate*	*Ethanol*	*Hexane*	*Water*
0	15.0 ± 0.0	25.0 ± 0.0	25.0 ± 5.0	25.0 ± 5.0	55.0 ± 5.0	$>10^3$
3	15.0 ± 0.0	20.0 ± 5.0	25.0 ± 5.0	25.0 ± 5.0	45.0 ± 5.0	>10
9	15.0 ± 0.0	20.0 ± 5.0	25.0 ± 0.0	25.0 ± 0.0	55.0 ± 0.0	$>10^3$
21	15.0 ± 0.0	25.0 ± 5.0	25.0 ± 5.0	25.0 ± 5.0	50.0 ± 5.0	$>10^3$
39	15.0 ± 0.0	25.0 ± 5.0	25.0 ± 0.0	25.0 ± 0.0	55.0 ± 0.0	$>10^3$

Neither the appearance nor the solubility of the extract in different solvents changed appreciably with storage (0-39 months). At every stage, the extract was most soluble in acetone, and least in water. The solubility of the extract in ethanol, ethylacetate and methanol were similar to one another. The solutions of the extract in the different organic solvents were clear and greenish-brown, but that in water was brown, and slightly cloudy, with no tinge of green. The results represent means ± SDs from 2 experiments in duplicate or triplicate (n=2–6).

Effect of Storage on Light Absorption and TLC Features of the Herb and Extract

Table 12.5 presents the effect of storage on light absorption and TLC features of the herb and extract. It shows the following: that the difference in absorbance between 0th month and the $21^{st}/39^{th}$ months was insignificant for the herb ($P > 0.05$). By contrast, the corresponding difference for the extract was significant ($P < 0.05$). Table 5 also shows that the number of TLC spots observed for the herb and extract at every stage of storage was unchanged up to the 39th month.

Effect of Storage on pH and Foaming Indices of Herb and Ethylacetate Extract

Table 12.6 shows that the pH of a 5 per cent w/v mixture of the herb in water was 5.4–5.9. No consistent or significant variation in the pH occurred with storage for up to 39 months. The corresponding results for extract was 6.1–6.9, with no significant variation with storage for up to 39 months. Although the 5 per cent w/v mixture of the fresh herb in water foamed a little; this property rapidly diminished, and was totally lost after the 3rd month of storage. By contrast, the ethylacetate extract never foamed at any stage of storage.

Discussion

NIPRIFAN® is an antifungal phytomedicine under development at NIPRD, Abuja, Nigeria. It contains the ethylacetate extract of the aerial parts of *Mitracarpus scaber* Zucc; Family: Rubiaceae. The Voucher specimen number of the herb, preserved in the Institute's Herbarium is: NIPRD/H/4208. The plant is an annual weed. It grows erect, with 4-angled pubescent stems, up to 55 cm high, usually branched; the leaves are lanceolate, 2-4 cm long, with the upper surface bearing minute hairs. The

Table 12.5: Real time stability of herb and ethylacetate extract of *Mitracarpus scaber* as monitored by light absorption and TLC.

Months of Storage in Capped Glass Bottles at RTH	*Herb*							*Extract*		
	Abs. at λ227 nm	*Types of solvent/number of TLC spots*						*Abs. at λ227 nm*	*TLC Spots*	
		A	*B*	*C*	*D*	*E*	*F*		*G*	*H*
0	2.172 ± 0.104 (5)	5	2	5	3	3	4	2.330 ± 0.094 (5)	7	10
3	2.221 ± 0.114 (6)	5	2	5	3	3	4	2.174 ± 0.107 (5)	7	10
9	2.144 ± 0.098 (5)	5	2	5	3	3	4	2.104 ± 0.070 (5)	7	10
21	2.322 ± 0.117[a](5)	5	2	5	3	3	4	2.039 ± 0.104[b](5)	7	10
39	2.233 ± 0.114[a] (5)	5	2	5	3	3	4	2.084 ± 0.111[b](6)	7	10

Solutions of herb and extract were made by thoroughly mixing 1 part of the solute and with 100 parts of solvent (MeOH: H_2O [80:20, v/v]), filtering, and diluting the filtrates by 150x with the same solvent. Absorbencies were measured at λ227 nm, using the solvent as the blank. For the herb, the difference in absorbance between 0th month and the 21st/39th months, denoted by ([a]), was insignificant ($P > 0.05$). By contrast, for the extract, the difference in absorbance between the 0th or 3rd month and the 21st or 39th month, denoted by ([b]), was significant ($P < 0.05$). The number of TLC spots observed for both the herb and extract at every stage remained unchanged up to the 39th month. For examples: the herb in acetone (A) gave 5 spots; the hexane extract (G) re-dissolve in hexane gave 7 spots; while the ethylacetate extract (H) re-dissolved in ethylacetate gave 10 spots, irrespective of storage for up to 39 months. Solvents B, C, D, E and F were: water, ethanol, ethylacetate, hexane and methanol respectively. G and H were dry extracts re-dissolved in hexane and ethylacetate respectively.

Table 12.6: Real time stability of herb and ethylacetate extract of *Mitracarpus scaber* as monitored by pH and foaming index.

Months of Storage in Capped Glass Bottles at RTH	*pH of mixture (5% w/v)*		*Foaming index*	
	Herb in Water	*Ethylacetate Extract in Water*	*Herb in water*	*Ethylacetate Extract in water*
Within 1st day of harvest or preparation	5.7 ± 0.3 (5)	6.9 ± 0.3 (5)	Foam: Slight < 100 (5)	Foam: Nil (5)
0	5.9 ± 0.2 (6)	6.1 ± 0.3 (5)	Foam: Slight < 100 (5)	Foam: Nil (6)
3	5.6 ± 0.2 (5)	6.2 ± 0.4 (6)	Foam: Slight < 100 (5)	Foam: Nil (5)
9	5.4 ± 0.2 (5)	6.1 ± 0.3 (5)	Foam: Nil (6)	Foam: Nil (5)
21	5.9 ± 0.3 (7)	6.7 ± 0.3 (5)	Foam: Nil (5)	Foam: Nil (5)
39	5.9 ± 0.3 (5)	6.5 ± 0.3 (5)	Foam: Nil (5)	Foam: Nil (5)

Both ([a]) and ([b]) indicate that the pH of the 5 per cent w/v aqueous mixtures at every stage fell within the mean values of 5.8 ± 0.2[a] and 6.4 ± 0.4[b] - they indicate that any deviations from these mean values were insignificant ($P > 0.05$). The freshly harvested samples foamed measurably, but the ability was totally lost after the 3rd month of dry storage. In all cases however, the dry ethylacetate extract was virtually insoluble in water, and did not foam at all.

plant manifests dense clusters of inflorescence, 6-14 mm across, with minute white flowers. The fruits are dehiscent capsules, about 0.5-1 mm long. The plant reproduces by seeds. Although the plant can grow at any time of the year if planted close to a water course, in NIPRD's botanical garden and in the wild, the plant usually begins to sprout as from about May-July, flowering about 3 to 4 months later. Both the fresh plant and air-dried weed are odourless but possess a slight warm taste. Both the weed and its ethylacetate extract contain tannins, terpenoids and anthraquinones. In our laboratory, the weed, but not the extract, contained detectable amounts of saponins; while tests for alkaloids were negative for both weed and ethylacetate extract. The well documented antimicrobial properties of *Mitracarpus scaber* (Nasipuri, 1999; Ogundiani, 1999; Cimanga *et al.*, 2004; Ogundiani, 2005) have been confirmed in NIPRD's department of Microbiology, Biotechnology and Human Virology (MBHV). The fungal species against which the MBHV, under the headship of Professor Kolo Ibrahim, had demonstrated activities were: *Epidermophyton flocossum* ATCC 10227; *Microsporum canis* ATCC 11622; *Trichophyton rubum* ATCC 28941; and *Trichophyton mentgrophytes* ATCC 4808. The aim of this present study was to apply the methods of WHO (1998) to assess the effect of real time storage on some selected parameters of the air-dried NIPRIFAN herb and its extract, with a view to estimating their shelf lives – a prerequisite for registration and GMP production. The results collectively show that the herb tended to be more stable than the extract, but a shelf life of a year or more can be ascribed to either material.

References

Agrawal, S. S. and Paridhavi, M. (2007). Herbal Drug Technology. University Press (India) Private Limited, Hydrabad, India, pp. 625-644.

Ameh SJ, Obodozie OO, Inyang US, Abubakar MS and Garba Magaji. (2010). Current phytotherapy – An inter-regional perspective on policy, research and development of herbal medicine. *Journal of Medicinal Plants Research* – accepted for publication on 19 July, 2010.

Ameh, S. J., Obodozie, O. O., Inyang, U. S., Abubakar, M. S. and Garba Magaji. 2010. Current phytotherapy - A perspective on the science and regulation of herbal medicine. *J. Med. Plants Res.*, **4(2):** 072-081.

Ann Godsell Regulatory (2008). Pharmaceutical Good Manufacturing Practice for Herbal Drug Substances. http: //www.pharmaceutical-int.com/article/category/treatment-herbal-medicines. Accessed 2010 April 8.

Cimanga, R. K., Kambu, K., Tona, L., Bruyne, T., Sandra, A., Totte, J., Pieters, L., Vlietinck, A. J. (2004) Antibacterial and antifungal activities of some extracts and fractions of *Mitracarpus scaber* Zucc. (Rubiaceae). *Journal of Natural Remedies*, **4(1):** 17-25.

De Smet, P. N. 2005. Herbal medicine in Europe – relaxing regulatory standards. *New England Journal of Medicine*, **352(12):** 1176-78.

DSHEA (1994). Dietary Supplements Health Education Act of 1994. http: //fda/Food/DietarySupplements/ucm109764.htm. Accessed 2010 April 8].

Nasipuri, R. N. (1999). R and D Priorities and Progress: Department of Pharmaceutical Technology, NIPRD. Nasipuri Papers, National Institute for Pharmaceutical Research and Development (NIPRD), Abuja, Nigeria. p. 1-6.

Ogundaini A. O. (1999) Antimicrobial agents from some Nigerian plants. *Nigerian Journal of Natural Products and Medicine*, **3:** 26-27.

Ogundaini, A. O. (2005). From Greens into Medicine: Taking a Lead From Nature. Inaugural Lecture Series 176. Tuesday 26, April, 2005. Obafemi Awolowo University Press Limited. Ile-Ife, Nigeria.

WHO (1998). Quality Control Methods for Medicinal Plant Materials. World Health Organization, Geneva, 128 p. http: //aps.who.int/medicinedocs/collect/medicin edocs/pdf/h1791e.pdf. Accessed 2009 Oct 2.

Traditional and Folk Herbal Medicine: *Pages* **289–300**
Recent Researches Vol. 1 (2012)
Editor: **V.K. Gupta**
Published by: **DAYA PUBLISHING HOUSE, NEW DELHI**

13

Effect of a Cysteine Proteinase from *Bauhinia forficata* Leaves on Thrombin-Induced Human Platelet Aggregation

Sheila S. Andrade[1], Mariana C.C. Silva[1], Iuri E. Gouvea[2], Maria A. Juliano[2], Adriana K. Carmona[2], Misako U. Sampaio[1] and Maria Luiza V. Oliva[1]*

ABSTRACT

B. forficata is trivially known as cow paw, and popularly used in Brazilian folk medicine for treatment of diabetes mellitus. From its leaves a cysteine proteinase was purified being the activity detected in the acetone precipitate 80 per cent (v/v). As the proteinase was shown to be related to papain, it was denominated baupain. Like a bromelain and papain, baupain is antithrombotic and anticoagulant in vitro. In this study, we investigated the baupain effect on aggregation of isolated human platelets in vitro and the results show that Baupain inhibits thrombin - but not ADP- and collagen- induced platelet aggregation. Using synthetic quenched-fluorescent peptides the kinetics of the cleavage site of human proteinase-activated receptor 1/2/3 and 4 [PAR-1/2/3 and 4] by baupain was determined. In conclusion, baupain activates human platelets, probably through a "thrombin-like" activity on proteinase-activating receptors (PARs).

Keywords: *Bauhinia*, Cysteine proteinase, Papain, Plant enzyme, Platelet aggregation, Thrombin, Trypsin.

1 Department of Biochemistry
2 Department of Biophysics,
Universidade Federal de Sao Paulo-Escola Paulista de Medicina, 04044-020 Sao Paulo, SP, Brazil.
* *Corresponding author*: E-mail: olivaml.bioq@epm.br, Tel: Phone: +5511- 55764444; Fax: +5511-55723006.

Introduction

Bauhinia forficata is the species most used in popular therapy for diabetes mellitus treatment in Brazil. A tree that can reach 3 meters, it is known as cow paw, due to the characteristic bilobed aspect of its leaves from which the homemade extract is prepared and ingested as an infusion (Khalil *et al.*, 2008).

Due to its folk medicinal application, pharmacological studies have been performed as part of the search for antidiabetic activity, and a significant antioxidant activity, beneficial in the prevention of diabetes complications associated with oxidative stress, was reported. Indeed, *B. forficata* has been the subject of several other studies (Oliveira *et al.*, 2005; Lim *et al.*, 2006; Khalil *et al.*, 2008; Volpato *et al.*, 2008). and recently the purification and characterization of a proteinase activity from its leaves was described (Andrade *et al.*, 2011).

The role of proteinases is not restricted to protein degradation, since their involvement as signaling molecules has been described (Salvesen and Dixit, 1997; Timmer *et al.*, 2007; Timmer and Salvesen, 2007).

Proteinase signaling pathways are strictly regulated, and the deregulation of proteinase activity can lead to cardiovascular and inflammatory diseases, cancer, osteoporosis and neurological disorders. Several small molecule drugs targeting proteinases are already available and many more are in development (Premzl *et al.*, 2003; Reuning *et al.*, 2003; De Clercq, 2004; Abbenante and Fairlie, 2005; Turk, 2006).

The network of enzyme activation is a complex process which occurs via at least two consecutive proteolytic steps, with one proteinase in a zymogen (inactive enzymatic precursors) form being the substrate for an active proteinase. In principle, all zymogen activations that are not autocatalytic fall into this category (Neurath, 1994; Neurath, 1999) and the most studied is the blood coagulation cascade, which involves the sequential activation of a series of serine proteinases with thrombin activation as a final step, leading to fibrinogen activation and blood clot formation (Davie *et al.*, 1991; Colman and Schmaier, 1997; Davie, 2003; Turk, 2006).

The serine proteinase thrombin plays pivotal roles in the activation of additional serine proteinase zymogens, cofactors, and cell-surface receptors. Thrombin generation is closely regulated in order to locally achieve rapid hemostasis after injury, without causing uncontrolled systemic thrombosis (Bode and Kubler, 1992; Bode *et al.*, 1992; Kim *et al.*, 2004a; Kim *et al.*, 2004b; Tanaka *et al.*, 2009). By triggering cellular responses at the site of vascular damage by extrinsic and intrinsic coagulation cascades, thrombin is an important agonist for platelet activation.

Platelet activation process and subsequent accumulation at sites of vascular injury are the first steps in hemostasis (Gachet, 2001; Kunapuli *et al.*, 2003; Hardy *et al.*, 2005). As in the process of aggregation, platelets adhere to each other at sites of vascular injury, aggregation has long been recognized as critical for hemostatic plug formation and thrombosis. Platelet aggregation is the consequence of multiple adhesion processes involving distinct receptors and adhesive ligands such as von Willebrand factor (vWF), fibrinogen, and fibronectin (McNicol and Israels, 2003). Furthermore, multiple pathways are involved in platelet activation, as those involving adenosine diphosphate (ADP), thromboxane A2 (TXA2), serotonin, collagen and thrombin (Adam *et al.*, 2005; Rivera *et al.*, 2009).

Thrombin mediates its cellular effects primarily through a family of proteinase-activated receptors (PARs), a family of G-protein coupled to seven transmembrane domain receptors that are activated by proteolytic cleavage of their amino-terminal exodomain. PARs are responsible for a variety of physiological roles that proteinases seem to play in as signaling molecules (Vu *et al.*, 1991; Fox *et al.*, 2002; Kim *et al.*, 2004b; Nieman *et al.*, 2005; Kim *et al.*, 2006). Activated upon hydrolysis after specific Arg or Lys residues, PARs 1–4 are widely expressed in the central nervous system and is important also in platelet activation by thrombin (Vu *et al.*, 1991; Striggow *et al.*, 2001; Angelo *et al.*, 2006), and are activated in several diseases (Ossovskaya and Bunnett, 2004; Bradesi, 2009), particularly PAR 2 and PAR 4 in inflammation (Steinhoff *et al.*, 2005). It is known that PARs couple to Gq and G12/13 pathways, and that thrombin and thrombin receptor–activating peptides cause Gi stimulation through P2Y12 receptor activation by secreting ADP (Kim *et al.*, 2006).

In vivo, a central role among platelet-activating factors is played by ADP, which induces multiple platelet responses and potentiates platelet aggregation caused by other agonists (Gachet, 2001; Hardy *et al.*, 2005). ADP is released from platelet-dense granules upon activation by agonists such as thrombin and collagen, and binds to purinergic receptors (P2Y1 and P2Y12) to reinforce platelet aggregation and thrombus formation (Hollopeter *et al.*, 2001; Kunapuli *et al.*, 2003). Collagen is the main platelet agonist at sites of vascular damage and is therefore central to the process of hemostasis. The collagen receptor comprises the recently cloned glycoprotein VI (GPVI) and the Fc receptor (FcR) g-chain. Within its cytoplasmic domain the FcR g-chain contains an ITAM that is tyrosinephosphorylated upon collagen stimulation, leading to multiple signaling pathways activation, platelet secretion and aggregation (Clark *et al.*, 1999; Bodor *et al.*, 2003).

Extracts from the latex of *Carica papaya* are the source of several proteinases therapeutically used in humans to treat thrombosis, rheumatic arthritis, adult periodontal disease, atherosclerosis and cancer, among other diseases (Metzig *et al.*, 1999). These enzymes (EC 3.4.22), papains, are the archetypal of the endopeptidyl hydrolases with a cysteine residue in their active center, and are usually identified on the basis of the effect of their active site inhibitors (iodoacetate, iodoacetamide and E64), and activation by thiol compounds, expressed in viruses, bacteria, animals and plants (Grudkowska and Zagdanska, 2004). The family of cysteine proteinases includes the mammalian lysosomal cathepsins, the cytosolic calpains (*i.e.* calcium activated cysteine proteinases) as well as several parasitic proteinases (Barrett and Rawlings, 2007; Rawlings *et al.*, 2008; Kos *et al.*, 2009).

From plants, it was reported that bromelain displays antiplatelet and anticoagulant properties and reduces the platelet sensitivity to ADP induced aggregation *in vitro* (Metzig *et al.*, 1999). Here, we report the anti-aggregating property of baupain, a cysteine proteinase found in *B. forficata* leaves and the PARs cleavage site, by using FRET peptides comprising the sequences of the proteinase activated receptors that span the cleavage sites for receptor activation.

Material and Methods

Enzyme Preparation

Baupain was prepared as described by Andrade *et al.* (2011). Briefly, leaves (50 g) collected from wild *Bauhinia forficata* trees were homogenized in a blender with 0.15 M NaCl (700 ml). The proteins in the crude extract were precipitated by 80 per cent (v/v) acetone at 4 °C. The sediment separated by centrifugation was dried at room temperature, dissolved in 0.1 M sodium phosphate buffer, pH 6.3 and applied on a size exclusion chromatography (Sephadex G-25) equilibrated with 0.1 M sodium phosphate buffer, pH 6.3. Further purification was achieved by affinity chromatographies on a Canecystatin-Sepharose, equilibrated with the same phosphate buffer and on Con A Sepharose, equilibrated with 0.1 M sodium phosphate buffer, pH 6.3, used for removal of pigments. The whole purification procedure was carried out at 4 °C.

Platelet Aggregation Assays

Venous blood, obtained from apparently healthy donors, was collected into plastic tubes containing 3.8 per cent trisodium citrate 1:10 (v/v). The effect of baupain (2, 4, and 10 μM) on the aggregation was assayed with washed human platelets or platelet-rich plasma (PRP). PRP was obtained by centrifugation at 141 x*g* for 12 min at room temperature. Washed platelets (WP) were obtained by centrifugation of PRP added of 2 per cent EDTA (EDTA/PRP 1:20) at 880 x*g* for 15 min. The procedure of washing platelets was repeated twice, with washing buffer (140 mM NaCl, 10 mM $NaHCO_3$, 2.5 mM KCl, 0.9 mM Na_2HPO_3, 2.1 mM $MgCl_2$, 22 mM $C_6H_5Na_3O_7$, 0.055 mM Glucose, 0.053 mM BSA, pH 6.5), and finally platelet pellets were resuspended in Tyrode's solution containing calcium (10 mM Hepes, 134 mM NaCl, 1 mM $CaCl_2$, 12 mM $NaHCO_3$, 2.9 mM KCl, 0.34 mM Na_2HPO_4, 1 mM $MgCl_2$, 0.055 mM glucose, pH 7.4). The cells were counted in Counter KX-21N System and adjusted for a final concentration of $2.6 \times 10^5 - 3 \times 10^5$ platelets with Tyrode buffer and platelet aggregation was performed in a Chrono-log Corporation, Whole-Blott following the method described by Born and Cross (Born and Cross, 1963). A 400 μl aliquot of the washed platelet suspension was placed in the aggregometer cuvettes, and allowed to warm to 37°C for 5 min. Then, the agonist solution containing baupain was added and the aggregation curve was recorded for 5 min. Thrombin (1 UNIH/500 μl) was used as agonist for aggregation in washed platelet suspensions. Control was run at the beginning and at the end of the experiment.

Statistical Analysis

Differences between mean values were calculated using one-Way ANOVA, followed by Turkey's correction test and considerated to be significant at $p<0.05$.

Peptide Synthesis

The FRET peptide substrates were synthesized by the solid-phase method (Korkmaz *et al.*, 2008). The molecular mass and purity of synthesized peptides were assessed by analytical HPLC and by MALDI-TOF using a Microflex - LT (Bruker – Daltonics, Billerica, MA, USA) mass spectrometer. Stock solutions of peptides were prepared in DMF and their concentrations were determined spectrophotometrically using the 2,4-dinitrophenyl goup (Dnp) molar extinction coefficient of 17.300 $M^{-1}cm^{-1}$ at 365 nm.

Hydrolysis of PARs Substrate by Baupain

Baupain (0.23 µM) hydrolysis of synthetic substrates comprising the sequences of the proteinase activated receptors, PAR1-4 (9.8 to 39.2 mM) was monitored. The substrates consisted of ten and fifteen amino acid residues spanning the receptor activation cleavage site, carrying an N-terminal 2-aminobenzoyl (Abz) group and a C-terminal 2,4-dinitrophenyl (EDDnp). PAR-1:Abz-TLDPRSFLLR-[EDDnp]-NH_2, PAR-2:Abz-SSKGRSLIGK-[EDDnp]-NH_2, PAR-3:Abz-TLPIKTFRGA-[EDDnp]-NH_2, PAR-4:Abz-LPAPRGYPGQ-[EDDnp]-NH_2. Longer substrates used were PAR 1 Abz-KATNATLDPRSFLLQ-[EDDnp]-NH_2, PAR 2 Abz-QGTNRSSKGRSLIGQ- [EDDnp]-NH_2, PAR 3 Abz-NLAKPTLPIKTFRGQ-[EDDnp]-NH_2, and PAR 4 Abz-STPSILPAPRGYPGQ-[EDDnp]-NH_2. The reaction was monitored fluorimetrically with an excitation wavelength of 320 nm and an emission wavelength of 420 nm, on a Hitachi F-2000 (Tokyo, Japan) spectrofluorimeter.

Determination of the Substrate Cleavage Sites

The scissile bonds of hydrolyzed FRET peptides were identified by isolation of the fragments using analytical HPLC. The molecular weight of each product was determined by LC/MS using an LCMS-2010 mass spectrometer equipped with an ESI-probe (Shimadzu, Japan) or by amino acid analysis.

Results and Discussion

Baupain activity was detected in the acetone precipitation (80 per cent v/v) of *B. forficata* leaves extract by Z-Phe- Arg-MCA hydrolysis. Mr was estimated at 30 kDa by SDS PAGE. Baupain hydrolyzes Z-Phe-Arg-MCA and Bz-Arg-pNan with K_m values 1.25 µM and 0.70 mM, respectively. With optimal activity at pH 6.3 and 40 °C, baupain is activated by dithiotreitol and inhibited by E-64(IC50 = 7.6 µM) and iodoacetamide (IC50 = 17.5 uM). No effective inhibition occurs with benzamidine, TLCK, TPCK and EDTA. The plant serine peptidase inhibitors EcTI (*Enterolobium contortisiliquum* trypsin inhibitor) and SBTI (soybean trypsin inhibitor) do not block baupain activity. On the other hand, baupain is inhibited by HMWK (high molecular weight kininogen) (K_{iapp} = 1.9×10^{-8} M) and by canecystatin (K_{iapp} = 0.18×10^{-9} M), which lodges the sequence motif QVVA, but not by BbCI, *Bauhinia bauhinioides* cruzipain inhibitor (Oliva and Sampaio, 2009). The CD spectra and the N-terminal sequence similarity include baupain as a member of papain (clan CA) family. By following the procedure described (Andrade *et al.*, 2011), baupain was isolated through a size exclusion chromatography on Sephadex G-25 and multiple chromatographic steps and purified to homogeneity. While the N-terminal sequence similarity, molecular mass, circular dichroism spectra and intrinsic fluorescences profiles points to a close structural relationship to papain, baupain activity on FRET and MCA peptides indicates substrate specificity more related to the mammalian cathepsin L enzyme. Baupain also shares with cathepsin L and papain the capacity of releasing bradykinin from high molecular weight kininogen (HMWK) (Andrade *et al.*, 2011). In our studies enough baupain was obtained for the studies of platelet aggregation to compare with the effect described for bromelain, a cysteine proteinase from the stem of *pineapple* (Rowan and Buttle, 1994; Metzig *et al.*, 1999; Maurer, 2001). The homogeneity of the preparation was followed by reverse phase chromatography on HPLC system C18 column (*Andrade et al.*, 2011).

As shown in Figure 13.1 A, addition of baupain at final concentrations up to 2 μM to a washed human platelet suspensions, led to a dose-dependent inhibition of extent of aggregation. Overall, the maximal extent of aggregation in thrombin stimulated platelets decreased from 62.2 per cent to 26.4 per cent in 4 μM of baupain ($P<0.01$) (Figure 13.1B). However, baupain does not affect ADP and collagen induced human platelet aggregation (Figures 13.1C and D), showing similar specificity of action to bromelain in which the effect on platelet aggregation induced by ADP and collagen was significantly lower (Metzig *et al.*, 1999).

Table 13.1: Kinetic parameters for the hydrolysis of the FRET peptides derived from sequences that span the cleavage sites for activation of proteinase-activated receptors (PAR) 1–4 by baupain.

Peptide	*Abz-peptidyl-EDDnp*	k_{cat}/K_m $(10^6 M^{-1}s^{-1})$
PAR 1	Abz TLDPRSF↓LLQ-EDDnp	1.4
PAR 2	Abz SSKGRSLIG↓Q-EDDnp	0.3
PAR 3	Abz TLPIKTF↓RGQ-EDDnp	1.0
PAR 4	Abz LPAP**R**↓GYPGQ-EDDnp	1.3

The following conditions for substrate hydrolysis were used: 0.1 M sodium phosphate buffer, pH 6.3 containing 2.0 mM DTT (dithiothreitol). All the reactions were performanced at 37 °C. ↓ indicates the cleaved peptide bond.

Protease-activated receptors unique mechanism of activation require proteases to cleave part of their extracellular N-terminal sequence, allowing the tethered ligand that is formed from the truncated portion of the N-terminus to bind to the receptor itself, leading to activation (Vu *et al.*, 1991; Kawabata *et al.*, 1997; Fox *et al.*, 2002). While PAR2 can be activated by trypsin and mast cell tryptase, PAR1 and 4 are considered as thrombin receptors. However, PAR4 has low affinity for thrombin, requiring high concentration of this agonist for activation. Protease-activated receptor-4 can also be activated by trypsin, cathepsin G, the activated factor X of the coagulation cascade and trypsin IV (Ramachandran and Hollenberg, 2008). The induction of platelet aggregation by baupain seems to be linked to the proteolysis of PARs, as seen by the hydrolysis of the synthetic substrates derived from sequences that span the cleavage sites for activation of these proteinase-activated receptors (Table 13.1). The efficiency of receptor hydrolysis by baupain is in the same order of magnitude of that of trypsin on PAR-1/2/3 and 4 substrates (k_{cat}/K_m values of 5.18×10^6, 2.08×10^7, 1.72×10^6 and 7.31×10^6 $M^{-1} s^{-1}$ respectively) and thrombin PAR-1/3 and 4 substrates (k_{cat}/K_m values of 2.72×10^5, 9.49×10^4 and 7.22×10^5 $M^{-1}s^{-1}$) (Fox *et al.*, 2002). Moreover, the extension of the substrate did not influence efficiency of baupain hydrolysis ($kcat/K_m$ PAR 1 Abz-**KATNATLDPRSFLLQ**[EDDnp]-NH_2 $0.98 \times 10^6 M^{-1} s^{-1}$, PAR 2

Figure 13.1: Inhibition of thrombin-induced human platelet aggregation by baupain. Washed human platelets or platelet-rich plasma (PRP) ($2.7 \times 10^5 - 3.0 \times 10^5$/mL) **(A)** was placed in the aggregometer cuvettes, and allowed to warm to 37°C for 5 min. Then, the agonist solution containing baupain (2.0, 4.0, and 10.0 μM) was added and the aggregation curve was recorded for 5 min. Thrombin (1 UNIH/500 μL) was used as agonist control. Tracings are representative results of a typical experiment. **(B)** Data are expressed as mean ± SEM for 3 independent experiments. * $p<0.01$ vs baupain 2 μM; *$p<0.01$ vs baupain 4 μM and **$p<0.001$ vs baupain 10 μM. Effect of baupain on ADP (5 μM) (**C**) and collagen (2 μg/ml) (**D**)-induced platelet aggregation.

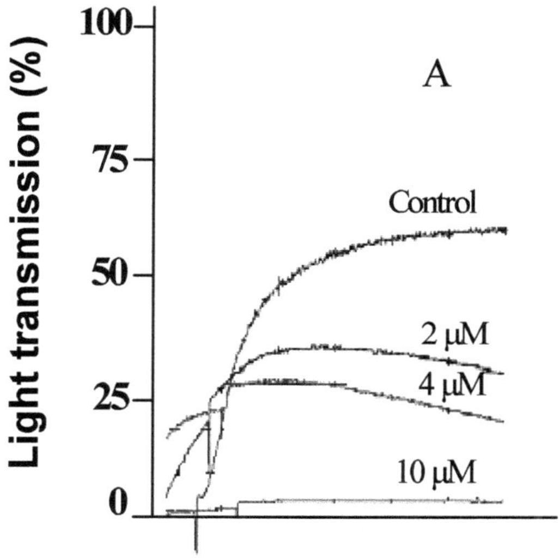

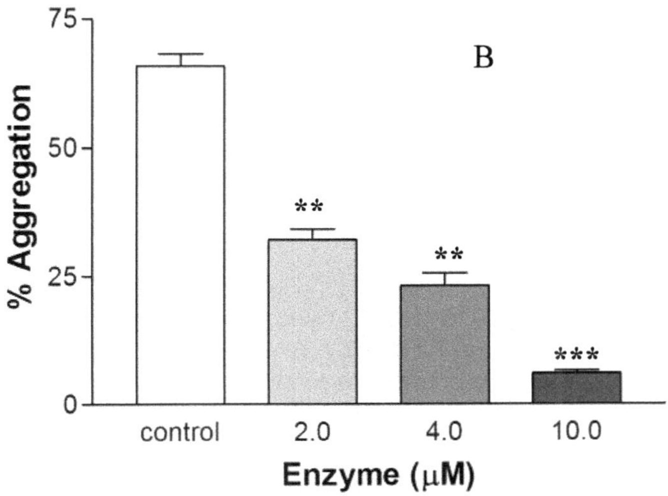

Contd...

Figure 13.1–*Contd...*

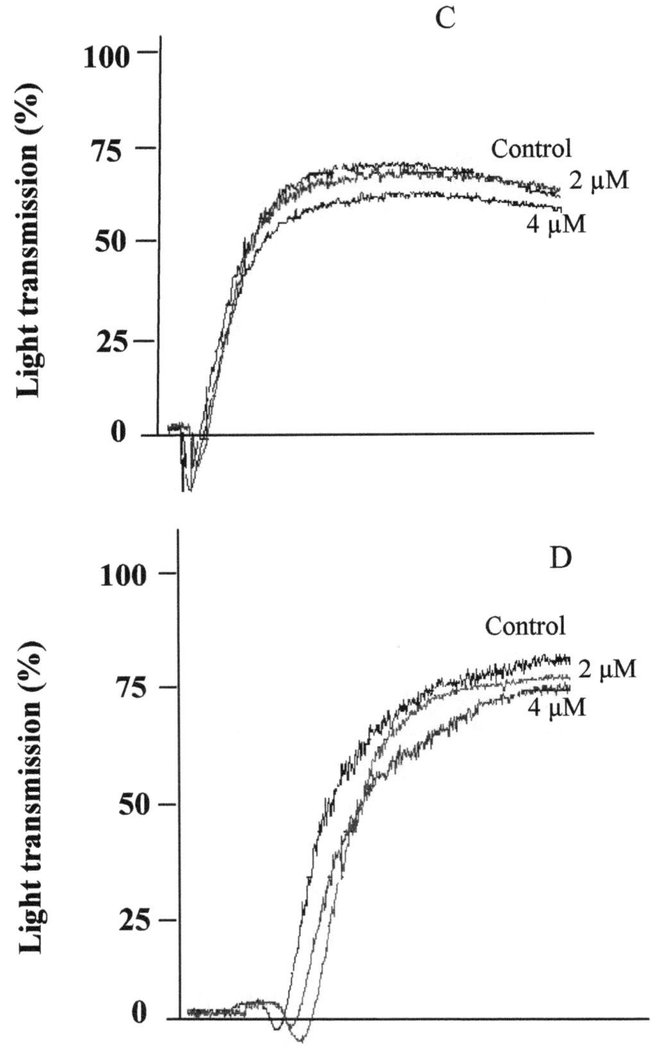

Abz-**QGTNR**SSKGRSLIGQ- [EDDnp]-NH_2 0.37 × $10^6 M^{-1} s^{-1}$, PAR 3 Abz-**NLAKP**TLPIKTFRGQ-[EDDnp]-NH_2 1.04 × 10^6 $M^{-1} s^{-1}$, and PAR 4 Abz-**STPSI**LPAPRGYPGQ-[EDDnp]-NH_2 0.31 × 10^6 $M^{-1} s^{-1}$ extended in five amino acid residues from the N-terminal region. By this action, the proteolytic activity of baupain might affect cell surface receptors leading to the hinders of thrombin mediated aggregation. Finally, of key interest in model studies, is the ability of baupain, like thrombin and trypsin, to regulate cell function by cleaving and activating the proteinase activated receptor (PAR1-4) pathway. As shown in Table 13.1, only the peptide derived from PAR 4 was hydrolyzed by baupain in the activated position upon hydrolysis after specific Arg residue, as observed in platelet activation by thrombin (Vu *et al.*, 1991; Striggow *et al.*, 2001, Angelo *et al.*, 2006). In this sense, hydrolysis of PAR1, 2 and 3 by baupain would not activate the PARs signaling pathways, but would inactivate the cleaved receptor. Particularly, our results indicate that hydrolysis of PAR-4 by baupain could be one of the factors responsible for interference on thrombin induction of platelet aggregation, observed at Figure 13.1. This action on PARs by baupain would confirm that this or other cysteine proteinases are part of a classical ligand-receptor signaling cascade. Recently, Reddy *et al.* (2010) showed that human cathepsin S evokes itch and activates human protease-activated receptors 2 and 4. Baupain and other endogenous or exogenous cysteine proteases may activate PARs as part of additional inflammatory processes. Taken together, we suggest that the exogenous cysteine proteinase baupain interacts with PAR-4 as part of signaling cascades in homeostasis and platelet activation. The usefulness of baupain in human diseases should be investigated.

Acknowledgments

This work was supported in Brazil by Fundaçao de Amparo à Pesquisa do Estado de Sao Paulo (FAPESP) and Conselho Nacional de Desenvolvimento Científico e Tecnológico (CNPq).

References

Abbenante, G., and Fairlie, D. P. (2005). Protease inhibitors in the clinic. *Med Chem.*, **1:** 71-104.

Adam, F., Zheng, S., Joshi, N., Kelton, D. S., Sandhu, A., Suehiro, Y., Jeimy, S. B., Santos, A. V., Masse, J. M., Kelton, J. G., *et al.* (2005). Analyses of cellular multimerin 1 receptors: *in vitro* evidence of binding mediated by alphaIIbbeta3 and alphavbeta3. *Thromb Haemost.*, **94**: 1004-1011.

Andrade, S. S., Silva-Lucca, R. A., Santana, L. A., Gouvea, I. E., Juliano, M. A., Carmona, A. K., Araújo, M. S., Sampaio, M. U., and Oliva, M. L. (2011). Biochemical characterization of a cysteine proteinase from *Bauhinia forficata* leaves and its kininogenase activity. *Process Biochemistry*, **46**: 572-578.

Angelo, P. F., Lima, A. R., Alves, F. M., Blaber, S. I., Scarisbrick, I. A., Blaber, M., Juliano, L., and Juliano, M. A. (2006). Substrate specificity of human kallikrein 6: salt and glycosaminoglycan activation effects. *J Biol Chem.*, **281**: 3116-3126.

Barrett, A. J., and Rawlings, N. D. (2007). 'Species' of peptidases. *Biol Chem.*, **388:** 1151-1157.

Bode, W., Turk, D., and Karshikov, A. (1992). The refined 1.9-A X-ray crystal structure of D-Phe-Pro-Arg chloromethylketone-inhibited human alpha-thrombin: structure analysis, overall structure, electrostatic properties, detailed active-site geometry, and structure-function relationships. *Protein Sci.*, **1**: 426-471.

Bodor, E. T., Waldo, G. L., Hooks, S. B., Corbitt, J., Boyer, J. L., and Harden, T. K. (2003). Purification and functional reconstitution of the human P2Y12 receptor. *Mol Pharmacol.*, **64:** 1210-1216.

Born, G. V., and Cross, M. J. (1963). The Aggregation of Blood Platelets. *J Physiol.*, **168:** 178-195.

Bradesi, S. (2009). PAR4: a new role in the modulation of visceral nociception. *Neurogastroenterol Motil.*, **21:** 1129-1132.

Clark, R. B., Knoll, B. J., and Barber, R. (1999). Partial agonists and G protein-coupled receptor desensitization. *Trends Pharmacol Sci.*, **20:** 279-286.

Colman, R. W., and Schmaier, A. H. (1997). Contact system: a vascular biology modulator with anticoagulant, profibrinolytic, antiadhesive, and proinflammatory attributes. *Blood*, **90:** 3819-3843.

Davie, E. W. (2003). A brief historical review of the waterfall/cascade of blood coagulation. *J Biol Chem.*, **278:** 50819-50832.

Davie, E. W., Fujikawa, K., and Kisiel, W. (1991). The coagulation cascade: initiation, maintenance, and regulation. *Biochemistry* (*Mosc*), **30:** 10363-10370.

De Clercq, E. (2004). Antivirals and antiviral strategies. *Nat Rev Microbiol.*, **2:** 704-720.

Fox, M. T., Greer, B., Lawson, J., Healy, A., and Harriott, P. (2002). Kinetic analysis of the cleavage of human protease-activated receptor-1/2/3 and 4 using quenched-fluorescent peptide substrates. *Protein Pept Lett.*, **9:** 387-397.

Gachet, C. (2001). ADP receptors of platelets and their inhibition. *Thromb Haemost.*, **86:** 222-232.

Grudkowska, M., and Zagdanska, B. (2004). Multifunctional role of plant cysteine proteinases. *Acta Biochim Pol.*, **51:** 609-624.

Hardy, A. R., Conley, P. B., Luo, J., Benovic, J. L., Poole, A. W., and Mundell, S. J. (2005). P2Y1 and P2Y12 receptors for ADP desensitize by distinct kinase-dependent mechanisms. *Blood*, **105:** 3552-3560.

Hollopeter, G., Jantzen, H. M., Vincent, D., Li, G., England, L., Ramakrishnan, V., Yang, R. B., Nurden, P., Nurden, A., Julius, D., and Conley, P. B. (2001). Identification of the platelet ADP receptor targeted by antithrombotic drugs. *Nature*, **409:** 202-207.

Kawabata, A., Saifeddine, M., al-Ani, B., and Hollenberg, M. D. (1997). Protease-activated receptors: development of agonists selective for receptors triggered by either thrombin (PAR1) or trypsin (PAR2). *Proc West Pharmacol Soc.*, **40:** 49-51.

Khalil, N. M., Pepato, M. T., and Brunetti, I. L. (2008). Free radical scavenging profile and myeloperoxidase inhibition of extracts from antidiabetic plants: *Bauhinia forficata* and *Cissus sicyoides*. *Biol Res.*, **41:** 165-171.

Kim, J. S., Kluskens, L. D., de Vos, W. M., Huber, R., and van der Oost, J. (2004a). Crystal structure of fervidolysin from *Fervidobacterium pennivorans*, a keratinolytic enzyme related to subtilisin. *J Mol Biol.*, **335:** 787-797.

Kim, S., Jin, J., and Kunapuli, S. P. (2004b). Akt activation in platelets depends on Gi signaling pathways. *J Biol Chem.*, **279:** 4186-4195.

Kim, S., Jin, J., and Kunapuli, S. P. (2006). Relative contribution of G-protein-coupled pathways to protease-activated receptor-mediated Akt phosphorylation in platelets. *Blood*, **107**: 947-954.

Korkmaz, B., Attucci, S., Juliano, M. A., Kalupov, T., Jourdan, M. L., Juliano, L., and Gauthier, F. (2008). Measuring elastase, proteinase 3 and cathepsin G activities at the surface of human neutrophils with fluorescence resonance energy transfer substrates. *Nat Protoc.*, **3:** 991-1000.

Kos, J., Jevnikar, Z., and Obermajer, N. (2009). The role of cathepsin X in cell signaling. *Cell Adh Migr.*, **3:** 164-166.

Kunapuli, S. P., Ding, Z., Dorsam, R. T., Kim, S., Murugappan, S., and Quinton, T. M. (2003). ADP receptors–targets for developing antithrombotic agents. *Curr Pharm Des.*, **9:** 2303-2316.

Lim, H., Kim, M. K., Lim, Y., Cho, Y. H., and Lee, C. H. (2006). Inhibition of cell-cycle progression in HeLa cells by HY52, a novel cyclin-dependent kinase inhibitor isolated from *Bauhinia forficata. Cancer Lett.*, **233:** 89-97.

Maurer, H. R. (2001). Bromelain: biochemistry, pharmacology and medical use. Cell Mol *Life Sci.*, **58:** 1234-1245.

McNicol, A., and Israels, S. J. (2003). Platelets and anti-platelet therapy. *J Pharmacol Sci.*, **93:** 381-396.

Metzig, C., Grabowska, E., Eckert, K., Rehse, K., and Maurer, H. R. (1999). Bromelain proteases reduce human platelet aggregation *in vitro*, adhesion to bovine endothelial cells and thrombus formation in rat vessels *in vivo. In Vivo*, **13:** 7-12.

Neurath, H. (1994). Proteolytic enzymes past and present: the second golden era. Recollections, special section in honor of Max Perutz. *Protein Sci.*, **3:** 1734-1739.

Neurath, H. (1999). Proteolytic enzymes, past and future. *Proc Natl Acad Sci U S A.*, **96:** 10962-10963.

Nieman, M. T., Pagan-Ramos, E., Warnock, M., Krijanovski, Y., Hasan, A. A., and Schmaier, A. H. (2005). Mapping the interaction of bradykinin 1-5 with the exodomain of human protease activated receptor 4. *FEBS Lett.*, **579:** 25-29.

Oliva, M. L., and Sampaio, M. U. (2009). Action of plant proteinase inhibitors on enzymes of physiopathological importance. *An Acad Bras Cienc.*, **81:** 615-621.

Oliveira, C. Z., Maiorano, V. A., Marcussi, S., Sant'ana, C. D., Januario, A. H., Lourenco, M. V., Sampaio, S. V., Franca, S. C., Pereira, P. S., and Soares, A. M. (2005). Anticoagulant and antifibrinogenolytic properties of the aqueous extract from *Bauhinia forficata* against snake venoms. *J Ethnopharmacol.*, **98:** 213-216.

Ossovskaya, V. S., and Bunnett, N. W. (2004). Protease-activated receptors: contribution to physiology and disease. *Physiol Rev.*, **84:** 579-621.

Premzl, A., Zavasnik-Bergant, V., Turk, V., and Kos, J. (2003). Intracellular and extracellular cathepsin B facilitate invasion of MCF-10A neoT cells through reconstituted extracellular matrix *in vitro. Exp Cell Res.*, **283:** 206-214.

Ramachandran, R., and Hollenberg, M. D. (2008). Proteinases and signalling: pathophysiological and therapeutic implications via PARs and more. *Br J Pharmacol.*, **153** (Suppl 1): 263-282.

Rawlings, N. D., Morton, F. R., Kok, C. Y., Kong, J., and Barrett, A. J. (2008). MEROPS: the peptidase database. *Nucleic Acids Res.*, **36:** D320-325.

Reuning, U., Sperl, S., Kopitz, C., Kessler, H., Kruger, A., Schmitt, M., and Magdolen, V. (2003). Urokinase-type plasminogen activator (uPA) and its receptor (uPAR): development of antagonists of uPA/uPAR interaction and their effects *in vitro* and *in vivo*. *Curr Pharm Des.*, **9:** 1529-1543.

Rivera, J., Lozano, M. L., Navarro-Nunez, L., and Vicente, V. (2009). Platelet receptors and signaling in the dynamics of thrombus formation. *Haematologica*, **94:** 700-711.

Rowan, A. D., and Buttle, D. J. (1994). *Pineapple* cysteine endopeptidases. *Methods Enzymol.*, **244:** 555-568.

Salvesen, G. S., and Dixit, V. M. (1997). Caspases: intracellular signaling by proteolysis. *Cell*, **91**: 443-446.

Steinhoff, M., Buddenkotte, J., Shpacovitch, V., Rattenholl, A., Moormann, C., Vergnolle, N., Luger, T. A., and Hollenberg, M. D. (2005). Proteinase-activated receptors: transducers of proteinase-mediated signaling in inflammation and immune response. *Endocr Rev.*, **26:** 1-43.

Striggow, F., Riek-Burchardt, M., Kiesel, A., Schmidt, W., Henrich-Noack, P., Breder, J., Krug, M., Reymann, K. G., and Reiser, G. (2001). Four different types of protease-activated receptors are widely expressed in the brain and are up-regulated in hippocampus by severe ischemia. *Eur J Neurosci.*, **14:** 595-608.

Tanaka, K. A., Key, N. S., and Levy, J. H. (2009). Blood coagulation: hemostasis and thrombin regulation. *Anesth Analg.*, **108:** 1433-1446.

Timmer, J. C., Enoksson, M., Wildfang, E., Zhu, W., Igarashi, Y., Denault, J. B., Ma, Y., Dummitt, B., Chang, Y. H., Mast, A. E., *et al.* (2007). Profiling constitutive proteolytic events *in vivo*. *Biochem J.*, **407:** 41-48.

Timmer, J. C., and Salvesen, G. S. (2007). Caspase substrates. *Cell Death Differ.*, **14:** 66-72.

Turk, B. (2006). Targeting proteases: successes, failures and future prospects. *Nat Rev Drug Discov.*, **5:** 785-799.

Volpato, G. T., Damasceno, D. C., Rudge, M. V., Padovani, C. R., and Calderon, I. M. (2008). Effect of *Bauhinia forficata* aqueous extract on the maternal-fetal outcome and oxidative stress biomarkers of streptozotocin-induced diabetic rats. *J Ethnopharmacol.*, **116:** 131-137.

Vu, T. K., Hung, D. T., Wheaton, V. I., and Coughlin, S. R. (1991). Molecular cloning of a functional thrombin receptor reveals a novel proteolytic mechanism of receptor activation. *Cell*, **64:** 1057-1068.

Traditional and Folk Herbal Medicine:
Recent Researches Vol. 1 (2012)
Editor: **V.K. Gupta**
Published by: **DAYA PUBLISHING HOUSE, NEW DELHI**

Pages **301–314**

14

Ajwain (*Tracyspermum ammi*) Seeds in Health Promotion and Disease Prevention

B. Andallu[1]*, C.U. Rajeshwari[1] and A.V. Vinay Kumar[2]

ABSTRACT

In Ayurvedic medicines, ajwain is used as a medicinal plant for its antispasmodic, stimulant, tonic and carminative properties. The ajwain volatile oil, rich in thymol, acts as a powerful antiseptic, has agreeable odour and exhibited a broad fungitoxic spectrum by inhibiting the mycelial growth of all tested fungi and especially it is useful in controlling a variety of fungal infections of skin. The methanolic extracts of ajwain seeds possess natural antioxidant properties. Besides, the volatile oil and acetone extract have been proven to be alternative sources of natural antioxidants and more efficacious than various synthetic antioxidants, viz., butylated hydroxy anisole (BHA) and butylated hydroxy toluene (BHT). Ajwain oil and thymol are known to possess a number of functional properties viz., antimicrobial, antiflatulent and antispasmodic, antirheumatic, diuretic, stimulant, carminative and expectorant.

Keywords: Ajwain (*Tracyspermum ammi*), Antimicrobial, Antiflatulent and anti spasmodic, Antirheumatic, Diuretic, Stimulant, Carminative and expectorant.

1 Department of Home Science, Sri Sathya Sai University, Anantapur, A.P., India

2 The Institute of Environmental and Human Health, Texas Tech University, Lubbock, USA.

* *Corresponding author*: E-mail: bandallu@gmail.com; Ph: 91-92474-50990.

Abbreviations

BHA: Butylated hydroxy anisole; BHT: Butylated hydroxy toluene; TBA: Thiobarbituric acid; AN: *Aspergillus niger*; AF: *Aspergillus flavus*; AO: *Aspergillus oryzae*; AO': *Aspergillus ochraceus*; FG: *Fusarium graminearum*; FM: *Fusarium monoliform*; PC: *Penicillium citrium*; PV: *Penicillium viridicatum*; PM: *Penicillium madriti*; CL: *Curvularia lunata*; FID: Flame ionization detection; tr: traces.

Introduction

Currently there has been an increasing interest globally to identify antioxidant compounds that are pharmacologically potent and have low or no side effects for use in preventive medicine and in the food industry. As several metabolic diseases and age-related degenerative disorders are closely associated with oxidative processes in the body, the use of herbs and spices as a source of natural antioxidants warrants further attention. The antioxidant activities in herbs may be attributable to active components of the essential oil and also due to synergistic interactions between them. Therefore, supplementation of a balanced diet with herbs may have beneficial health effects (Lee and Shibamoto, 2002). With time, a greater body of scientific evidence supporting the benefits of herbs and spices in the overall maintenance of health and protection from disease will be seen. Immediate studies should focus on the validation of the antioxidant capacity of herbs and spices after harvest, as well as testing their effects as markers of oxidative stress (Ali *et al.*, 2008). Ajwain, commonly known as 'omum' is such a herb mentioned in the Indian traditional medicine but not much exploited scientifically.

Botanical Description and Nutrient Composition

Ajwain (*Tracyspermum ammi* L.) is an annual aromatic and herbaceous plant.

English name: Ajwain seeds

Family name: Apiaceae

Common Names

Ajwain, carom, Ethiopian cumin, wild parsley, bishop's weed. It is also called netch (white) azmad (Amharic), ajwan, kamun al-mulaki, taleb el koub (Arabic), joni-gutti (Assamese), jowan, yamani (Bengali), yan-jhon-wuihheung (Cantonese), nanava (Farsi), ajowan (Dutch, French, German, Italian), ayamo, yavan (Gujarati), ajwain, carom omum (Hindi), ajamoda, oma (Kannada), ayowan (Korean), ajowan (Japanese), ayamodakam (Malayalam), yin-dou-zeng-hui-xiang (Mandarin), javano (Nepali), oregano-semente, ajowan (Portugese), ajavain (Punjabi), assamodum (Singhalese), ajowan (Spanish), omam (Tamil), omu (Telugu), chilan (Thai) and misir anason (Turkish).

Parts Used

Seeds

Important Chemical Constituents

Beta-pinene, para-cymene, alpha-pinene limonene with gamma and beta-terpinenes.

Ajwain seeds contain variable amounts of nutrients and crude fibre (Table 14.1) and are considered as diuretic, antimicrobial, antiflatulent, antispasmodic, antirheumatic, diuretic, stimulant, carminative and expectorant. Stimulating chief constituent of essential oil is thymol and is used for flavouring, culinary, household and cosmetic purposes (The Wealth of India, 1976).

Table 14.1: Nutrient composition of ajwain seeds (The Wealth of India, 1976).

Parameter		*g per cent*
Moisture	–	8.9
Protein	–	15.4
Fat	–	18.1
Fibre	–	11.9
Carbohydrates	–	38.6
Total Mineral	–	7.1
Calcium	–	1.42
Phosphorus	–	0.30
Iron	–	14.6 mg/100g

Cultivation and Usage

Ajwain is cultivated extensively as a cold season crop in the plains and as a summer crop in the hills. It requires a warm and long frost – free growing season. It can be grown in any soil type loamy to sandy loam and even in black soils. In India, it is grown under both rain-fed and irrigated cultivation systems (Agarwal *et al.*, 2009).

Ajwain, *Trachyspermum ammi* L. Sprague, is an erect annual herb with striate stem, originated in eastern regions of Persia and India. Its fruits are grayish, minute and shaped like and egg. Its seeds are used in Indian, Greek, Unani and Egyptian systems of medicines (Figure 14.1). Ajwain is cultivated in the Mediterranean region and south-west Asian countries. Ajwain originated in the Middle East, possibly in Egypt. It is now primarily grown and used in the Indian subcontinent, but also in Iran, Egypt and Afghanistan. It is sometimes used as an ingredient in berbere, a spice mixture favored in Eritrea and Ethiopia. In India, the major ajwain producing states are Rajasthan and Gujarat, where Rajasthan produces about 90 per cent of India's total production. The most utilized part of ajwain is the small caraway like fruit, which is particularly popular in Indian savory recipes, savory pastries and snacks (http://en.wikipedia.org/wiki/ajwain).

Chemical Composition and Medicinal Properties

Ajwain is highly valued in India as a gastrointestinal medicine and as an antiseptic. It is combined with salt and hot water and taken after meals to relieve pain in bowel or colic pain, and to improve digestion. Ajwain is also a traditional remedy for cholera and fainting spells (Nadkarni, 1976). Westerners generally use it against coughs and throat issues. Ajwain is an ingredient in mouthwashes and toothpastes because of its antiseptic properties and also valued for the following properties *viz.*, flatulence, indigestion, polyuria, asthma, bronchitis, common cold, toothache, various other gastro-intestinal disorders, cardialgia, ear-ache, pain in throat, arthritic and rheumatic pains, migraine and also used as an aphrodisiac. It is also widely used to

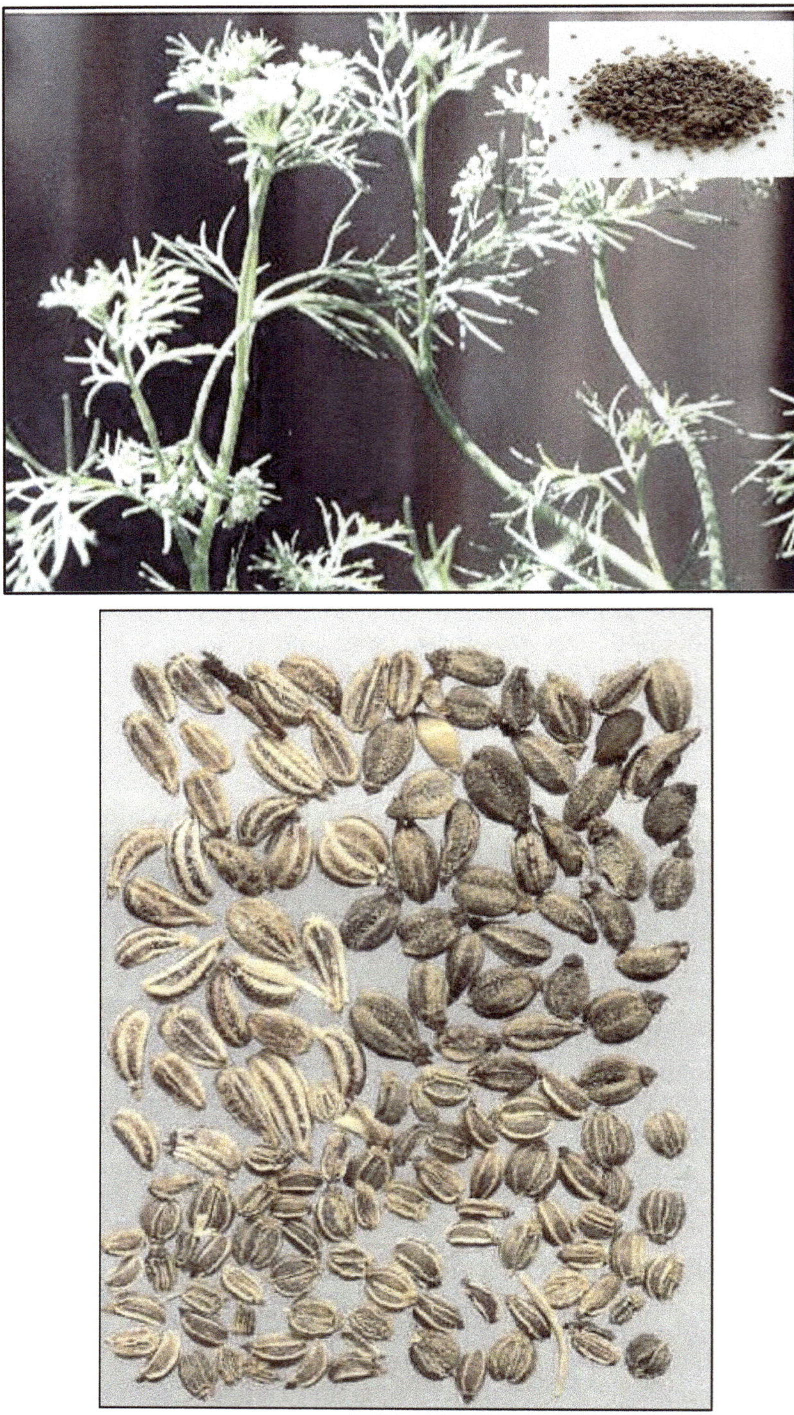

Figure 14.1: Ajwain (*Tracyspermum ammi*) leaves and seeds. (www.kitchenwonders.blogsspot.com)

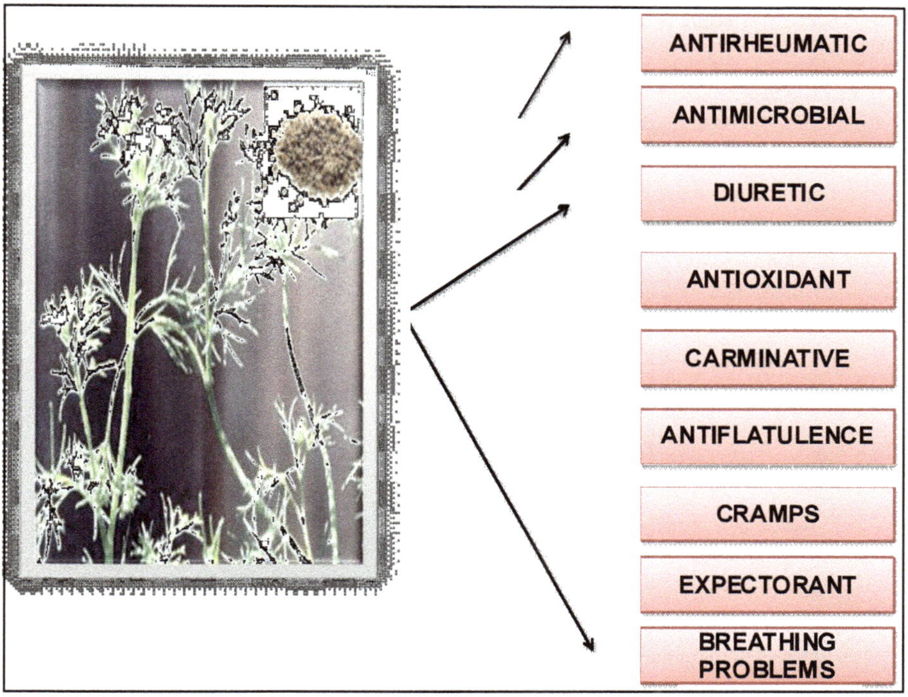

Figure 14.2: Traditional uses of Ajwain (*Tracyspermum ammi*).

enhance virility and premature ejaculation, besides use of its grayish brown seeds and fruits as spice (www.webindia123.com/spices) (Figure 14.2).

Chemical Composition

The volatile oil present in the seeds of ajwain is one of the principal constituents responsible for providing a typical flavour owing to the presence of thymol. It also contains a cumene and terpene – "thymene". The fruits of ajwain yield 2-4 per cent of essential oil, containing thymol. The remainder of the oil consists of p-cymene, β-pinene, α-pinene and carvacrol (Pruthi, 2001).

The investigation of aroma compounds of the essential oils and acetone extract of dried fruits of ajwain was carried out by means of GC and GC-MS to identify the odorous target component responsible for characteristic odor of valuable spice and food flavoring products. The analysis of oil and acetone extract of ajwain was undertaken by using previously stated HP-5 series. The volatile oil, on analysis showed the presence of 26 identified components (Table 14.2), which account for 96.3 per cent of the total amount. Thymol (39.1 per cent) was found as a major component along with *p*-cymene (30.8 per cent), γ-terpinene (23.3 per cent), α-pinene (1.7 per cent), terpinene-4-ol (0.8 per cent), and several other components in minor percentages. Moreover, acetone extract showed the presence of 18 identified components, which account for 68.8 per cent of the total amount. The major component was thymol (39.1 per cent), followed by oleic acid (10.4 per cent), linoleic acid (9.6 per cent), γ-terpinene

Table 14.2: Chemical composition of ajwain essential oil and acetone extract[a] (Singh *et al.*, 2004).

Compound	*Essential Oil*		*% FID of Acetone Extract*
	Per cent FID	*KIb*	
2-methyl but-3-ene-2-ol	tr	0807	
2-methyl(methyl butanoate)	tr	0872	
alpha-thujene	0.2	0931	
alpha-pinene	0.2	0941	tr
camphene	tr	0953	
sabinene	tr	0975	
beta-pinene	1.7	0980	0.2
myrcene	0.4	0993	
alpha-phellandrene	tr	1007	
delta-3-carene	tr	1013	
alpha-terpinene	0.2	1020	
p-cymene	30.8	1026	1.6
beta-phellandrene	0.6	1030	1.3
1,8-cineole	tr	1035	
gamma-terpinene	23.2	1064	2.6
terpinolene	0.2	1088	
trans-sabinenhydrate	0.1	1098	
linalool	0.1	1099	
terpinen-4-ol	0.8	1177	tr
alpha-terpineol	0.1	1189	tr
thymol	39.1	1291	39.1
carvacrol	0.3	1298	0.3
beta-selinene	0.1	1492	
alpha-selinene	tr	1499	
caryophyllene oxide	tr	1583	
delta-dodecalactone	tr		
4-methylpent-3-ene-2-one			0.5
4-hydroxy-4-methylpent-2-one			1.1
xylene (isomer not identified)			tr
undecane			0.1
thymoquinone			0.2
2,3,4,6-tetramethyl phenol			0.1
palmitic acid			1.6
linoleic acid			9.6
oleic acid			10.4
Total	96.3 per cent		68.8 per cent

a: Percentages are the mean of three runs and were obtained from electronic integrations measurements using flame ionization detection (FID) trace, <0.01 per cent.

b: The retention indices were calculated for all volatile constituents using a homologous series of *n*-alkanes C8-C16.

(2.6 per cent), *p*-cymene (1.6 per cent), palmitic acid (1.6 per cent), and 4-hydroxy-4-methylpenta-2-one (1.1 per cent). It is interesting to note that thymol content in both the oil and acetone extract was almost the same (Singh *et al.*, 2004). Chialva *et al.* (1993) already reported that thymol is a major component of this oil and there is a compositional variation between a steam distilled oil and oleoresin.

Therapeutic Uses and Biological Activities

Ajwain (omum) with its characteristic aromatic smell and pungent taste is widely used as a spice in curries. The seeds are used in small quantities for flavouring numerous foods, as preservative in medicines and for the manufacture of essential oil for ultimate use in perfumery (Pruthi, 1992). Chemical constituents and the biological activities of *Trachyspermum ammi* are given in Figure 14.3 (www.ars-grin.gov/duke). In literature, there are many reports available on chemical composition and antimicrobial studies (Singh *et al.*, 2002) of volatile oil of ajwain. Ajwain oil and thymol are known to possess a number of functional properties (Meera and Sethi, 1994) *viz.*, antimicrobial, antiflatulent, antispasmodic, antirheumatic, diuretic, stimulant, carminative and expectorant (Singh *et al.*, 2004).

In Indian system of medicine, *T. ammi* is administered as a household remedy for stomach disorders, a paste of crushed fruits is applied externally for relieving colic pains and a hot and dry fermentation of the fruits applied on chest, is used as a common remedy for asthma (The Wealth of India, 1995). *T. ammi* has been shown to possess anti-aggregatory effects (Srivastava, 1988), anthelmintic (Lateef *et al.*, 2006), antihyperlipidaemic (Javed *et al.*, 2006), antifilarial (Mathew *et al.*, 2008), insecticidal (Chaubey, 2008), kidney stone inhibitory (Kaur *et al.*, 2009), molluscicidal (Singh *et al.*, 1997, 1999; Singh and Singh, 2000), mosquito repellent (Pandey *et al.*, 2009) and nematicidal activities (Park *et al.*, 2007).

Antioxidative Activity

The methanolic extracts of ajwain seeds possess natural antioxidant properties. However, the acetone extract showed better antioxidative activity for linseed oil as compared with synthetic antioxidants such as butylated hydroxy toluene and butylated hydroxy anisole (Frag *et al.*, 1990). Figure 14.4 shows absorbance values measured for oil, acetone extract, BHA, and BHT, along with the control for 12 days. These results are well correlated with those obtained previously by peroxide value and thiobarbituric acid (TBA) methods. It is known that the effectiveness of added antioxidants varies depending on the food, on the processing and storage conditions (Yanishlieva *et al.*, 2002). BHT is very effective in animal fat but less effective in vegetable oils. Volatile oil and BHT may also be lost during heating because of their volatility. BHA is known to be very effective antioxidant for vegetable oils and it is more stable at high temperatures than BHT (Gordon and Kourimska, 1995).

Most likely, acetone extracts are more effective than BHA and BHT at high temperature. In terms of retarding the formation of primary and secondary oxidation products, the effectiveness of samples added to the concentration of 200 ppm can be put into the following order: acetone extract > oil > BHA > BHT > control. Several factors can induce changes in the antioxidant activity during the distillation of volatile

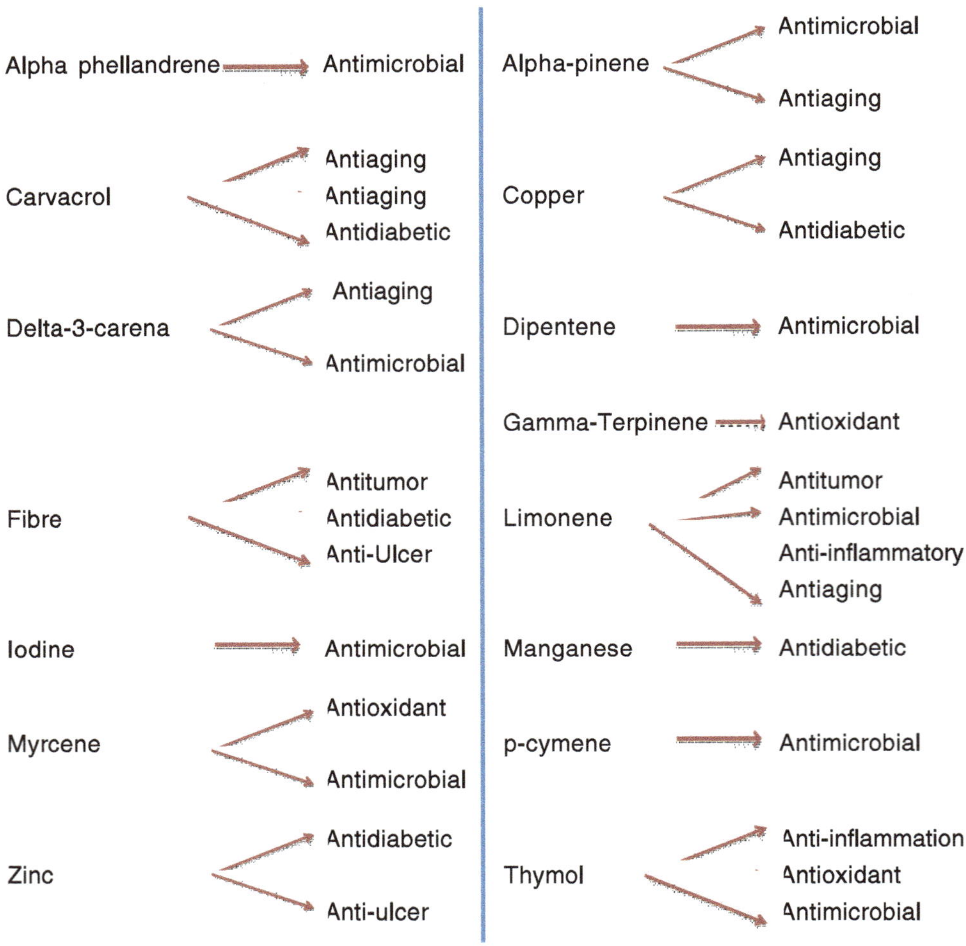

Figure 14.3: Chemicals in *Trachyspermum ammi* (ajwain) seeds and their biological activities (www.ars_grin.gov/duke).

constituents from spices. The volatile oil and acetone extract both contain thymol (39.1 per cent) as a major component which refers in general, to the antioxidative properties of these extracts. In acetone extract, oleic acid (10.4 per cent) and linoleic acid (9.6 per cent) do not possess antioxidant activity. Probably other substances which could be present in acetone extract and not identified by GC can contribute to improve antioxidant activity of acetone extract, which needs further investigations. Moreover, heat- and water-induced chemical reactions can also change the activity of a complex extract system consisting of numerous compounds with different chemical and physical properties (Singh *et al.*, 2004).

Fungicidal Activity

Ajwain oil exhibited a broad spectrum of fungitoxic behavior against all tested fungi such as *Aspergillus niger, Aspergillus flavus, Aspergillus oryzae,*

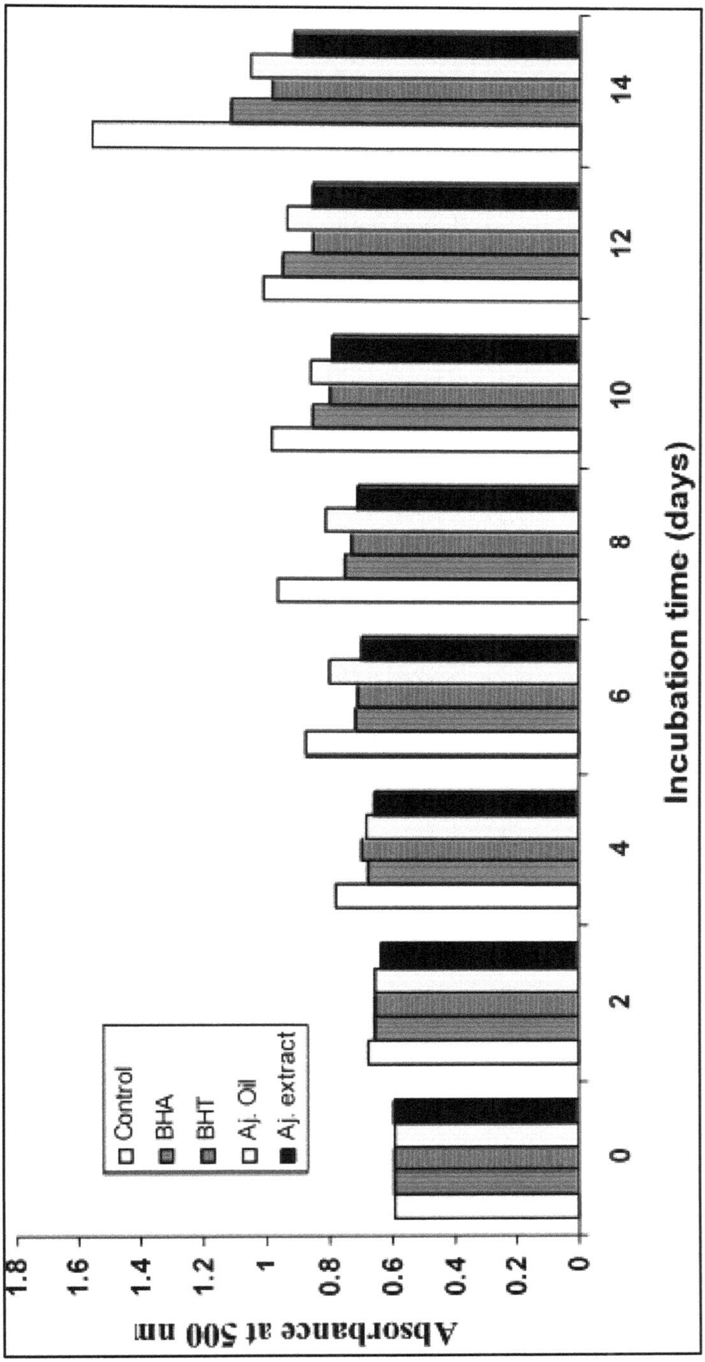

Figure 14.4: Determination of antioxidant activity by linoleic acid method at 40°C (Singh *et al.*, 2004).

Aspergillus ochraceus, Fusarium monoliforme, Fusarium graminearum, Pencillium citrium, Penicillium viridicatum, Pencillium madriti, and *Curvularia lunata* as absolute mycelial zone inhibition was obtained at a 6-µL dose of the oil using Inverted Petriplate technique, and the oil was found to be 100 per cent antifungal against all the tested fungi (Table 14.3). It was highly effective in controlling mycelial growth of tested Penicillium species even at a 2-µL dose of the oil. Moreover, the oil was also found to be active against AO′, PC, and CL at a 2-µL dose as more than 80 per cent mycelial zone inhibition was obtained, whereas using the same method, acetone extract was found less effective than the volatile oil, as only more than 50 per cent mycelial zone inhibition of AF, AO′, FG, FM, and CL was obtained even at a 6-µL dose. Moreover, using food poison technique (Table 14.4), the oil was found highly effective against AN, AF, AO, AO′, FM, and PV as absolute mycelial zone inhibition was obtained at 6 µL. It is very interesting to note that 100 per cent activity of the fungi AN, FM and PV was obtained even at a 2-µL dose of the oil, whereas using the same method acetone extract was found to be comparatively less effective than the volatile oil, as only more than 70 per cent mycelial zone inhibition was obtained for FG, FM, PC and PM. Various researchers have already reported that this oil exhibited a broad fungitoxic spectrum, inhibiting the mycelial growth of a number of fungi at 100, 200, and 300 ppm (Diwivedi and Singh, 1999). The oil was thermostable and more efficacious than various synthetic fungicides. Moreover, the data was found to be highly significant ($p < 0.05$) (Meera and Sethi, 1994).

Table 14.3: Antifungal investigations of ajwain essential oil and acetone extract by using inverted petriplate technique (Singh *et al.*, 2004).

Fungus[a]	*Per cent Mycelial Zone Inhibition at Different Dose of Sample*[b]					
	Ajwain Essential Oil			*Ajwain Acetone Extract*		
	2 µL	*4µL*	*6µL*	*2µL*	*4µL*	*6µL*
Aspergillus niger (AN)	68.7	87.5	100	12.5	25.0	37.5
Aspergillus flavus (AF)	37.5	56.3	100	12.5	50.0	52.5
Aspergillus oryzae (AO)	75.0	87.5	100	25.0	13.7	21.3
Aspergillus ochraceus (AO¢)	87.5	100	100	11.3	62.5	87.5
Fusarium graminearum (FG)	50.0	75.0	100	0.0	50.0	75.0
Fusarium monoliforme (FM)	75.0	87.5	100	12.5	37.5	50.0
Penicillium citrium (PC)	87.5	100	100	12.5	18.7	43.8
Penicillium viridicatum (PV)	100	100	100	6.25	12.5	25.0
Penicillium madriti (PM)	100	100	100	18.7	26.3	38.7
Curvularia lunata (CL)	87.5	93.7	100	43.7	50.0	75.0

a: For all tested fungi, the data was found to be highly significant ($p < 0.05$).

b: Average of three replicates.

Antimicrobial Activities

Ajwain oil exhibited a remarkable antibacterial activity against *Staphylococcus aureus, Escherichia coli, Salmonella typhi, Shigella dysenteriae* and *Vibrio cholera* (Syed *et al.*, 1986; Anonymous, 1995). The essential oils extracted from the seeds of *T. ammi* showed antibacterial activity (Singh *et al.*, 2002; Mayaud *et al.*, 2008). Extracts prepared in different solvents exhibited variable activity against *E. coli, P. aeruginosa, S. typhi* and *S. aureus* (Ahmad *et al.*, 1998; Patel *et al.*, 2008) suggesting their centuries old usage in the treatment of gastrointestinal disorders. This historical use of ajwain (omum) seeds to cure various gastrointestinal disorders has also been scientifically proved in another study carried out by Kaur and Arora (2009) wherein aqueous and organic extracts of omum seeds have also shown their antibacterial effect. Methanol extracts of *T. ammi* showed significant *in vitro* inhibitory effect on hepatitis C virus (HCV) protease at a concentration 100 µg/ml (Hussein *et al.*, 2000).

Table 14.4: Antifungal investigations of ajwain essential oil and acetone extract by using food poison technique (Singh *et al.*, 2004).

Fungus[a]	*Per cent Mycelial Zone Inhibition at Different Dose of Sample*[b]					
	Ajwain Essential Oil			*Ajwain Acetone Extract*		
	2 µL	*4µL*	*6µL*	*2µL*	*4µL*	*6µL*
Aspergillus niger (AN)	100.0	100.0	100.0	18.7	26.3	38.7
Aspergillus flavus (AF)	87.5	100.0	100.0	27.5	41.3	46.3
Aspergillus oryzae (AO)	87.5	100.0	100.0	25.0	43.8	62.5
Aspergillus ochraceus (AO¢)	25.0	37.5	100.0	25.0	31.3	50.0
Fusarium graminearum (FG)	18.7	50.0	56.3	6.3	50.0	75.0
Fusarium monoliforme (FM)	100.0	100.0	100.0	62.5	77.5	87.5
Penicillium citrium (PC)	6.2	13.8	40.0	25.0	62.5	87.5
Penicillium viridicatum (PV)	100.0	100.0	100.0	50.0	52.5	60.0
Penicillium madriti (PM)	50.0	75.0	87.5	37.5	50.0	75.0
Curvularia lunata (CL)	18.7	26.3	46.3	18.8	26.3	46.3

a: For all tested fungi, the data was found to be highly significant ($p < 0.05$).

b: Average of three replicates.

Antibacterial Activity

The antibacterial activity of homogenized seed oil and residues indicates that the powdered seeds of *C. copticum* exhibit antibacterial activity against *Staphylococcus aureus* only and not *E. coli*. The oil extracted with *n*-hexane exhibited antibacterial activity against both organisms, while the oil extracted with ethanol and *n*-hexane did not. The residue left after ethanol extraction exhibits antibacterial activity against *E. coli* only (Qasim and Khan, 2001). The antimicrobial activity of ajwain oils against *Aspergillus niger, S. cerevisiae, Mycoderma* sp., *Lactobacillus acidophilus* and *B. cereus*, as estimated by the paper disc agar diffusion method, has been reported by Meera and Sethi (1994).

Antifilarial Activity

The methanolic extract of *T. ammi* has exhibited promising macrofilaricidal activity against adult bovine filarial worm *S. digitata* and the active principle responsible for this has been identified as thymol. The position isomer of thymol namely carvacrol also showed *in vitro* antifilarial activity but to a lesser extent than thymol. The active molecule isolated from *T. ammi* has promising *in vivo* antifilarial activity against human filarial worm *B. malayi* in terms of adult worm mortality as well as female worm sterility (Mathew *et al.*, 2008).

Conclusions

Ajwain is valued for the following properties *viz.*, flatulence, indigestion, polyuria, asthma, bronchitis, common cold, toothache, various other gastro-intestinal orders, cardialgia, ear-ache, pain in throat, arthritic and rheumatic pains, migraine, and also used as an aphrodisiac. Ajwain is highly valued in India as a gastrointestinal medicine and as an antiseptic. Ajwain seed oil and thymol are known to possess a number of functional properties *viz.*, antioxidant, antimicrobial, antiflatulent, antispasmodic, antirheumatic, diuretic, stimulant, carminative and expectorant. Ajwain seed oil also possesses fungicidal, antibacterial, antifilarial properties.

References

Agarwal, B.B., Michelle, E., Kuiken, V., Laxmi, H. Iyer., Kuzhuvelil, B., Harikumar. and Sung, B. (2009). Molecular targets of nutraceuticals derived from dietary spices: Potential role in suppression of inflammation and tumorgenesis. *Experimental Biology and Medicine*, 1-35.

Ahmad, I., Mehmood, J. and Mohammad, F. (1998). Screening of some Indian medicinal plants for their antimicrobial properties. *Journal of Ethnopharmacology*, **62**: 183-193.

Ali, S. S., Kasoju, N., Luthra, A., Sing, A., Sharanabasava, H., Sahu, A. and Bora, U. (2008). Indian medicinal herbs as source of antioxidants. *Food Research International*, **141**: 1-15.

Anonymous, (1995). In: The Wealth of India, A dictionary of Indian Raw Materials and Industrial Products Publications and Information Directorate (CSIR), New Delhi. Vol. XXI,

Chaubey, M.K. (2008). Fumigant toxicity of essential oils from some common spices against pulse beetle, *Callosobruchus chinensis* (Coleoptera: Bruchidae). *Journal of Oleoresin Science*, **57**: 171-179.

Chialva, F., Monguzzi, F., Manitto, P. and Akgul. (1993). Essential oil constituents of *Trachyspermum capticum* (L.) Link fruits. *Journal of Essential Oil Research*, **5**: 105-106.

Diwivedi, S. K. and Singh, K. P. (1999). Fungitoxicity of some higher plant products against *Macrophomina phaseolina* (Tassi) Goid. *Flavour and Fragrance Journal*, **14**: 315-318.

Frag, R. S., Ali, M. N. and Taha, S. H. (1990).Use of some essential oils as natural preservatives for butter. *Journal of American Oil Chemists Society*, **68**: 188-191.

Gordon, H. H.and Kourimska, L. (1995). The effect of antioxidants on changes in oils during heating and deep frying. *Journal of Science Food Agriculture*, **68**: 347-353.

http: //en.wikipedia.org/wiki/ajwain (accessed on February 2010).

Hussein, G., Miyashiro, H., Nakamura, N., Hattori, M., Kakiuchi, N. and Shimotohno K (2000). Inhibitory effects of Sudanese medicinal plant extracts on hepatitis C virus (HCV) protease. *Phytotherapy Research*, **14**: 510-516.

Javed, I., Iqbal, Z., Rahman, Z.U., Khan, F.H., Muhammad, F., Aslam, B. and Ali, L. (2006). Comparative antihyperlipidaemic efficacy of *Trachyspermum ammi* extracts in albino rabbits. *Pakistan Veterinary Journal* **26**: 23-29.

Kaur, G.J. and Arora, D.S. (2008). *In vitro* antibacterial activity of three plants belonging to the family Umbelliferae. *International Journal of Antimicrobiology Agents*, **31**: 393-395.

Kaur, T., Bijarnia, R.K., Singla, S.K. and Tandon, C. (2009). Purification and characterization of an anti-calcifying protein from the seeds of *Trachyspermum ammi* (L.). *Protein and Peptide Letters*, **16**: 173-181.

Lateef, M., Iqbal, Z., Akhtar, M.S., Jabbar, A., Khan, M.N. and Gilani, A.H. (2006). Preliminary screening of *Trachyspermum ammi* (L.) seed for anthelminthic activity in sheep. *Tropical Animal Health Production*, **38**: 491-496.

Lee, K. G. and Shibamoto, T. (2002). Determination of antioxidant potential of volatile extracts isolated from various herbs and spices. *Journal of Agricultural Food Chemistry*, **50**: 4947-4952.

Mathew, N., Misra-Bhattacharya, S., Perumal, V. and Muthuswamy, K. (2008). Antifilarial lead molecules isolated from *Trachyspermum ammi*. *Molecules*, **13**: 2156-2168.

Mayaud, L., Carricajo, A., Zhiri, A. and Aubert, G. (2008). Comparison of bacteriostatic and bactericidal activity of 13 essential oils against strains with varying sensitivity to antibiotics. *Letters of Applied Microbiology*, **47**: 167-173.

Meera, M. R. and Sethi, V. (1994). Antimicrobial activity of essential oils from spices. *Journal of Food Science and Technology*, **31**: 68-70.

Nadkarni, K.M. (1976). *Indian Materia Medica*, Vol 1, 3rd Edn, Popular Prakashan, pp 11-1243.

Pandey, S.K., Upadhyay, S. and Tripathi, A.K. (2009). Insecticidal and repellent activities of thymol from the essential oil of *Trachyspermum ammi* (Linn) Sprague seeds against *Anopheles stephensi*. *Parasitology Research*, **105**: 507-512.

Park, I.K., Kim, J., Lee, S.G. and Shin, S.C. (2007). Nematicidal activity of plant essential oils and components from Ajowan (*Trachyspermum ammi*), Allspice (*Pimenta dioica*) and Litsea (*Litsea cubeba*) Essential oils against Pine Wood Nematode (*Bursaphelenchus xylophilus*). *Journal of Nematology*, **39**: 275-279.

Patel, J.D., Patel, D.K. and Shrivastava A, Kumar, V. (2008) Screening of plant extracts used in traditional antidiarrhoeal medicines against pathogenic *Escherichia coli*. *Scientific World*, **6**: 63-67.

Pruthi, J. S. (1992) Spices and Condiments. 4th Ed. National Book trust, New Delhi.

Pruthi, J. S. (2001). Spices and condiments: chemistry, microbiology, technology. New York: Academic Press, pp 251-265.

Qasim, M. and Khan, M.R. (2001) Biochemical and antibacterial studies of Ajowan (*Carum copticum*) oil. *Pakistan Journal of Scientific and Industrial Research*, **44**(3), 184–185.

Singh, K. and Singh, D.K. (2000). Effect of different combinations of MGK-264 or piperonyl butoxide with plant derived molluscicides on snail reproduction. *Archives of Environmental Contamination and Toxicology*, **38**: 182-190.

Singh, S., Singh, V.K. and Singh, D.K. (1997). Molluscicidal activity of common spice plants. *Biological Agricultural Horticulture*, **14**: 237-249.

Singh, V.K., Singh, S. and Singh, D.K. (1999). Effect of active molluscicidal component of spices on different enzyme activities and biogenic animal levels in the nervous tissue of *Lymnae acuminate*. *Phytotherapy Research*, **13**: 649-654.

Singh, G., Kapoor, I. P. S., Pandey, S. K., Singh, S. K., Singh,U. K. and Singh, R. K. (2002). Studies on essential oils, Part 10: Antibacterial activity of volatile oils of some spices. *Phytotherapy Research*, **16**: 680-682.

Singh, G., Maurya, S., Catalan, C., and De, Lampasona. M. P. (2004). Chemical constituents, antifungal and antioxidative effects of ajwain essential oil and its acetone extract. *Journal of Agricultural Food Chemistry*, **11**: 52: 3296-3292.

Srivastava, K.C. (1988). Extracts of a spice-omum (*Trachyspermum ammi*)- shows anti-aggregatory effects and alters arachidonic acid metabolism in human platelets. *Prostaglandins Leukotrienes and Essential Fatty Acids*, **33**: 1-6.

Syed, M., Sabir, A.W., Chaudhary, F.M. and Bhatty, M.K. (1986). Antimicrobial activity of essential oils of umbelliferae part II- *Trachyspermum ammi, Daucus carota, Anethum graveolens* and *Apium graveolens*. *Pakistan Journal of Science and Industrial Research*, **28**: 189-192.

The Wealth of India (1976). A dictionary of Indian raw materials and industrial products materials. Council of Scientific and Industrial Research, Hill Side, Road. New Delhi. **11**: 80-285.

www.ars-grin.gov/duke (accessed on February 2010).

www.kitchenwonders.blogsspot.com (accessed on February 2010).

www.webindia123.com/spices (accessed on February 2010).

Yanishlieva, N. V., Eldin, A. K., Marinova, E. M. and Toheva, A. S. G. (2002). Kinetics of antioxidant action of R- and ç-tocopherols in sunflower and soybean triacyl glycerols. *European Journal of Lipid Science and Technology*, **104**: 262-270.

Traditional and Folk Herbal Medicine: *Pages* **315–328**
Recent Researches Vol. 1 (2012)
Editor: **V.K. Gupta**
Published by: **DAYA PUBLISHING HOUSE, NEW DELHI**

15

Quality Control, Standardization and Chemo Profiling of Herbal Drugs and ISM Preparations

R.K. Khajuria[1*] and D.K. Gupta[1]

ABSTRACT

According to World Health Organization (WHO) estimates 4 billion people use herbal medicine. WHO currently encourages, recommends and promotes traditional herbal remedies in national healthcare programmes because such drugs are available at reasonable price, comparatively safe and people have faith in such remedies. At the same time, WHO in its number of resolutions have emphasized the need to ensure quality, safety and efficacy of the traditional herbal drugs/remedies. Herbal medicine is a major component of all traditional medicine such as ayurvedic, homeopathic, naturopathic and native American Indian Medicine. A standardized extract means that the manufacturer has verified the active ingredient believed to be present in the herb as well as in the formulation and at the same time potency and the amount of the active ingredient is assured. The action of the herb may be from a number of constituents and not from just one or two ingredients. This article describes the various techniques which are used in the quality control and standardization of herbal drugs and ISM preparations.

Keywords: World Health Organization (WHO), Traditional medicines, Ayuvedic, Homeopathic, Naturopathic, Native American Indian medicine, HPLC, GC, GC-MS, LC-MS.

1 Indian Institute of Integrative Medicine (CSIR), Canal Road, Jammu – 180 001, India.

* *Corresponding author*: E-mail: rkhajuria@iim.ac.in

Introduction

The primary health care of 70-80 per cent of the world's population is based on the use of medicinal plants derived from traditional systems of medicine and local health practices. Besides developing countries and under-developed countries, even western countries are changing over to alternative medicines due to the more holistic approach and with substantially fewer side effects - as opposed to a depleted immune system due to the prolonged use of allopathic medicine.

The recent years have witnessed resurgence of interest in herbal medicines as more and more people throughout world are turning to use medicinal plant products in healthcare system. The sales for herbal medicine products have plateaued to such an extent that these products have become available to consumers as positive healthcare just like vitamins. They are now found in supermarkets, pharmacies and numerous other main stream retail outlets as over-the-counter drug products. Herbal medicines are in great demand in the developed as well as developing countries for primary healthcare because of their wide biological activities, higher safety margins and lesser costs. In India. the herbal drug market is about $ one billion and the export of plant based crude drugs is around $ 80 million. But the most important challenges faced by these formulations arise because of their lack of complete standardization. Herbal medicines are prepared from materials of plant origin which are prone to contamination, deterioration and variation in composition. Therefore, quality control of herbal medicines offers a host of problems. To solve this problem, first and foremost task is the selection of the right kind of plant material which is therapeutically efficacious. Fundamentally, a better approach would be the one in which some direct correlation of marker compounds is generated with respect to the biological activity of the extract for a particular therapeutic area or disease pattern. The commonly used analytical methods like chromatography have narrow scope in the analysis of heterogeneous botanical extracts. Most often a desired biological response is due to not one, but a mixture of bioactive plant components and the relative proportions of single bioactive compound can vary from batch to batch, while the bioactivity still remains within tolerable limits. New trends are emerging in the standardization of herbal raw materials whereby it is carried out to reflect the total content of phytoconstituents like polyphenols which can be correlated with biological activity like antioxidant activity which many times has a direct or indirect correlation to the pathophysiological disorders like diabetes, cancer, inflammatory and age related disorders.

Botanical medicines, similar to all pharmacologically active substances have the potential to contribute positively, neutrally, or negatively to the health status of the people. World wide need of alternative medicine has resulted in growth of natural product markets and interest in traditional systems of medicine. The use of plants for health or medicinal purposes has been a part of every culture or region but now this science of herbal drugs is passing through the age of renaissance.

Herbal drugs are now being produced at the large scale and many of the issues such as availability of good quality raw material, authentication of raw material, availability of standards, proper standardization methodology of single drugs and formulations, quality control parameters, etc. are of greater importance. Through out the world, the major problem that hinders the prospects of herbal products in becoming the main stay of any treatment modality is the difficulty in standardizing a formulation. System of Ayurveda is well respected globally and many of the herbals used in Ayurveda, are covered in WHO monographs on selected medicinal plants, thus confirming the efficacy of the Ayurvedic, Siddha and Unani (ASU) System.

There are about 10,000 plant species which are being used in Indian System of Medicines (ISM)/traditional medicines in Indian subcontinent. Out of these, 450-500 species are mostly utilized in over 85 per cent of the Ayurvedic, Unani and Siddha formulations and about 40 plant species are used in modern drugs. Twenty five of these are cultivated while other plant species are collected by proper/destructive harvesting. Important parameters for collection of the plant materials are proper botanical identification, proper chemo-type/ecotype, desired state of maturity, post-harvest processing technology and proper storage conditions. These herbals account for a large chunk of global drug market, therefore assurance of their quality, safety and efficacy needs additional attention.

Safety Parameters

Standardization of herbal drugs is gaining momentum in India and as a result, it is also proposed to include the safety parameters as per International norms. So far as drugs are concerned one must remember that "Safety comes first and quality afterwards". Therefore, In addition to proper botanical identification and physico-chemical parameters, test for heavy metals (Mercury, Lead, Cadmium, Arsenic), microbial contaminants (Total viable aerobic count, Total *Enterobacteriaceae*, Total fungal count), specific pathogens (*E. coli*, *Salmonella* spp., *S. aureus*, *Pseudomonas aeruginosa*), pesticide residue (organochlorine, organophosphorus, pyrethroids and others), aflatoxin (B1, B2, G1, G2) should also be an important part of any Quality control protocol. State-of-art facilities have been installed at Quality Control Laboratory of I. I. I. M., Jammu to test such parameters.

Botanical Concerns

Identification of true ASU herbal drugs, Taxonomic characterization, Collection, Post-Harvest Processing and Storage Many a time similar names have been very frequently used for several related umbelliferous fruits, mint leaves and various other plant materials and thus, different plant species are sold in the market under the same name in different parts of Indian sub-continent. We would like to mention a few to draw attention. "Ajamoda" (*Trachspermum roxbrghianum*), a reputed drug of the Ayurvedic and Unani systems of medicine but in the market *Apium graveolens*, *Apium leptophyllum*, *Trachspermum roxbrghianum* and *T. ammi* are sold under one trade name *i.e.*, Ajamoda (Table 15.1) in different regions (Dutt, 1974). The major chemical constituents of all these four-plant spp. are different and can easily be characterized on the basis of chemical markers.

Many species of *Mentha* such as *M. spicata*, *M. virides*, *M. longifolia*, *M. piperita*, *M. arvensis*, *M. aquaicta* and *M. pulegium* are sold under one name *i.e.* Pudina in various parts of the country but all the spp. contain different monoterpenes. In some cases

Table 15.1: Different plants sold under the name Ajmoda.

Sl.No.	*Species*	*Volatile Oil*	*Important Constituents*
1.	*Apium graveolens* (Celery)	1-2 per cent	Limonene (~85 per cent), pentyl benzene, 3-n-butyl phthalide and other phthalides
2.	*Apium leptophyllum*	1-2 per cent	Monoterpenes, coumarins
3.	*T. roxberghianum*	4-5 per cent	Limonene (~15 per cent), Cadinener (~ 24 per cent), β-cyclolavandulala/acid is 15-25 per cent, Seselin (~12-15 per cent)
4.	*T. ammi*	2-5 per cent	p-cymene, γ-terpinene, thymol

Table 15.2: Different species of Mentha sold under the same name as Pudina.

Sl.No.	*Species*	*Volatile Oil*	*Important Constituents*
1.	*M. spicata*	l-limonene, l-carvone	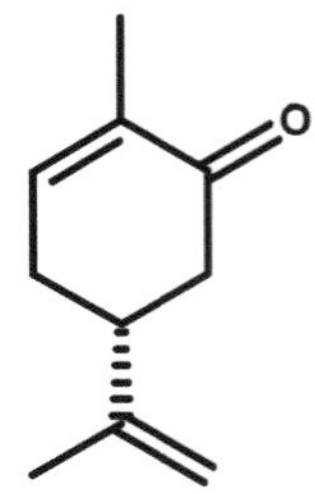
2.	*M. virides*	Piperitenone oxide	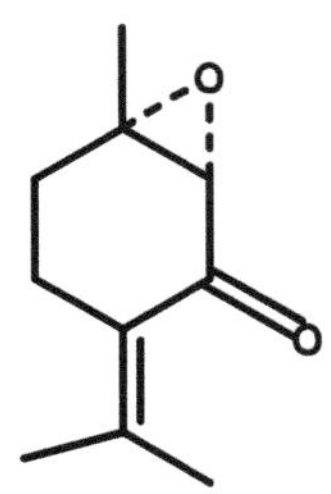
3.	*M. longifolia*	l-limonene, l-carvone	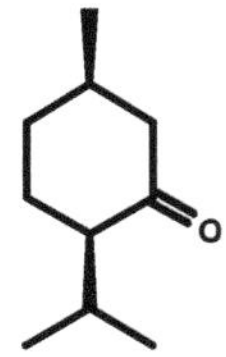
4.	*M. piperita*	l-menthol (~ 50 per cent), isomenthone others	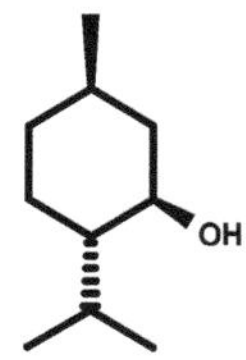

Contd...

Table 15.2–*Contd...*

Sl.No.	Species	Volatile Oil	Important Constituents
5.	*M. arvensis*	l-menthol (~75 per cent), menthone, others	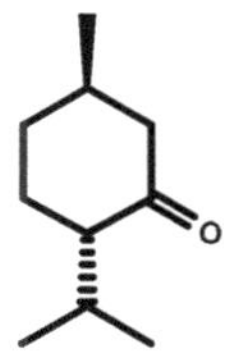
6.	*M. aquatica*	Isopinocamphone	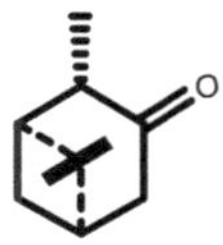
7.	*M. pulegium*	Pulegone (~80 per cent), piperitone, menthone	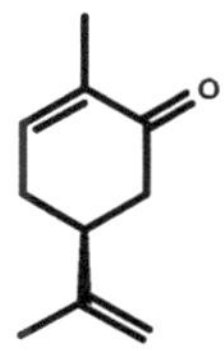

constituting monoterpenes are same; they are present in different amounts in different species (Agnihotri and Agarwal, 2005). Three plant species such as *Carum carvi*, *Bunium persicum* and *Bunium cylindricum* are sold under one name *i.e.*, Kalazeera and all the three contain entirely different constituents. Fruit of another plant *i.e. Cuminum cyminum* is sold under the name of Safed zeera has constituents similar to one of Kalazeera *i.e., Bunium persicum* (Table 15.3) (Thappa and Agarwal, 1991; Agarwal and Vashist, 1974; Agarwal and Thappa, 1979). Similarly different ecotypes yield different constituents with various proportions (Table 15.4)

On the other hand, many a time different plant species contain similar chemical markers and same is true with *Mucuna puriens, M. utilis* and *M. cochchinensis*, which all contain L-dopa as marker compound, although these are very much different in size and appearance. Our own studies (Ahmed *et al.*, 2006) on three *Tinospora* spp. which are sold in the market under the name of "Guduchi" have revealed that there is lot of variation in the chemical constituents of three spp of the plant *i.e. Tinospora cordifolia, T. malabarica* and *T. crispa*.

Further, chemical composition of many plant species is directly related to the phenological stages of plant growth and changes with season. The best example is commonly used vasaka (*Adhatoda vasica*) for cough and cold. At vegetative stage, the total alkaloid content are highest with vasicine (90 per cent) and vasicinone (5 per cent); but at full bloom stage alkaloid content falls with simultaneous variations in quantitative composition with vasicine (45 per cent) and of vasicinone (25 per cent) of the total alkaloids along with several other water soluble glycosylated, oxides and other derivatives of vasicine/vasicinone (Pandita *et al.*, 1993). The post harvest processing of *Crocus sativus* stigma by traditional methods which takes 3-5 days, the amount of total crocins content is 7-9 per cent and if the same stigma is processed at 45°C ± 5°C with tray load of $1M^3$ which requires 6-7 hours of drying, the amount of crocins (Table 15.5) from the plant increases to more than 14 per cent (Raina *et al.*, 1996).

Table 15.3: Different plants sold under the name Kalazeera and Safed Zeera.

Sl.No.	*Species*	*Volatile Oil*	*Important Constituents*
		KALA ZEERA	
1.	*Carum Carvi*	2.5-4 per cent	d-limonene, d-carvone
2.	*Bunium persicum*	4-12 per cent	p-cymene, γ-terpinene, cuminaldehyde, p-mentha-1,3- dienals, p-mentha-1,4-dienals
3.	*Bunium cylindricum*	1-2 per cent	Myristicin, Dillapiole, elemenes
4.	*Nigella sativa*	–	Carvone, limonene, p-cymene, thymol, nigellone
		SAFED ZEERA	
5.	*Cumin cyminum*	1.5-3 per cent	p-cymene, γ-terpinene, Cuminaldehyde, p-mentha-1,3-dienals, p-mentha-1,4-dienals

Table 15.4: Different ecotypes giving different constituents in different ratios.
Ecotype

AJWAIN			
(*Trachyspermum ammi*)	*Oil*	*Thymol*	*p-cymene + γ-terpinene*
Moti Ajwain	1-3 per cent	85-90 per cent	10-15 per cent
Choti Black	3-5 per cent	35-45 per cent	40-60 per cent
KHUS			
(*Vetiveria zizaniodes*)		North Indian oil	**l -Vetiverol** 63-78 per cent
		South Indian oil	**d-Khusinol** (major)

Table 15.5: Yields of crocins by using different drying techniques.

Crocus sativus	
Postharvest Processing	*Crocins Pigment Content*
Traditional processing 3-5 days	7-9 per cent
Processing at 45°C ± 5°C (6 hrs drying)	14-17 per cent

Thus, there is a need to remove ambiguities related to proper herbals to be used in ASU System of medicine followed by proper taxonomical identification, harvesting time, post-harvest processing, storage conditions etc. as these are of utmost importance in maintaining the quality of an herbal product. Therefore, "Standard Operating

Procedures" should be adopted for collection, processing storage and sampling of herbs and herbal products. Further, ASU drugs should be categorized separately, from the EU Herbal drugs, as these have evidence based documented records and are widely used in Indian sub-continent for centuries. Additionally, ASU drugs belong to organized system of healthcare and practiced by experts in the field.

Chemical Standardization and Chemo Profiling of ASU Drugs

There are always chances of wide variations with respect of their chemical contents in crude drugs/raw materials of plant origin due to varied reasons such as climatic conditions, geographical distribution, source and season of collection and lack of scientific methods of post-harvest processing, storage and preservation. Keeping these facts in view for acceptance or rejection criteria, minimum- maximum limits for marker compounds, based on chemical analysis of large number of samples for each drug/herbal collected from different agro-climatic zones with passport data has to be worked out, to have reproducible results.

Therefore, the only solution to place ASU System of medicines on global market, is to chromatographically standardize the herbal products and wherever possible quantitate them for active/or phytochemical marker(s). Thin Layer chromatography is a powerful and simple analytical tool, used for this purpose, there were situations where this tool of analysis did not give satisfactory results because of its own limitations (Singh and Pandey, 2003). Quantitative and qualitative HPLC, HPLC/MS, HPLC/MS/MS, GC, GC/MS, GC/MS/MS, HPLC-UV–DAD and HPTLC are well-suited analytical methods of choice to control the quality of phyto-pharmaceuticals.

In modern times the issues of quality, safety and efficacy of medicines are interrelated. The quality of drug is of paramount importance as it can affect the issues of efficacy and safety. Long history of use and better patient tolerance as well as acceptance, renewable source, cultivation and processing, environmental friendly, local availability especially in the third world and several important recent break-through are the major factors responsible for the resurgence of interest in plant based drugs. The people all over the world have realized that vast plant wealth has much to offer in the shape of new remedies which are efficacious, safe and accessible to masses. By laying guidelines for the assessment of herbal medicines under the programme on traditional medicines, WHO has recognized the role of traditional system of medicines in healthcare programmers all over the world. We in India therefore, have to gear up to the challenge especially when we possess wealth of natural drugs second to none in the world.

According to world statistics developing one new drug needs screening of 10,000 molecules with 10-15 years of time and an expenditure of more than $ 900 million. But if we screen plant drugs from known natural sources with assured therapeutic efficacy, the percentage of hits be increased, thus saving time and money. WHO currently encourages, recommends and promotes traditional herbal remedies in national healthcare programmes because such drugs are available at reasonable price, comparatively safe and people have faith in such remedies. At the same time, WHO in its number of resolutions have emphasized the need to ensure quality, safety and efficacy of the traditional herbal drugs/remedies.

In the present global market scenario related to traditional and herbal drugs, People Republic of China is leading with 50 per cent market share followed by Japan (20 per cent) and Russia (16 per cent).Unfortunately India's share is only 2.5 per cent although the internal market has crossed over Rs.4000 crore per year. The reason is very simple and to ensure fair share in world market the quality control and scientific evaluation data for Indian products has to be generated/presented to global community. Development of ASU drugs with proper efficacy has been undertaken by Department of Ayush, ICMR and CSIR through Golden Triangle Programme utilizing traditional knowledge, latest safety, chemo-profiling and reverse pharmacology route to arrive at standardized formulation (preferably with 3 to 5 ingredients) with IPR, for placing at global market.

Why Standardization

Reproducible assays of the plant preparation generate confidence in the mind of the user and prescriber. To ensure consistent quality of the preparation, the qualitative and quantitative chromatographic fingerprint on the basis of characteristic substance(s) for raw materials and finished products should be provided. Thus, the standardization of the herbal drugs and preparation thereof is not just an analytical operation that ends with the identification and assay of an active principle; rather it embodies total information (passport data of raw material, quality parameters, process control) and controls that are necessary to guarantee consistency of composition. The question of quality acquires greater relevance today than ever before. Regulatory agencies and consumer groups on one hand and increasingly stringent quality norms for global positioning of the products on the other hand, demand that new materials and herbal products there from, used for medicinal, nutritional and cosmetic purposes should be of consistent reproducibility; scientifically (chemically and biologically) validated and supplemented by data on documented evidence of efficacy and safety through limited human clinical trials.

However, with rapid advancement in the field of instrumentation where limits of detection (LOD) and limits of quantification (LOQ) have routinely been upgraded to femto and atto levels, the chemical standardization of herbal products by hyphenated techniques (HPTLC, HPLC, LC/MS and LC/MS^n, GC/MS) has become more credible than biological evaluation of data.

Although liquid and gas chromatography are powerful tools for chemical profiling of the herbals, but these in combination with mass spectrometric detection has become more effective and convenient. The techniques are very useful for the characterization and quantification of the individual constituents in the plant extracts. LC-MS and GC-MS based multi component hyphenated methods allow the determination of the components in the mixture. These techniques are sensitive, selective, fast with inexpensive clean up for sample preparation; where simultaneous separation as well as identification of the components in a mixture is possible. LC/GC perform the function of separation whereas MS performs the function of identification of the components in the mixture on the basis of molecular mass and fragmentation pattern. It gives us two dimensional information, the information from the UV/VIS or diode array detector in LC and FID/ECD/NPD detectors in GC is the a first dimensional information in the form of retention time whereas the second dimension of information comes from the mass detector in the form of molecular mass and fragmentation pattern. Techniques are very sensitive, selective and specific and allow the detection of the compounds even in picogram amounts.

LC/MS has been playing a more significant role in plant medicine research (Cai *et al.*, 2002) because the technique is capable of characterizing active components ranging from small molecules to macromolecules and recent scientific results and publications show that application of LC/MS has been rapidly expanding into the area of structure elucidation and characterization of active components, in addition to valuable quantitative analysis. The LC/GC-MS instruments are of two types, Ion trap and linear quadrupoles. Linear quadrupole instruments are very sensitive and are used for quantitative analysis whereas Ion trap technology is multi step fragmentation technology and the instruments based on this technology are used for characterization of the molecule. Under CSIR's NMITLI Project, our studies on quantification of four chemical markers in different cultivars (cultivated/wild accessions) of *Withania somnifera* (Khajuria *et al.*, 2004) have provided us a very valuable and significant data. These studies have revealed that an accession cultivated at Indian Institute of Integrative Medicine (CSIR), Jammu and coded, as AGB002 can be the best source for withaferin-A. The PCR based RAPD and AFLP studies on this accession have revealed that this accession has entirely different DNA sequence when compared with other accessions. Gradual improvements in the LC-MS methods led to the development of methods for quantification on the basis of five (Figure 15.1), seven (Figure 15.2) and finally eight withanolides and withanosides (Figure 15.3).

LC/MS methods are very useful in determining active components and their metabolites in pre-clinical studies. Studies on the metabolites of piperine, an alkaloid constituent of *Piper nigrum* and *Piper longum* led to the characterization (Bajad *et al.*, 2003) of two metabolites in rat urine. These metabolites were characterized on the basis of LC/MS/MS and LC/NMR/MS/MS data.

Over recent decade, a number of lead compounds and new natural products derived from medicinal herbs have been successfully isolated and identified. Chemical analysis plays a central role in development and modernization of herbal or herbal-based medicines and mass spectrometry coupled to LC is emerging as the technique of choice in the identification of active ingredients, compositional analysis and chemical finger printing studies. Tandem mass spectrometry (MS/MS) is a powerful technique for detecting a target compound in complex matrices, including plant extracts and herbal drugs. The strength of this technique lays in the selectivity, high sensitivity and fast screening capabilities compared with many other separation and identification techniques. We have developed a rapid LC/MS/MS screening method for detection of camptothecin in an endophytic fungus *Entrophosphere infrquens*, which resides in the plant *Nothpodytes foetida* (Amna and Khajuria, 2006). For preparation of marker grade hyperforin from *Hypericum perforatum* a preparative HPLC method (Anand *et al.*, 2003) has been developed by our group and similarly a very sensitive

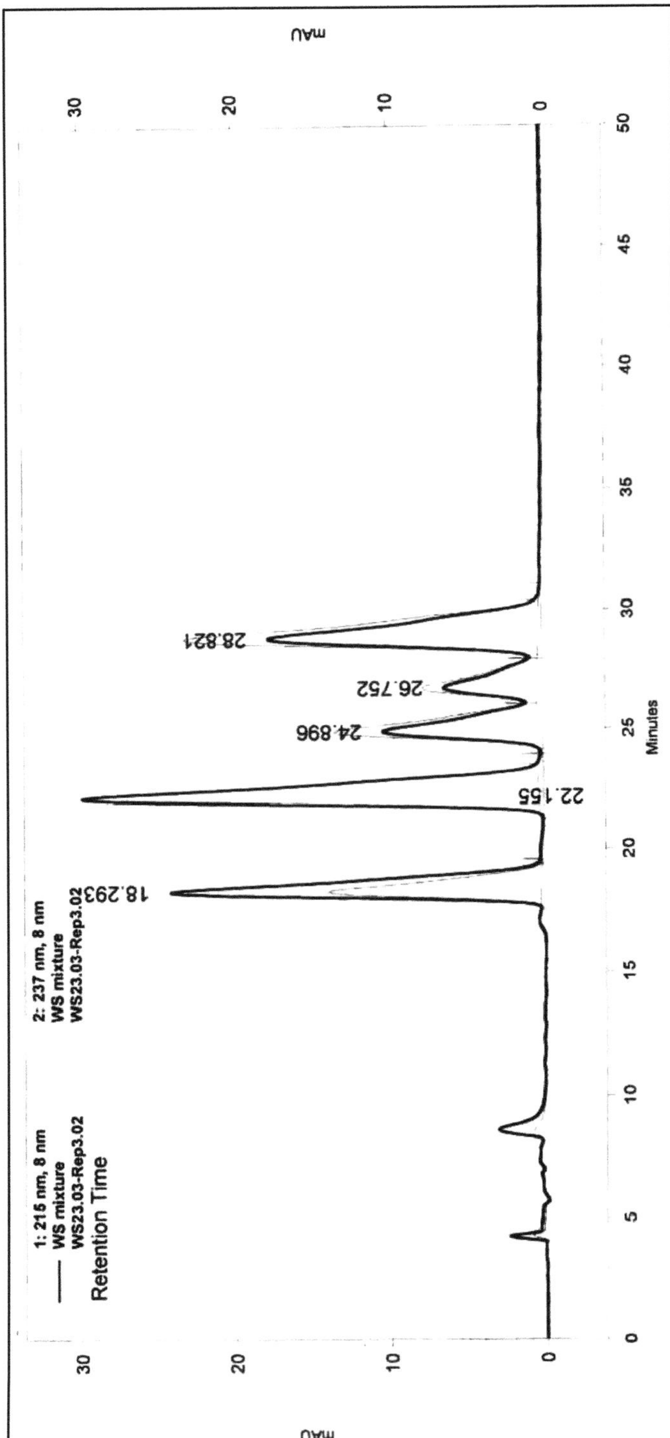

Figure 15.1: HPLC chromatogram of five markers of *Withania somnifera*.

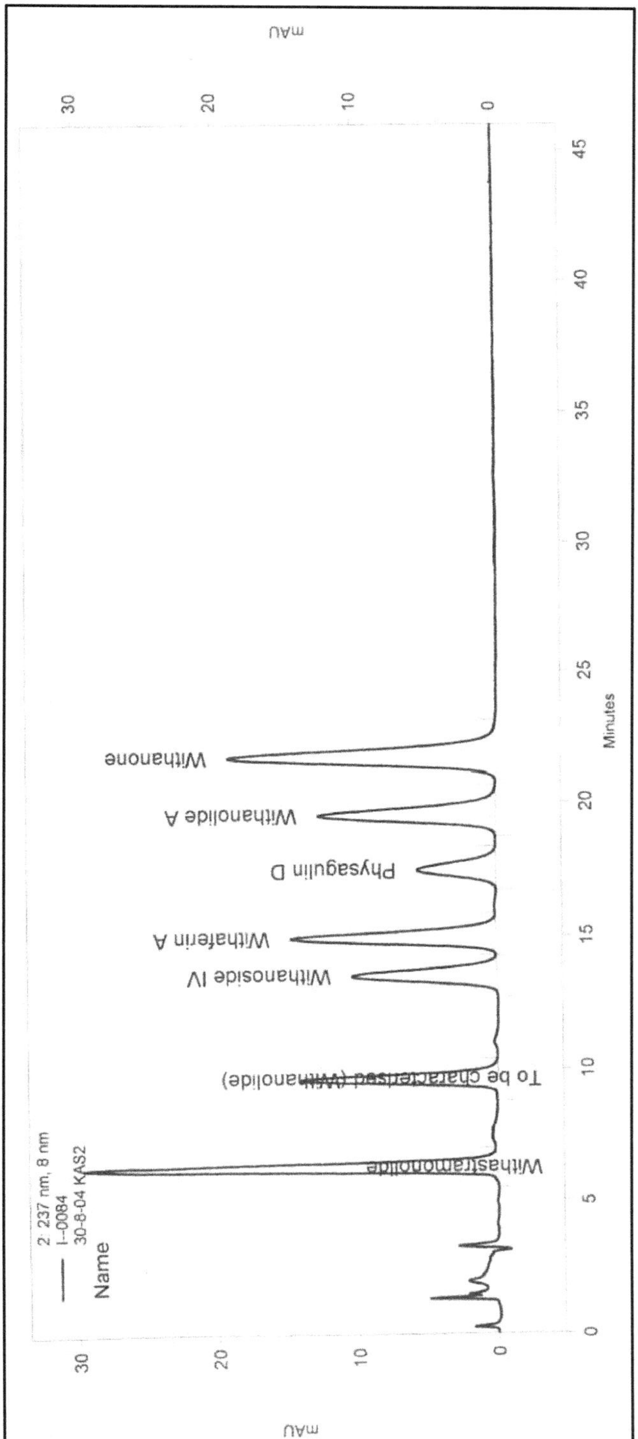

Figure 15.2: HPLC chromatogram of seven markers of *Withania somnifera.*

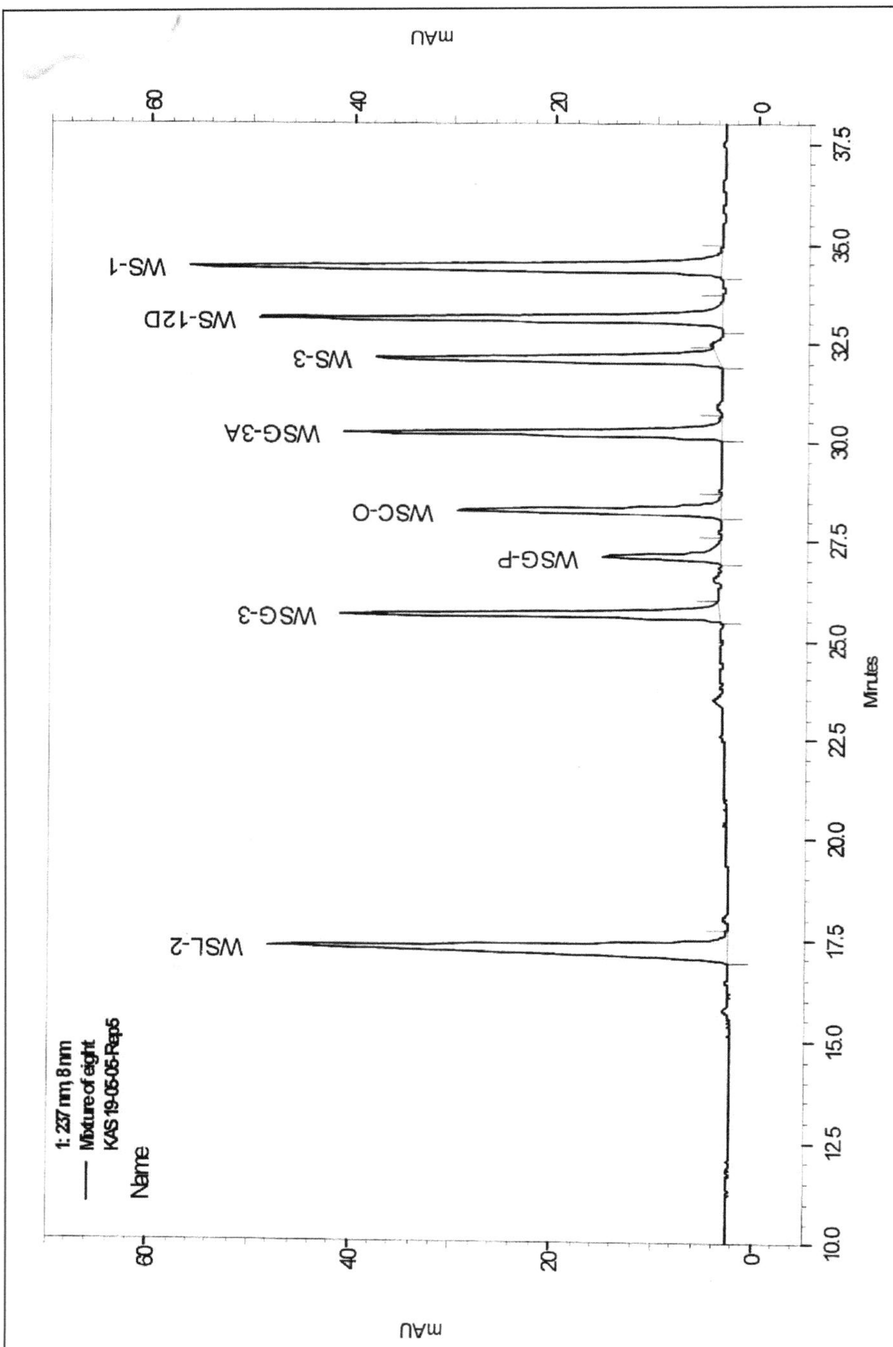

Figure 15.3: HPLC chromatogram of eight markers of *Withania somnifera*.

LC-MS method was developed to check the purity and stability of the same. LC/MS methods for the detection of different compounds like ß, ß-dimethylacrylshikonin in tissue cultures of *Onosoma echiodes* etc. have been developed (Lattoo *et al.*, 2005). These methods are also used to detect the same compounds in the plant extracts.

Conclusions

A golden triangle consisting of Ayurveda, modern medicine and science has converged to form a real discovery engine that can result in newer, safer, cheaper and effective therapies. Globally there is a positive trend towards holistic health, integrative sciences, systems biology approaches in drug discovery and therapeutics which is the unique feature of Ayurveda. Ayurveda and modern medicine techniques must be coupled in order to bring out high quality herbal products with rapid onset of action and good bioavailability. Herbal drug development is possible only through the development of standardized herbal products with reference to their active phytoconstituents present for commercialization, correct identification and supply of raw material and to avoid adulteration. Since some botanicals have undergone changes in their physical characteristics, the concept of active markers needs a flexible approach in favour of the complex nature of these material.The ultimate goal of ethnopharmacology must be to identity drugs to eliminate human illness by a thorough analysis of plants through out the world.

Proper botanical identification, proper chemo type/ecotype, proper/desired stage of maturity, post harvest processing, storage, isolation of marker compounds (preferably the therapeutic ones), validated analytical (GC, GCMS, LC, LC-MS and LC-MS/MS) methodologies for the quantification of the extracts/fractions are the important milestones in the herbal drug standardization.

Ever increasing discovery costs and increased failures at the end of the discovery line makes medicines unaffordable to the developing countries. This makes new approaches such as system biology and reverse pharmacology more attractive that provide innovation opportunities that are based on experiential wisdom and holistic view point of 'Traditional Medicine'. It is known to all of us that drug discovery pipeline in modern drug discovery is getting dry and modern world is looking towards the herbal world with great expectations.

References

Agarwal, S. G., Vashist, V. N., Thappa, R. K., Singh, K., Ghosh, S. C. and Atal, C.K. (1974). Terpenes and other components from *Bunium cylindricum* seeds. *Phytochemistry*, **13:** 2024-2025.

Agarwal, S. G., Thappa, R. K., Dhar, K. L., Atal, C. K. (1979). Essential oils of the seeds of *Bunium bulbocastanum* Linn. *Carum gracile* Lindle and *Cuminum cyminum* Linn. *Indian Perfumer*, **23:** 34-37.

Agnihotri, V. K., Agarwal, S.G., Dhar, P.L., Thappa, R.K., Meena, B., Kapahi, B.K., Saxena, R. K., Qazi, G.N. (2005). Essential oil composition of *Mentha pulegium* L. *Flavour and Fragrance Journal*, **20:** 607-610.

Ahmed, S. M., Manhas, L. R., Verma, V., and Khajuria, R. K. (2006). Quantitative determination of four constituents of *Tinospora* sps. by reverse phase HPLC-UV(DAD) method. Broad based studies revealing variation in content of four secondary metabolites in the plant in different eco-geographical regions in India. *Journal of Chromatographic science*, **44:** 504.

Anand, A., Puri, S. C., Verma, V., Handa, G., Khajuria, R. K., Gupta, V. K., Suri, O. P., and Qazi, GN. (2003). A simple and reliable semi preparative HPLC technique for the isolation of marker grade Hyperforin from *Hypericum Perforatum* L. *Journal of Chromatographic Science*, **41(8):** 444-446.

Bajad, S., Coumar, M., Khajuria, R. K., Bedi, K. L. and Suri, O. P. (2003). Characterization of a new major metabolite 3,4-methylenedioxy phenyl-2E.4E-pentadienoic acid - N-(3yl-propionic acid) amide of piperine, an ominipresent food component by LC-NMR-Positive ESI-MS. *European Journal of Pharmaceutical Sciences*, **19(5):** 413-421.

Bajad, S., Khajuria, R. K., Bedi, K. L., Suri, O. P. and Singla, A. K. (2003). Characterisation of a new minor urinary metabolite 2E, 4E-1-[5- (3 – methoxy –4-hydroxy – phenyl)-1-oxo-2,4-pentadienyl]- piperidine of piperine, an ominipresent food component, by LC-MS/MS. *J. of separation Science*, **26 (9-10):** 943-946.

Caig, Z., Lee F. S. C., Wang, X. R. and Yu, W. J.(2002). A capsule review of recent studies on the application of mass spectrometry in the analysis of Chinese medicinal herbs. *Journal of Mass Spectrometry*, **37:** 1013-1024.

Dutt, A. K. (1974).The Botanical Identity of Ajmoda. *Jour. Res. Ind. Med.*, **9(1):** 96-98.

Lattoo, S. K., Kaul, S., Dhar, M. K., Khajuria, R. K., Gupta, D. K. and Qazi, G. N. (2005). Production of b, b-dimethylacryl shikonin in callus cultures of *Onosoma hispidum*, Var. *hispidum*. *J. of Plant Biochemistry and Biotechnology*, **14:** 193-196.

Pandita, K., Bhatia, M.S., Thappa, R.K., Agarwal, S.G., Dhar, K. L. and Atal, C. K. (1983). Seasonal variation of the alkaloids of *Adhatoda vasica* and the detection of glycosides and N-oxides of vasicine and vasicinone. *Planta Medica*, **48 :** 81-82.

Singh, S., Pandey, S. C., Srivastava, S., Gupta, V. S., Paetro, B. and Ghosh, A. C. (2003). Chemistry and medicinal properties of *Tinospora cordifolia* (GUDUCHI). *Indian Journal of pharmacology*, **35:** 83-91.

Thappa, R. K., Agarwal, S.G., Singh, K. and Ghosh, S. (1991). Comparative studies on the major volatiles of Kalazira (*Bunium persicum* Seed) of wild and cultivated sources. *Food Chemistry*, **41:** 129-134.

Touseef, A., Khajuria, R. K., Puri, S. C., Verma, V. and Qazi, G. N. (2006). Determination and quantification of camptothecin in an endophytic fungus by liquid chromatography–Positive mode electrospray ionisation tandem mass spectrometry (LC-MS/MS). *Current Science*, **91(2) :** 208-212.

Traditional and Folk Herbal Medicine: Recent Researches Vol. 1 (2012) *Pages* **329–348**
Editor: **V.K. Gupta**
Published by: **DAYA PUBLISHING HOUSE, NEW DELHI**

16

Ficus gnaphalocarpa, Ficus cordata and *Ficus chlamydocarpa:* Potential Sources of Bioactive Compounds

Donfack J. Hubert[1*], Ngueguim T. Florence[2], Bathelemy Ngameni[3], Tchana N. Angele[1], Daniela Buonocore[5], P. Vita Finzi[4], G. Vidari[4], Fulvio Marzatico[5], Moundipa F. Paul[1] and Ngadjui T. Bonaventure[3]

ABSTRACT

Ficus gnaphalocarpa, Ficus cordata and Ficus chlamydocarpa are plants used in Cameroonian folk medicine to treat certain diseases including toxic hepatitis. The in vitro hepatoprotective effect of the crude extracts from these plants on the CCl_4-induced liver cell damage as well as the possible antioxidant mechanisms involved in this protective effect, were investigated. The phytochemical investigation of these crude extracts led to the isolation of

1 Laboratory of Pharmacology and Toxicology, Department of Biochemistry, Faculty of Science, University of Yaounde I, P.O. Box. 812 Yaounde - Cameroon.

2 Laboratory of Animal Physiology, Department of Aanimal Biology and Physiology, Faculty of Science, University of Yaounde I,P.O. Box. 812 Yaounde – Cameroon.

3 Department of Pharmacy and Traditional Pharmacotherapeutics, Faculty of Medicine and Biomedical Science, University of Yaounde I, P.O. Box 1364 Yaounde - Cameroon.

4 Dipartimento di Chimica Organica, Università degli Studi di Pavia and Centre of Ethnobiopharmacy (CISTRE) – Via Taramelli 10 – 27100 – Italy.

5 Dipartimento di Medicina Legale, SC.Forensi e Farmacotossicologiche Sezione di Sc. Farmacologiche e Tossicologiche via Ferrata 9-27100, Università Degli Studi Di Pavia – Italy.

* Corresponding author: E-mail: donfackhubert@gmail.com;donfackhubert@yahoo.fr; Ph: (+237) 75 13 32 42, (+237) 22 10 15 81.

some compounds. The hepatoprotective activity of these compounds was tested in vitro against CCl_4-induced damage in rat hepatoma cells. Some of these compounds showed significant hepatoprotective activity as indicated by their ability to prevent liver cell death and LDH leakage during CCl_4 intoxication. In addition, 2,4-dinitrophenyl-1-picrylhydrazyl (DPPH) radical scavenging activity, β-Carotene-Linoleic Acid Model System (β-CLAMS), Ferric-Reducing Antioxidant Parameter (FRAP) and microsomal lipid peroxidation assays were used to measure antioxidant activity of crude extracts and isolated compounds. The obtained results suggest that the protective effects of crude extracts from these plant extracts against the CCl_4-induced hepatotoxicity possibly involve the antioxidant effect of some isolated compounds.

Keywords: Antioxidant activity, Carbon tetrachloride, *Ficus chlamydocarpa, Ficus cordata, Ficus gnaphalocarpa*, Hepatoprotective effects, Plant compounds, Sylimarine.

Introduction

Lipid peroxidation has gained more importance today because of its involvement in pathogenesis of many diseases like atherosclerosis, cancer, hepatitis, diabetes mellitus, myocardial infarction, and also ageing. Free radicals or reactive oxygen species (ROS) are produced *in vivo* from various biochemical reactions and also from the respiratory chain as a result of occasional leakage. These free radicals are the main agents in lipid peroxidation. Antioxidants thus play an important role of protecting the human body against damage by reactive oxygen species (Lollinger, 1981; Cheeseman and Scater, 2003). Based on growing interest in free radical biology and lack of effective therapies for most chronic diseases, the usefulness of antioxidants in protection against these diseases is warranted (Rubinstein, 1962; Jacob and Sotoudeh, 2002; Yi-Hang *et al.*, 2009; Bo Huang *et al.*, 2010).

A number of *Ficus* species are used as food and for medicinal properties (Mousa *et al.*, 1994; Alziro de Amorin *et al.*, 1999; Lansky *et al.*, 2008). The biological activities of several plants of *Ficus* species have already been proven. Some few examples include: *Ficus glabrata* (Anders *et al.*, 1986), *Ficus racemosa* (Subhash *et al.*, 2000), *Ficus platyphylla* (Chindo, 2003), *Ficus bengalensis* (Rimi *et al.*, 2004), *Ficus insipida* (Anders *et al.*, 2005), *Ficus sycomorus* (Sandabe *et al.*, 2006), *Ficus asperifolia* (Kofi and Peter, 2008), *Ficus capensis* (Mpiana *et al.*, 2008), *Ficus glomerata* (Rao *et al.*, 2008), *Ficus bengalensis* (Rakesh, 2009), *Ficus religiosa* (Damanpreet and Rajesh, 2009), *Ficus ovata* (Kuete *et al.*, 2009), *Ficus exasperata* (Bafor and Igbinuwen, 2009) and *Ficus hirta* (Yihui *et al.*, 2009).

Ficus gnaphalocarpa (Miq.) Steud. ex A. Rich (Moraceae) is a sub species of *Ficus sycomorus*, which takes the form of tree or shrub reach 10 to 30 meters high. It has often been a stocky, up to 3 meters in diameter, straight, cylindrical and slightly wide at the base. Her crown is wide open, with spreading branches (Maydell, 1983). The fruits are figs solitary or in pairs in the axils of leaves. They are 2 to 6 centimetres in diameter, more or less hairy, yellow or reddish when ripe. Fruiting takes place in early dry season and rainy season (Cusset, 1985). The bark of the trunk is smooth and scaly in places. It abounds in white or yellow latex (Maydell, 1983).

Ethnopharmacological investigations show that *Ficus gnaphalocarpa* is widely used in treatment of many diseases such as hepatitis, jaundices (Maydell, 1983), sickle-cells diseases (Nongonierma, 2005), malaria (Asase, 2005) dysentery, cholera, hemorrhagic diarrhoea, gingivitis (Kerharo, 1974), cough, throat inflammation, tuberculosis and some urogenital infections (Adjanohoun, 1989). This plant is also used as a tonic, stimulant, diuretic and in the cases of difficult childbirth (Adjanohoun, 1989).

Ficus cordata Thunb (Moraceae) is a savana tree of around ten meters height found in Senegal, Angola, South Africa and Cameroon (Sabatie, 1985). The leaves of this plant are used against hyperaesthesia, ataxia, muscle tremor and padding motions and can kill heifers 48h after ingestion (Poumale *et al.*, 2008). Additional ethnopharmacological investigations showed that the stem bark of this plant can be used to treat several liver related diseases. The antibacterial activity of the crude extract and some isolated compounds from the stem barks of this plant have been also investigated (Kuete *et al.*, 2008).

Ficus chlamydocarpa is traditionally used in Cameroon for the treatment of different diseases, from filarial to diarrhoeal infections and tuberculosis (Kuete *et al.*, 2008). Further ethnopharmacological surveys have revealed that decoction of the stem bark of this plant is used in the West Cameroon folk medicine for the treatment of jaundice, which could be a symptom of liver-related diseases.

In the present study, the effects of the crude extracts from *Ficus gnaphalocarpa, Ficus cordata, Ficus chlamydocarpa* and some isolated metabolites were examined for their *in vitro* hepatoprotective and antioxidant properties.

Materials and Methods

Plant Material

Ficus gnaphalocarpa

The stem bark of *Ficus gnaphalocarpa* was collected in Touloum (North region of Cameroon). The botanical identification of the plant was carried out at the Cameroon National Herbarium, where the voucher specimen was conserved under the reference number 28054/HNC.

Ficus cordata

The stem bark of *Ficus cordata* was collected in the Center Region of Cameroon. The botanical identification of the plant was carried out at the Cameroon National Herbarium, where the voucher specimen was conserved under the reference number 8613/SRF/CAM.

Ficus chlamydocarpa

Samples of *Ficus chlamydocarpa* stem bark were collected in Bahouan (West province of Cameroon) in July 2004. The plant was identified by Nana Victor of the Cameroon National Herbarium where a voucher specimen is conserved under the reference number 35446/HNC.

Extraction and Isolation of Compounds

Crude Extract and Isolated Compounds from *Ficus gnaphalocarpa*

Air-dried and powdered stem barks of *Ficus gnaphalocarpa* (2 Kg) were macerated in methanol (5 L) for 48 h at room temperature. The macerate was filtered through Whatman no.1 filter paper. The filtrate was concentrated under vacuum to a paste that constituted the crude extract (CE1, 300 g). This extract was subjected to differential solubility on ethyl acetate to yield 60 g of ethyl acetate extract. These extracts were subjected to silica gel 60 (0.063-0200 mm) column chromatography using hexane/ethyl acetate and ethyl acetate/methanol gradients as eluents at 2 ml/min. Fractions of 150 ml each were collected as follows: pure hexane (1 to 5), hexane/ethyl acetate 5 per cent (6 to 12), hexane/ethyl acetate 10 per cent (13 to 25), hexane/ethyl acetate 15 per cent (26 to 33), hexane/ethyl acetate 20 per cent (34 to 41), hexane/ethyl acetate 30 per cent (42 to 56), hexane/ethyl acetate 40 per cent (57 to 70), hexane/ethyl acetate 50 per cent (71 to 90), hexane/ethyl acetate 75 per cent (91 to 102), ethyl acetate (103 to 115), ethyl acetate/methanol 10 per cent (116-125) and methanol (126 to 132). These fractions were pooled on the basis of the thin layer chromatography analysis on five sub-fractions namely F1 (1 to 5), F2 (6 to 35), F3 (36 to 69), F4 (70 to 118) and F5 (119 to 132) in the respective quantities of 7 g, 18 g, 11 g, 9 g and 13 g. Two sub-fractions (F2 and F3) were separately subjected to silica gel 60 (0.063-0200 mm) column chromatography and preparative TLC using hexane/ethyl acetate gradients as eluent while dichloromethane/methanol gradients eluent were used to purify sub-fractions F4 and F5. Fractions of 50 ml each were collected. Purification of subfraction **F3** yields **betulinic acid** [compound (**1**); in hexane/ethyl acetate 20 per cent; white powder; 153 mg; m.p. 316-318 °C; *MW*: 456; Rf: 0.5 (in hexane/ethyl acetate 15 per cent)] (Mahato and Kundu, 1994). Purification of subfraction **F4** afforded two compounds: **catechin** [compound (**2**); in dichloromethane/methanol 15 per cent; pale-yellow powder; 52 mg; m.p. 240-242 °C; *MW*: 290; Rf: 0.5 (in dichloromethane/methanol 20 per cent)] (Benavides *et al.*, 2006) and **quercetin** [compound (**3**); in dichloromethane/methanol 10 per cent; yellow powder; 22 mg; m.p. 316 °C; *MW*: 302; Rf: 0.5 (in dichloromethane/methanol 15 per cent)] (Dutta *et al.*, 2007).

Crude Extract and Isolated Compounds from *Ficus cordata*

The powdered stem bark of *Ficus cordata* (1.8 kg) was extracted with CH_2Cl_2/MeOH (1:1) at room temperature for 24 H. After removing the solvents by evaporation under reduce pressure, the crude extract (CE2;155 g) was chromatographed on silica gel and eluted using hexane followed by hexane/ethyl acetate mixture; ethyl acetate and ethyl acetate/methanol gradients as eluent. Sixty fractions of 250 mL each were collected, then combined together on the basis of the analytic TLC in two fractions named A and B. Fraction A (subfractions 1-15; 17 g) was chromatographed on silica gel and eluted with a mixture of hexane/ethyl acetate of increasing polarity to yield amount other **β-amyrin acetate** [(**4**) 97 mg; m/z 426; amorphous powder; Rf: 0.80 using hexane-AcOEt: 191] and **lupeol** [(**5**) 10 mg; m/z; m.p; Rf: 0,71 using hexane-AcOEt: 19-1] (Swift *et al.*, 1952). Fraction B (subfractions 30-60; 36 g) was chromatographed on silica gel and eluted with a mixture of CH_2Cl_2/MeOH of increasing polarity to yield **catechin** [**(2)** 57 mg; m/z 290; m.p: 214; Rf: 0.36 using CH_2Cl_2/MeOH: 3-1)] (Wan *et al.*, 2004)

Crude Extract and Isolated Compound from *Ficus chlamydocarpa*

Sun-dried and powdered stem bark of *F. chlamydocarpa* Mildbr. and Burr. (2 kg) was macerated in methanol (15 L) at room temperature for 48 h. The filtrate was concentrated under vacuum to give a dark green methanol crude extract (CE 3) (150 g). The CE3 (140 g) were subjected to flash column chromatography (FCC) on silica gel 60 and eluted with a hexane-EtOAc gradient, followed by EtOAc-MeOH mixtures and MeOH to afford six fractions: fractions A (10 g, hexane), B (3 g, hexane-EtOAc, 3:1), C (1 g, hexane-EtOAc, 1:1), D (2 g, hexane-EtOAc, 1:3), E (5 g, EtOAc), F (50 g, EtOAc-MeOH, 9:1, followed by MeOH). Fraction A was subjected to FCC on silica gel 60. Elution with a hexane-EtOAc gradient yielded **α-amyrin acetate** [(**6**), 100 mg, M^+ m/z 448, m.p 228-230 °C; $R_f = 0.80$ using hexane-EtOAc, 19:1] (Sami *et al.*, 1983).

The structures of the isolated compounds were established using NMR spectral analysis such as 2D experiments, COSY, HMQC and HMBC, and direct comparison with published information and with authentic specimens obtained by our research group. NMR spectra were measured on varian Unity 300 (300.125 MHz) and various Inova 500 (499.876 MHz) spectrometers. ESIMS was recorded on a Finnigan LCQ with a quaternary pump Rheos 4000 (Flux instrument). ESIHR mass spectra were recorded on Bruker FTICR 4.7 T mass spectrometer. EI.MS were recorded on a Finnigan MAT 95 spectrometer (70 eV) with perfluor and kerosene as reference substance for HREI MS. IR spectra were recorded on a Perkin-Elmer 1600 series FT-IR spectrometer from films. Flash chromatography was carried out on silica gel (230-400 mesh). Thin layer chromatography (TLC) was performed on silica gel 60 F254 (Merck) precoated aluminum sheets and spots were visualized using both UV light (254 and 366 nm) and 50 per cent H_2SO_4 spray reagent.

Experimental Design

Chemicals

All reagents used in the study were obtained from SIGMA Chemicals Co. (Dorset, UK) and Prolabo (Paris, France).

Screening for Hepatoprotective Activity Using Hepatoma Cells

Hepatoma Cell Culture

The effect of isolated compounds and crude extracts from *Ficus gnaphalocarpa, Ficus cordata* and *Ficus chlamydocarpa* on cell viability was assessed in a cell culture system, using cells from the rat Morris hepatoma cell line BS TCL 41. The antihepatotoxic effects of the crude extract and isolated compounds were assayed on CCl_4-treated cells. The CCl_4 concentration used for cell culture treatment was previously determined and chosen because of its ability of induce up to 75 per cent cell culture mortality (Chen *et al.*, 1996; Rodeiro *et al.*, 2008; Donfack *et al.*, 2010). The crude extract and isolated compounds from *Ficus cordata* were not completely soluble in aqueous medium and had to be emulsified in dimethylsulfoxyde (DMSO) prior to their addition to the cell culture medium. A 3 per cent solution of DMSO was used in cell culture medium. The cells were grown in Ham's F. 10/F10 medium supplemented with 20 per cent (V/V) inactivated fetal calf serum, 2 mM L-glutamine and 1 per cent penicillin/streptomycin 100x solution to prevent microbial contamination. Cells were maintained in a humidified atmosphere with 5 per cent CO_2 at 37 °C. The cell medium was changed twice a week. At 70–80 per cent confluence, cells were trypsinized and seeded on 96-well plates at a cell density of 30.000 cells/well in serum-free culture medium. Twenty-four hours after cell seeding, cell were simultaneously exposed to 12.5, 25, 50, 100 and 200 µg/mL of the plant extracts and isolated compounds and 2.5 mM CCl_4 in fresh serum-free medium. The positive control was a set of cells maintained in a 2.5mM CCl_4 test compounds free culture medium; while the negative control was a set of cells maintained in DMSO culture medium.

MTT Assay

Twenty-four hours after cell seeding, cells were treated with CCl_4 (2.5 mM) and incubated with 12.5, 25, 50, 100 and 200 µg/mL crude extracts and isolated compounds for 24 hrs at 37°C. Following removal of the supernatants from each well, cells were washed with phosphate-buffered saline solution and incubated with 0.05 per cent MTT (3-(4,5-Dimethylthiazol-2-yl)-2,5-diphenyltetrazolium bromide) in culture medium for 4 h. These supernatants were used to determine the LDH leakage. Subsequently, the medium was removed, and the cells washed with phosphate-buffered saline solution were incubated for 15 min in acidic isopropanol to dissolve the formazan crystals. The absorbance of the MTT formazan was determined at 570 nm using a multiwell plate reader. Viability was defined as the ratio (expressed as a percentage) of the absorbance of treated cells to that of untreated cells.

Lactate Dehydrogenase Assay

For the LDH (lactate dehydrogenase) assay, 30.000 cells were seeded per well of 96 well plates in serum-free culture medium. Then, 24 hrs after cell seeding, cells were exposed to 2.5 mM CCl_4 and 12.5, 25, 50 100 or 200 µg/mL of the plant extract or isolated compounds. After 24 hrs incubation, the supernatant was collected from each well. The LDH activity was measured in each supernatant using a cytotoxicity assay kit (Colorimetric Assay for Cytotoxicity Product NO. LK100 Oxford Biomedical Research), in accordance with the manufacturer's instructions. In this test, the intensity of the colour obtained from reaction is proportional to the LDH activity. The absorbance was determined at 490 nm using a plate reader. The percentage of LDH released from the cells was determined as the ratio (expressed as a percentage) of the absorbance of treated cells to that of untreated cells.

Screening for Antioxidant Activity

Four model systems 2,4-dinitrophenyl-1-picryl hydrazyl (DPPH) radical scavenging activity, β-Carotene-Linoleic Acid Model System (β-CLAMS), Ferric-Reducing Antioxidant Power (FRAP) assay and microsomal lipid peroxidation (IPL) were used to measure the antioxidant activities of the crude extract and isolated compounds from *Ficus cordata*. In each assay, the crude extract and isolated compounds were tested at doses of 12.5, 25, 50, 100 or 200 µg/mL. Trolox was used as a standard antioxidant. The EC_{50} values were then calculated for the crude extract and each individual compound.

Free Radical-scavenging Activity

The free radical-scavenging activity of the crude extracts and isolated compounds was evaluated by assessing the decrease in absorbance of 3,4-DPPH at 517 nm according to Brand-Williams *et al.* (1995). The decrease in absorbance was monitored at 517 nm, exactly 30 seconds after addition of the appropriate volume of the extract or methanol to the blank.

β-Carotene-linoleic Acid Model System (β-CLAMS) Assay

The β-CLAMS method is based on the discoloration of β-carotene by the peroxides generated during the oxidation of linoleic acid (a free radical chain reaction) at high temperature (Miller, 1971). In brief, 1mL β-carotene (0.02 per cent w/v) dissolved in $CHCl_3$ was introduced in a 250 mL round-bottom flask. Linoleic acid (20 µL) and 200 mg Tween 20 were added to the mixture and the $CHCl_3$ was removed using a rotary evaporator. The fifty mL of distilled water was added and the flask was shaken vigorously until all the material dissolved. This test mixture was prepared fresh and used immediately. Three millilitres of test mixture plus 10 µL of extract/compound solution or water (used as a blank) were added to each spectrophotometric cuvette. The spectrophotometric cuvettes were incubated at 50 °C for 5 min and readings were carried out at 470 nm immediately after and then every 10 min for 3 hours.

Ferric-reducing Antioxidant Power "FRAP"

The ferric-reducing antioxidant power (FRAP) assay measures the potential of antioxidants to reduce the Fe^{3+}/2,4,6-tripyridyl-s-triazyne (TPTZ) complex present in stoichiometric excess to the blue coloured Fe^{2+} form which increases the absorption at 593 nm. This method was carried out as described by Benzie and Strain (1996).

Lipid Peroxidation Assay

Inhibition of lipid peroxidation was investigated using rat liver microsomes isolated by the calcium aggregation procedure as described by Garl and Fry (1989). Lipid peroxidation was non-enzymatically initiated using ascorbate as described by Ulf *et al.* (1989) and assayed for thiobarbituric acid-reactive substances (TBA-RS) were assayed according to Wills (1989).

Calculations and Statistical Analyses

Results are presented as means ± S.D. The total variation present in a set of data was analysed by one-way analysis of variance (ANOVA) using the Graph Pad Prism software. A value of $P<0.05$ was taken as statistically significant (* $p \leq 0.05$, ** $p \leq 0.01$).

The EC_{50} values, taken as the concentrations of the sample required to scavenge 50 per cent DPPH or to inhibit 50 per cent of another oxidant mechanism, were estimated using Graph Pad Prism 3.0 software.

Results

Structures of Isolated Compounds

The structures of the isolated compounds and sylimarin (used as an antihepatotoxic reference compound in this study) are shown in Figure 16.1.

Figure 16.1: Chemical structures of isolated compounds and silymarin (mixture of silybine, silydianine and silychristine).

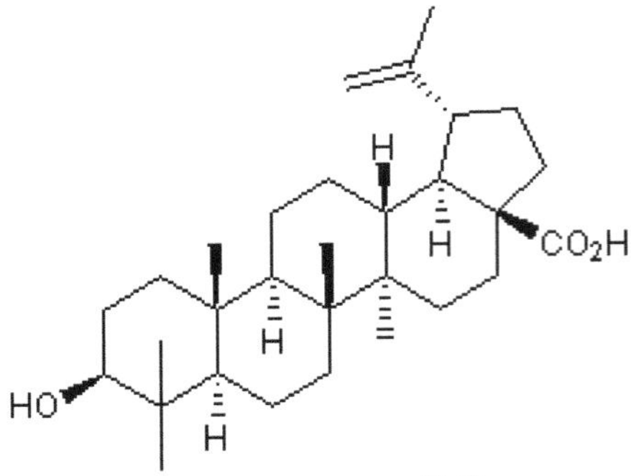

Betulinic acid (1)
Source: *Ficus gnaphalocarpa*

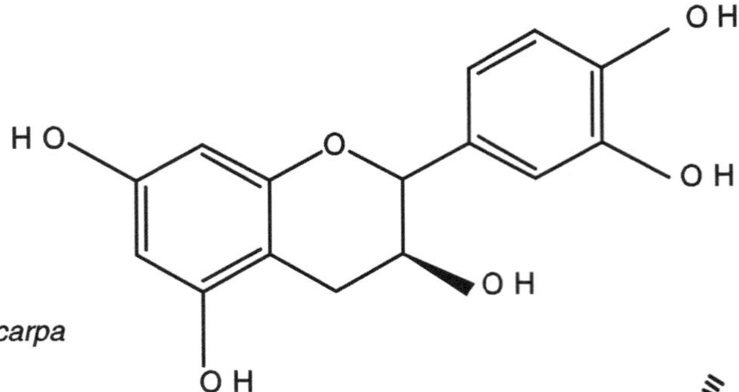

Catechin (2)
Source: *Ficus gnaphalocarpa* and *Ficus cordata*

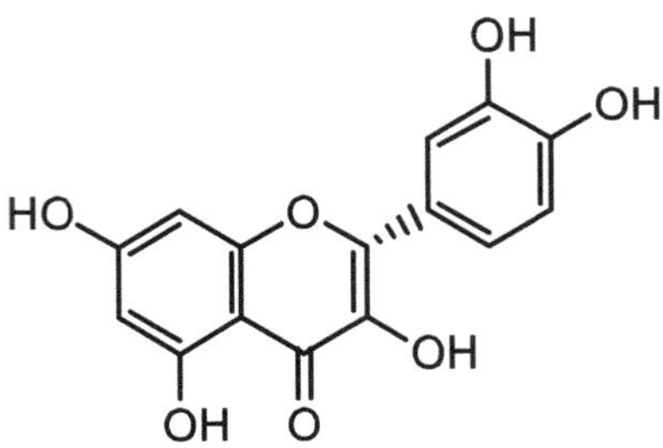

Quercetin (3)
Source: *Ficus gnaphalocarpa*

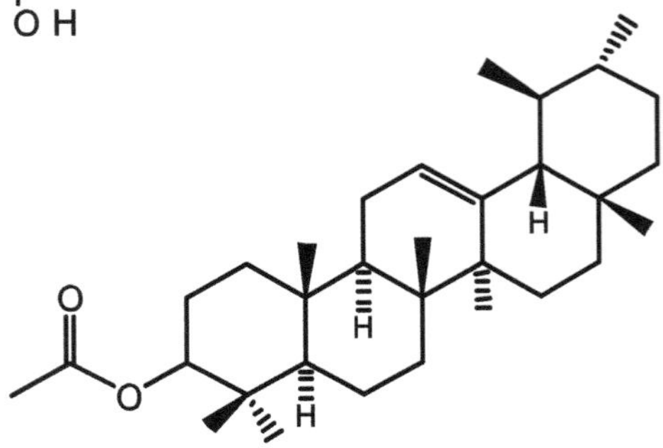

β-amyrin acetate (4)
Source: *Ficus cordata*

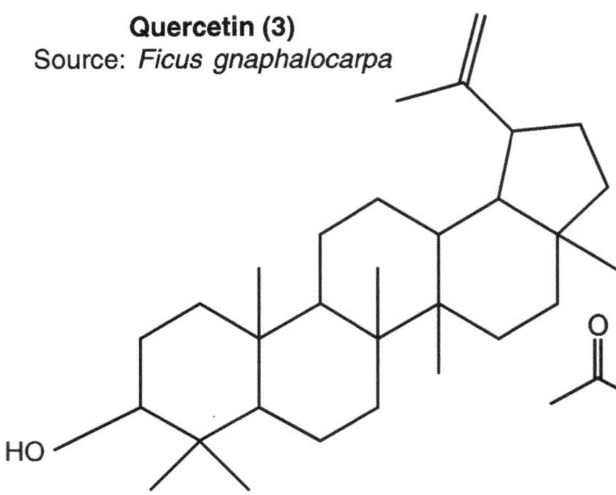

Lupeol (5)
Source: *Ficus cordata*

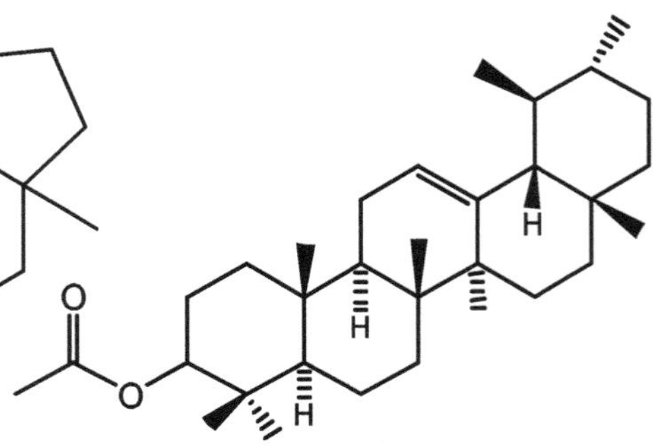

α-amyrin acetate (6)
Source: *Ficus chlamydocarpa*

Contd...

Figure 16.1–*Contd...*

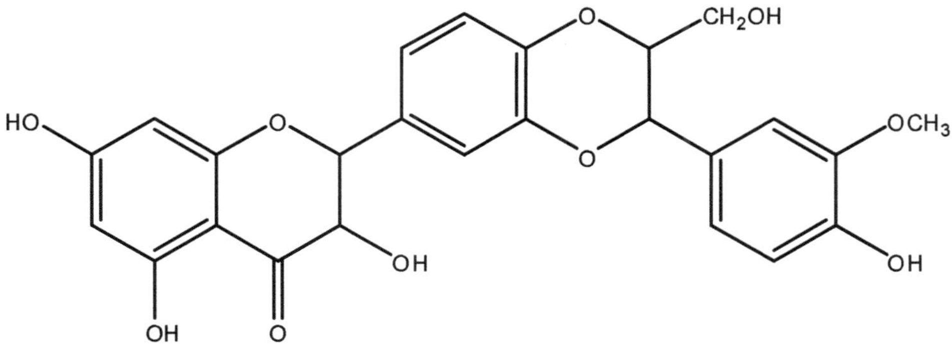

Silybine

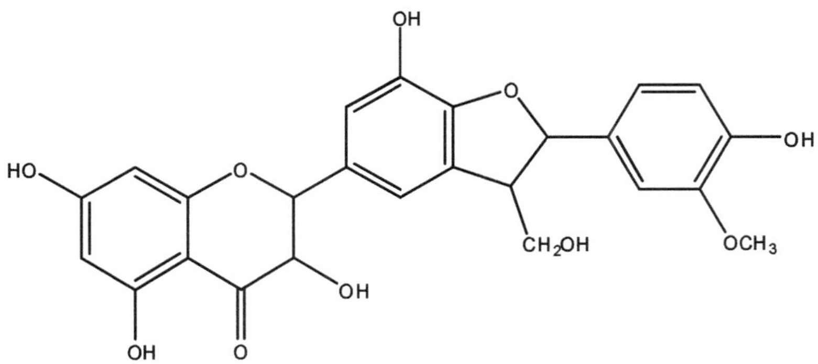

Silychristine

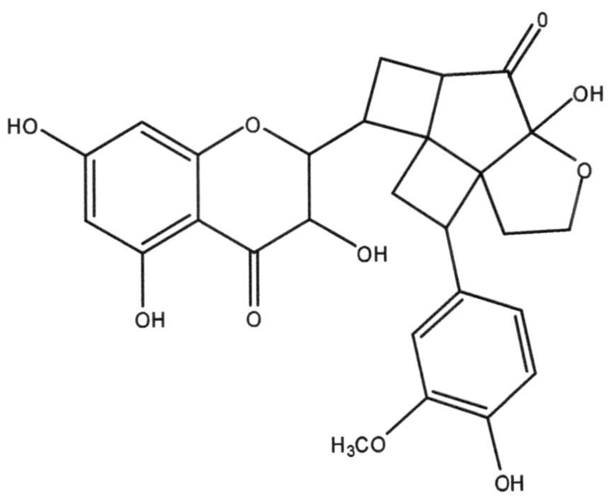

Silydianine

In vitro Hepatoprotective Activities of Extract and Isolated Compounds

Metabolic activity can be evaluated by measuring the activity of the mitochondrial enzyme succinate dehydrogenase using the MTT test. MTT was designed for the quantification of both cell proliferation and cell viability in a cell population using a 96-well plate format. This test is widely used for the *in vitro* evaluation of bioactivity of plant extracts. In addition, membrane integrity can be evaluated by measuring lactate dehydrogenase (LDH) activity. LDH, an enzyme located in the cytoplasm, catalyses the interconversion of lactate and pyruvate. The presence of LDH in a medium can either be interpreted as a result of cell death or cell leakage. When cells are disrupted, LDH activity is increased. In the present study MTT and LDH assays were carried out to evaluate the *in vitro* hepatoprotective activity of the crude extract and isolated compounds. The results of these tests are summarized in Tables 16.1 and 16.2. These results showed that until the dose of 100 µg/mL, cell viability increased with increasing amounts of the crude extract and isolated compounds. These compounds exhibited the lowest percentage of LDH leakage and display an important hepatotoprotective activity, which was characterized by their ability to preserve cell viability during CCl_4 intoxication.

Antioxidant Activity of Crude Extracts and Isolated Compounds

In order to determine the antioxidant effects of the crude extracts and isolated compounds in terms of the mechanism of their hepatoprotective effects, the anti-lipid peroxidation in rat liver microsomes, the bleaching inhibition (measured by peroxydation of β-carotene), reducing power and scavenging activities were investigated. The results are shown in Table 16.3. These results show that apart of compound **(4)** and **(6)**, all compounds showed antioxidant activities for the four studied mechanisms.

Discussion

The liver is exposed to absorbed drugs and xenobiotics in concentrated form. Detoxification reactions (phase I and phase II) metabolize xenobiotics aiming to increase substrate hydrophilicity for excretion. Drug-metabolizing enzymes detoxify many xenobiotics but bioactivate or increase the toxicity of others (Jaeschke *et al.*, 2002). In case of bioactivation, the liver is the first organ exposed to the damaging effects of the newly formed toxic substance. Therefore, protective mechanisms relevant to the liver are of particular interest because free radicals and reactive oxygen species play a central role in liver diseases pathology and progression, dietary antioxidants have been proposed as therapeutic agents to counteract liver damage (Hensley *et al.*, 2000; Higuchi and Gores, 2003). Natural antioxidants may act as protectors against several compounds but, more importantly, may exert modulatory effects in cells through actions in antioxidant, drug metabolizing and repairing enzymes as well as working as signalling molecules in important cascades for cell survival (Nandita *et al.*, 2008; Ning *et al.*, 2008). In the past decade, the bioactivities of flavonoids on human health have given rise to much attention, especially the antioxidant activity. Flavonoids are important phytochemicals that cannot be synthesized by humans.

Table 16.1: MTT assay in rat hepatoma cells after 24 h incubation with 12.5, 25, 50, 100, and 200 μg/mL extracts and isolated compounds.

Compounds and Crude Extracts	*Concentration of Compounds and Percentages of Cell Viability*				
	12.5 μg/mL	*25 μg/mL*	*50 μg/mL*	*100 μg/mL*	*200 μg/mL*
Betulinic acid (1)	06.78 ± 03.48	11.78 ± 03.87	21.98 ± 04.89	29.38 ± 05.09	22.78 ± 08.97
Catechin (2)	04.18 ± 03.17	06.78 ± 05.89	19.57 ± 06.48	24.28 ± 03.89	20.78 ± 08.79
Quercetin (3)	08.28 ± 06.12	13.27 ± 03.79	27.48 ± 04.79	50.17 ± 07.08	40.17 ± 08.09
β-amyrin acetate (4)	05.59 ± 02.15	09.28 ± 01.89	28.89 ± 03.78	36.02 ± 01.27	37.23 ± 04.48
Lupeol (5)	07.58 ± 03.79	14.28 ± 04.09	30.78 ± 03.47	48.72 ± 02.87	26.48 ± 04.89
α-amyrin acetate (6)	4.87 ± 03.54	08.43 ± 03.71	31.14 ± 02.57	39.57 ± 05.57	38.57 ± 06.27
CE1	06.21 ± 01.28	09.58 ± 03.48	12.48 ± 03.17	22.57 ± 05.79	15.37 ± 08.79
CE2	08.71 ± 03.17	31.54 ± 02.78	41.05 ± 0.50	52.59 ± 04.28	56.88 ± 07.80
CE3	04.74 ± 02.54	07.05 ± 02.27	26.75 ± 03.07	34.81 ± 03.17	33.78 ± 02.78
Si	12.35 ± 02.25	16.30 ± 03.28	33.50 ± 01.25	62.39 ± 02.54	81.12 ± 01.35

The absorbance of the MTT formazan was determined at 570 nm in an ELISA/plate reader. Cell viability was defined as the ratio (expressed as a percentage) of the absorbance of treated cells to untreated cells. Reported values are the mean ± SD (standard deviations) of three independent experiments carried out in triplicate. **(1) - (6)**: isolated compounds; **CE**: crude extract of *Ficus gnaphalocarpa*; **CE2**: crude extract of *Ficus cordata*, **CE3**: crude extract of *Ficus chlamydocarpa*, **Si**: silymarin taken as a hepatoprotective reference compound.

Table 16.2: LDH leakage from rat hepatoma cells after 24 h incubation with 12.5, 25, 50, 100 and 200 µg/mL extract and isolated compounds.

Compounds and Crude Extracts	*Concentration of Compounds and LDH Leakage from Rat Hepatoma Cells*				
	12.5 µg/mL	*25 µg/mL*	*50 µg/mL*	*100 µg/mL*	*200 µg/mL*
Betulinic acid (1)	60.10 ± 04.78	50.76 ± 03.56	30.70 ± 05.67	28.78 ± 06.72	30.50 ± 05.89
Catechin (2)	66.58 ± 04.89	62.58 ± 03.60	50.38 ± 06.18	58.48 ± 03.49	44.27 ± 04.09
Quercetin (3)	48.59 ± 06.28	40.48 ± 05.47	35.28 ± 06.27	24.57 ± 03.48	38.27 ± 04.79
β-amyrin acetate (4)	60.45 ± 03.58	57.89 ± 05.48	48.26 ± 02.81	31.25 ± 05.19	40.18 ± 06.71
Lupeol (5)	53.48 ± 06.58	46.58 ± 05.87	35.48 ± 04.09	27.28 ± 07.89	53.89 ± 03.71
α-amyrin acetate (6)	58.17 ± 03.45	55.89 ± 05.78	45.78 ± 02.49	33.27 ± 04.89	38.27 ± 05.89
CE1	63.28 ± 03.45	42.15 ± 05.29	36.28 ± 06.28	36.37 ± 02.78	52.48 ± 03.57
CE2	40.81 ± 04.70	34.87 ± 03.47	22.89 ± 06.89	30.47 ± 05.78	63.80 ± 05.71
CE3	20.10 ± 09.71	52.48 ± 06.51	48.79 ± 04.88	40.84 ± 03.17	53.87 ± 07.57
Si	46.70 ± 03.89	33.60 ± 03.76	22.50 ± 04.75	16.60 ± 03.04	10.60 ± 02.97

LDH activity was measured in the supernatants. Reported values are the mean ± SD (standard deviations) of three independent experiments carried out in triplicate. **(1) - (6)**: **CE1**: crude extract of *Ficus gnaphalocarpa*; **CE2**: crude extract of *Ficus cordata*, **CE3**: crude extract of *Ficus chlamydocarpa*, **Si**: silymarin taken as a hepatoprotective reference compound.

Phenolics have been known to possess a capacity to scavenge free radicals. They are commonly found in both edible and nonedible plants, and have multiple biological effects, including antioxidant activity (Káhkónen *et al.*, 1999; Valenzuela *et al.*, 2003). The antioxidant activity of phenolics is principally due to their redox properties, which allow them to act as reducing agents, hydrogen donors and singlet oxygen quenchers. In addition, they have a metal chelation potential (Rice-Evans *et al.*, 1996).

Table 16.3: Antioxidant activities of crude extracts and isolated compounds. Values are EC_{50} ± SD of tree experiments in triplicate.

Compounds and Crude Extracts	*Biochemical Antioxidant Parameters (EC_{50} μg/mL)*			
	DPPH	*IPL*	*β-CLAMS*	*FRAP*
Betulinic acid (1)	∂	∂	31.95±0.64***	27.52 ± 0.82***
Catechin (2)	08.01 ± 02.78^{ns}	24.89 ± 01.78***	16.58 ± 03.75*	13.47 ± 04.79*
Quercetin (3)	05.45 ± 03.89^{ns}	14.66 ± 3.66*	07.25 ± 04.28	05.45 ± 03.25
β-amyrin acetate (4)	∂	∂	∂	∂
Lupeol (5)	23.48 ± 04.48***	24.18 ± 05.87***	22.47 ± 04.57***	16.27 ± 01.47*
α-amyrin acetate (6)	∂	∂	∂	∂
CE1	06.10 ± 02.87^{ns}	18.98 ± 02.55**	10.79 ± 01.78	06.58 ± 04.27
CE2	17.23 ± 03.87**	26.78 ± 02.87***	37.13 ± 02.90***	11.87 ± 02.83
CE3	07.45 ± 02.09^{ns}	15.78 ± 04.98*	10.98 ± 02.08	09.89 ± 04.78
Trolox	03.52 ± 02.74	06.19 ± 01.04	05.74 ± 03.89	04.37 ± 01.89

DPPH: radical scavenging activity; **IPL**: Lipid peroxidation assay; **FRAP**: ferric-reducing antioxidant power. **β-CLAMS**: inhibition of degradative oxidation of β-carotene. **(1) - (6)** are isolated compounds **CE1**: crude extract of *Ficus gnaphalocarpa*; **CE2**: crude extract of *Ficus cordata*, **CE3**: crude extract of *Ficus chlamydocarpa*. For each antioxidant mechanism, compounds/extracts were tested at five concentrations: 12.5, 25, 50, 100 and 200 μg/mL. ∂: No activity was shown. ***significant, compared with reference compound P<0.001. **significant, compared with reference compound P<0.01. *significant, compared with reference compound (P<0.05). ^{ns}not significant, compared with reference compound (P>0.05).

Study of any herbal drug becomes more significant when it ameliorates some diseases conditions. The current research on plant based medications focuses on isolation of biologically active substances from potent plants, their characterization and commercialization. Research in this direction has been greatly facilitated by modern physico-chemical techniques of isolation and structural elucidation (Suresh and Vandana, 2008; Rajesh *et al.*, 2009). CCl_4 is a well-known hepatotoxic agent and the preventive action of drugs on liver damage by CCl_4 has been widely used as an indicator of their liver protective activity. Changes associated with CCl_4-induced liver damage are similar to that of acute viral hepatitis (Rubinstein, 1962; Avijeet *et al.*, 2009; Yi-Hang *et al.*, 2009; Bo Huang *et al.*, 2010). CCl_4 is a hepatotoxic haloalkane whose mechanism has been studied intensively over the past years. It is one of the best studied solvent in regards to liver toxicity. It is well known that CCl_4 undergoes metabolic activation by a cytochrome P-450 dependent step to free radical products which can initiate lipid peroxidation. The toxicity induced by CCl_4 *in vivo* and in cultured hepatocytes, involves the stimulation of lipid peroxidation. CCl_4 treatment caused a significant decrease in cell viability. In addition, the toxin treatment initiated lipid peroxidation, caused leakage of enzymes like alanine transaminase and lactate dehydrogenase (LDH) levels (Gowri *et al.*, 2008; Nandita *et al.*, 2008; Rajesh *et al.*, 2009). In the present study, we applied the MTT and LDH tests to evaluate the *in vitro* hepatoprotective activity of the crude extracts and isolated compounds in rat hepatoma cell cultures. The tetrazolium dye, MTT, is widely used to assess the viability and/or the metabolic state of cells. This colorimetric assay is based on the conversion of the yellow tetrazolium bromide salt (MTT) to the red formazan derivative by mitochondrial succinate dehydrogenase in viable cells. In the lactate dehydrogenase (LDH) assay, leakage of the cytoplasmic located enzyme LDH into the extracellular medium is measured. The presence in the cell culture medium of the exclusively cytosolic enzyme LDH is indicative of cell membrane damage (Saad *et al.*, 2003; Rajesh *et al.*, 2009). We established that simultaneous treatment of hepatoma cells with the crude extracts and isolated compounds prevented the toxic effects of CCl_4 as judged by cellular viability and LDH leakage. The hepatoprotection, showed by the aptitude of these molecules to preserve cellular viability and to inhibit the leakage of LDH in extracellular medium was particularly pronounced with lupeol, quercetin and betulinic acid. Silymarin, which is an active constituent of the fruit of the milk thistle (*Silybum marianum*, Compositae) was used as standard reference and exhibited the best hepatoprotective activity against CCl_4-induced haptotoxicity in rat Morris hepatoma cell.

Quercetin is one of the ubiquitous flavonol-type flavonoids predominant in edible vegetables and fruits. As a member of the flavonol family, quercetin exhibited multiple biological effects on human health. It has been shown to prevent cardiovascular disease (Sesso *et al.*, 2003), inhibit platelet aggregation and prevent atherosclerosis and thrombosis (Formica and Regelson, 1995), have anticancer, anti-inflammatory responses, as well as antiulcer, antiallergic, and antiviral activities (Brown, 1980; Middleton and Kandaswami, 1992). Most of the pharmacological effects of quercetin are ascribed in part to its antioxidant activity. The antioxidant properties of quercetin might be due to its ability to chelate transition metal ions. Moreover, quercetin inhibited divalent cation-mediated lipid peroxidation, such as Fe^{2+} and Cu^{2+} (da Silva *et al.*, 1998). Quercetin could reduce oxidative damage of macromolecules such as lipids and DNA (Negre-Salvayre and Salvayre, 1992).

It was demonstrated that Lupeol ameliorates aflatoxin B(1)-induced peroxidative hepatic damage in rats (Kanniappan *et al.*, 2006). Hepatoprotective effect of lupeol on tissue antioxidant defence system in cadmium-induced hepatotoxicity in rats is known (Sunitha *et al.*, 2001). This pentacyclic triterpene also have some effect on the lipid peroxidation and antioxidant status in rat kidney after chronic cadmium exposure (Nagaraj *et al.*, 2000). Lupeol also prevents free radical mediated macromolecular damage and alleviates benzoyl peroxide induced biochemical alterations in murine skin (Sultana *et al.*, 2003).

The fact that antioxidant agents inhibit carbon tetrachloride-induced liver damage (Muriel and Murelle, 1990; Hussain *et al.*, 2009; Naseem *et al.*, 2009; Yi-Hang *et al.*, 2009) prompted us to study the antioxidant effect of the crude extract and isolated compounds. The DPPH radical-scavenging activity, β-CLAMS, FRAP assays, and microsomal lipid peroxidation are frequently used by researchers for a rapid evaluation of antioxidant activity. These systems were used in our study as well, with trolox as standard antioxidant compound. As shown by the results, apart of compound **(4)** and **(6)**, all compounds tested in the present study exhibit antioxidant activities.

Conclusions

This study provides a scientific basis for the ethnomedical claims that *Ficus gnaphalocarpa; Ficus cordata* and *Ficus chlamydocarpa* are effectives against certain liver-related diseases. Taken together, our findings provide evidence that the crude extracts and some isolated compounds from these plants exhibit hepatoprotective and/or antioxidant activities. This crude extracts and isolated compounds might be useful for the prevention of toxic-induced and free radical-mediated liver diseases since it has been suggested that antioxidant compounds may be used as prophylactic agent.

Acknowledgements

This research was supported by the University Centre for International Cooperation and Development (CICOPS) and by the Institute for Right to University Studies (ISU University of Pavia, Italy). The authors are also grateful to the members of NGO *Help for African Students* (HAS).

References

Adjanohoun, E.J., Adjakidje, V., Ahyi, M.R.A. *et al.* (1989). Contribution aux études ethnobotaniques et floristiques en République Populaire du Bénin. Collection "Médecine traditionnelle et pharmacopée". *Agence de Coopération Culturelle et Technique,* **111:** 339-405.

Alziro de, Amorin., Helcio, R.B., Jorge, P.P., Carauta, D.L., and Maria, A.C.K. (1999). Anthelmintic activity of the latex of *Ficus* species. *Journal of Ethnopharmacology,* **64:** 255-258.

Anders, H., Graciela, V., Cesar, N., Maud, A., Miguel, A., and Guillermo, A. (1986). Preclinical and clinical studies with latex from *Ficus glabrata* hbk, a traditional intestinal anthelminthic in the amazonian area. *Journal of Ethnopharmacology,* **17:** 105-138.

Anders, H., Julio, C., Zelada, H., and Noriega, P. (2005). Reevaluation of risks with the use of latex as a traditional anthelmintic remedy in the Amazon. *Journal of Ethnopharmacology,* **98:** 251-257.

Anwarul, H.G., Malik, H.M., Khalid, H.J., Arif-ullah, K., Sheikh, A.S. (2008). Ethnopharmacological studies on antispasmodic and antiplatelet activities of *Ficus carica. Journal of Ethnopharmacology,* **119:** 1-5.

Asase, A., Oteng-Yeboah, A.A., Odamtten, G.T. *et al.* (2005). Ethnobotanical study of some Ghanaian anti-malarial plants. *Journal of Ethnopharmacology,* **99:** 273–279.

Avijeet, J., Manish, S., Lokesh, D. *et al.* (2008). Antioxidant and hepatoprotectiveactivity of ethanolic and aqueous extracts of *Momordica dioica* Roxb. Leaves. *Journal of Ethnopharmacology*, **115:** 61-66.

Bafor, E.E., and Igbinuwen, O. (2009). Acute toxicity studies of the leaf extract of *Ficus exasperata* on haematological parameters, body weight and body temperature. *Journal of Ethnopharmacology*, **123:** 302-307.

Benavides, A., Montoro, P., Bassarello, C. *et al.* (2006). Derivatives in *Jatropha macrantha* stems: Characterization and LC/ESI/MS quail-quantitative analysis. *Journal of pharmaceutical and Biomedical Analysis*, **40:** 639-647.

Benzie, I., and Strain, J.J. (1996). The ferric reducing ability of plasma (FRAP) as a measure of "antioxidant power". The FRAP assay. *Analytical Biochemistry*, **239:** 70-76.

Bo, Huang., Xiaoquan, B., and Jingsheng, H. (2010). Hepatoprotective and antioxidant activity of ethanolic extracts of edible lotus (*Nelumbo nucifera* Gaertn.) leaves. *Food Chemistry*, **120:** 873–878.

Brand-Williams, W., Cuvelier, M.E., and Berset, C. (1995). Use of free radical method to evaluate antioxidant activity. *Lebensm Wiss Technol.*, **28:** 25-30.

Brown, J.P. (1980). A review of the genetic effects of naturally occurring flavonoids, anthroquinones and related compounds. *Mutat. Res.*, **75:** 243–277.

Cheeseman, K.H., and Scater, T.F. (2003). Free radical in medicine. In *British Medical Bulletin*. Churchill Livingstone, London **49:** 479-724.

Chen, S., Morimoto, S., Tamatani, M. *et al.* (1996). Calcitonin prevents CCl_4-induced hydroperoxide generation and cytotoxicity possibly through C1b receptor in rat hepatocytes. *Biochem Biophys Res Commun.*, **26:** 865-871

Chindo, B.A., Amos, S., Odutola, A.A., Vongtau, H.O., Abbah, J.C., and Wambebe, K.S. (2003). Gamaniel central nervous system activity of the methanol extract of *Ficus platyphylla* stem bark. *Journal of Ethnopharmacology*, **8:** 131-137.

Cusset, C. (1985). «Flore du Cameroun», *Muséum National d'Histoire Naturelle*, Paris 5, **18**: 108-132.

Damanpreet, S., and Rajesh, K.G. (2009). Anticonvulsant effect of *Ficus religiosa*: Role of serotonergic pathways. *Journal of Ethnopharmacology*. **123:** 330-334.

da Silva, E.L., Piskula, M.K., Yamamoto, N., *et al.* (1998). Quercetin metabolites inhibit copper ion–induced lipid peroxidation in rat plasma. *FEBS Lett.*, **430:** 405–408.

Donfack, J.H., Wabo, F.G., Ngameni, B. *et al.* (2010). *In vitro* hepatoprotective and antioxidant activities of the crude extract and isolated compounds from *Irvingia gabonensis*. *Asian Journal of Traditional Medicine*, **5:** 79-88.

Dutta, N.K., Mazumdar, K., Mishra, U.S. *et al.* (2007). Isolation and identification of flavones (quercetin) from *Butea frondosa* bark. *Springer. New-York.*, **41:** 269-271.

Formica, J.V., and Regelson, W. (1995). Review of the biology of quercetin and related bioflavonoids. *Food Chem. Toxicol.*, **33:** 1061–1080.

Garle, M.J., and Fry, J.R. (1989). Detection of reactive metabolites *in vitro*. *Toxicology*, 154: 101-110.

Hensley, K., Robinson, K., Gabbita, S.P. *et al.* (2000). Reactive oxygen species, cell signaling, and cell injury. *Free Radic. Biol. Med.*, **28:** 1456.

Higuchi, H., and Gores, G.J. (2003). Mechanisms of liver injury: an overview. *Curr. Mol. Med.*, **3:** 483-491.

Hussain, Z., Amresha, G., Satyawan, S. *et al.* (2009). Hepatoprotective and antioxidant activity of *Amaranthus spinosus* against CCl_4 induced toxicity. *Journal of Ethnopharmacology*, **125:** 364–366.

Jaeschke, H., Gores, G.J., Cederbaum, A.I. *et al.* (2002). Mechanisms of hepatotoxicity. *Toxicol. Sci.*, **65:** 166-172.

Jacob, R.A., and Sotoudeh, G. (2002). Vitamin C function and status in chronic disease. *Nutr. Clin. Care.*, **5:** 66–74.

Káhkónen, M.P., Hopia, A.I., Vuorela, H.J., Rauuha, J.P., Pihlaja, K., Kujala, T.S., and Heinonen, M. (1999). Antioxidant activity of plant extracts containing phenolic compounds. *J Agrie Food Chem.*, **47:** 3954-3962.

Kanniappan, M.E., Selvakumar, M. N., and Varalakshmi, P. (2006). Lupeol ameliorates aflatoxin B_1-induced peroxidative hepatic damage in rats. *Comparative Biochemistry and Physiology Part C: Toxicology and Pharmacology*, **143(3)**.

Kerharo, J., and Adam, J.G. (1974). «La pharmacopée sénégalaise traditionnelle. Plantes médicinales et toxiques», Editions Vigot Frères, 1011. Paris.

Kofi, A., and Peter, J.H. (2008). Antibacterial, antioxidant and fibroblast growth stimulation of aqueous extracts of *Ficus asperifolia* Miq. and *Gossypium arboreum* L., wound-healing plants of Ghana. *Journal of Ethnopharmacology*, **119:** 141-144.

Kuete, V., Ngameni, B., Fotso, S.C.C., Kengap, T.R., Tchaleu, N.B., Meyer, J.J.M., Lall, N., and Kuiate, J.R. (2008). Antimicrobial activity of the crude extracts and compounds from *Ficus chlamydocarpa* and *Ficus cordata* (Moraceae). *Journal of Ethnopharmacology*, **120:** 17–24.

Kuete, V., Frédéric, N., Bathélémy, N., Armelle, T.M., Félix, K., and Bonaventure, T.N. (2009). Antimicrobial activity of the crude extract, fractions and compounds from stem bark of *Ficus ovata* (Moraceae). *Journal of Ethnopharmacology*, **124:** 556-561.

Lansky, E.P., Paavilainen, H.M., Pawlus, A.D., and Newman, R.A. (2008). *Ficus* spp. (fig): Ethnobotany and potential as anticancer and anti-inflammatory agents. *Journal of Ethnopharmacology*, **119:** 95–213.

Lollinger, J. (1981). Free radicals and food additives. Taylor and Francis London 21.

Mahato SB and Kundu AP (1994). ^{13}C-NMR spectra of pentacyclic triterpenoids. A compilation and some salient features. *Phytochemistry*, **37**: 1517-1575.

Maydell, H.J.V. (1983). «Arbres et arbustes du sahel, leurs caractéristiques et leurs utilisations», *GTZ, Eschborn*. **147:** 250-251.

Middleton, E.J., and Kandaswami, C. (1992). Effects of flavonoids on immune and inflammatory cell functions. *Biochem. Pharmacol.*, **43:** 1167–1179.

Miller, H.E. (1971). A simplified method for the evaluation of anti-oxidant. *J. Am. Oil Chem. Soc.*, 48-91.

Mousa, O., Vuorela, P., Kiviranta, J., Abdel, S.W., Hiltunen, R., and Vuorela, H. (1994). Bioactivity of certain Egyptian *Ficus* species. *Journal of Ethnopharmacology*, **41:** 71-76.

Mpiana, P.T., Mudogo, V., Tshibangu, D.S.T., Kitwa, E.K., Kanangila, A.B., Lumbu, J.B.S., Ngbolua, K.N., Atibu, E.K., and Kakule, M.K. (2008). Antisickling activity of anthocyanins from *Bombax pentadrum*, *Ficus capensis* and *Ziziphus mucronata*: Photodegradation effect. *Journal of Ethnopharmacology*, **120:** 413-418.

Muriel, P., and Murelle, M. (1990). Prevention by silymarin of membrane alterations in acute CCl_4 liver damage. *J. Appl. Toxicol.*, **10:** 275-279.

Nagaraj, M., Sunitha, S., and Varalakshmi, P. (2000). Effect of lupeol, a pentacyclic triterpene, on the lipid peroxidation and antioxidant status in rat kidney after chronic cadmium exposure. *J. Appl. Toxicol.*, **20:** 413-417.

Nandita, S., Vasudeva, K., Narasimhamurthy, K., *et al.* (2008). Protective effect of potato peel extracts against carbon tetrachloride-induced liver injury in rats. *Environmental Toxicology and Pharmacology*, **26:** 241–246.

Naseem, N.Q., Bhanudansh Nadeem, S.K. *et al.* (2009). Antioxidant and hepatoprotective activity of *Cordia macleodii* leaves. *Saudi Pharmaceutical Journal*, **17:** 299– 302.

Negre-Salvayre, A., and Salvayre, R. (1992). Quercetin prevents the cytotoxicity of oxidized LDL on lymphoid cell lines. *Free Radic. Biol. Med.*, **12:** 101–106.

Ning, W., Peibo, L., Yonggang, W. *et al.* (2008). Hepatoprotective effect of *Hypericum japonicum* extract and its fractions. *Journal of Ethnopharmacology*, **116:** 1–6.

Nongonierma, R.B., Ndiaye, L., Thiam, D. *et al.* (2005). «Activité antidrépanocytaire de la fraction F3 de l'extrait acétonique des écorces des tiges de *Ficus gnaphalocarpa* (Mig) steud (Moraceae)». *Pharmacien d'Afrique*, **187:** 3-6.

Poumale, H.M.P., Kengap, R.T., Tchouankeu, J.C., Keumedjio, F., Laatsch, H., and Ngadjui, B.T. (2008). Pentacyclic Triterpenes and Other Constituents from *Ficus cordata* (Moraceae). *Z. Naturforsch.*, **636:** 1335-1338.

Rajesh, K., Ramasamy, M., Ramtej, J. *et al.* (2009). Isolation, characterization and antioxidative effect of phyllanthin against CCl_4-induced toxicity in HepG2 cell line. *Chemico-Biological Interactions*, **181:** 351–358.

Rakesh, K.S., Shikha, M., Dolly, J., Prashant, K.R., and Geeta, W. (2009). Antidiabetic effect of *Ficus bengalensis* aerial roots in experimental animals. *Journal of Ethnopharmacology*, **123:** 110-114.

Rao, C.V., Arti, R.V., Vijayakumar, M., and Rastogi, S. (2008). Gastroprotective effect of standardized extract of *Ficus glomerata* fruit on experimental gastric ulcers in rats. *Journal of Ethnopharmacology*, **115:** 323-326.

Rice-Evans, C.A., Miller, N.J., Paganga, G. (1996). Structure-antioxidant activity relationships of flavonoids and phenolic acids. *Free Radical Biol Med.*, **20:** 933-956.

Rimi, S., Shweta, G., Gambhir, J.K., Prabhu, K.M., and Murthy, P.S. (2004). Antioxidant effect of aqueous extract of the bark of *Ficus bengalensis* in hypercholesterolaemic rabbits. *Journal of Ethnopharmacology*, **92:** 47-51.

Rodeiro, I., Donato, M.T., Martínez, I. *et al.* (2008). Potential hepatoprotective effects of new Cuban natural products in rat hepatocytes culture. *Toxicology in Vitro*, **22:** 1242-1249.

Rubinstein, D. (1962). Epinephrine release and liver glycogen levels after carbon tetrachloride administration. *American Journal of Physiology*, **203:** 1033–1037.

Saad, B., Abu-Hijleh, G., and Suter, U.W. (2003). Polymer biocompatibility assessment by cell culture techniques. In: Arshady R (ed). The PMB Series Volume 1: Introduction Polymeric Biomaterials. The Citus Books.

Sabatie, B. (1985). Flore du Cameroun, Moraceae (incl; Cecropiaceae), Ministère de l'Enseignement Supérieur et de la Recherche Scientifique (MESRES) (Ed), Yaoundé- Cameroun. pp 154-155.

Sami, A.K., and Petter, G.W. (1983). Thonningine-A and Thonningine-B. Two 3-phenylcoumarins from the seeds of *Milletia thonningii*. *Phytochemistry*, **22:** 1001-1003.

Sandabe, U.K., Onyeyili, P.A., Chibuzo, G.A. (2006). Phytochemical screening and effect of aqueous extract of *Ficus sycomorus* L. (Moraceae) stembark on muscular activity in laboratory animals. *Journal of Ethnopharmacology*, **104:** 283-285.

Sesso, H.D., Gaziano, J.M., Liu, S. *et al.* (2003). Flavonoid intake and the risk of cardiovascular disease in women. *Am. J. Clin. Nutr.*, **77:** 1400–1408.

Subhash, C.M., Tapan, K.M., Das, J., Saba, B.P., and Pal, M. (2000). Anti-inflammatory evaluation of *Ficus racemosa* Linn. leaf extract. *Journal of Ethnopharmacology*, **72:** 87-92.

Sunitha, S., Nagaraj, M., and Varalakshmi, P. (2001). Hepatoprotective effect of lupeol and lupeol linoleate on tissue antioxidant defence system in cadmium-induced hepatotoxicity in rats. *Fitoterapia*, **72:** 516-523.

Sultana, S., Saleem, M., Sharma, S., and Khan, N. (2003). Lupeol, a triterpene, prevents free radical mediated macromolecular damage and alleviates benzoyl peroxide induced biochemical alterations in murine skin. *Ind. J. Exp. Biol.*, **41:** 827-831.

Suresh, R.N., and Vandana, S.P. (2008). Hepatoprotective effect of Ginkgoselect Phytosome® in rifampicin induced liver injurym in rats: Evidence of antioxidant activity. *Fitoterapia*, **79:** 439–445.

Swift, L.I. (1952). Isolation of β-sitosterol, β-sitosterol-3-O-β-D-glucopyranoside from the juice of *Flurida Valencia* oranges (*Citrus sinensis* L.). *Journal of American Chemical Society*, **74:** 99-100.

Ulf, A.N., Lars-Inge, O., Gunnar, C. *et al.* (1989). Inhibition of lipid peroxidation by spin labels. *The Journal of Biological Chemistry*, **264:** 11131-11135.

Valenzuela, A., Sanhueza, J., and Nieto, S. (2003). Cholesterol oxidation: Health hazard and the role of antioxidants in prevention. *Biol Res.*, **36:** 291-302.

Wan, B.S., and Chan, H.T. (2004). Enantioselective synthesis of afzelzchin and epiafzelechin. *Tetrahedron*, **60**: 8207-8211.

Wills, E.O. (1987). Evaluation of lipid peroxidation and biological membranes. In: K. Snell and B. Mullock, (Eds), Biochemical Toxicology. A practical approach. IRL Press Oxford.

Yihui, Li., Juping, D., Tingting, G., Wei, X., Shineng, Y., Bo, Li., Yanqin, Z., and Yuxiang, C. (2009). *In vivo* pharmacokinetics comparisons of icariin, emodin and psoralen from Gan-kang granules and extracts of Herba Epimedii, Nepal dock root, *Ficus hirta* yahl. *Journal of Ethnopharmacology*, **124:** 522-529.

Yi-Hang, W., Xiao-Meng, Z., Ming-Hui, H. *et al.* (2009). Effect of *Laggera alata* on hepatocyte damage induced by carbon tetrachloride *in vitro* and *in vivo*. *Journal of Ethnopharmacology*, **126:** 50–56.

Traditional and Folk Herbal Medicine: *Pages* **349–354**
Recent Researches Vol. 1 (2012)
Editor: **V.K. Gupta**
Published by: **DAYA PUBLISHING HOUSE, NEW DELHI**

17

Effects of *Holarrhena floribunda* Leaf Extract on Gastrointestinal Functions in Male Rats

Bamidele V. Owoyele[1*], Adebisi, A. Adebayo[1], Oluwakemi T. Alabi[1] and John I. Okwusidi[2]

ABSTRACT

An aqueous extract of the dried leaves of Holarrhena floribunda *(false rubber tree) was investigated for its effect on gastric transit time and purgative properties in male Wistar rats using the indicator displacement method and production of wet faeces. Transit time was evaluated using the indicator dilution method while the purgative test was carried out by observing the number of wet faeces produced by each animal following the administration of the extract. Extract was administered in a dose range of 25-200mg/kg p.o. The results show that the extract (25- 200mg/kg) produced a dose related increase in intestinal transit as well as a dose related increase in number of wet faeces produced per hour by the same animals. The result from this study demonstrated the purgative effect of* Holarrhena floribunda *leaves and its potential use by traditional medical practitioners.*

Keywords: *Holarrhena floribunda*, Intestinal motility, Indicator displacement, Purgative.

1 Department of Physiology, Faculty of Basic Medical Sciences College of Health Sciences, University of Ilorin, P. M. B. 1515, Ilorin, Nigeria.

2 Department of Physiology, College of Health Sciences, Ladoke Akintola University of Technology, Ogbomosho, Nigeria.

* *Corresponding author*: E-mail: deleyele@yahoo.com, deleyele@gmail.com; Ph: +234 803 5065190.

Introduction

Holarrhena floribunda (G. Don) Durand and Schinz (Apocynaceae) is found in tropical regions of Africa and south East Asia. In Africa the plant is distributed from the republic of Guinea to Angola (Iwu, 1983). Different parts of the plant are used for the treatment of many types of ailments. The bark macerated in palm wine is used in the treatment of fever, amoebic dysentery, as a febrifuge and as a tonic (Watt and Breyer, 1962; Iwu, 1983). The seeds are used for the treatment of diarrhoea, fever and chronic chest pain (Gill, 1991). The leaves, bark and roots are used in many parts of African continent as remedy for malaria (Iwu, 1983). Locally the traditional medical practitioners commonly prescribe the decoction of the leaves for the treatment of digestive disorders such as constipation.

Previous works on *Holarrhena floribunda* shows that it contains anti tumor and anti-malarial agents (Abreu, 1999; Fotie, 2001). Other studies also reveal the presence of long- chain fatty esters, trichthenes, alkaloids, conessine, holarrhenine, holadienine, holamine, holaphylline, holaphyllamine and kurchicine are well known. Progesterone and some flavonoids (Foucaud, 1959; Leboeuf *et al.*, 1964; Paris and Duret, 1973; Tamboura *et al.*, 2005). The focus of the present study is to evaluate the anecdotal claims that *Holarrhena floribunda* can be used for the treatment of gastric disorders such as constipation. Specifically, this study was undertaken to investigate the effect of dried *Holarrhena floribunda* leaf extract on the gastrointestinal transit time and its purgative effects.

Materials and Methods

Plant Material and Extraction

The plant material used in this study was collected from Igosun, (in Kwara state, Nigeria) and identified by T.K. Odewo at Forestry Research Institute of Nigeria. The leaves were dried in the early morning sun and reduced to powder with mortar and pestle. 500g of the powder was soaked in 5L of water for 48 hours. The macerated mixture was filtered and evaporated in a carefully regulated water bath maintained at 50°C to yield a dark brown residue weighing 15g the extract was stored in a refrigerator at 4°C. The residue of the extract was re-suspended in saline for administration to animals.

Animals

Male Wistar rats (weighing 150g±10) were used for the study. They were bred and housed in the Faculty of Basic Medical Sciences Animal house, College of Health Sciences, University of Ilorin, Nigeria, under standard conditions of temperature and humidity. They were fed mouse cubes (Bendel Feeds). Feed and water were available *ad libitum*.

Phytochemical Screening

Phytochemical analysis of the plant extract was carried out using the methods described by Trease and Evans (1989).

Effects of the Extract on Intestinal Transit in Rats

Thirty rats were randomly divided into six groups of five animals each. The animals were fasted for 24 hours after which they were then forced fed orally with 5ml of Leishman's stain food mixture (3g of Leishman stain dissolved in 200ml of distilled water containing 50g of powdered growers marsh). Saline (50ml/kg) and atropine (3mg/kg) was administered to the animals in the control and reference groups respectively after 30 minutes of leishman's stain administration. The rats in the test groups received 25, 50, 100 and 200 mg/kg of the extract respectively in addition to the leishman's stain.

The treatment regimens were allowed to circulate for 1 hr after which the rats were killed by decapitation. The intestines were carefully removed and detached from the stomach at the pyloric junction. The distance traveled (in centimeter) by the lead stain (*i.e.* the food stain mixture) from the pylorus was measured (Schwarz *et al.*, 1999).

Purgative Test in Rats

This was carried out according to the method described by Akah *et al.* (1997). Rats were housed in metal cages without food and water for 6 hours before the experiment. The rats were divided into six groups of five rats each, the groups includes the extract (25, 50, 100 and 200 mg/kg), Bisacodyl (50mg/kg) or saline (50mg/kg) groups. The extract, reference drug (Bisacodyl) or saline was orally administered to the animals in their respective groups.

The animals were then housed individually in cages lined with white absorbent paper. Water was restored and the animals were observed for 24 hours. The total number of wet faeces produced by each group was noted and normalized to hourly output.

Statistical Analysis

Results were expressed as means ± SEM. The significance of difference between means was determined by the Student's t-test and results were regarded as significant when $p < 0.05$.

Results and Discussion

Phytochemical Analysis

The preliminary phytochemical screening of the leaves of *H. floribunda* revealed the presence of saponins, reducing sugars, glycosides, tannins, alkaloids and flavonoids.

Intestinal Transit Time and Purgative Tests

The aqueous extract produced a marked increase in propulsive movement of the leishman's stain-food mixture in the intestine compared with the control group (Table 17.1). This increase in intestinal transit occurred in a dose related fashion. The results of the purgative tests show that higher doses of the extract significantly reduced the number of wet faeces compared with the reference drug (bisacodyl) (Table 17.2).

Table 17.1: Intestinal transit effect of aqueous extract of *Holarrhena floribunda* leaves.

Groups	*Dose (mg/kg)*	*Mean Distance Traveled (cm)*
Control (saline)	–	54.2 ± 3.50
Holarrhena floribunda	25	66.2 ± 3.02*
Holarrhena floribunda	50	77.0 ± 4.56**
Holarrhena floribunda	100	83.4 ± 2.04***
Holarrhena floribunda	200	89.4 ± 3.98***
Atropine	3	42.2 ± 3.26

a: Each value is the mean ± S.E.M. of 5 rats,
*: P < 0.05; **: P < 0.001; ***: P < 0.001 compared with control; Student's t-test.

Table 17.2: Purgative effect of aqueous extract of *Holarrhena floribunda* leaves.

Groups	*Dose (mg/kg)*	*Number of Wet Faeces*
Control (saline)	–	–
Holarrhena floribunda	25	0.11 ± 0.07[c]
Holarrhena floribunda	50	0.15 ± 0.06*[c]
Holarrhena floribunda	100	0.30 ± 0.03***[b]
Holarrhena floribunda	200	0.33 ± 0.08***[b]
Bisacodyl	50	1.53 ± 0.26***

a: Each value is the mean ± S.E.M. of 5 rats,
*: P < 0.05; **: P < 0.001; ***: P < 0.001, compared with control;
b: P< 0.01; c: P<0.001 compared with Bisacodyl; Student's t-test.

In the present study, the purgative and motility effects of *H. floribunda* aqueous extract was investigated using standard animal models. In the purgative test the number of wet faeces produced served as the index of purgative activity and this agrees with standard practices (Yagi *et al.*, 1997). Likewise in the motility test the Leishman stain used is one of the standard stains commonly employed for such purposes. The distance travelled by the food and stain mixture serves as the index of the motor/propulsive action of the gastrointestinal tracts (Eghianruwa *et al.*, 2006).

The results obtained from the purgative test show that the extract has significant purgative effects compared with the untreated control. However, the purgative effect recorded with the extract was significantly lower than the effect observed from bisacodyl (a standard purgative agent) application. In the motility test, the extract dose dependently decreases transit time for the stain/food mixture compared with the untreated control and the reference drug (atropine). The results obtained from the two test generally indicated that the extract probably has secretomotor activities. The administration of the extract may be useful in conditions of food poisoning or constipation.

Furthermore, the activities of the extract may not be unconnected with some of the bioactive constituents which the phytochemical analysis revealed. Such phytochemical constituents include saponnins, alkaloids, tannins and glycosides. The presence of some of these constituents such as alkaloid and flavonoids has been reported previously (Iwu, 1983; Leboeuf *et al.*, 1964; Paris and Foucand, 1959). Specifically, some alkaloids have been shown to possess purgative effects while some flavonoids have been shown to have the capacity to increase gastrointestinal motility. Previous report on the bark of *H. floribunda* has shown that it contains alkaloids such as conessine which inhibits amoebiasis (Tamboura *et al.*, 2005). However, none of such studies on gastrointestinal activities have been extended to the leaves of this plant.

In conclusion, the present study has shown that the aqueous extract of the leaves of *H. floribunda* enhances gastric motility and emptying. These observations also justify the traditional uses of the plant leaves for the treatment of constipation and as purgative. The focus of further studies will include employment of *in vitro* models, characterization of active ingredients as well as toxicity studies.

References

Abreu, P.M., Martins, E.S., Kayser, O., Bindseil, K.U., Siems, K., Seemann, A., and Frevert, J. (1999). Antimicrobial, antitumor and antileishmania screening of medicinal plants from Guinea-Bissau. *Phytomedicine,* **6:** 187-195.

Akah, P.A., Oli, A.N., Enwerem, N.M., and Gamaliel, K. (1997). Preliminary studies on the purgative effect of *Carica papaya* root extract. *Fitoterapia,* **68**: 327-331.

Duez, P., Chamart, S., Lejoly, J., Hanocq, M., Zeba, B., Sawadogo, M., Guissou, P., and Molle, L. (1987). Changes in conessine in stem bark of *Holarrhena floribunda* in Burkina Faso. *Ann Pharm Fr.,* **45(4):** 307-313.

Eghianruwa, K.I., Ogunleye, O.A., Saba, A.B., Famakinde, S.A., Ola-Davies, E.O., and Abu, H,H. (2006). Influence of atropine and loperamide on reduced intestinal transit induced by *Calotropic procera* latex in rats. *Afr J Biomed Res.,* **9:** 125-128.

Fotie, J., Bohle, D,S., Leimanis, M,L., Georges, E., Rukunga, G., and Nkengfack, A,E. (2006). Lupeol long-chain fatty acid esters with antimalarial activity from *Holarrhena floribunda. J Nat Prod.,* **69:** 62-67.

Gill, L.S. (1992). Ethnomedical uses of plants in Nigeria. Uniben press. Pp 133

Iwu, M.M. (1993). Handbook of African Medicinal plants. *C.R.C Press,* LL.C. USA: pp 191.

Leboeuf, M., Cave, A., and Goutarel, R. (1964). Steroid alkaloids. Presence of progesterone in the leaves of *Holarrhena floribunda* (G. Don) Duer. and Schinz. *C R Hebd Seances Acad Sci.,* **9:** 3401-3403.

Leboeuf, M., Cavé, A., and Goutarel, R. (1969). Steroid alkaloids. LXXXII. (16). Chemical composition of the leaves of *Holarrhena floribunda* Dur. and Schinz. Isolation of progesterone and 4 new alkaloids: methylholaphylline, holaphyllinol, holaphyllidine and dihydroholaphyllamine. *Ann Pharm Fr.,* **27**: 2172-2228.

Paris, R.R., and Duret, S. (1973). On poly phenols of different species of Holarrhena. Plantes *Medicinales and Phytotherapie,* **7:** 145-150.

Paris, R.R., and Foucaud, A. (1959). Flavonoids in the leaves of *Holarrhena floribunda* G. Don (Durand and Schinz): isolation of a flavonoside identified as isoquercitroside. *C R Hebd Seances Acad Sci.*, **4:** 2634-2635.

Schwarz, R.D., Callahan, M.J., Coughenour, L.L., Dickerson, M.R., Kinsora, J.J., Lipinski W.J, Raby, C.A., Spencer, C.J., and Tecle, H. (1999). Milameline (CI-979/RU35926): A muscarinic receptor agonist with cognition-activating properties: biochemical and *in vivo* characterization. *J Pharmacol Exp Ther.*, **291:** 812–822.

Trease, G.E., and Evans, W.C. (1989). *A textbook of Pharmacognosy*, 13th edn. Bailliere-Tyndall Ltd., London. pp 5-53.

Tamboura, H. H., Bayala, B., Lompo, M., Guissou, I.P., and Sawadogo, L. (2005). Ecological distribution, morphological characteristics and acute toxicity of aqueous extracts of *Holarrhena floribunda* (G. Don) Durand and Schinz, *Leptadenia hastata* (pers.) Decne and *Cassia sieberiana* (D C) used by veterinary healers in Burkina faso. *Afr. J. Trad. CAM.*, **2:** 13 – 24.

Watt, J.M., and Breyer-Brandwijk, M.G. (1962). The Medicinal and poisonous plants of Southern and Eastern and Eastern Africa, 2nd edn. Edinburgh: E and S Livingstone, pp131.

Yagi, T., Yamauchi, K., and Kuwano, S. (1997). The synergistic purgative action of aloe – emodin in mice: synergism in large intestinal propulsion and water secretion. *J. Pharm. Pharmacol.*, **49** (1): 22-25.

Traditional and Folk Herbal Medicine: Recent Researches Vol. 1 (2012) *Pages* **355–364**
Editor: **V.K. Gupta**
Published by: **DAYA PUBLISHING HOUSE, NEW DELHI**

18

Insulin Releasing Effects of Cumin (*Cuminum cyminum* L.) Seeds in Diabetic Rats

Akram Eidi[1*], Maryam Eidi[2], Alireza Sadeghipour[3] and Davood Magsoodi[1]

ABSTRACT

To determine the insulin releasing effect of cumin seeds, we investigated the effects of ethanolic extract of the plant on healthy and streptozotocin-induced diabetic rats. The ethanolic extract (0.1, 0.2, 0.4 and 0.6 g/kg body wt) was injected intraperitoneally. Blood samples were obtained from retro-orbital sinus before administration and 1, 3 and 5 h after administrations. The serum glucose and insulin levels were measured. After blood sampling, the pancreas of animals were excised. Pancreatic sections were processed for examination of insulin-releasing activity using an immunocytochemistry kit. The results showed that the plant extract significantly decreased serum glucose while increased serum insulin levels in diabetic rats but not in healthy rats. Also, our results showed that the extract increased the number of beta cells with insulin secretory activity. The present data indicate that cumin extract has hypoglycemic and insulin releasing effect in diabetic animals.

Keywords: Cumin, *Cuminum cyminum* L., Hypoglycemia, Immunocytochemistry, Pancreatic β-cells, Rat.

1. Department of Biology, Science and Research Branch, Islamic Azad University, Tehran, Iran.
2. Department of Biology, Varamin Branch, Islamic Azad University, Varamin, Iran.
3. Department of Pathology, Rasoul Akram Hospital, Tehran, Iran.

* *Corresponding author*: E-mail: eidi@srbiau.ac.ir.

Introduction

Diabetes mellitus is a disease due to abnormality of carbohydrate metabolism and it is mainly linked with low blood insulin level or insensitivity of target organs to insulin. It is the most prevalent chronic disease in the world affecting nearly 25 per cent of the population. From literature review it has been revealed that 15-20 per cent of diabetic patients are suffering from insulin-dependent diabetes mellitus or type-I (Urger and Foster, 1998; Muhlhauser *et al.*, 2000). In modern medicine, no satisfactory effective therapy is till available to cure the diabetes mellitus. Though insulin therapy is also used for the management of diabetes mellitus but there are several drawbacks like insulin resistance (Piedrola *et al.*, 2001), anorexia nervosa, brain atrophy and fatty liver after chronic treatment (Yaryura-Tobias *et al.*, 2001).

Renewed attention in recent decades to alternative medicines and natural therapies has stimulated a new wave of research interest in traditional practices. The plant kingdom has become a target for the search for new drugs and biologically active compounds (Evans, 1996). Ethnobotanical information indicates that more than 800 plants are used as traditional remedies for the treatment of diabetes (Pushparaj *et al.*, 2000; Alarcon-Aguilara *et al.*, 1998), but only a few have received scientific scrutiny.

Spices increase the taste and flavor of food. These esoteric food adjuncts have been in use for thousands of years. Spices have also been recognized to possess several medicinal properties and have been effectively used in the indigenous systems of medicine in different countries (Nadkarni and Nadkarni, 1976). Apart from the traditional use, a host of beneficial physiological effects have been brought to the fore by extensive animal studies during the past three decades (Chandrasekhara and Srinivasan, 1999; Srinivasan, 2000; Kempaiah and Srinivasan, 2006). Several spices and herbs have been shown to have hypoglycemic action in animals and humans (Brinker, 1998). Among the spices cumin seeds (*Cuminum cyminum* L.) belonging to the Apiaceae family are consumed in fairly large quantities by Asian people. Cumin seed is generally used as a food additive in the form of powder for imparting flavor to different food preparations. It also has a variety of medicinal properties. Cumin is widely used in Ayurvedic medicine for the treatment of dyspepsia, diarrhea and jaundice. It also has stomachic, diuretic, carminative, emmenagogic and antispasmodic properties (Joshi, 2000). Literature survey shows that a decoction of cumin seed possesses hypoglycemic effect. Roman-Ramos in preliminary studies on the effect of cumin seeds observed that they produce hypoglycemia (Roman-Ramos, 1995).

Since there are some recommendations concerning the hypoglycemic effects of the plant in Iranian folk medicine (Zargari, 1997), it became worthwhile to evaluate the influence of administration of alcoholic extract of cumin seeds in healthy and streptozotocin (STZ)-induced diabetic rats.

Materials and Methods

Plant Material

Cumin seeds were collected from Kerman area and scientifically approved in the department of botany of Islamic Azad University (Voucher number: 5032, deposited in: IAU Herbarium). The seeds plant was cleaned, shed dried at 25°C, and the dried seeds were ground by a blender, and the powder was kept in nylon bags in a deep freezer until the time of experiments.

Extraction of Ethanolic Extract

Dried and ground seeds (about 60 g) were submitted to extraction with 300 ml ethanol (80 per cent) in a Soxhlet apparatus for 72 h. After extraction, the solvent was filtered and then evaporated under reduced pressure by Rotavapor. The obtained cumin alcoholic extract was stored at -20° C until being used. The extract yield was 18 per cent.

Animals

Male Wistar rats weighing 200–250 g were housed in clean cages with temperature (22–24 °C), 12-h light:12-h dark cycle and relative air humidity 40–60 per cent. The animals were housed in groups of five per cage with free access to standard laboratory chow (35 per cent carbohydrates, 25 per cent proteins, 7 per cent lipids, and 3 per cent vitamins) and tap water. The diet was purchased from Pars-Dam food service, Tehran, Iran. The study was conducted in accordance with ethical procedures and policies approved by the Animal Care and Use Committee of Islamic Azad University, Iran.

Preparation of Diabetic Rat

Animals were injected with STZ (70 mg/kg, i.p.). Five days after injection, the rats with fasting blood glucose higher than 300 mg/dl were used for the experiments. Eight rats were used in each experiment. Each animal was used once only in all of experiments. The food was removed from cages 12 h before testing.

Experimental Design

Cumin extract was dissolved in saline and administered intraperitoneally. Animals were randomly divided into the following eight experimental groups. The volume of administration was 0.5 ml. Each group included 8 animals.

Group I included healthy rats and administrated intraperitoneally with saline.

Groups II-V included healthy rats and administrated intraperitoneally cumin extract (0.1, 0.2, 0.4 and 0.6 g/kg body weight [wt], respectively).

Group VI included diabetic control rats and administrated intraperitoneally with saline.

Groups VII-X included diabetic rats and administrated intraperitoneally cumin extract (0.1, 0.2, 0.4 and 0.6 g/kg body wt, respectively).

Group XI included diabetic rats and administrated intraperitoneally glibenclamide (600 μg/kg body wt).

Biochemical Assays

Before administration and 1, 3 and 5 h after administrations of the ethanolic extract, glibenclamide or saline, blood samples were drawn from the retro-orbital

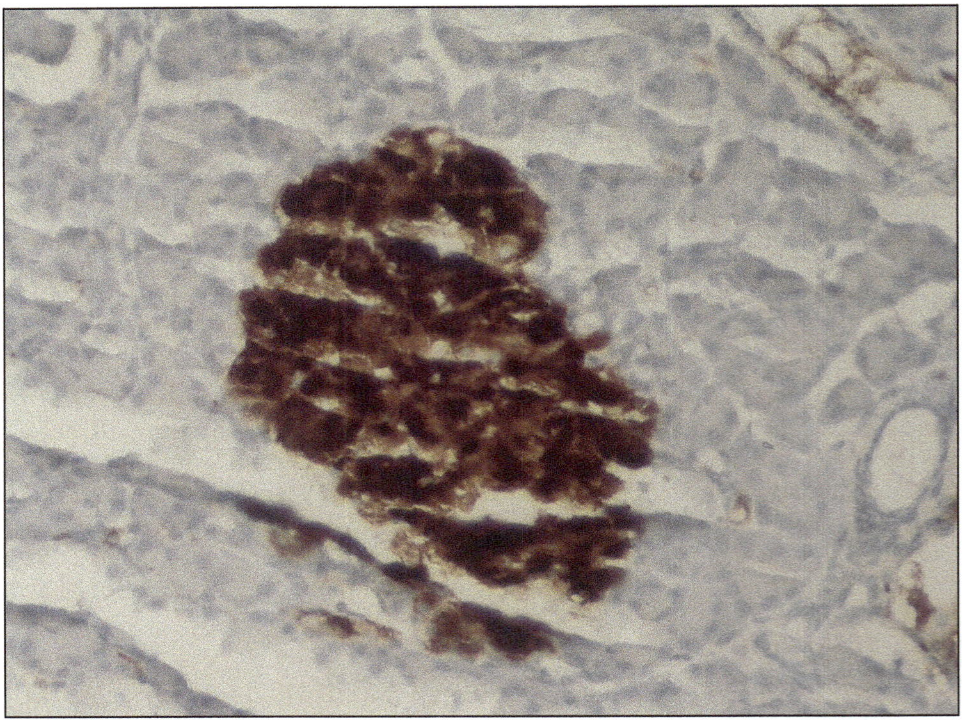

Figure 18.1: Immunoreaction of insulin in pancreatic β-cells after staining with the anti-insulin antibody (×400).

sinus. Serum glucose levels were determined by enzymatic method of glucose oxidase (Barham and Trinder, 1972). The serum insulin levels were estimated by using the radioimmunoassay kit (Diasorin, Saluggia, Italy). Serum glucose and insulin levels were considered as control values before administration.

Determination of Insulin Releasing Activity

To determine the insulin releasing activity of the extract, 5 h after extract treatment, the animals were anaesthetized with diethyl ether. The pancreases were excised, fixed in 10 per cent formaldehyde for 24 h, and embedded in paraffin for sectioning. Sections of 5 µm were processed for insulin-releasing activity using an immunocytochemistry kit. Briefly, endogenous peroxidase activity was blocked with methanol-H_2O_2 (30 per cent H_2O_2; diluted 1:30 in methanol), and to reduce non-specific binding to cellular immunoglobulin (Fc)-receptors, the specimens were treated with buffered rat serum (dilution 1:10 in PBS buffer). For insulin detection, a rat polyclonal insulin-antibody (DAKO-company) was used as primary antibody (dilution 1:200 in PBS-buffer). Biotinylated anti-rat-F (ab') 2-fragment (dilution 1:200 in PBS-buffer) served as the secondary antibody. The detection of the specific antibody bonding was accomplished using the avidin–biotin–peroxidase-complex method (DAKO-company) with 3,3'-diaminobenzidine. Finally, the background was stained with haematoxylin. For the morphometric semi-quantitative analysis of stained β-cells in pancreas of rats, the stained Langerhan's islets in 13 ocular fields per specimen were measured at 10 × magnification (Wehner, 1999).

Statistical Analysis

Statistical analyses were carried out by SPSS 10 program for windows. All data were expressed as mean ± S.E.M. Statistical analysis was carried out using one-way ANOVA followed by Tukey post hoc test. The criterion for statistical significance was $p<0.05$.

Results

Effect of Cumin Extract on Serum Glucose in Healthy and Diabetic Fasted Rats

The results showed that serum glucose levels are increased, while serum insulin levels are decreased in diabetic control rats when compared with healthy rats. Table 18.1 showed that the effect of the cumin extract on serum glucose in healthy fasted rats. The results showed that the intraperitoneal administration of cumin ethanolic extract at doses of 0.1, 0.2, 0.4 and 0.6 g/kg body wt did not exhibit any significant alterations in serum glucose levels after administration.

Table 18.2 showed that the effect of the cumin extract on serum glucose in STZ-induced diabetic fasted rats. The results showed that the intraperitoneal administration of cumin ethanolic extract at doses of 0.2, 0.4 and 0.6 g/kg body wt produces a statistically significant decrease in serum glucose levels after administration.

Effect of Cumin Extract on Serum Insulin in Healthy and Diabetic Fasted Rats

Table 18.3 showed that the effect of the cumin extract on serum insulin in STZ-induced diabetic fasted rats. The results showed that the intraperitoneal administration of cumin ethanolic extract at doses of 0.4 and 0.6 g/kg body wt produces a statistically significant increase in serum insulin levels after administration.

Table 18.4 showed that the effect of cumin extract on number of beta cells with insulin releasing activity in STZ-induced diabetic fasted rats. The result showed that intraperitoneal administration of cumin ethanolic extract at doses of 0.4 and 0.6 g/kg body wt produces a statistically significant increase in number of active beta cell after administration.

Discussion

A wide variety of spices including cumin is added to food during processing, and is consumed by human population daily in variable quantities through diet. These spices improve the quality of food. The health benefits of spice consumption are well known. Among these are their beneficial influence on lipid metabolism (Srinivasan *et al.*, 2004), efficacy as anti-diabetics (Kempaiah and Srinivasan, 2006; Srinivasan, 2005), ability to stimulate digestion, antioxidant property and anti-inflammatory potential. Because of its stomachic, diuretic, carminative, emmenagogic

Table 18.1: Effect of intraperitoneally administration of cumin seeds ethanolic extract at doses of 0.1, 0.2, 0.4 and 0.6 g/kg body wt on serum glucose in healthy fasted rats for 1, 3 and 5 h after administration.

Experimental Groups	*Treatment and Doses*	*Serum Glucose Levels (mg/dl) ± S.E.M.*			
		0 h	*1 h*	*3 h*	*5 h*
I	Control (-) healthy (Saline)	102.5 ± 4.28	108.33 ± 2.01	102.6 ± 4.25	97.83 ± 4.2
II	Cumin extract (0.1 g/kg)	97.83 ± 4.16	102.16 ± 3.68	91.66 ± 2.11	90.66 ± 3.19
III	Cumin extract (0.2 g/kg)	107.16 ± 3.79	100.5 ± 3.91	94.16 ± 4.72	90.83 ± 4.9
IV	Cumin extract (0.4 g/kg)	102.16 ± 5.19	94.83 ± 4.7	96.83 ± 4.54	92.5 ± 3.81
V	Cumin extract (0.6 g/kg)	100.5 ± 5.73	97.33 ± 4.9	90.83 ± 2.38	85.83 ± 2.01

Each point represents mean ± SEM for eight rats. Serum glucose levels at zero time or before administrations were considered as control values. Control group was administered with saline as a vehicle.

Table 18.2: Effect of intraperitoneally administration of cumin seeds ethanolic extract at doses of 0.1, 0.2, 0.4 and 0.6 g/kg body wt on serum glucose in diabetic fasted rats for 1, 3 and 5 h after administration.

Experimental Groups	*Treatment and Doses*	*Serum Glucose Levels (mg/dl) ± S.E.M.*			
		0 h	*1 h*	*3 h*	*5 h*
VI	Control (+) diabetic (Saline)	398 ± 33	361 ± 56	332 ± 30	340 ± 29
VII	Cumin extract (0.1 g/kg)	348.16 ± 10.25	330 ± 12.11	315 ± 8.46	312.5 ± 6.55
VIII	Cumin extract (0.2 g/kg)	334.83 ± 9.98	324.16 ± 10.03	313.33 ± 8.81	294.16 ± 5.54 *
IX	Cumin extract (0.4 g/kg)	343.33 ± 7.14	316 ± 13.16	300 ± 11.54 *	284.16 ± 6.11**
X	Cumin extract (0.6 g/kg)	346.66 ± 8.43	311 ± 4.01 **	287.5 ± 4.42 ***	275 ± 5.62 ***
XI	Glibenclamide (600 µg/kg)	351.66 ± 6.54	330 ± 10.32	305.83 ± 5.23 **	285 ± 6.7 ***

Each point represents mean ± SEM for eight rats. Serum glucose levels at zero time or before administrations were considered as control values. Control group was administered with saline as a vehicle. Glibenclamide used as standard hypoglycemic drug. $^{*}p<0.05$, $^{**}p<0.01$, $^{***}p<0.001$ different from zero time.

Table 18.3: Effect of intraperitoneally administration of cumin seeds ethanolic extract at doses of 0.1, 0.2, 0.4 and 0.6 g/kg body wt on serum insulin in diabetic fasted rats for 1, 3 and 5 h after administration.

Experimental Groups	*Treatment and Doses*	*Serum Insulin Levels (IU/l) ± S.E.M.*			
		0 h	*1 h*	*3 h*	*5 h*
VI	Control (+) diabetic (Saline)	1.11 ± 0.039	1.08 ± 0.03	1.13 ± 0.05	1.15 ± 0.08
VII	Cumin extract (0.1 g/kg)	1.07 ± 0.02	1.06 ± 0.03	1.14 ± 0.04	1.2 ± 0.073
VIII	Cumin extract (0.2 g/kg)	1.1 ± 0.02	1.17 ± 0.05	1.2 ± 0.03	1.23 ± 0.04
IX	Cumin extract (0.4 g/kg)	1.1 ± 0.03	1.24 ± 0.04	1.27 ± 0.02*	1.35 ± 0.06***
X	Cumin extract (0.6 g/kg)	1.15 ± 0.02	1.27 ± 0.03*	1.31 ± 0.01**	1.4 ± 0.04***
XI	Glibenclamide (600 µg/kg)	1.13 ± 0.04	1.27 ± 0.03*	1.29 ± 0.03*	1.36 ± 0.04***

Each point represents mean ± SEM for eight rats. Serum insulin levels at zero time or before administrations were considered as control values. Control group was administered with saline as a vehicle. Glibenclamide used as standard hypoglycemic drug. *$p<0.05$, **$p<0.01$, ***$p<0.001$ different from zero time.

and antispasmodic activity, cumin is used for treating common aliments such as dyspepsia, flatulence, colic, irregularly menstrual cycle and spasm (Joshi, 2000).

Table 18.4: Effect of intraperitoneally administration of cumin seeds ethanolic extract at doses of 0.4 and 0.6 g/kg body wt on number of active beta cells in diabetic fasted rats for 5 h after administration.

Experimental Groups	*Treatment and Doses*	*Number of Active Beta Cells*
I	Control (-) healthy (Saline)	60 ± 12
VII	Control (+) diabetic (Saline)	21 ± 7.5 ***
IX	Cumin extract (0.4 g/kg)	34 ± 9.2 ++
X	Cumin extract (0.6 g/kg)	42 ± 11.3 +++

Each point represents mean ± SEM for eight rats. Control group was administered with saline as a vehicle. ***$p<0.001$ different from control healthy rats. ++ $p<0.01$, +++ $p<0.001$ different from control diabetic rats.

The present data indicated that extract of cumin seeds had a statistically significant decrease in serum glucose levels whereas it increased serum insulin only in STZ-induced diabetic fasted rats but not in healthy fasted rats. Also, our results showed that the extract increased the activity of beta cells in diabetic rats. The hypoglycemic effect of the extract was similar to that observed for glibenclamide.

By STZ, most islet cells are impacted to death and remaining islet cells almost exhibit a significant decrease in the level of insulin compared to normal rats (Platel and Srinivasan, 2004). Therefore, STZ-induced diabetes in rats has been frequently used as a model for type 1 diabetes or for type 2 diabetic complications. Glibenclamide, one of the most widely used oral hypoglycemic agents in the treatment of diabetes mellitus, exerted its beneficial effects on extracellular site by opening Ca^{2+} channels to stimulate insulin secretion and also duodenal insulin-releasing agent (Srivivasan *et al.*, 2003).

The possible mechanism by which cumin brings about its hypoglycemic action may be, by potentiating the insulin effect, either by increasing the pancreatic secretion of insulin from the cells of islets of Langerhan's or its release from bound insulin (Davis and Granner, 1996). Moreover, our result indicates that part of the anti-hyperglycemic activity of these plants is through release of insulin from the pancreas. In addition, since cumin extract showed some anti-hyperglycemic effect even in severely diabetic rats, there is a possibility that it also exerts a direct insulinomimetic effect.

From the polar portion of the methanolic extract of cumin (fruit of cumin), two sesquiterpenoid glucosides, cuminoside A and B, and two alkyl glycosides were isolated together with five known compounds (Dhandapani *et al.*, 2002). It is reported that sesquiterpene glycoside produces a significant antihypoglycemic effect when administered orally to alloxan- diabetic mice (Takayanagi *et al.*, 2003).

Moreover, weight gain was lesser in the diabetic rats as compared to the control. Al-Shamaony *et al.* (1994) have also observed a similar result. Administering cumin improved the weight gain in alloxan diabetic rats (Chen *et al.*, 2008). The ability of cumin to protect body weight loss seems to be due to its hypoglycemic and hypolipidemic effect.

Summarizing, it could be proofed that the traditional use of cumin seeds as a hypoglycemic agent is justified and that extracts from this plant show a dose-dependent activity which is comparable to the standard hypoglycemic drug glibenclamide.

Acknowledgments

We would like to thank Deputy Research of the Science and Research Branch, Islamic Azad University, for financial support of the project.

References

Alarcon-Aguilara, F.J., Roman-Ramos, R., and Perez-Gutierrez, S. (1998). Study of the anti-hyperglycemic effect of plants used as anti-diabetics. *Journal of Ethnopharmacology,* **61**: 101-110.

Al-Shamaony, L., Al-Khaznaji, S.M., and Twaij, H.A. (1994). Hypoglycemic effect of *Artemisia herba alba*. II. Effect of valuable extract on some blood parameters in diabetic animals. *Journal of Ethnopharmacology*, **43**: 167-171.

Barham, D., and Trinder, P. (1972). An improved color reagent for the determination of blood glucose by the oxidase system. *Analyst*, **97**: 42-45.

Brinker, F. (1998). Herb contraindications and drug interactions. 2nd ed, Sandy, OR: Eclectic Medical Publications, pp. 36-82.

Chandrasekhara, N., and Srinivasan, K. (1999). Health benefits of spices. In: Recent trends in food science and technology-Global perspective. Mysore: Association of Food Scientists and Technologists, pp. 725-737.

Chen, J., Li, W.L., Wu, J.L., Ren, B.R., and Zhang, H.Q. (2008). Hypoglycemic effects of a sesquiterpene glycoside isolated from leaves of loquat (*Eriobotrya japonica* (Thunb.) Lindl.). *Phytomedicine*, **15**: 98-102.

Davis, S.N., and Granner, D.K. (1996). Insulin, oral hypoglycemic agents and the pharmacology of endocrine pancreas. In: Hardman JG, Limberd LE, Malinoff PB, Ruddon RW, Gilman AG, Goodman and Gilman's the pharmacological basis of therapeutics, 9th ed. New York: McGraw Hill, pp. 1487-1518.

Dhandapani, S., Subramanian, V.R., Rajagopal, S., and Namasivayam, N. (2002). Hypolipidemic effect of Cuminum cyminum L. on alloxan-induced diabetic rats. *Pharmacological Research*, **3**: 251-255.

Evans, W.C. (1996). Trease and Evans's Pharmacognosy, 14th ed. London: Saunders, pp. 124-143.

Joshi, S.G. (2000). Medicinal plants: family Apiaceae. 1st ed, Oxford and IBH Publishing Company, pp. 34-35.

Kempaiah, R.K., and Srinivasan, K. (2006). Beneficial influence of dietary curcumin, capsaicin and garlic on erythrocyte integrity in high-fat fed rats. *Journal of Nut Biochem.*, **17**: 471-478.

Muhlhauser, I., Sawicki, P.T., Blank, M., Overmann, H., Bender, R., and Berger, M. (2000). Prognosis of persons with type 1 diabetes on intensified insulin therapy in relation to nephropathy. *Journal of International Medicine*, **248**: 333-341.

Nadkarni, K.M., and Nadkarni, A.K. (1976). In Indian materia medica. Mumbai: Popular Prakashan.

Piedrola, G., Novo, E., Escober, F., and Garcia-Robles, R. (2001). White blood cell count and insulin resistance in patients with coronary artery disease. *Annual of Endocrinology*, **62**: 7-10.

Platel, K., and Srinivasan, K. (2004). Digestive stimulant action of spices: a myth or reality? *Indian Journal of Medical Research*, **119**: 167-179.

Pushparaj, P., Tan, C.H., and Tan, B.K. (2000). Effects of *Averrhoa bilimbi* leaf extract on blood glucose and lipids in streptozotocin-diabetic rats. *Journal of Ethnopharmacology*, **72**: 69-76.

Roman-Ramos, R. (1995). Antihyperglycaemic effect of some edible plants. *Journal of Ethnopharmacology*, **48**: 25-32.

Srinivasan, K. (2005). Plant foods in the management of diabetes mellitus: spices as potential antidiabetic agents. *International Journal of Food and Science Nutrition*, **56**: 399-414.

Srinivasan, K., Sambaiah, K., and Chandrasekhara, N. (2004). Spices as beneficial hypo-cholesterolemic food adjuncts: a review. *Food Reviews International*, **20**: 187-220.

Srivivasan, A., Menon, V.P., Periaswamy, V., and Rajasekaran, K.N. (2003). Protection of pancreatic beta-cell by the potential antioxidant bis-o-hydroxycinnamoyl methane, analogue of natural curcuminoid in experimental diabetes. *Journal of Pharmacological Science*, **6**: 327-333.

Srinivasan, K. (2000). Spices: Valued for more than taste and flavour of foods. In De AK (ed), Recent trends in spices and medicinal plants research, New Delhi: Associated Publishing Company, pp. 31-38.

Takayanagi, T., Ishikawa, T., and Kitajima, J. (2003). Sesquiterpene lactone glucosides and alkyl glycosides from the fruit of cumin. *Phytochemistry*, **63**: 479-484.

Urger, R.H., and Foster, D.W. (1998). Diabetes mellitus. In: Wilson JD, Foster DW, Kronenserg HM, Larson PR (eds). Williams's Text Book of Endocrinology, Philadelphia: W.B Saunders, pp. 973-1060.

Wehner, F., Wehner, H.D., Schieffer, and M.C. Sube, J. (1999). Delimitation of the time of death by immunohistochemical detection of insulin in pancreatic β-cells. *Forensic Sci Int.*, **105**: 161-169.

Yaryura-Tobias, J.A., Pinto, A., and Neziroglu, F. (2001). *Anorexia nervosa*, diabetes mellitus, brain atrophy, and fatty liver. *Int J Eat Disord.*, **30**: 350-353.

Zargari, A. (1997). Medicinal Plant, Vol 4, Iran: Tehran University Press, pp. 59-64.

Traditional and Folk Herbal Medicine: *Pages* **365–389**
Recent Researches Vol. 1 (2012)
Editor: **V.K. Gupta**
Published by: **DAYA PUBLISHING HOUSE, NEW DELHI**

19

Phytochemical Analysis and *in vitro* Antioxidant Activity of Essential Oils and Methanolic Extracts of *Ocimum* spp. (Indian Basils) Growing in Uttarakhand Regions of Central Himalaya

S. Sethi[1], M. Chandra[1], Om Prakash[1*] and A.K. Pant[1]

ABSTRACT

In continuation to our research on plants of family Lamiaceae from central Himalaya region of Uttarakhand, we have collected various species of genus Ocimum (Indian basils) with the objective to study their chemical composition, antioxidant activity and total phenolic assay. Ocimum basilicum, Ocimum kilimandscharicum, Ocimum gratissimum, Ocimum sanctum syn. Ocimum tenuiflorum and Ocimum americanum syn. Ocimum canum collected from different ecological niches were analyzed by the combination of GC and GC/MS. More than 30 compounds contributing over 80 per cent of the total oil in eight collections of Ocimum were identified. All the oils were dominated by oxygenated monoterpenoids. The major compounds identified in various species were linalool, camphor, eugenol, methyl cinnamate, chavicol and methyl chavicol beside other minor constituents. Linalool/methyl chavicol/germacrene-D, linalool/β-selinene, germacrene-D/eugenol/geranyl acetate/Z-methyl cinnamate and β-selinene/germacrene-D/eugenol/geranyl acetate chemotypes were identified among Ocimum sanctum syn. Ocimum tenuiflorum (OS/OT) while linalool/neral/geranial and linalool/methyl chavicol chemotypes were identified in Ocimum basilicum. All the oils

1 Department of Chemistry, College of Basic Sciences and Humanities, G.B. Pant University of Agriculture and Technology, Pantnagar, Uttarakhand, India

* *Corresponding author*: E-mail: op_orgchem@yahoo.co.in; oporgchem@gmail.com; Ph.: +919411159551; Fax: +91-5944-233473.

and methanolic extracts of all the Ocimum spp. exhibited good to moderate in vitro antioxidant activity when tested for reducing power activity, 2, 2',-diphenyl-1-picrylhydrazyl (DPPH) radical scavenging activity and chelating activity of Fe^{2+} *ions. The total phenolic contents among all the methanolic extracts varied from 0.181-3.00 mg/100mg, flavonols from 0.020-0.12 mg/100mg and ortho dihydric phenols from 0.025-1-15mg/100mg.*

Keywords: Antioxidant activity, Chavicol, Essential oil, Eugenol, Linalool, Methyl chavicol, *Ocimum* spp., Total phenols.

Introduction

More than 45,000 plant species are reported from India and among them several thousand have been claimed to possess medicinal properties. The use of plants as a source of medicine is an important component of the indigenous health care system in India (Grover *et al.*, 2002). Due to the increased adverse drug reactions and economic reasons of allopathic medicines the public, academic and government interest is growing exponentially in alternative and traditional medicines (Dubey *et al.*, 2004). Since practitioners formulate and dispense their own recipes in Indian system of medicine hence proper documentation and research is required to check their chemical diversity (Bent and Ko, 2004).

Aromatic plants have traditionally been used in perfumery as well as in the treatment of illnesses and control of insects. Their essential oils have been known for their various therapeutic, insecticidal and antifungal activities (Murillo *et al.*, 2002). Essential oils exhibits antifungal activity (Reuveni *et al.*, 1984; Sartoratto *et al.*, 2004; Dharmagadda *et al.*, 2005; Ferreira *et al.*, 2006; Kumar *et al.*, 2007; Bansod and Rai, 2008; Zhang *et al.*, 2009), antimicrobial activity (Ahonkhai and Ayinde, 2005; Bozin *et al.*, 2006; Stefanini *et al.*, 2006) and antibacterial activity (Nakamura *et al.*, 1999; Dharmagadda *et al.*, 2005). The oils also possess the antioxidant activity (Bozin *et al.*, 2006; Folarin and Akinhanmi, 2006; Lertsatitthanakorn *et al.*, 2006; Stefanini *et al.*, 2006; Pereira and Maia, 2007; Leelapornpisid *et al.*, 2008).

Family Lamiaceae a rich source of terpenoids consists of 264 genera and 6,990 species distributed world over (Thorne, 2007). The Indian sub-continent represents 72 genera and 435 species (Kartikeyan and Singh, 2000). Indian Himalayan region is a big repository of the family Lamiaceae. Western Himalaya represents 60 genera and 225 species (Rana *et al.*, 1997). Himachal Pradesh represents 42 genera and 111 species (Chowdhary and Wadhwa, 1984), while 49 genera and 151 species of Lamiaceae are reported from Uttarakhand region out of total species reported from India (Uniyal *et al.*, 2007).

Ocimum spp. are native throughout the old World tropics and widespread as cultivated plants and also as escaped weeds. *Ocimum sanctum* (Tulsi or Holy basil) is cultivated for religious and medicinal purposes, and for its essential oil. It is widely known across South Asia as a medicinal plant and an herbal tea, commonly used in Ayurveda. It has an important role within the Vaishnavite tradition of Hinduism, in which devotees perform worship involving tulsi plants or leaves (Staples and Michael, 1999). Tulsi's extracts are used in ayurvedic remedies for common colds, headaches, stomach disorders, inflammation, heart disease, as antidotes against poisoning, and in treatment of malaria. Traditionally and indigenously tulsi is used in many forms *viz.*, as herbal tea, dried powder, fresh leaf, or mixed with ghee. Essential oil extracted from Karpoora tulsi is mostly used for medicinal purposes and in herbal cosmetics, and is widely used in skin preparations due to its antibacterial activity. For centuries the dried leaves of tulsi had also been mixed with stored grains to repel insects (Biswas and Biswas, 2005).

Voluminous research work on essential oils, extracts and antioxidant activity of various *Ocimum* spp. growing in different parts of the world has been reported in literature viz; *Ocimum basilicum* L. (Sweet Basil) (Keita *et al.*, 2000; Juliani *et al.*, 2002; Kruger *et al.*, 2002; Slougui *et al.*, 2003; Javanmardi *et al.*, 2003; Feng *et al.*, 2004; Ozcan and Chalchat, 2004; Lee *et al.*, 2005; Hinneburg *et al.*, 2006; Gulcin *et al.*, 2007; Siddiqui *et al.*, 2007; Hussain *et al.*, 2008; Modnicki and Balcerek, 2009; Javanmardi, 2009; Zhang *et al.*, 2009). *Ocimum sanctum* L. syn *Ocimum tenuiflorum* L. (Holy Basil, Tulsi) (Kelm *et al.*, 2000; Gill *et al.*, 2002; Jadhav *et al.*, 2005; Awasthi and Dixit, 2007; Mondal *et al.*, 2007; Bhowmik *et al.*, 2008; Hakkim *et al.*, 2008; Leelapornpisid *et al.*, 2008; Prakash *et al.*, 2009). *Ocimum americanum* L. syn *Ocimum canum* Sims (basilic camphor, Indian Basil) (Chagonda *et al.*, 2000; Yayi and Moudachirou, 2001; Silva *et al.*, 2003; Ngassoum *et al.*, 2004; Pereira and Maia, 2007; Hakkim *et al.*, 2008). *Ocimum grattissimum* L. (Clove Basil, Shrubby basil)(Keita *et al.*, 2000; Cimanga *et al.*, 2002; Jirovetz *et al.*, 2003; Jirovetz *et al.*, 2005; Orafidiya *et al.*, 2007; Matiucci and Feirra, 2007; Hakkim *et al.*, 2008; Akinmoladun *et al.*, 2007; Oboh *et al.*, 2008; Benitez *et al.*, 2009; Kumar *et al.*, 2010). *Ocimum kilimandscharicum* Guerke (Kilimanjaro basil, camphor basil, hoary basil, feverplant) (Ntezurubanza *et al.*, 1985; Codignola, 1986; Charles and Simon, 1992; Hakkim *et al.*, 2008; Kumar *et al.*, 2009).

Herbal products are developed from phytochemicals, the active chemical ingredients which are responsible for broad spectrum of biological activities. The herbal products mainly include medicines, cosmetics and perfumery. At present India ranks sixth in the essential oil production (6.1 per cent) with China (30.7 per cent), Brazil (13.1 per cent), Turkey (10.4 per cent), Indonasia (98.7 per cent) and Morocco (99.1 per cent) in the decreasing order. Similar is the situation in terms of the export of medicinal plant material (Varshney, 2001). Biodiversity is bestowed with immense chemical diversity even within the same species. Because of wide range of chemical diversity, the medicinal plants have served and are still used as a valuable source for new drug molecules. Since ancient times Himalayan region had been big repository of medicinal and aromatic plants. Of nearly 3,000 herbal drugs in "*Ayruveda*" and 50 per cent of species in Indian System of Medicine have their origin in Himalaya (Mathela, 1995). In continuation to our research on plants of family Lamiaceae in present investigation we have reported chemical composition, total phenolic assay and *in vitro* antioxidant activity of the all *Ocimum* spp. growing in central Himalaya region of Uttarakhand.

Material and Methods

Source of Plant Material

Plants material (whole herbs) was collected from the different natural habitats of the Kumaun and Garhwal region, in the month of August, September and was identified by Dr. D.S.Rawat (Asstt. Professor and taxonomist), Department of Biological Science, G.B. Pant University of Agriculture and Technology, Pantnagar. Herbaria are maintained in department of Botany G.B., Pant University of Agriculture and Technology, Pantnagar.

Chemical Investigation

Isolation of Essential Oils

Fresh aerial parts of each collection of *Ocimum* were subjected to hydro distillation in Clevenger's type apparatus for 8 hours separately. Extraction of distillate by diethyl ether followed by drying over anhydrous Na_2SO_4 and removal of solvent under vacuum yielded essential oils in different quantities. The yields are recorded in Table 19.1.

Table 19.1: Yield of essential oils in different *Ocimum* spp.

Plant's Name	*Place of Collection*	*Yield (%) in Fresh Weight Basis*
Ocimum basilicum (Lemon basil) OB (LTP)	Pantnagar, Uttarakhand	0.68
O.basilicum (Van Tulsi) OB (VTP)	Pantnagar, Uttarakhand	0.99
O. sanctum/O.tenuiflorum OS/OT (PNR)	Pantnagar, Uttarakhand	0.37
O. sanctum/O.tenuiflorum OS/OT (ROR)	Roorkee, Uttarakhand	0.2
O. sanctum/O.tenuiflorum OS/OT(JNP)	Jawaharnagar (Pantnagar), Uttarakhand	0.18
O. kilimandscharicum OK (PNR)	Haldi (Pantnagar), Uttarakhand	0.4
O. *gratissimum* (Clove basil) OG (CBP)	Pantnagar, Uttarakhand	0.56
O. americanum OA (DH)	Dwarahat, Uttarakhand	0. 075

GC and GC-MS Analyses

The essential oils were analyzed on Nucon 5765 GC Rtx-5MS column (30m × 0.32mm i.d. FID), and ThermoQuest Trace GC 2000 coupled with Finnigan Mat PolarisQ ion trap mass spectrometer fitted with Rtx-5MS fused silica capillary column (30m × 0.25 mm i.d. × 0.25 μm film thickness). The column temperature was 60°C to 210°C at 3°C/min, with He as carrier gas at 1ml/min., and the injector temp. was 210°C. MS were recorded under EI condition (70 eV) for a mass range of 40-450 amu, injection vol. 0.1 μl with split mode of 1:40. The compounds were identified based on kovat index, co-injections with authentic samples, mass spectral library searches (NIST-Wiley) and comparing the spectra with literature (Adams, 1995).

Antioxidant Activity

Antioxidant potential of the essential oils were evaluated in terms of 2,2'-diphenyl-1-picrylhydrazyl (DPPH) radical scavenging ability, chelation of Fe^{+2} and the reducing power in comparison to standard antioxidants *viz.*, butylated hydroxytoluene (BHT), gallic acid and catechin. All determinations were performed in triplicate.

Reducing Power Activity

The reducing power of the essential oils was determined according to the methods reported earlier (Yen and Duh, 1993; Singh *et al.*, 2005). Absorbance at 700 nm was plotted against the different concentrations of the oils. Increase in absorbance indicated increase in reducing power.

Chelation of Fe^{+2}

The chelating activity of essential oils on ferrous ions (Fe^{+2}) were measured according to the method of Decker and Welch, 1990 and Hsu *et al.*, 2003. The chelating activity of the oils on Fe^{+2} were compared with that of EDTA at a level of 0.01 mM and citric acid at a level of 0.025 M and calculated according to equation (1)

$$\text{Chelating activity (per cent)} = [1 - (A_t/A_0)] \times 100 \qquad \text{(eq. 1)}$$

(where A_t is the absorbance of the sample at 562 nm and A_0 is the absorbance of the control at 562 nm)

DPPH Radical Scavenging Activity

The scavenging effect on the DPPH radical was determined according to the methods reported earlier (Yen and Dun, 1993; Cuendet *et al.*, 1997; Singh *et al.*, 2005) and calculated according to the equation (2).

$$\text{DPPH scavenging activity (per cent)} = [1 - (A_t/A_0)] \times 100 \qquad \text{(eq. 2)}$$

(where A_t is the absorbance of the sample at 515 nm and A_0 is the absorbance of the control at 515 nm)

Biochemical Assay

The biochemical assay of methanolic extracts of all the species of *Ocimum* were studied quantitatively by Folin–Ciocalteu method in terms of total phenols, flavonols and ortho dihydroxy phenols and by the methods developed earlier (Shetty *et al.*, 1995; Mahadevan and Sridhar, 1986 a,b).

Results and Discussion

GC and GC-MS Analysis of Essential Oils

In the present study the essential oils of the eight collections of *Ocimum* belonging to five species *viz*; *Ocimum basilicum* (Lemon basil collected from Pantnagar [OB(LTP)], *Ocimum kilimandscharicum* (Camphor basil collected from Pantnagar [OK(PNR)], *Ocimum gratissimum* (Clove basil collected from Pantnagar [OG(CBP)], *Ocimum sanctum* syn. *Ocimum tenuiflorum* collected from Jawaharnagar Pantnagar [OS/OT(JNP)], *Ocimum basilicum* (Van Tulsi collected from Pantnagar [OB(VTP)], *Ocimum sanctum* syn. *Ocimum tenuiflorum* collected from Roorkee[OS/OT(ROR)], *Ocimum sanctum* syn. *Ocimum tenuiflorum* collected from Pantnagar[OS/OT(PNR)] and *Ocimum americanum* syn. *Ocimum canum* collected from Dwaharhat[OA(DH)] were analyzed by the combination of GC and GC-MS. Several constituents have been identified in different oils which contribute to 90.2 per cent, 85.1 per cent, 98.1 per cent, 97.2 per cent, 96.4 per cent, 90.7 per cent, 95 per cent and 94.9 per cent of the total oils respectively. All the collections were dominated by monoterpenoids contributing to 89.21 per cent, 56.4 per cent, 79.9 per cent, 93.9 per cent, 93.2 per cent, 52.4 per cent, 71.4 per cent and 50.3 per cent of the total oil respectively. The detailed comparative essential oil composition is presented in Table 19.2 and the structures of major identified constituents in Figure 19.1. Every collection was rich in oxygenated monoterpenes linalool(32.0 per cent), neral(14.5 per cent), geranial(21.4 per cent), citronellol (7.4 per cent) were major constituents in OB(LTP), camphor (39.2 per cent), methyl cinnamate (6.2 per cent), methyl eugenol (4.3 per cent) in OK(PNR), eugenol(77.5 per cent) in OG(CBP), methyl chavicol (75.1 per cent), linalool (16.9 per cent) in OS/OT(JNP), methyl chavicol (62.7 per cent), linalool (24.7 per cent), eugenol (2.6 per cent), geranial (1.9 per cent), neral (1.3 per cent) in OB(VTP), geranial acetate (24.5 per cent), *Z*-methyl cinnamate (10.2 per cent), *E*- methyl cinnamate (9.6 per cent), methyl eugenol (1.5 per cent) in OS/OT(ROR), eugenol (41.9 per cent), geranyl acetate (12.3 per cent), methyl eugenol (11.7 per cent), *E*-methyl cinnamate (4.6 per cent) in OS/OT(PNR) and eugenol (24.2 per cent), methyl chavicol (10.5 per cent), geranyl acetate (2.5 per cent), methyl eugenol (1.9 per cent), 10-*epi*-γ-eudesmol (1.0 per cent) in OA(DH) respectively.The highest percentage (8.9 per cent) of monoterpene hydrocarbons were detected in OA(DH). The major constituents were *Z*-β-ocimene (7.9 per cent) and *E*-β-ocimene(1 per cent). Only 0.9 per cent monoterpene hydrocarbons were detected in OS/OT(PNR). None of the monoterpene hydrocarbon could be detected in the collections of OB(VTP) and OS/OT(ROR).

The oxygenated sesquiterpenoids could not be detected in the collections except in OB (LTP) and OA (DH) collections which contribute 0.2 per cent and1 per cent of the total oil respectively. However the highest percentage of sesquihydrocarbons could be detected in OA(DH) (43.6 per cent) followed by OS/OT(ROR) (38.3 per cent), OK(PNR)(28.7 per cent), OS/OT (PNR) (23.6 per cent), OG (CBP) (18.2 per cent), OS/OT (JNP) (3.3 per cent), and OB(VTP) (3.2 per cent). Only 0.8 per cent sesquiterepene hydrocarbons were detected in OB(LTP). The major sesquiterpenoids detected were caryophyllene, germacrene-D, β-selinene, germacrene-B, γ-cadinene, bisabolene etc. In the present investigation we could identified the three chemotypes (Tables 19.3 and 19.4) in *Ocimum sanctum* and two chemotypes in *Ocimum basilicum* based on qualitative and quantitative differences in their chemical make up.

Antioxidant Activity

Reducing Power Activity

The essential oils and methanolic extracts of eight different collections of *Ocimum* comprising five species were studied for their reducing power activities. The presence of reductants (antioxidant) in the tested sample would result, in reduction of Fe^{3+} complex to ferrous (Fe^{2+}) ion can be measured by UV-Vis spectrophotometer. All the

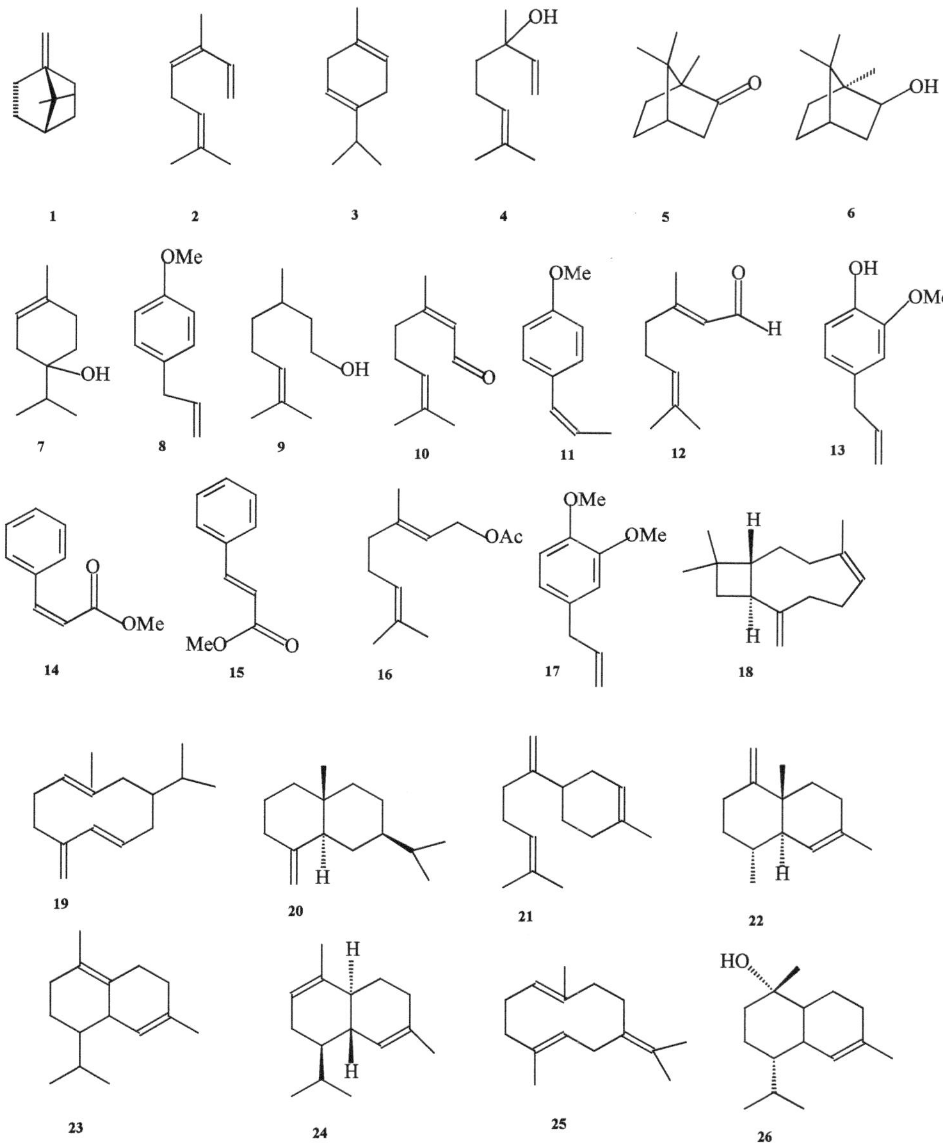

Figure 19.1: Structures of some mono and sesquiterpenoids present in different species of *Ocimum*.

tested essential oils of *Ocimum* from eight collections exhibited reducing power activity. The maximum reducing power was observed in OA (DH) with absorbance at 700nm (A_{700}=4.000-4.000 (±0.0000-±0.0000)), comparable to that of the standard catechin and gallic acid (A_{700}=4.000±0.000) and OB(VTP) (A_{700}=1.956-3.593, (±0.0019-±0.0005)) Figure 19.2. The essential oils from OS/OT(ROR)(A_{700}=1.729-3.121 (±0.0000-±0.1287)), OG(CBP) (A_{700} =1.692-2.639 (±0.0033-±0.0000)), OS/OT(PNR) (A_{700} = 0.852-1.962

Table 19.2: Comparative chemical composition (FID per cent) of essential oils of different species of *Ocimum* collected from different regions of Uttarakhand.

Sl.No.	Compounds	FID %							
		OB (LTP)	OK (PNR)	OG (CBP)	OS/OT (JNP)	OB (VTP)	OS/OT (ROR)	OS/OT (PNR)	OA (DH)
1.	α-thugene	0.1	–	–	–	–	–	–	–
2.	α-pinene **(1)**	0.4	–	–	–	–	–	–	–
3.	sabinene	0.1	–	–	–	–	–	–	–
4.	β-pinene	0.5	–	–	–	–	–	–	–
5.	α-phellandrene	0.5	–	–	–	–	–	–	–
6.	α-terpinene	0.2	–	–	–	–	–	–	–
7.	p-cymene	0.5	–	–	0.7	–	–	0.9	–
8.	β-phellendrane	0.3	–	–	–	–	–	–	–
9.	Z-β-ocimene **(2)**	–	4.1	2.1	–	–	–	–	7.9
10.	E-β-ocimene	–	–	–	–	–	–	–	1.0
11.	γ-terpinene **(3)**	0.2	–	–	1.2	–	–	–	–
12.	linalool **(4)**	32.0	–	–	16.9	24.7	0.8	–	0.5
13.	cis-p-mentha-2-en-1-ol	0.4	–	–	–	–	–	–	–
14.	camphor **(5)**	–	39.2	–	–	–	–	–	–
15.	borneol **(6)**	–	1.5	–	–	–	–	–	–
16.	terpinen-4-ol **(7)**	3.9	1.1	–	–	–	–	–	–
17.	α-terpineol	0.5	–	–	–	–	–	–	–
18.	methyl chavicol **(8)**	–	–	–	75.1	62.7	–	–	10.5
19.	citronellol **(9)**	7.4	–	–	–	–	–	–	–
20.	neral **(10)**	14.5	–	–	–	1.3	–	–	–
21.	cis-anethole **(11)**	4.6	–	–	–	–	–	–	–
22.	geranial **(12)**	21.4	–	–	–	1.9	–	–	–
23.	α-cubebene	0.4	–	–	–	–	–	–	–
24.	Z-methyl cinnamate **(14)**	–	–	–	–	–	10.2	–	–
25.	eugenol **(13)**	–	–	77.5	–	2.6	5.8	41.9	24.2
26.	α-copaene	1.3	–	–	–	–	–	–	–
27.	E-methyl cinnamate **(15)**	–	6.2	–	–	–	9.6	4.6	1.8
28.	geranial acetate **(16)**	–	–	–	–	–	24.5	12.3	2.5
29.	methyl eugenol **(17)**	–	4.3	0.3	–	–	1.5	11.7	1.9
30.	β-caryophyllene **(18)**	0.2	1.3	6.2	–	0.1	–	–	–
31.	γ-gurjunene	0.6	–	–	–	–	–	–	–
32.	germacrene-D **(19)**	–	10.3	12.0	–	1.6	23.8	11.2	1.6
33.	β-selinene **(20)**	–	8.4	–	–	1.5	12.6	12.4	–

Contd...

Table 19.2–*Contd...*

Sl.No.	*Compounds*	*FID %*							
		OB (LTP)	*OK (PNR)*	*OG (CBP)*	*OS/OT (JNP)*	*OB (VTP)*	*OS/OT (ROR)*	*OS/OT (PNR)*	*OA (DH)*
34.	β-bisabolene **(21)**	–	7.6	–	–	–	–	–	–
35.	γ-cadinene **(22)**	–	1.1	–	–	–	1.9	–	–
36.	δ-cadinene **(23)**	–	–	–	–	–	–	–	26.8
37.	α-cadinene **(24)**	–	–	–	–	–	–	–	15.2
38.	germacrene-B **(25)**	–	–	–	3.3	–	–	–	–
39.	10-epi-γ-eudesmol	0.1	–	–	–	–	–	–	–
40.	α-muurolol **(26)**	–	–	–	–	–	–	–	1.0
41.	β-eudesmol	0.1	–	–	–	–	–	–	–
	Total	**90.2**	**85.1**	**98.1**	**97.2**	**96.4**	**90.7**	**95.0**	**94.9**
	Monoterpene Hydro carbons	**4.5**	**4.1**	**2.1**	**1.9**	**–**	**–**	**0.9**	**8.9**
	Monoterpene Oxygenated	**84.7**	**52.3**	**77.8**	**92.0**	**93.2**	**52.4**	**70.5**	**41.4**
	Sesquiterpene Hydro carbons	**0.8**	**28.7**	**18.2**	**3.3**	**3.2**	**38.3**	**23.6**	**43.6**
	Sesquiterpene Oxygenated	**0.2**	**–**	**–**	**–**	**–**	**–**		**1.0**

Table 19.3: Chemotypic variation among *Ocimum sanctum* collected from different locations.

OS/OT(JNP)	*OS/OT(ROR)*	*OS/OT(PNR)*
linalool/methyl chavicol/ germacrene-D type	linalool/β-selinene, germacrene-D/eugenol/geranyl acetate/Z-methyl cinnamate type	β-selinene/germacrene-D/eugenol/geranyl acetate type

Table 19.4: Chemotypic variation between *Ocimum basilicum* collected from Pantnagar.

OB(LTP)	*OB(VTP)*
linalool/neral/geranial type	linalool/methyl chavicol type

(±0.0005-±0.0000)), OS/OT(JNP) (A_{700} = 0.537-0.963 (±0.0012-±0.0203)) showed significant variation in their reducing power activity in a dose dependant manner lower than BHT, gallic acid and catechin. The minimum reducing power activity was observed in OB (LTP) (A_{700}=0.425-0.677 (±0.0000-±0.0017)) and in OK (PNR) (A_{700}=0.373-0.408 (±0.0008-±0.0009)). Note that the reducing power (in terms of increase in A_{700}) of all the essential oil except OA(DH)(Dwaharhat) was significantly lower as compared to gallic acid (A_{700} =4.000-4.000 (±0.0000-0.0000)) and catechin (A_{700} =4.000-4.000 (±0.0000-0.0000))and greater than that of BHT(A_{700} =2.303-2.508 (±0.0005-0.0014)).The above mentioned result reveals that essential oils under study are electron donor and react with free radical chain reactions as suggested earlier (Yen and Chen, 1995). The reducing power of the oils might be the result of their

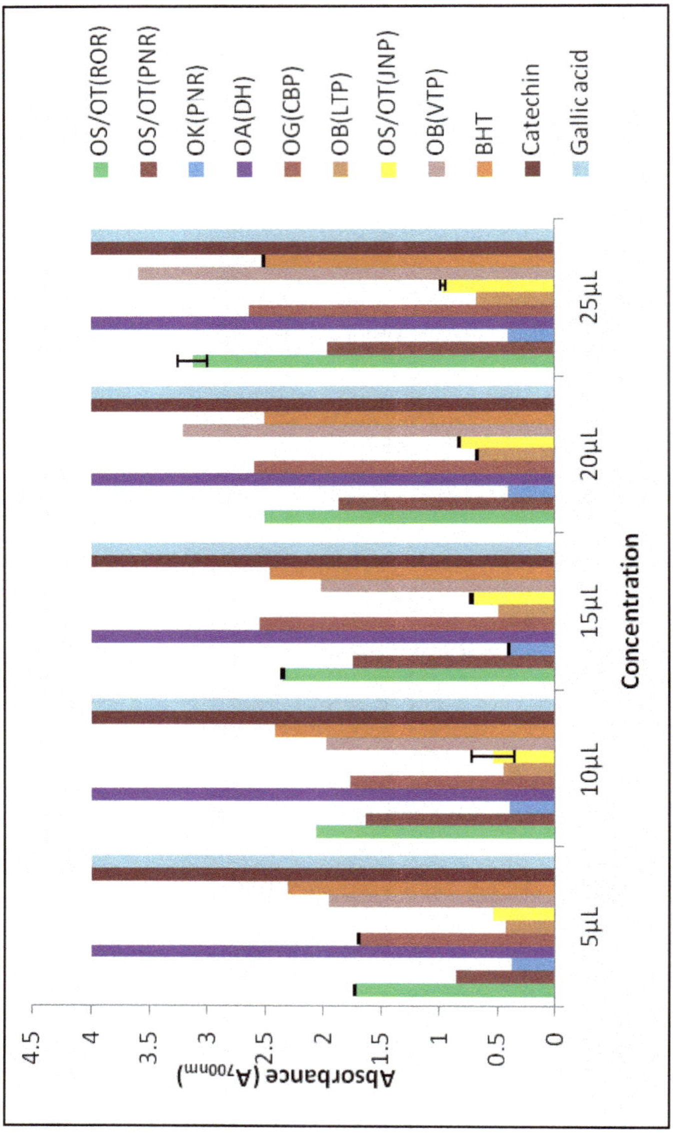

Figure 19.2: Reducing power activity of the essential oils of different *Ocimum* at different concentrations.

hydrogen donating ability (Shimada *et al.*, 1992) which is generally associated with the presence of reductance (Pin-Der, 1998).

The maximum reducing power in the methanolic extract of different *Ocimum* species were observed in OB (LTP) with absorbance 700nm (A_{700}=3.517-4.000 (±0.0000-±0.0000)) and OS/OT (JNP) (A_{700}=2.178-4.000, (±0.0000-±0.0000)) comparable to that of the standard catechin and gallic acid (A_{700}=4.000±0.000) Figure 19.3. The methanolic extract of OB(VTP) (A_{700}=2.178-4.0000 (±0.0000-±0.0000)), OS/OT (A_{700}=1.654-4.0000 (±0.0000-±0.0000)), OK(PNR) (A_{700}=2.047-3.015 (±0.0014-±0.0005)), OG(LTP) (A_{700} =0.972-4.000 (±0.0043-±0.0000)),OS/OT(ROR) (A_{700}=0.834-2.517 (±0.0009-±0.0000)), showed significant variation in their reducing power activity in a dose dependant manner lower than dose of BHT, gallic acid and catechin. The minimum reducing power activity was observed in OA(DH) (A_{700}=0.490-1.327 (±0.0009-±0.0000)). Note that the reducing power (in terms of increase in A_{700}) of all the extract except OA(DH)(Dwaharhat) was significantly lower as compared to those of the natural antioxidant gallic acid (A_{700}=4.000-4.000 (±0.000)) and catechin (A_{700}=4.000-4.000 (±0.000)) and greater than that of synthetic standard BHT(A_{700}=2.303-2.508 (±0.0005-0.0014)).

DPPH Scavenging Activity

The 2,2'-diphenyl-1-picrylhydrazyl (DPPH) radical scavenging activity was studied both for essential oils and methanolic extracts of all the eight collection of five different species at selected dose levels. The essential oils differed in their activity to scavenge 2, 2-diphenyl picryl-1- hydrazyl (DPPH) free radicals. The essential oil of OB (VTP) and OG (CBP) exhibited maximum scavenging activity (98.69±0.0000 per cent and 92.37±0.0075 per cent respectively) at concentration of 25µL of the oil per 5mL methanol containing 40ppm of DPPH in comparision to the standards. With the increase in the amount of essential oils an increase in scavenging activity was observed for all the oils. Sample of OS (ROR) also exhibited good scavenging activity (96.61±0.0016 per cent) than others at 25µL. The variation in the radical scavenging activity of essential oils of different *Ocimum* spp. samples has been presented in Figure 19.4.

The methanolic extracts also differed in their activity to scavenge DPPH radicals. The methanolic extracts of OB (VTP), OS(ROR) and OS/OT(JNP) exhibited maximum scavenging activity (98.82±0.0000 per cent, 96.61±0.0016 per cent and 86.09±0.0000 respectively) at concentration of 25µL of the extract per 5mL methanol containing 40ppm of DPPH. With the increase in the amount of methanolic extracts an increase in scavenging activity was observed for all the extracts. Sample of collection OG (CBP) also showed good scavenging activity (85.40±0.0000 per cent) than others at 25µL (Figure 19.5). It is well known that free radicals cause auto-oxidation of unsaturated lipids in food (Kaur and Perkins, 1991). In addition, antioxidants are known to interrupt the free radical chain of oxidation and to donate hydrogen from phenolic hydroxyl groups, thereby forming stable free radicals, which do not initiate or propagate further oxidation of lipids (Sherwin, 1978). Many medicinal plants contain large amounts of antioxidants other than vitamin C, vitamin E and carotenoids (Velioglu *et al.*, 1998). Recent reports have described antioxidants and compounds

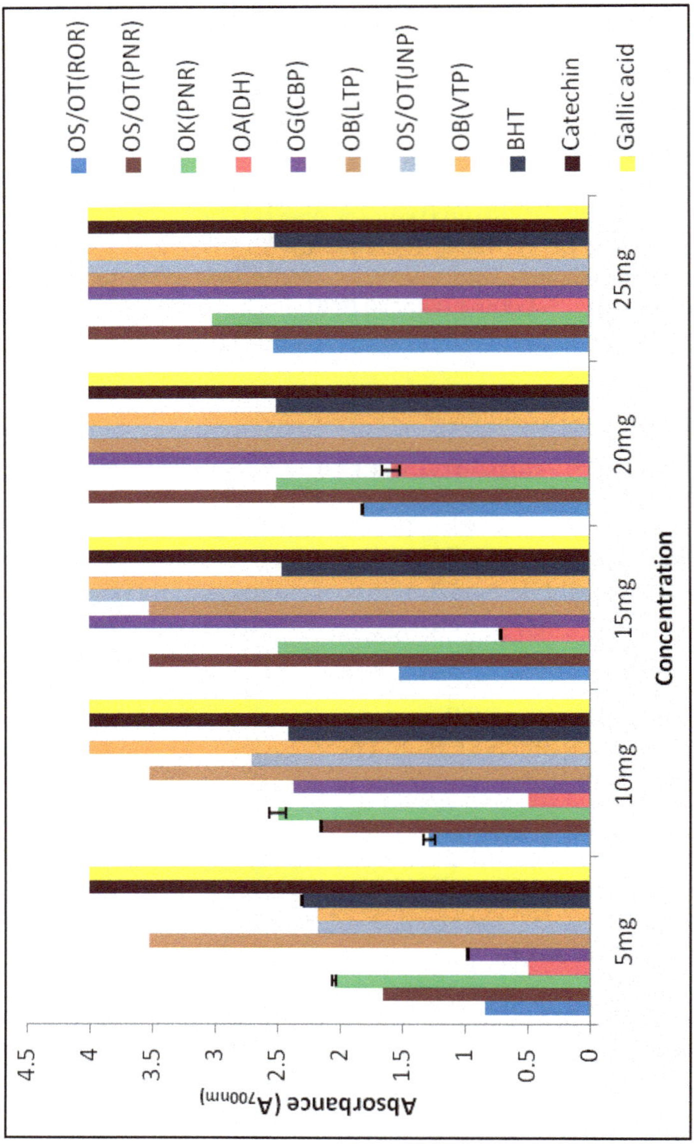

Figure 19.3: Reducing power activity of the methanolic extracts of different *Ocimum* spp. at different concentration.

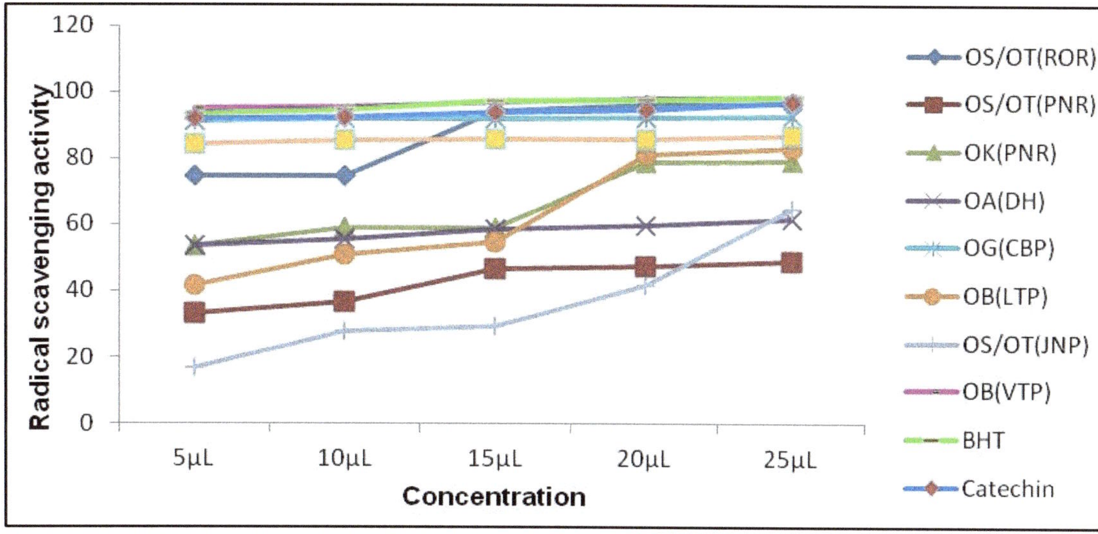

Figure 19.4: Percent 2,2'-diphenyl-1-picrylhydrazyl radical scavenging activity of the essential oils of different *Ocimum* spp. at different concentrations.

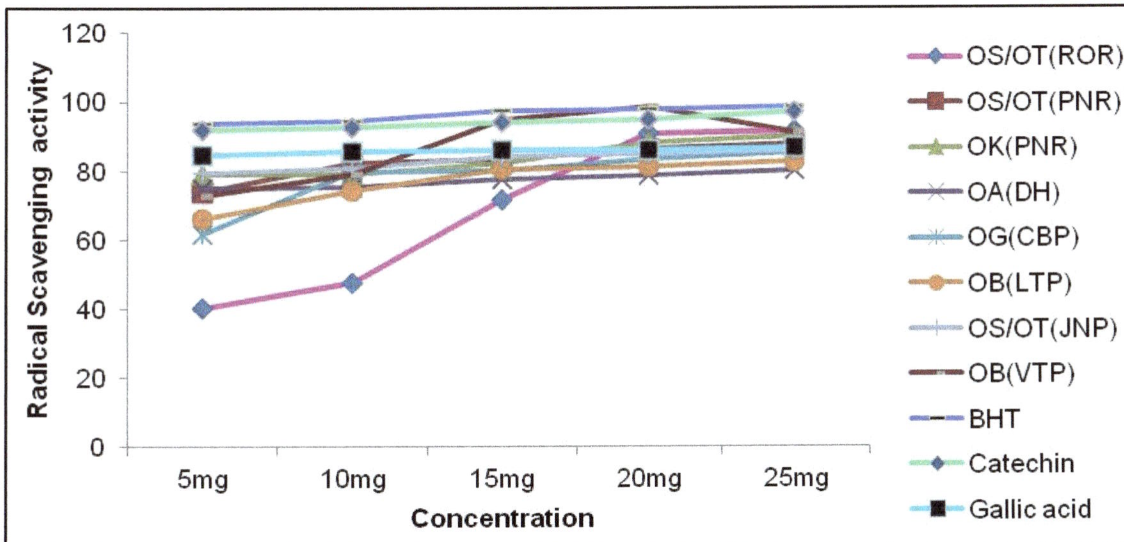

Figure 19.5: Percent 2,2-diphenyl-1-picrylhydrazyl radical scavenging activity of the methanolic extracts of different *Ocimum* spp. at different concentrations.

with free radical scavenging activity present in onion, garlic, sage and thyme extracts (Miura *et al.*, 2002; Nuutila *et al.*, 2003). For the determination of scavenging activity DPPH radical was a very fast method to evaluate the antioxidant activity. With this method it was possible to determine the antiradical power of an antioxidant activity by measuring the decrease in absorbance of the DPPH radical at 517nm. In the radical form this molecule had an absorbance at 517nm which disappeared after acceptance of an electron or hydrogen radical from an antioxidant compound to become a stable diamagnetic molecule (Matthaus, 2002). All these oils and extracts showed variation in their antioxidant potential it is known that most natural antioxidant often were syngertically to produce a broad spectrum of anti oxidative activity that create an effective defense system against free radical attack. Therefore we assume that the observed variation in antioxidant potential is related to different chemical composition and concentration levels. These oils and extracts could, thus be used as natural antioxidant in place of synthetic antioxidant.

Effect on the Chelating Activity of Fe^{2+}

All the essential oils and methanolic extracts in $MeOH/H_2O$ were tested for their chelating activity on Fe^{2+} in the presence of $FeCl_2$ and ferrozine in a dose dependent manner. Variation in the chelating activity of *Ocimum* spp. with increasing in concentrations of the essential oils has been presented in Figure 19.6. The chelating activity was found maximum in OS/OT(JNP) followed by OB(VTP) (65.22±0.0000 per cent) and in OB(LTP) (61.57±0.0009 per cent).

Variation in the chelating activity of methanolic extracts with increasing in concentrations is shown in Figure 19.7. The maximum chelating activity was observed in OS/OT(ROR) (92.50±0.0000 per cent) followed by OS/OT(PNR) (88.91±0.0000 per cent) and OS/OT(JNP) (82.01±0.0009 per cent). Methanolic extract of OS/OT (ROR) showed comparable chelating activity with that of citric acid at 0.025M (92.54±0.0009 per cent).

Among the transition metals, iron is known as the most important lipid oxidation pro-oxidant due to its high reactivity. The ferrous state of iron accelerates lipid oxidation by breaking down hydrogen and lipid peroxide to reactive free radicals like Fenton type reaction. Fe^{3+} ions produce radicals from peroxides, although the rate is tenfold less than that of Fe^{2+} (Miller, 1996). Fe^{2+} ions are the most powerful pro-oxidant among various species of metal ions (Halliwell and Gutteridge, 1984). Ferrozine, a chelating reagent, was used to indicate the presence of chelator in the reaction system. Ferrozine forms a complex with free Fe^{2+} but not with Fe^{2+}. In presence of chelating agents, the complex formation between ferrous and ferrozine is disturbed, resulting in decrease the colour of the complex. Measurement of colour reduction therefore allows estimating the metal chelating activity of the coexisting chelator (Yamaguchi *et al.*, 2000).

The mentioned data shows that essential oils and extract of different *Ocimum* spp. collected from different ecological niches exhibits moderate to good Fe^{2+}chelating activity and could thus function as Fe^{2+} chelator. These oils may either directly bind to the metal ions or indirectly suppress their chelating activity by occupying their coordinate sites (Mahoney and Graf, 1986). Therefore it can be concluded that essential

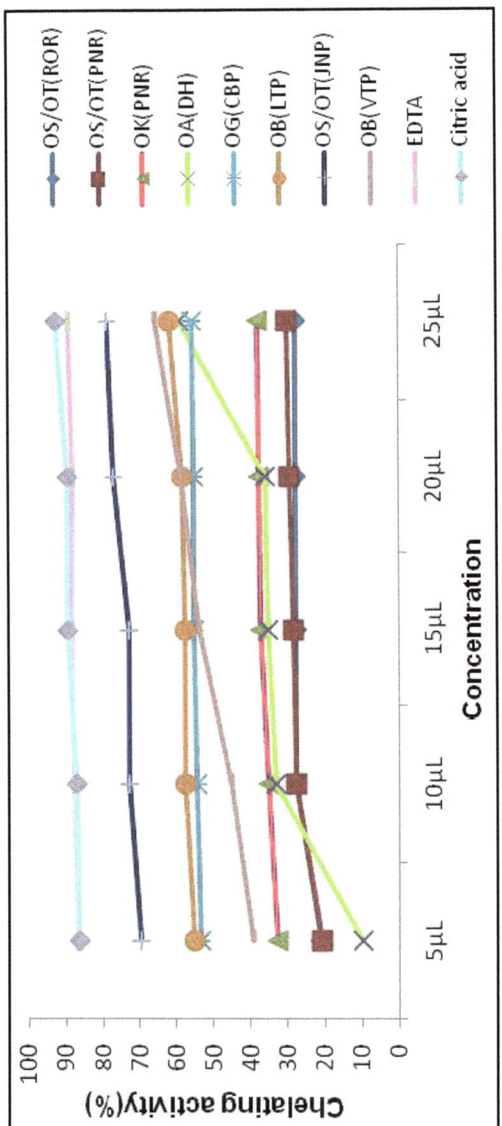

Figure 19.6: Percent chelating activity of the essential oils of different *Ocimum* spp. at different dose levels.

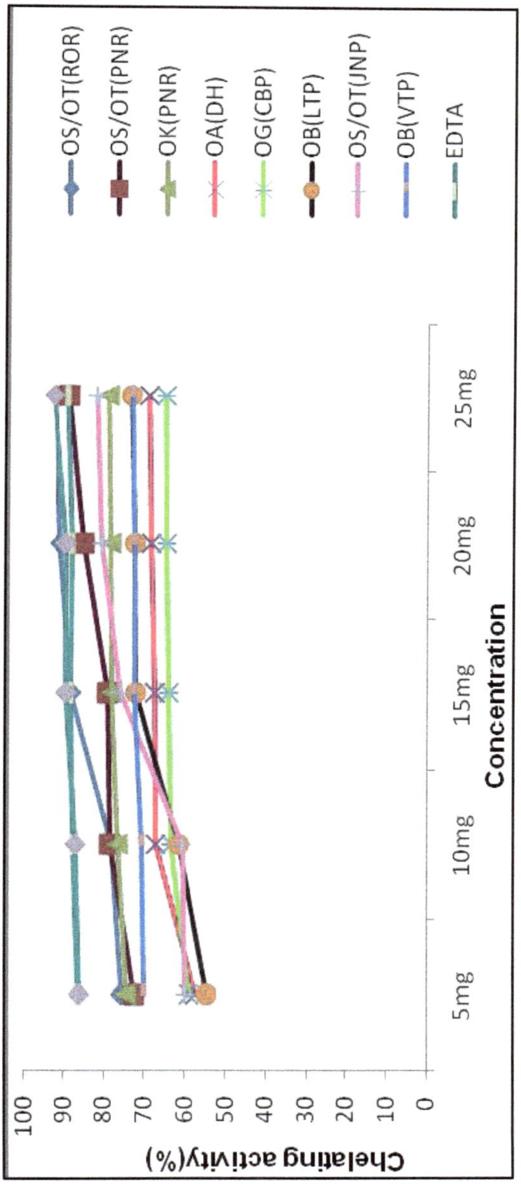

Figure 19.7: Percent chelating activity of the methanolic extracts of different *Ocimum* spp. at different dose levels.

oils and methanolic extracts of *Ocimum* from eight different collections comprising of five different species are an effective chelating agent and could effort protection against oxidizing agent.

Biochemical Assay

Biochemical study of the methanolic extracts of different *Ocimum* spp. were studied quantitatively in terms of their total phenols, flavonols and ortho dihydroxy phenolic content with the help of their respective calibration curves obtained calorimetrically. OB(VTP), OS/OT(JNP), OA(DH), OK(PNR) and OS/OT(ROR) contained the total phenols more than 1mg/100mg followed by OB(LTP),OG(CBP) and OS/OT(PNR) in which the quantity was less than 1mg/100mg. Ortho dihydric phenolic contents was found to be more than 1mg/100mg in OB(VTP) while the content was less than 1mg/100mg in all the other species. The flavonol content was observed less than 1mg/100mg in all the species under study. The differences in the antioxidant activity of methanolic extracts of different species of *Ocimum* may possibly be due to the different biochemical make up of the extracts in terms of phenols, flavonols and orthodihydroxy phenolic compounds and their concentration in the extracts. The detailed comparative and quantitative make up of total phenols,flavonols and ortho dihydric phenols has been presented in Table 19.5.

Table 19.5: Total phenols, flavonols and orthodihydric phenols present in methanolic extracts from different species of *Ocimum*

Sl.No.	*Plant Extracts*	*Total Phenol (mg/100mg)*	*Flavonols (mg/100mg)*	*Orthodihydric Phenol (mg/100mg)*
1.	OB(VTP)	3.000	0.125	1.15
2.	OS/OT(JNP)	1.071	0.05857	0.121
3.	OB(LTP)	0.181	0.02061	0.025
4.	OG(CBP)	0.625	0.0325	0.540
5.	OA(DH)	1.000	0.08571	0.100
6.	OK(PNR)	2.200	0.110	0.550
7.	OS/OT(PNR)	0.625	0.065	0.125
8.	OS/OT(ROR)	1.400	0.086	0.170

Acknowledgements

Authors are thankful to UGC, New Delhi and UCOST, Dehradun for financial assistance and Dr. C.S. Mathela, Professor Emeritus, Kumaun University, Nainital for GC/MS facilities. Thanks are also due to G.B. Pant University of Agriculture and Technology, Pantnagar for providing necessary research facilities.

References

Adams, R.P. (1995). Identification of Essential Oil Components by Gas Chromatography/Mass Spectrometry. Allured Publishing Corporation, Carol Stream, Illinois, USA.

Ahonkhai, I., and Ayinde, B.A. (2005). Antimicrobial activities of some volatile oils against some pathogenic organisms. *Journal of Pharmacy and Bioresources,* **2(2):** 56-59.

Akinmoladun, A.C., Ibukun, E.O., Afor, E., Obuotor, E.M., and Farombi, E.O. (2007). Phytochemical constituent and antioxidant activity of extract from the leaves of *Ocimum gratissimum. Scientific Research and Essays,* **2(5):** 163-166.

Awasthi, P.K., and Dixit, S.C. (2007). Chemical compositions of *Ocimum sanctum* Shyama and *Ocimum sanctum* Rama oils from the plains of northern. *Indian Journal of Essential Oil Bearing Plants,* **10(4):** 292-296.

Bansod, S., and Rai, M. (2008). Antifungal activity of essential oils from Indian medicinal plants against human pathogenic *Aspergillus fumigatus* and *A. niger. World Journal of Medical Sciences,* **3 (2):** 81-88.

Benitez, N.P., Melendez, L.E.M., and Stashenko, E.E. (2009). Eugenol and methyl eugenol chemotypes of essential oil of species *Ocimum gratissimum* L. and *Ocimum campechianum* Mill. from Colombia. *Journal-of-Chromatographic-Science,* **47(9):** 800-803.

Bent, S., and Ko, R. (2004).Company used herbal medicines in the United States : a review. Am. J.Med, **116**: 478-85.

Bhowmik, S., Chowdhury, S.D., Kabir, M.H., and Ali, M.A. (2008). Chemical composition of some medicinal plant products of indigenous origin. *Bangladesh-Veterinarian,* **25(1):** 32-39.

Biswas, N.P., and Biswas, A.K. (2005). Evaluation of some leaf dusts as grain protectant against rice weevil Sitophilus oryzae (Linn.). *Environment and Ecology,* **23(3):** 485–488.

Bozin, B., Mimica, D.N., Simin, N., and Anackov, G. (2006). Characterization of the volatile composition of essential oils of some Lamiaceae spices and the antimicrobial and antioxidant activities of the entire oils. *Journal of Agricultural and Food Chemistry,* **54(5):** 1822-1828.

Chagonda, L.S., Makanda, C.D., and Chalchat, J.C. (2000). The essential oils of *Ocimum canum* Sims (basilic camphor) and *Ocimum urticifolia* Roth from Zimbabwe. *Flavour and Fragnance Journal,* **15(1):** 23-26.

Charles D.J., and Simon, J.E. (1992). Essential oil constituents of *Ocimum kilimandscharicum* Guerke. *Journal of Essential Oil Research,* **4(2):** 125-128.

Chowdhary, H.J., and Wadhwa, M.B. (1984). *Flora of Himachal Pradesh* vol.-II, BSI, Howrah.

Cimanga, K., Kambu, K., Tona, L., Apers, S., Bruyne, T. de., Hermans, N., Totte, J., Pieters, L., and Vlietinck, A.J. (2002). Correlation between chemical composition and antibacterial activity of essential oils of some aromatic medicinal plants growing in the Democratic Republic of Congo. *Journal-of-Ethnopharmacology,* **79(2):** 213-220.

Codignola, A. (1986). A camphorated basil of pharmaceutical interest: *Ocimum Kilimandscharicum* Gurke. *Essenze Derivati agrumari*, **54(1):** 91-101.

Cuendet, M., Hostettmank, K., Potterat, O., and Dyatmiko, W. (1997). Iridoid glucosides with free radical scavenging properties from Fagraea blumei. *Helv.Chem. Acta.*, **80**: 1144-1152.

Decker, E.A., and Welch, B. (1990). Roll of ferritin as alipid oxidation catalyst in muscles food. *J. Agric. Food Chem.*, **38:** 674-677.

Dharmagadda, S.S.V., Tandonb, M., and Vasudevan, P. (2005). Biocidal activity of the essential oils of *Lantana camara, Ocimum sanctum* and *Tagetes patula. Journal of Scientific and Industrial Research*, **64**: 53-56.

Dubey, N.K., Kumar, R., and Tripathi, P. (2004). Global promotion of herbal medicine. *India's Opportunity Curr.Med*, **86**: 37-41.

Feng, Y., Hong, H.L., and Chang, L.F. (2004). Chemical studies on the herb of *Ocimum basilicum* L. *Chinese Journal of Natural Medicines*, **2(1):** 20-24.

Ferreira, R.S., Yassumoto, L., Roberto, J., Souza, P.D., Ishikawa, N.K., and Barbosa, M.A. (2006). Antifungal activity of essential oil isolated from *Ocimum gratissimum* L. (eugenol chemotype) against phytopathogenic fungi. *Brazilian Archives of Biology and Technology an International Journal*, **49(6):** 867-871.

Folarin, O.M., and Akinhanmi, T.F. (2006). Antioxidant activity of essential oils of some aromatic plants. *Biosciences- Biotechnology Research Asia*, **3(1a)**: 87-90.

Gill, K., Malik, O.P., Kalidhar, S.B., Mishra, R.C., and Malik, M.S. (2002). Honey bee repellent activity of leaf extract and essential oil of *Ocimum sanctum* and isolation of a new compound. *Journal of Medicinal and Aromatic Plant Sciences*, **24(4):** 951-956.

Grover, J.K., Yadav, S., and Vats, V. (2002). Medicinal plants of India with antidiabetic.*Journal of EthnoPharmacology*, **81(1):** 81-100.

Gulcin, I., Elmastas, M., and Aboul, E.H.Y. (2007). Determination of antioxidant and radical scavenging activity of basil (*Ocimum basilicum* L. family Lamiaceae) assayed by different methodologies. *Phytotherapy Research*, **21(4):** 354-361.

Hakkim, F.L., Girija, A., and Boopathy, R. (2008). Antioxidant property of selected *Ocimum* species and their secondary metabolite content. *Journal of Medicinal Plants Research*, **2(9):** 250-257.

Halliwell, B., and Gutteridge, J.M.C. (1984). Oxygen Toxicology, oxygen radicals, transition metals and disease. *Biochemical Journal*, **219**: 1-4.

Hinneburg, I., Dorman, H.J.D., and Hiltunen, R. (2006). Antioxidant activities of extracts from selected culinary herbs and spices. *Food Chemistry*, **97(1):** 122-129.

Hsu, C.L., Chen, W., Weng, Y.M., and Tseng, C.Y. (2003). Chemical composition, physical properties and antioxidant activities of yam flours as affected by different drying methods. *Food Chemistry*, **83(1):** 85–92.

Hussain, A.I., Anwar, F., Sherazi, S.T.H., and Przybylski, R. (2008). Chemical composition, antioxidant and antimicrobial activities of basil (*Ocimum basilicum*) essential oils depend on seasonal variations. *Food-Chemistry*, **108(3):** 986-995.

Jadhav, H.R., Singh, A., and Bhutani, K.K. (2005). Rationale for immunomodulatory and anti-inflammatory effects of *Ocimum sanctum:* radical scavenging potential and effect on nitric oxide production. *Acta-Horticulturae*, **678**: 159-162.

Javanmardi, J. (2009). Profiling volatile oil compounds using HS-SPME, GC-MS detection, a fast and reliable tool for classification of basil (*Ocimum basilicum* L.) accessions. *Acta-Horticulturae*, **826**: 297-300.

Javanmardi, J., Stushnoff, C., Locke, E., and Vivanco, J.M. (2003). Antioxidant activity and total phenolic content of Iranian *Ocimum* accessions. *Food-Chemistry*, **83(4):** 547-550.

Jirovetz, L., Buchbauer, G., Ngassoum, M.B., Ngamo, L.T., and Adjoudji, O. (2005). Combined investigation of the chemical composition of essential oils of *Ocimum gratissimum* and *Xylopia aethiopica* from Cameroon and their insecticidal activities against stored maize pest *Sitophilus zeamais*, **29(2):** 55-60.

Jirovetz, L., Buchbauer, G., Shafi, M.P., and Kaniampady, M.M. (2003). Chemotaxonomical analysis of the essential oil aroma compounds of four different *Ocimum* species from southern India. *European Food Research and Technology*, **217(2):** 120-124.

Juliani, H.R., Simon, J.E., Ramboatiana, M.M.R., Behra, O., Garvey, A.S., and Raskin, I. (2004). Malagasy aromatic plants: essential oils, antioxidant and antimicrobial activities. *Acta-Horticulturae*, **629**: 77-81.

Kartikeyan, S., and Singh, N.P. (2000). *Flora of India*, introductory volume part-II, *Botanical survey of India*, 201-217.

Kaur, H., and Perkins, J. (1991). *The free radical-chemistry of food additives*.In: O.I. Aruoma, B. Halliwell (Eds), free radical and food additives. New York, 17-35.

Keita, S.M., Vincent, C., Schmit, J.P., and Belanger, A. (2000). Essential oil composition of *Ocimum basilicum* L., *O. gratissimum* L. and *O. suave* L. in the Republic of Guinea. *Flavour and Fragrance Journal*, **15(5):** 339-341.

Kelm, M.A., Nair, M.G., Strasburg, G.M., and DeWitt, D.L. (2000). Antioxidant and cyclooxygenase inhibitory phenolic compounds from *Ocimum sanctum* Linn. *Phytomedicine*, **7(1):** 7-13.

Kruger, H., Wetzel, S.B., and Zeiger, B. (2002).The chemical variability of *Ocimum* species. *Journal of Herbs, Spices and Medicinal Plants*, **4**: 335-344.

Kumar, A.; Singh, G.R.; Chauhan, N., and Chhokar, S.S. (2010). Essential oil composition and antimicrobial activity of *Ocimum gratissimum.Indian Perfumer*, **54**: 43-46.

Kumar, N., Shukla, R., Chanotiya, C.S., Yadav, A., Singh, A.K., and Bagchi, G.D. (2009). Effect of seasons and drying on the essential oil composition of leaf in *Ocimum kilimandscharicum* Guerke. *Journal of Essential Oil Research*, **21(5):** 400-402.

Kumar, R., Dubey, N.K., Tiwari, O.P., Tripathi, Y.B., and Sinha, K.K. (2007). Evaluation of some essential oils as botanical fungitoxicants for the protection of stored food commodities from fungal infestation. *Journal of the Science of Food and Agriculture*, **87(9):** 1737-1742.

Lee, S.J., Umano, K., Shibamoto, T., and Lee, K.G. (2005). Identification of volatile components in basil (*Ocimum basilicum L.*) and thyme leaves (*Thymus vulgaris L.*) and their antioxidant properties. *Food-Chemistry*, **91(1):** 131-137.

Leelapornpisid, P., Chansakaow, S., Chaiyasut, C., and Wongwattananukul, N. (2008). Antioxidant activity of some volatile oils and absolutes from Thai aromatic plants. *Acta-Horticulturae*, **786**: 61-65.

Lertsatitthanakorn, P., Taweechaisupapong, S., Aromdee, C., and Khunkitti, W. (2006). In vitro bioactivities of essential oils used for *acne* control. *International Journal of Aromatherapy*, **16(1):** 43-49.

Mahadevan, A., and Sridhar, R. (1986a). *In Methods in physiological plant pathology* (3rd end.) Sivakami Publication, Chennai, pp 190-192.

Mahadevan, A., and Sridhar, R. (1986b). *In Methods in physiological plant pathology* (3rd end.) Sivakami Publication, Chennai, pp.183-184.

Matiucci, C.A., and Ferreira, M.J. (2007). Study of the antioxidant activity and essential oil from wild basil (*Ocimum gratissimum L.*) leaf. *Ciencia-e-Tecnologia de Alimentos*, **27(3):** 624-632.

Matthaus, B. (2002). Antioxidant activity of extracts obtained from the residue of different oilseeds. *Journal of Agriculture Food Chemistry*, **50**: 3444-3452.

Mathela, C. S. (1995). New Terpenoids from Inian Medicinal Plants Seminar lecture in California State Polytechnic University, Pomona, USA.

Miller, D. D. (1996). *Mineral in Food Chemistry*. O.R. fennema (Ed.), New York, Marcel Deckker, pp. 618-649.

Miura, K., Kikuzaki, H., and Nakatani, N. (2002). Antioxidant activity of chemical compounds form sage (*Salvia officinalis* L.) and thyme (*Thymus vulgaris* L.) measured by the oil stability index method. *Journal of Agriculture Food Chemistry*, **50**: 1845-1851.

Modnicki, D., and Balcerek, M. (2009). Estimation of total polyphenols contents in *Ocimum basilicum L., Origanum vulgare L.* and *Thymus vulgaris L.* commercial samples. *Herba Polonica*, **55(1):** 35-42.

Mohaney, J. R., and Graf, E. (1986). Citric acid and EDTA as oxidant in a model system. *Journal of Food Science*, **51**: 1293-1296.

Mondal, S., Mahapatra, S.C., Mirdha, B. R., and Naik, S.N. (2007). Antimicrobial activities of essential oils obtained from fresh and dried leaves of *Ocimum sanctum* (L.) against enteric bacteria and yeast. *Acta-Horticulturae*, **756**: 267-269.

Murillo, E., Vina, A., and Linares, M. (2002). Chemical composition, insecticidal and antifungal activity of *Ocimum micranthum Willd. Revista-Colombiana-de-Entomologia*, **28(1):** 109-113.

Nakamura, C.V., Nakamura, T.V., Bando, E., Melo A.F.N., Cortez, D.A.G., and Filho, B.P. (1999). Antibacterial activity of *Ocimum gratissimum* L. essential oil. *Mem. Inst. Oswaldo Cruz*, **94**: 675-678.

Ngassoum, M.B., Ousmaila, H., Ngamo, L.T., Maponmetsem, P.M., Jirovetz, L., and Buchbauer, G. (2004). Aroma compounds of essential oils of two varieties of the spice plant *Ocimum canum Sims* from northern Cameroon. *Journal of Food Composition-and-Analysis*, **17(2):** 197-204.

Ntezurubanza, L., Scheffer, J.J.C., Looman, A., and Svendsen, A.B. (1985). Composition of essential oil of *Ocimum kilimandscharicum* grown in Rwanda. *Planta Med.m*, **50(5):** 385-388.

Nuutila, A.M., Pimia, R.P., Aarni, M., and Cadenty, K.M.O. (2003). Comparison of antioxidant activities of onion and garlic extracts by inhibition of lipid per oxidation and radical scavenging activity. *Food Chemistry*, **81**: 485-493.

Oboh, G., Raddatz, H., and Henle, T. (2008). Antioxidant properties of polar and non-polar extracts of some tropical green leafy vegetables.*Journal-of-the-Science of Food and Agriculture*, **88(14):** 2486-2492.

Orafidiya, L.O., Kaloustian, J., Agbani, E.O., Portugal, H., Abudu, A., Igbeneghu, O.A., Abou, L., Mikail, C., Ivaldi, B., and Adetogun, G.A.A. (2007). A study on the composition, autoxidation and antimicrobial activity of the leaf essential oil of *Ocimum gratissimum* Linn. against *Helicobacter pylori* and *Neisseria sicca*. *International Journal of Essential Oil Therapeutics*, **1(3):** 103-109.

Ozcan, M., and Chalchat, J.C. (2004). Essential oil composition of a new chemotype of Basil (*Ocimum basilicum* L.) cultivating in Turkey. *Journal of Essential Oil Bearing Plants*, **7(2):** 155-159.

Pereira, C.A.M., and Maia, J.F. (2007). Study of the antioxidant activity and essential oil from wild basil (*Ocimum gratissimum* L.) leaf. *Ciencia-e-Tecnologia de Alimentos*, **27(3):** 624-632.

Pin-Der, D. (1998). Antioxidant activity of Budrock (*Arctium laooa* Linn), its scavenging effect on free radical and active oxygen. *Journal of American Oil Research Society*, **75**: 455-461.

Prakash, V., Kishor, M.P., and Mishra, M. (2009). Screening of medicinal plant extracts for antioxidant activity. *Journal-of-Medicinal-Plants-Research*, **3(8):** 608-612.

Rana, T.S., Nath, B., and Datt, B. (1997). A census of the family lamiaceae syn labiateae in Western Himalaya, Higher Plant. *Indian Subcast*, **6**: 31-55.

Reuveni, R.A., Fleisher, N., and Putievsky, E. (1984). Fungistic activity of essential oils from *Ocimum basilicum* chemotypes. *Phytopath.*, **110**: 20-22.

Sartoratto, A., Machado, M.L.A., Delarmelina, C., Figueira, G.M., Duarte, T.M.C., and Rehder G.V.L. (2004). Composition and antimicrobial activity of essential oils from aromatic plants used in brazil. *Brazilian Journal of Microbiology*, **35**: 275-280.

Sherwin, E.R. (1978).Oxidation and antioxidants in fat and oil. *Journal of the American Oil Chemist's Society*, **55**: 809-841.

Shetty, K., Curtis, O.F., Levin, R.E., Wikowsky, R. and Ang, W. (1995). Prevention of verification associated with in vitro shoot culture of oregano (*Origanum vulgare*) by Pseudomonas spp. Journal of Plant Physiology, **147**: 447–451.

Shimada, K., Fujikawa, K., and Namakura, T. (1992). Antioxidative properties of xanthan on autooxidation of soyabean oil in cyclodextrin emulsion. *Journal of Agriculture and Food Chemistry*, **40**: 945-948.

Siddiqui, B.S., Huma, A., Ali, S.T., Sabira, B., and Nasima, K. (2007). Two new triterpenoids and a steroidal glycoside from the aerial parts of *Ocimum basilicum*. *Chemical and Pharmaceutical Bulletin*, **55(4):** 516-519.

Silva, M.G.V., Santos, R.N.D., Matos, F.J.A., and Machado, M.I.L. (2003). Volatile constituents from leaf, inflorescence and root oils of *Ocimum americanum* L. grown in north-eastern Brazil. *Flavour and Fragrance Journal*, **18(4):** 303-304.

Singh, G., Marimuthu, P., Murali, H.S., and Bawa, A.S. (2005). Antioxidative and antibacterial potential of the essential oils and extract isolated from various spices material. *Journal of Food Safety* **25**: 130-145.

Slougui, N., Hadj-Mahammed, M., Changriha, N., and Rolando, C. (2003). Analysis of essential oil of basilicum by gas chromatography (GC) and gas chromatography combined with mass spectrometry (GC-MS). *Rivista Italiana EPPOS*, **36**: 15-20.

Staples, G., and Michael S.K. (1999). *Ethnic Culinary Herbs*.University of Hawaii Press. pp. 73.

Stefanini, I., Piccaglia, R., Marotti, M., and Biavati, B. (2006). Characterization and biological activity of essential oils from fourteen Labiatae species. *Acta-Horticulturae*, **723**: 221-226.

Throne, R.F. (2007). An updated classification of the class Mangnoliopsida (Anigospermas). *The Botanical Review*, **73**(2): 67-182.

Uniyal, B.P., Sharma, J.R., Choudhary, U., and Singh, D.K. (2007). *Flowering plants of Uttarakhand* (A Check list) Bishan Singh Mahendra Pal Singh, Dehradun.

Varshney, S.C. (2001). Value addition in Indian essential oils-The key to success in International Conference on Globalizxation of Indial essential oils Agra, India.

Velioglu, Y.S., Mazza, G., Gao, L., and Oomah, B.D. (1998). Antioxidant activity and total phenolics in selected fruits, vegetables and grainproducts. *Journal of Agriculture Food Chemistry*, **46**: 4113-4117.

Yamaguchi, F., Ariga, T., Yoshimira,Y., and Nakazawa, H. (2000). Antioxidant and anti-glycation activity of carcinol from *Garcinia indica* Fruit Rind. *Journal of Agri. Food Chem.*, **49**: 180-185.

Yayi, E., and Moudachirou, M. (2001). Chemotyping of three *Ocimum* species from Benin: *O. basilicum, O. canum and O. gratissimum*. *Journal of Essential Oil Research*, **13(1):** 13-17.

Yen, G.C., and Chen, H.Y. (1995). Antioxidant activity of various tea extracts in relation to their mutagenicity test. *Journal of Agriculture and Food Chemistry*, **40**: 27-32.

Yen, G.C., and Duh, H.Y. (1993). Antioxidative propertiesmethanolic extracts from peanut hulls. *Journal of American Oil Chemist's Society*, **70**: 383-386.

Zhang, J.W., Li, S.K., and Wu, W.J. (2009). The main chemical composition and in vitro antifungal activity of the essential oils of *Ocimum basilicum* Linn. var. *pilosum* (Willd.). *Benth. Molecules*, **14(1):** 273-278.

Index

B

C

Q

www.ingramcontent.com/pod-product-compliance
Ingram Content Group UK Ltd.
Pitfield, Milton Keynes, MK11 3LW, UK
UKHW022216051026
15624UKWH00001B/187

9 789351 242154